Advances in Neurobiology

Volume 18

Series Editor
Arne Schousboe

More information about this series at http://www.springer.com/series/8787

Michael Aschner • Lucio G. Costa

Editors

Neurotoxicity of Metals

 Springer

Editors
Michael Aschner
Department of Molecular Pharmacology
Albert Einstein College of Medicine
Bronx, NY, USA

Lucio G. Costa
Department of Environmental and
  Occupational Health Sciences
University of Washington
Seattle, WA, USA

ISSN 2190-5215   ISSN 2190-5223   (electronic)
Advances in Neurobiology
ISBN 978-3-319-60188-5   ISBN 978-3-319-60189-2   (eBook)
DOI 10.1007/978-3-319-60189-2

Library of Congress Control Number: 2017953381

Printed on acid-free paper

This Springer imprint is published by Springer Nature
The registered company is Springer International Publishing AG
The registered company address is: Gewerbestrasse 11, 6330 Cham, Switzerland

Contents

Part I
Developmental Neurotoxicity

Developmental Neurotoxicity of Lead

Samuel Caito and Michael Aschner

Abstract Lead exposure is a major concern for the developing nervous system. Environmental exposures to lead, predominantly from contaminated water or lead paint chips, account for the majority of exposures to children. In utero and early life exposures to lead have been associated with lower IQ, antisocial and delinquent behaviors, and attention-deficit hyperactivity disorder. In this review, we will discuss sources of developmental lead exposure and mechanisms of lead neurotoxicity. We will highlight both human epidemiological studies showing associations between lead exposure and behavioral abnormalities as well as experimental data from animal studies. Finally, we will discuss the effects of lead on neurological endpoint past childhood, namely, development of Alzheimer's disease in old age.

Keywords Air quality criteria • Behavioral impairments • Permissible exposure limit (PEL) • Encephalopathy • Metalloproteins • Alzheimer's disease

Introduction

Lead is a widely used metal that has a long history of toxicity in humans. It is a very common element in the Earth's crust but not very prevalent on the Earth's surface. Therefore, developmental exposures to lead are the result of industrial processes/products. Lead is not an essential metal, such as iron, cobalt, or copper, as there are no physiological processes in humans that are dependent on lead. Exposure to lead does not confer any benefits and can result in toxicity.

Lead has many uses in industry and consumer products, primarily due to its chemical properties: softness, low melting temperature, malleability, ductility, poor conductivity, and resistance to corrosion and easily combined with other metals to form alloys. Currently lead is used in batteries, ammunition, electrodes for electrolysis, radiation shielding and reactor coolant, semiconductors, polyvinyl chloride (PVC) plastics, and sailing ballasts. However, lead is a persistent contaminant in our

S. Caito • M. Aschner (✉)
Department of Molecular Pharmacology, Albert Einstein College of Medicine, 1300 Morris Park Ave, Forchheimer Rm 209, Bronx, NY 10461, USA
e-mail: michael.aschner@einstein.yu.edu

© Springer International Publishing AG 2017
M. Aschner and L.G. Costa (eds.), *Neurotoxicity of Metals*, Advances in Neurobiology 18, DOI 10.1007/978-3-319-60189-2_1

environment. It has been estimated that the Greeks and Romans deposited around 400 tons of lead into the environment, which can still be measured in the polar regions of Greenland (Needleman 2004). Lead has been used in the past in pipes, kitchen utensils and tableware, ceramic pigment, cosmetics, and, due to its sweet taste, a wine sweetener. More recently, lead was used in paints, stain glass, book printing, and antiknock agents (Hernberg 2000).

The toxicity of lead has been appreciated since ancient times, with the Greek physician Dioscorides describing how "lead makes the mind give way" and that by the eighteenth century, Benjamin Franklin remarked how he did not understand how lead poisonings could still be occurring (Major 1931). Lead affects every organ system, but the nervous system is the most sensitive. The toxicity of lead has been expertly reviewed extensively (White et al. 2007; Neal and Guilarte 2010; Winneke et al. 1996); herein we give an overview of its effects on the developing nervous system. The US Environmental Protection Agency has performed a series of assessments on the safety of lead concluding that the developing organism is of greatest risk, there is no evident threshold that has been found for lead's effects on the nervous system, and behavioral impairments of developmental exposures persist into childhood and adulthood (US Environmental Protection Agency 1977, 1986, 2006). Blood lead levels are the predominant biomarker used in both human and animal studies, where blood lead levels of 80–100 µg/dl result in encephalopathy, 30–80 µg/dl disrupt cognitive function, and 30–50 µg/dl lower IQ in humans (US Environmental Protection Agency 1986, 2006). Rat and monkey studies have demonstrated that blood lead levels as low as 10 µg/dl can cause neurobehavioral deficits and learning impairments that can persist into adulthood (US Environmental Protection Agency 2006).

The developing brain is susceptible to a variety of toxins, including lead, due to its unique characteristics. The developing brain of children is highly susceptible to lead and more vulnerable than the adult brain. The high susceptibility and vulnerability as compared to adults are due to differences in exposure and toxicokinetics. Children under the age of 5 absorb triple the amount of lead from their GI tract than adults (Chamberlain et al. 1978). Brain development is a long process, with waves of cell division, migration, synaptogenesis, cellular pruning, and myelination. These processes occur at varying rates and persist into childhood. For this reason, lead has been shown to interrupt trimming and pruning of synapses, migration of neurons, and formation of neuron-glia interactions, all of which can result in failure to establish the proper connections between structures and lead to functional deficits. The duration and time of exposure are important determinants to the extent of damage. As different brain areas mature at varying rates, exposure to lead in utero can have different effects on the developing brain than a pediatric exposure.

Environmental exposures to lead have multiple effects on the developing nervous system. Some effects manifest early in an individual's life; however, recent research has associated developmental exposure with lead in the development of neurological and psychological diseases later in life. In this review, we will discuss exposure to lead in children and fetuses, mechanisms of lead toxicity, and effects of lead on cognition, attention, IQ, behavior, and the development of Alzheimer's disease (AD).

Developmental Exposures to Lead

Environmental exposures to lead compromise the major source of lead for developmental exposures in children. Pregnant mothers can be exposed to lead occupationally if they are involved in the manufacture of lead-containing products or battery recycling. Regulations from Occupational Safety and Health Administration (OSHA) have reduced lead exposure in the workplace, setting their permissible exposure limit (PEL) at 30 $\mu g/m^3$ averaged over 8 hours and a reduction in the PEL for shifts longer than 8 hours. However, the PEL is set for adults and may not be low enough for fetuses in utero.

Geographic location influences a child's exposure to lead. Lead contaminates the soil and groundwater around sites of its use, including mines, industrial sites, power plants, incinerators, and hazardous waste sites (Mielke and Reagan 1998). Food crops grown in areas of lead contamination will absorb lead from the groundwater and incorporate into the vegetable or fruits. Public drinking water contains only trace levels of lead; however, acidic (soft) water is corrosive to older lead pipes and solder, which results in lead dissolving into the water. This has been observed recently in Flint, Michigan, in 2014 and Washington, D.C., in 2001, where there was a 9.6-fold increase of elevated blood lead levels in children (Edwards et al. 2009).

Living in older homes and low-income urban dwellings that contain lead paint increases one's exposure to lead. Lead paints were desirable for their vibrant colors and durability; however, due to health concerns, leaded paints and dyes have been phased out of use to minimize lead's harmful effect on people. Ingestion of paint chips by children remains a major source of exposure in the United States. Young children are especially prone to hand-to-mouth behavior, which increases the likelihood of children eating lead paint chips. This is of great concern since children under the age of 5 absorb triple the amount of lead in their gastrointestinal tract than adults (Chamberlain et al. 1978). As paint peels and chips, paint can disintegrate into dust along friction surfaces. Lead dusts can be inhaled, and alarmingly it has been shown that particulate lead between 2 and 10 μm does not degrade but remains as a persistent contaminant (Mielke and Reagan 1998; Gasana and Chamorro 2002). The removal of lead from paints and gasoline has caused a remarkable reduction in exposures seen in the United States. In the 1970s, the median blood lead level of preschool children was 15 $\mu g/dl$, and 88% of children had a level exceeding 10 $\mu g/dl$ (Mahaffey et al. 1983), according to the current Centers for Disease Control screening guideline. Presently, the mean blood lead level of preschool children in the United States is less than 2 $\mu g/dl$, and less than 2% are above 10 $\mu g/dl$ (Bellinger and Bellinger 2006).

Children can be exposed to lead from its mother through the placenta in utero or through breast milk. Pregnancy allows for lead being released from bone stores. Studies using lead isotope ratios demonstrated that 80% of lead in fetal cord blood derives from maternal bone stores, whereas 20% derive from the more recent exposure (Gulson et al. 2003). Alcohol consumption late in pregnancy and high blood pressure have been shown to increase lead deposition from the mother into cord blood, while high hemoglobin content or sickle cell trait is associated with decreased cord blood levels of lead (Harville et al. 2005). Unlike in utero lead exposure, lead exposure through breast milk is more influenced by maternal blood lead concentrations than by maternal bone lead levels (Ettinger et al. 2006).

Neurobehavioral Effects of Lead

Extremely high levels of lead exposure (blood lead levels between 60 and 300 µg/dL) result in encephalopathy in children. Lead encephalopathy can present as hyperirritability, ataxia, convulsions, stupor, coma, and death. Pathologically, it is characterized by endothelial cell swelling and necrosis of the cerebral and cerebellar capillaries, capillary leakage and cerebral edema, loss of neuronal cells, cytoplasmic vacuolization, interstitial edema, and demyelination of nerve fibers. Nonfatal neurobehavioral effects occur at much lower blood lead levels than lead encephalopathy.

Lead exposure has been associated with delinquent behavior, attention deficit hyperactivity disorder (ADHD), and decrements in IQ. The effects of lead on delinquent behavior come from analyses on school-aged children, teenagers, and crime statistics. Bone lead levels have been associated with aggression, attention, and delinquency in children as well as with arrest and adjudication in the juvenile court system (Needleman et al. 2002). Prenatal lead exposure also has a positive correlation with delinquent behavior and drug use as a teenager (Dietrich et al. 2001). It has also been noted that in areas of lead air pollution from gasoline, there is increased incidence of homicides and violent crimes in the Unites States after adjusting for unemployment and percent of population in the high-crime age group (Nevin 2000; Stretesky and Lynch 2001). The mechanism behind lead-induced delinquency is not fully understood. Delinquent behavior in adolescence has been associated with alterations in the hypothalamus pituitary adrenal (HPA) axis (Popma et al. 2006). The HPA axis has been shown to be a target of lead toxicity. In rats exposed to lead either maternally or early in life, there were changes in the functioning of the HPA axis, including altered corticosterone levels (Virgolini et al. 2006; Cory-Slechta et al. 2004). One of the many functions of the HPA axis is to respond to and manage stress; rats exposed to lead had poor stress responses, which were worse in rats exposed to lead and stress (Virgolini et al. 2006).

ADHD is a chronic condition where those affected have difficulty with attention and concentration and are hyperactive and impulsive. An assessment of child behavior by classroom teachers using the Child Behavior Checklist and Disruptive Behavior Disorders Rating Scale of 279 Inuit children aged 11 years in Arctic Quebec found low levels of childhood lead exposure were associated with ADHD behaviors (Boucher et al. 2012). This population was characterized as a predominantly fish-eating population and had significant blood methylmercury levels, which may be a confounder. However, in a case-control study with 71 medically diagnosed

ADHD cases and 58 controls performed near a former lead refinery in Omaha, NE, there was a clear increase in the odds ratio that for each natural log unit of blood lead, there was an odds ratio of 2.52 having ADHD, after adjusting for maternal smoking, socioeconomic status, and environmental smoke exposure (Kim et al. 2013). No similar risk was found for blood mercury or cadmium levels (Kim et al. 2013). Furthermore, a performance and questionnaire study of children in Romania found correlations between blood lead levels, but not mercury or aluminum (Nicolescu et al. 2010). Lead exposure causes similar hyperactive and attention deficits in wild-type rodents (Luo et al. 2014; Sanchez-Martin et al. 2013) or in a genetic rat model prone to neuropsychiatric problems (Ruocco et al. 2015).

Intelligence is negatively affected by lead exposure. The amount of IQ points that are decreased in an individual lead-exposed child is small; however, the troubling effect of lead is on the population as a whole. Lead shifts the population's IQ, leading to fewer individuals in the higher end of the IQ spectrum and more individuals in the lower end. In a study performed from 1978 to 2007, researchers compared blood lead levels in Swedish children ages 7–12, school performance at age 16, and overall IQ at ages 18–19. Over the course of 29 years, there was a statistically significant negative association between school performance, IQ, and blood lead levels below 50 µg/dL. Low exposures of lead (<60 µg/dL blood lead level) early in life cause a decrease in IQ around the time children enter school even though at school age blood lead levels are lower than at the time of exposure (Chen et al. 2005). In a study, 780 children were followed from age 2 to age 7 after being treated for elevated blood lead levels (20–44 µg/dL), and serial IQ tests were administrated, showing decreased IQ, while average lead level at age 7 was 8 µg/dL (Chen et al. 2005). Furthermore, in utero exposures to lead as early as during the first trimester of pregnancy have been associated with decreases in intelligence scores (Hu et al. 2006). Areas of intelligence that have shown decrements include arithmetic skills, reading skills, nonverbal reasoning, reaction time, visual-motor integration, fine motor skills, attention, and short-term memory (White et al. 2007; US Environmental Protection Agency 1977, 1986, 2006). While multiple studies have examined the effects of lead on IQ in various locations around the globe, there are variables that affect the effects of lead. Socioeconomic status, prenatal smoking, maternal age, and prenatal alcohol use exacerbate the effects of lead on IQ (Lanphear et al. 2005).

Molecular Mechanisms of Neurotoxicity

Following inhalation or ingestion, lead enters the bloodstream and will find its way to the brain through both the blood-brain barrier (BBB) and the blood-cerebrospinal fluid (CSF) barrier. The endothelial cells in the BBB microvasculature and the choroid plexus cells that comprise the blood-CSF barrier accumulate lead, causing the barriers to become leaky. This can result in increased permeability of the barriers, brain swelling, herniation, ventricular compression, petechial and cerebral hemorrhages, thrombosis, and arteriosclerosis (Zheng et al. 2003). Lead mimics the action of both iron and calcium, altering these ions' homeostasis and signaling. In the BBB and blood-CSF cells, lead can bind to calcium-dependent protein kinase C (PKC) enzymes, activating the kinases and increasing endothelial permeability (Markovac and Goldstein 1988). Additionally, accumulation of lead by the choroid plexus causes a decrease in transthyretin production (Zheng et al. 1996), disrupting thyroid hormone signaling. The thyroid itself is also targeted by lead and, upon developmental exposure, shows abnormal architecture and decreased functioning (Kumar et al. 2016). Whether the effects on the thyroid are direct actions of lead or indirect due to altered transthyretin production remains to be determined.

Alteration in calcium signaling has important implications on learning and memory deficits in lead-exposed children. Neurotransmitter release through voltage-gated Ca^{2+} channels has been shown to either impede or spontaneously release neurotransmitters (Minnema et al. 1988; Atchison and Narahashi 1984). Cognitive function in rodent studies is usually measured by long-term potentiation (LTP) from hippocampal slices, which require presynaptic glutamate release and subsequent activation of the postsynaptic N-methyl-D-aspartate (NMDA) glutamate receptor (Sui et al. 2000a, b; Altmann et al. 1993). Chronic exposures to lead beginning in utero and continued past weaning as well as transient exposures to lead from in utero to weaning altered presynaptic release of glutamate in the hippocampus (Gilbert et al. 1996, 1999; Lasley et al. 1999). This data suggests that continual presence of lead is not necessary for neurochemical changes but that there is a window of exposure that can produce irreversible deficits. Acute exposures to lead in cell culture or brain tissue homogenates have demonstrated its ability to block the NMDA receptor (Neal et al. 2010; Lasley and Gilbert 1999; Alkondon et al. 1990). Blockade of the NMDA receptor by lead further disrupts calcium signaling. In altering both the presynaptic glutamate release and postsynaptic receptor signaling, lead significantly changes the ways in which the hippocampus produces LTPs. This has implications on memory function, as a neuronal mechanism for memory has different phases of LTPs (short term, intermediate, and long term) (Matthies et al. 1990; Reymann and Frey 2007). Lead exposure results in very long-lasting LTPs, which hinders the formation of short-term and intermediate phases (Gilbert and Mack 1998).

In addition to altering calcium homeostasis, lead can displace metals in metalloproteins and induce oxidative stress. Lead can substitute for physiologic metals in metalloproteins, leading to alteration in protein function. For example, lead binds to the zinc-binding site of the Cys_2/His_2 zinc finger transcription factors TFIIIA and Sp1 (Hanas et al. 1999; Rodgers et al. 2001), the function of which is important for the developing brain. Lead also substitutes for divalent metals present in Cu/Zn superoxide dismutase (SOD), MnSOD, and glutathione peroxidase (GPx) 1 and GPx4, inhibiting these enzymes which are responsible for scavenging reactive oxygen species (ROS). As with many heavy metal exposures, lead can bind to glutathione (GSH) and decrease the reactive thiol pool, increasing the oxidative stress. Disruption of mitochondrial calcium signaling can lead to the generation of ROS and loss of mitochondrial membrane potential, initiate apoptosis, and inhibit the Na^+/K^+ ATPase, decreasing cellular ATP levels (Baranowska-Bosiacka et al. 2011).

Developmental Exposures to Lead and Alzheimer's Disease

While several studies investigate the influence of lead exposure on IQ and cognition in children, developmental lead exposure in rodents and nonhuman primates has shown links to the development of Alzheimer's disease (AD) later in life. Alzheimer's disease is the most common neurodegenerative disease. It is characterized by dementia and loss of cognition, with a brain pathology comprised of proteinaceous plaques comprised of amyloid beta (Aβ). In postmortem human brains of AD patients, lead levels have been measured to be significantly higher in the globus pallidus, dentate gyrus, temporal cortex, and temporal white matter than in control healthy age-matched brains (Haraguchi et al. 2001a, b). An observational study of elderly individuals exposed to multiple heavy metals living near the volcano Etna in Sicily found increased lead in the blood of AD patients than in healthy controls (Giacoppo et al. 2014). Pb exposure increases amyloid precursor protein (APP) mRNA and aggregation of Aβ in rats, amyloidogenesis, and senile plaque deposition and upregulates APP proteins in nonhuman primates exposed to lead as infants (Basha et al. 2005a, b; Wu et al. 2008). After exposing mice to lead during different life span periods, Bihaqi et al. found that a window of vulnerability to lead toxicity exists in the developing brain, where cognitive impairment occurred only in mice exposed to Pb as infants, but not as adults (Bihaqi et al. 2014a). Early life exposure of mice to lead enhances the expression of AD-associated protein tau and alters epigenetic markers associated with the development of AD (Bihaqi et al. 2014b; Masoud et al. 2016). An epigenetic basis for the increased expression of AD-related proteins and cognitive decline is an emerging hypothesis to explain the link between early life exposure to lead and AD. Exposures that occur during fetal or early life stages can produce epigenetic changes in the brain leading reprogramming of genes. In a study of rats exposed in utero or postnatally to Pb, decreased DNA methyltransferase expression was found in the hippocampus of exposed females (Schneider et al. 2013), suggesting that less DNA methylation may be occurring and allowing for genes that are normally repressed to be expressed. Gene expression for DNA methyltransferases in this study was performed at postnatal day 55 (Schneider et al. 2013). Conversely, in a genome-wide expression and methylation profiling experiment carried out in infant Pb-exposed mice aged to postnatal day 700, there was a repression of a set of genes that are normally expressed in aged mice (Dosunmu et al. 2012). These genes were involved in the immune response, metal binding, and metabolism, repression of which due to developmental exposure to Pb compromises the brain's ability to defend against age-related stressors.

Conclusions

Lead is a highly toxic metal that poses great risks to the developing nervous system. Environmental exposures to lead are a major problem for children growing up communities with old homes and water systems, as the recent mass exposure in Flint, Michigan (USA), illustrates. While the effects of lead have been associated with behavioral and cognitive deficits in childhood, we are starting to understand the long-term effects of developmental lead exposure as the population ages.

References

Alkondon M, Costa AC, Radhakrishnan V, Aronstam RS, Albuquerque EX. FEBS Lett. 1990;261(1):124–30.

Altmann L, Weinsberg F, Sveinsson K, Lilienthal H, Wiegand H, Winneke G. Toxicol Lett. 1993;66(1):105–12.

Atchison WD, Narahashi T. Neurotoxicology. 1984;5(3):267–82.

Baranowska-Bosiacka I, Gutowska I, Marchetti C, Rutkowska M, Marchlewicz M, Kolasa A, Prokopowicz A, Wiernicki I, Piotrowska K, Baskiewicz M, Safranow K, Wiszniewska B, Chlubek D. Toxicology. 2011;280(1-2):24–32.

Basha MR, Murali M, Siddiqi HK, Ghosal K, Siddiqi OK, Lashuel HA, Ge YW, Lahiri DK, Zawia NH. FASEB J Off Publ Feder Am Soc Exp Biol. 2005a;19(14):2083–4.

Basha MR, Wei W, Bakheet SA, Benitez N, Siddiqi HK, Ge YW, Lahiri DK, Zawia NH. J Neurosci Off J Soc Neurosci. 2005b;25(4):823–9.

Bellinger DC, Bellinger AM. J Clin Invest. 2006;116(4):853–7.

Bihaqi SW, Bahmani A, Subaiea GM, Zawia NH. Alzheimers Dement. 2014a;10(2):187–95.

Bihaqi SW, Bahmani A, Adem A, Zawia NH. Neurotoxicology. 2014b;44:114–20.

Boucher O, Jacobson SW, Plusquellec P, Dewailly E, Ayotte P, Forget-Dubois N, Jacobson JL, Muckle G. Environ Health Perspect. 2012;120(10):1456–61.

Chamberlain A, Heard C, Little M. Philos Trans R Soc Lond A. 1978;290:557–89.

Chen A, Dietrich KN, Ware JH, Radcliffe J, Rogan WJ. Environ Health Perspect. 2005;113(5):597–601.

Cory-Slechta DA, Virgolini MB, Thiruchelvam M, Weston DD, Bauter MR. Environ Health Perspect. 2004;112(6):717–30.

Dietrich KN, Ris MD, Succop PA, Berger OG, Bornschein RL. Neurotoxicol Teratol. 2001;23(6):511–8.

Dosunmu R, Alashwal H, Zawia NH. Mech Ageing Dev. 2012;133(6):435–43.

Edwards M, Triantafyllidou S, Best D. Environ Sci Technol. 2009;43(5):1618–23.

Ettinger AS, Tellez-Rojo MM, Amarasiriwardena C, Peterson KE, Schwartz J, Aro A, Hu H, Hernandez-Avila M. Am J Epidemiol. 2006;163(1):48–56.

Gasana J, Chamorro A. J Expo Anal Environ Epidemiol. 2002;12(4):265–72.

Giacoppo S, Galuppo M, Calabro RS, D'Aleo G, Marra A, Sessa E, Bua DG, Potorti AG, Dugo G, Bramanti P, Mazzon E. Biol Trace Elem Res. 2014;161(2):151–60.

Gilbert ME, Mack CM. Brain Res. 1998;789(1):139–49.

Gilbert ME, Mack CM, Lasley SM. Brain Res. 1996;736(1-2):118–24.

Gilbert ME, Mack CM, Lasley SM. Neurotoxicology. 1999;20:57–70.

Gulson BL, Mizon KJ, Korsch MJ, Palmer JM, Donnelly JB. Sci Total Environ. 2003;303(1–2):79–104.

Hanas JS, Rodgers JS, Bantle JA, Cheng YG. Mol Pharmacol. 1999;56(5):982–8.

Haraguchi T, Ishizu H, Takehisa Y, Kawai K, Yokota O, Terada S, Tsuchiya K, Ikeda K, Morita K, Horike T, Kira S, Kuroda S. Neuroreport. 2001a;12(18):3887–90.

Haraguchi T, Ishizu H, Kawai K, Tanabe Y, Uehira K, Takehisa Y, Terada S, Tsuchiya K, Ikeda K, Kuroda S. Neuroreport. 2001b;12(6):1257–60.

Harville EW, Hertz-Picciotto I, Schramm M, Watt-Morse M, Chantala K, Osterloh J, Parsons PJ, Rogan W. Occup Environ Med. 2005;62(4):263–9.

Hernberg S. Am J Ind Med. 2000;38(3):244–54.

Hu H, Tellez-Rojo MM, Bellinger D, Smith D, Ettinger AS, Lamadrid-Figueroa H, Schwartz J, Schnaas L, Mercado-Garcia A, Hernandez-Avila M. Environ Health Perspect. 2006;114(11):1730–5.

Kim S, Arora M, Fernandez C, Landero J, Caruso J, Chen A. Environ Res. 2013;126:105–10.

Kumar BK, Reddy AG, Krishna AV, Quadri SS, Kumar PS. Vet World. 2016;9(2):133–41.

Lanphear BP, Hornung R, Khoury J, Yolton K, Baghurst P, Bellinger DC, Canfield RL, Dietrich KN, Bornschein R, Greene T, Rothenberg SJ, Needleman HL, Schnaas L, Wasserman G, Graziano J, Roberts R. Environ Health Perspect. 2005;113(7):894–9.

Lasley SM, Gilbert ME. Toxicol Appl Pharmacol. 1999;159(3):224–33.

Lasley SM, Green MC, Gilbert ME. Neurotoxicology. 1999;20:619–30.

Luo M, Xu Y, Cai R, Tang Y, Ge MM, Liu ZH, Xu L, Hu F, Ruan DY, Wang HL. Toxicol Lett. 2014;225(1):78–85.

Mahaffey KR, Annest JL, Murphy RS. N Engl J Med. 1983;307:573–9.

Major RH. Ann Med Hist. 1931;3:218–27.

Markovac J, Goldstein GW. Nature. 1988;334(6177):71–3.

Masoud AM, Bihaqi SW, Machan JT, Zawia NH, Renehan WE. J Alzheimers Dis. 2016;

Matthies H, Frey U, Reymann K, Krug M, Jork R, Schroeder H. Adv Exp Med Biol. 1990;268:359–68.

Mielke HW, Reagan PL. Environ Health Perspect. 1998;106(Suppl 1):217–29.

Minnema DJ, Michaelson IA, Cooper GP. Toxicol Appl Pharmacol. 1988;92(3):351–7.

Neal AP, Guilarte TR. Mol Neurobiol. 2010;42(3):151–60.

Neal AP, Stansfield KH, Worley PF, Thompson RE, Guilarte TR. Toxicol Sci. 2010;116(1):249–63.

Needleman H. Annu Rev Med. 2004;55:209–22.

Needleman HL, McFarland C, Ness RB, Fienberg SE, Tobin MJ. Neurotoxicol Teratol. 2002;24(6):711–7.

Nevin R. Environ Res. 2000;83(1):1–22.

Nicolescu R, Petcu C, Cordeanu A, Fabritius K, Schlumpf M, Krebs R, Kramer U, Winneke G. Environ Res. 2010;110(5):476–83.

Popma A, Jansen LM, Vermeiren R, Steiner H, Raine A, Van Goozen SH, van Engeland H, Doreleijers TA. Psychoneuroendocrinology. 2006;31(8):948–57.

Reymann KG, Frey JU. Neuropharmacology. 2007;52(1):24–40.

Rodgers JS, Hocker JR, Hanas RJ, Nwosu EC, Hanas JS. Biochem Pharmacol. 2001;61(12):1543–50.

Ruocco LA, Treno C, Gironi Carnevale UA, Arra C, Boatto G, Pagano C, Tino A, Nieddu M, Michel M, Prikulis I, Carboni E, de Souza Silva MA, Huston JP, Sadile AG, Korth C. Amino Acids. 2015;47(3):637–50.

Sanchez-Martin FJ, Fan Y, Lindquist DM, Xia Y, Puga A. PLoS One. 2013;8(11):e80558.

Schneider JS, Kidd SK, Anderson DW. Toxicol Lett. 2013;217(1):75–81.

Stretesky PB, Lynch MJ. Arch Pediatr Adolesc Med. 2001;155(5):579–82.

Sui L, Ruan DY, Ge SY, Meng XM. Neurotoxicol Teratol. 2000a;22(5):741–9.

Sui L, Ge SY, Ruan DY, Chen JT, Xu YZ, Wang M. Neurotoxicol Teratol. 2000b;22(3):381–7.

US Environmental Protection Agency. Air quality criteria for lead. Research Triangle Park: Health Effects Research Laboratory, Criteria and Special Studies Office; 1977.

US Environmental Protection Agency. Air quality criteria for lead, Office of Health and Environmental Assessment. Research Triangle Park: Environmental Criteria and Assessment Office; 1986.

US Environmental Protection Agency. Air quality criteria for lead, vol. I and II of II. Research Triangle Park: National Center for Environmental Assessment – RTP Office; 2006.

Virgolini MB, Bauter MR, Weston DD, Cory-Slechta DA. Neurotoxicology. 2006;27(1):11–21.

White LD, Cory-Slechta DA, Gilbert ME, Tiffany-Castiglioni E, Zawia NH, Virgolini M, Rossi-George A, Lasley SM, Qian YC, Basha MR. Toxicol Appl Pharmacol. 2007;225(1):1–27.

Winneke G, Lilienthal H, Kramer U. Arch Toxicol Suppl. 1996;18:57–70.

Wu J, Basha MR, Brock B, Cox DP, Cardozo-Pelaez F, McPherson CA, Harry J, Rice DC, Maloney B, Chen D, Lahiri DK, Zawia NH. J Neurosci Off J Soc Neurosci. 2008;28(1):3–9.

Zheng W, Shen H, Blaner WS, Zhao Q, Ren X, Graziano JH. Toxicol Appl Pharmacol. 1996;139(2):445–50.

Zheng W, Aschner M, Ghersi-Egea JF. Toxicol Appl Pharmacol. 2003;192(1):1–11.

Manganese and Developmental Neurotoxicity

Roberto Lucchini, Donatella Placidi, Giuseppa Cagna, Chiara Fedrighi, Manuela Oppini, Marco Peli, and Silvia Zoni

Abstract Manganese (Mn) is an essential metal that plays a fundamental role for brain development and functioning. Environmental exposure to Mn may lead to accumulation in the basal ganglia and development of Parkinson-like disorders. The most recent research is focusing on early-life overexposure to Mn and the potential vulnerability of younger individuals to Mn toxicity also in regard to cognitive and executive functions through the involvement of the frontal cortex.

Neurodevelopmental disturbances are increasing in the society, and understanding the potential role of environmental determinants is a key for prevention. Therefore, assessing the environmental sources of Mn exposure and the mechanisms of developmental neurotoxicity and defining appropriate biomarkers of exposure and early functional alterations represent key issues to improve and address preventive strategies. These themes will be reviewed in this chapter.

Keywords Basal ganglia • Vehicle emissions • Methylcyclopentadienyl Mn tricarbonyl (MMT) • Fungicides • Deposited dust • Revised Conners' Teacher Rating Scale • Wechsler Intelligence Scale for Children (WISC) • Olfactory loss

Manganese in the Environment

The assessment of Mn occupational or environmental exposure is a key factor in order to investigate Mn toxicity. While in the occupational contexts personal air monitoring to different particles' granulometry (respirable vs inhalable) is essential to control and prevent Mn excessive exposure, an increasing variety of natural and anthropogenic sources of Mn in the environment can increase pre- and postnatal early-life exposure:

R. Lucchini (✉)
Department of Environmental Medicine and Public Health, Icahn School of Medicine at Mount Sinai, New York, NY, USA

Department of Medical and Surgical Specialties, Radiological Sciences and Public Health, University of Brescia, Viale Europa, 11, 25123 Brescia, Italy
e-mail: roberto.lucchini@mssm.edu

D. Placidi • G. Cagna • C. Fedrighi • M. Oppini • M. Peli • S. Zoni
Department of Medical and Surgical Specialties, Radiological Sciences and Public Health, University of Brescia, Viale Europa, 11, 25123 Brescia, Italy

© Springer International Publishing AG 2017
M. Aschner and L.G. Costa (eds.), *Neurotoxicity of Metals*, Advances in Neurobiology 18, DOI 10.1007/978-3-319-60189-2_2

(i) groundwater contamination, as a consequence of the weathering and leaching of Mn-bearing minerals and rocks into aquifers; (ii) use of the fungicides maneb and mancozeb, which contain approximately 21% Mn by weight; and (iii) emission from ferromanganese and iron industry (FAO 1979; Gulson et al. 2006; Wasserman et al. 2006; Menezes-Filho et al. 2009a, 2009b; Bouchard et al. 2011; ATSDR 2012; Lucchini et al. 2012b; Borgese et al. 2013; Gunier et al. 2013). Early-life exposure to Mn may be caused also by contamination of houses and cars used by Mn workers in various occupations, including welders and agricultural workers (Gunier et al. 2013). Manganese concentration in airborne particles is also higher in areas with intense traffic (Poulakis et al. 2015), showing a contribution of vehicle emissions in relation to non-exhaust sources such as road dust resuspension, break and tire wear, and road-wear abrasion (Thorpe and Harrison 2008) and to exhaust of potential combustion of the gasoline additive methylcyclopentadienyl Mn tricarbonyl (MMT) (Walsh 2007).

In order to investigate all the different ways in which Mn moves from the surrounding environment inside the human body, many studies have been conducted in various locations around the world during the last 30 years, targeting different environmental matrices: airborne particles, drinking water, deposited dust, soil, and vegetables.

Airborne Particles

Industrial activities, agricultural use of fungicides (maneb, mancozeb), and vehicle traffic can increase the ambient levels of Mn. Average air concentrations have been reported to range between 220 and 300 ng/m^3 near industrial sources, whereas in urban and rural areas without point sources have been reported to range from 10 to 70 ng/m^3 (WHO 2011).

Sampling for ambient levels and personal exposures provides direct measures of Mn exposure, but it may be difficult to obtain many data since such samples are time-consuming and expensive. Additionally, sampling only captures levels of exposure associated with the location and environmental conditions during the time of sampling, and this represents a limitation when the concern is cumulative exposure. Proximity measures used as surrogates for Mn exposure are inexpensive and easy to obtain but are limited in their ability to capture other factors that impact ambient air Mn concentrations such as wind direction, precipitation, and terrain. Air dispersion modeling may provide a viable alternative to ambient Mn exposure assessment: the AERMOD model (USEPA 2005) from the US Environmental Protection Agency, for example, estimates ambient air Mn values accounting for Mn emissions, terrain, and weather within a spatial and temporal context, all factors that influence the magnitude of exposure to an air pollutant (Fulk et al. 2016).

The inhalation of airborne particulate matter is the primary source of early-life exposure to Mn in the USA. Populations living in close proximity to industrial and agricultural sources may be at higher risk for developmental effects among children.

Deposited Dust

Deposited dust in houses and schools is another relevant potential source of pediatric Mn exposure, through both ingestion of contaminated hands and foods and inhalation of resuspended particles (Lioy et al. 2002; Zota et al. 2011). Dust sampling yields information on Mn concentration and loading. Samples are generally collected from a given measured surface of the main living area and of furniture or floor, sieved to the desired particle size (e.g., <250 μm), microwave-digested with HNO_3, and analyzed using ICP-MS or XRF techniques. If the dust sample has sufficient mass (>0.1 g), this procedure has detection limits of 0.2 μg/g for Mn concentration. Dust loadings ($μg/m^2$) are usually calculated as [(concentration × mass collected)/area vacuumed].

Zota et al. (2016) evaluated 53 infants at the Tar Creek Superfund site (Oklahoma, USA) in two points in time corresponding to developmental stages before and during initial ambulation (0–6 and 6–12 months). They measured Mn, lead (Pb), arsenic (As), and cadmium (Cd) in indoor air, house dust, tap water, and yard soil and found that except for Cd, metals were detected in all dust samples and at a lesser extent in indoor air, tap water, and soil. They found hair Mn, Pb, and Cd associated to the dust levels, concluding that deposited dust may represent a better measurement of infant exposure to Mn and Pb, compared to air and soil. In fact, infants spend most of their time indoors, and therefore risk assessments and exposure mitigation strategies should prioritize intervention on house dust for early-life prevention.

Gulson et al. (2014) conducted a 5-year longitudinal study to assess potential changes to the environment and exposure of young children associated with the introduction of MMT into Australia in 2001 and its cessation of use in 2004. They evaluated a cohort of 108 children aged 0–5 in Sydney collecting longitudinal samples of Mn and Pb blood, soil, duplicate diet samples, and several types of house dust samples: interior house and day-care dust fall accumulation using Petri dishes, exterior dust fall accumulation, and exterior dust sweepings. Although they did not stratify their results for age, they found dust accumulation ($μg/m^2$/30 day) being the only significant predictor for blood Pb, while no medium predicted blood Mn. More recently, Menezes-Filho et al. (2016) measured Mn and Pb dust fall accumulation on Petri dishes in 15 elementary schools, located between 1.25 and 6.48 km from a Mn alloy production plant in the municipality of Simões Filho, Bahia, Brazil. Their sampling method was similar to the one adopted by Gulson et al. (2014), but they found that the interior school environments, located within a 2-km radius from the plant, showed loading rates on average 190 times higher than the Mn levels measured in the day-care centers in Sydney, while Pb loading rates were not associated with distance from the plant and were lower than the rates observed in the same day-care centers in Sydney.

Surface Soil

Exposure through soil is especially relevant for children playing in contaminated playground or environment, due to their hand-to-mouth behavior and maximized gastrointestinal absorption. Manganese is in soils both in organic and inorganic forms and oxidation states, i.e., 0, +2, +3, +4, +6, and +7. Its mobility is extremely sensitive to soil conditions such as acidity, wetness, organic matter content, and biological activity (Nadaska et al. 2012). Research emphasis has been placed on the toxic effects of compounds containing inorganic Mn^{2+}, Mn^{3+}, and Mn^{4+} ions since these are the forms most commonly encountered in biological systems (Millaleo et al. 2010). With decreasing pH, the amount of exchangeable Mn – mainly Mn^{2+} form – increases in the soil solution. This Mn form is available for plants and can be readily transported into the root cells and translocated to the shoots, where it is finally accumulated. In contrast, other forms of Mn predominate at higher pH values, such as Mn(III) and Mn(IV), which are not available and cannot be accumulated in plants (Millaleo et al. 2010). Different techniques have been applied to laboratory determination of total Mn in soil, including spectrophotometry, polarography, atomic absorption spectrometry (AAS), and inductively coupled plasma atomic emission spectrometry (ICP-AES) or mass spectrometry (ICP-MS) (Pearson and Greenway 2005). Field portable x-ray fluorescence (XRF) is an exemplary field method, offering extremely rapid, cost-effective screening of total heavy metal concentration in soil by in situ measurement.

The total Mn content in soils is variable. Some authors reported small amounts of Mn in soils, ranging from 20 to 10,000 mg/kg soil, whereas other authors have registered total Mn contents between 450 and ~4000 mg Mn/kg soil (Adriano 2001). Potential bioavailability may not be properly addressed by the measure of total soil and can be investigated by means of sequential chemical extraction procedures, where a soil sample is divided into its composing fractions: metal compounds present in the first fraction are those that are weakly bound at cation exchange sites in the matrix and hence chemically very labile. Subsequent processing steps typically extract metals from the carbonate phase, organic matter, etc. Metals in the water/ acid soluble and exchangeable fractions are considered the most mobile and potentially bioavailable forms present in soils and may best capture the anthropogenic contribution of greatest possible concern for children exposure, followed by the carbonate phase (Borgese et al. 2013).

Metal soil contamination resulting from anthropogenic activities is associated to increased health risks among children in the surrounding of smelters (Carrizales et al. 2006) and mines (Pruvot et al. 2006). A significant positive association between soil Mn exposure and both impaired motor coordination and odor discrimination was observed among Italian adolescents (Lucchini et al. 2012b,) and elderly (Zoni et al. 2012; Lucchini et al. 2014) residing near ferroalloy emission sites.

Edible Vegetables

Manganese occurs naturally in many food sources, such as leafy vegetables, nuts, grains, and animal products. For vegetables and vegetable products, mean concentrations range between 0.42 and 6.64 mg/kg (IOM 2002). The Food and Nutrition Board of the Institute of Medicine set adequate intake levels for Mn at 2.3 mg/day for adult men and 1.8 mg/day for adult women (IOM 2002). Adequate intake levels for Mn were also set for other age groups; the values were 0.003 mg/day for infants from birth to 6 months, 0.6 mg/day for infants from 7 months to 1 year, 1.2 mg/day for children aged 1–3 years, 1.5–1.9 mg/day for children aged 4–13 years, and 1.6–2.3 mg/day for adolescents and adults (WHO 2011). Higher levels on Mn were measured in lattice, but not in *Cichorium* spp. or turnip, cultivated in the vicinities of ferroalloy emissions (Ferri et al. 2012, 2015).

Drinking Water

The detection of Mn in groundwater in the USA is approximately 70% of the sites due to the ubiquity of Mn in soil and rock, although the levels detected are generally not considered of public health concern (USEPA 2002). ATSDR (2012) reported Mn levels from <11 to >51 µg/l in a river water survey in the USA. Higher levels found in aerobic waters are usually associated with industrial pollution. Concentrations in seawater have been reported from 0.4 to 10 µg/l, averaging 2 µg/l, whereas in freshwater, they typically range from 1 to 200 µg/l (ATSDR 2012).

Manganese intake from drinking water is substantially lower than the food intake. At the median drinking water level of 10 µg/l determined by the National Inorganic and Radionuclide Survey, the intake of Mn would be 20 µg/day for an adult, assuming a daily water intake of 2 liters. In Germany, the drinking water supplied to 90% of all households contains less than 20 µg/l of Mn (WHO 2011). Gonzalez-Merizalde et al. (2016) investigated the case of artisanal and small-scale gold mining activities performed in mountain areas of the southern Ecuadorian Amazon, which contaminated the aquatic system of the Nangaritza River Basin with mercury (Hg) and Mn, posing health risks for the populations living in the adjacent zones. Children living in alluvial areas showed the highest Mn concentrations in hair, whereas greater values of urinary Hg were found in children living in the high mountain areas, where the ore processing takes place inside or close to houses and schools. This suggests that Hg vapors impact directly the area where they are produced, while waterborne Mn can travel significant distances before impacting the population.

Studies about high level of Mn in drinking water in Quebec reported that the Revised Conners' Teacher Rating Scale oppositional and hyperactivity subscales (Bouchard et al. 2007) were inversely related to hair Mn. Among Bangladeshi children drinking tube-well water, it was found that IQ (Wasserman et al. 2006) and academic achievement (Khan et al. 2012) were inversely related to the level of Mn in drinking water. In the fourth edition of Guidelines for Drinking-Water Quality (WHO 2011), the 400-µg/L drinking-water guideline for manganese (Mn) was discontinued. Concentrations > 400 µg/L are found in a substantial number of countries worldwide and may have been too high to adequately protect public health (Ljung and Vahter 2007). Toxic effects and geographic distribution of Mn in drinking-water supplies require reevaluation by the WHO of its decision to discontinue its drinking-water guideline for Mn (Frisbie et al, 2012).

Toxicology and Biomarkers

Metabolism

Manganese is a naturally occurring trace element essential for human development and function of the brain and other biological processes. As a trace element, Mn is assumed with the diet (mainly with grains, fruits, vegetables, tea) and, once ingested, is absorbed through the small intestine in a proportion of about 3–4%. Gastrointestinal absorption is influenced by the iron metabolism: a deficiency of iron increases the absorption of Mn through some transport proteins, like DMT1 and TFr, which both of these metals have in common (DeWitt et al. 2013).

Manganese is highly needed for the developing brain, and therefore the transplacental absorption is maximized during pregnancy (Guan et al. 2014). When exposure occurs through inhalation, Mn is absorbed through the alveolar-capillary membrane in percentage between 40 and 70%, depending on the size of the particles, and by their water solubility. Another possible route of absorption is the olfactory tract, especially for small particles, that, through the olfactory mucosa, can reach directly sensory areas of the brain (Elder et al. 2006; Lucchini et al. 2012c). An important exposure route for the children is inhalation/ingestion of resuspended soil particulates (Harris and Davidson 2005) or deposited house and school dust (Pavilonis et al. 2014; Lucas et al. 2015) and consumption of contaminated locally grown vegetables (Hough et al. 2004; Ferri et al. 2012, 2015).

Manganese is subjected to an efficient homeostatic control of gastrointestinal absorption and urinary excretion, based on the ratio between absorbed amount and concentration of Mn in the tissues. The absorbed Mn is conveyed in the blood bound to proteins (transferrin, alpha-2-macroglobulin) and in the proportion of 85% to red blood cells. It is deposited mainly in the liver, pancreas, and kidneys and less in bone and adipose tissue. The brain has a small proportion of Mn deposit, but the retention times are long. The Mn values in the adult population not occupationally exposed are between 3.0 and 8.0 ug/l in whole blood and between 0.1 and 1.2 ug/l in the serum (SIVR 2011). Much higher levels are measured during pregnancy and at birth, as shown by Mn in umbilical cord, and gradually decrease postnatally (Claus Henn et al. 2010).

Manganese is eliminated via the gastrointestinal tract through the bile, the intestinal mucosa, and pancreatic secretion. The main excretion of Mn is carried in the feces regardless of route of introduction, while the portion excreted in the urine is low, about 6% of the total, but with a high individual variability depending on age, sex, smoking, and alcohol intake.

Pediatric Absorption

Children are exposed to Mn through the mother during pregnancy and after birth through breastfeeding or formulas; in the neonatal period, during childhood and adolescence, the primary sources of exposure are potentially through drinking water, inhalation of airborne particles, and ingestion of particles from dust and soil. Exposure to Mn by ingestion or inhalation poses higher risks compared to adults, in relation to the different mechanisms of absorption and elimination: the intestinal absorption rate of ingested Mn in children is higher; the high demand for Fe linked to growth can further enhance the absorption of ingested Mn; the excretion rate is lower than in adults because of the poorly developed biliary excretion mechanism; the ratio of inhaled air/body weight is substantially higher (Menezes-Filho et al. 2009a, b).

Biomarkers

A variety of potential biomarkers are available to evaluate Mn exposure in children, including maternal/cordonal blood, blood, serum, plasma, urine, nails, saliva, and hair (Zheng et al. 2011). Blood and urinary Mn reflect exposure over a short and recent period of time (from hours to days), whereas nails and hair longer periods up to several months (Smith et al. 2007; Zheng et al. 2011). Nails and in particular toenails show higher accumulation related to longer-term cumulative exposure (Laohaudomchok et al. 2011). Hair Mn is considered the most consistent and valid biomarker for pediatric exposure and has been found to be associated to intelligence quotient (IQ) decrement by most studies (Coetzee et al. 2016). Hair grows 1 cm per month and provides exposure information for a period of 1–6 months, with variability due to hair pigmentation and potential external contamination (Eastman et al. 2013; Haynes et al. 2015).

A new biomarker, Mn in deciduous teeth, can estimate the exposure windows during prenatal development and early childhood. Deciduous teeth accumulate metals and their mineralization proceeds in an incremental pattern spanning the prenatal and early postnatal periods, commencing gestational week 13–16 for incisors, and concluding postnatal age 10–11 months for molars. Therefore, the distribution of Mn in deciduous teeth can provide information on environmental Mn exposure during fetal development and early childhood. Usually, deciduous teeth are replaced by permanent teeth from the age of 6–12 years. The deposit of metals in teeth has been correlated with exposure in pre- and postnatal period, measuring Mn in the house dust and in the blood and bone of the mother prenatally and cord blood and blood in the postnatal period (Arora et al. 2012; Gunier et al. 2015). This biomarker can provide information about exposure timing and intensity over the fetal period, in particular second and third trimesters, and during early childhood and cumulative early-life exposure (Andra et al. 2015).

Effects on Cognitive Functions

An increasing number of studies have focused on the potential impact of early-life exposure to Mn on cognitive functions (Wasserman et al. 2006, 2011; Wright et al. 2006; Kim et al. 2009; Riojas-Rodríguez et al. 2010; Bouchard et al. 2011; Menezes-Filho et al. 2011; Khan et al. 2012; Torres-Agustín et al. 2013; Haynes et al. 2010, 2012, 2015). Manganese plays a vital role in brain growth and development, and therefore children are more vulnerable than adults to Mn dysfunction (Zoni and Lucchini 2013) in a U-shaped relationship where both insufficiency and excessive absorption can cause adverse effects (Claus Henn et al. 2010). Since brain susceptibility varies during the different phases of development, exposure windows are critical for neurotoxicity (Grandjean and Landrigan 2014). Prenatal and early postnatal periods are sensitive developmental windows for Mn exposure (Claus Henn et al. 2010; Lin et al. 2013; Liu et al. 2014) that can act as essential or toxic element as a function of exposure timing and dose (Sanders et al. 2015; Claus Henn et al. 2010; Chung et al. 2015).

Manganese is transported through the placenta (Erikson et al. 2007), and an elevated maternal exposure during pregnancy can lead to excessive fetal overload (Takser et al. 2004), with accumulation in the developing brain and changes in different neurological structures, which may be responsible of motor, cognitive, and behavioral impairment postnatally. Basal ganglia are the main target of Mn accumulation (Kim et al. 1999) and are involved in the regulation of inhibitory and disinhibitory processes at a cellular and behavioral level throughout the body, via dopaminergic pathways connecting to the frontal lobes (Lezak et al. 2012). These pathways are responsible for the coordination of higher-level cognitive functions including cognitive flexibility, response inhibition, and planning (Miller and Cummings 2007).

Intellectual Ability

A variety of tests and test batteries are used to detect and quantify cognitive effects of Mn exposure in children mostly with the Wechsler Intelligence Scale for Children (WISC) but also with other mental developmental indices (BSID, CDIIT, etc.). A relevant number of studies have now reported effects on the reduction of IQ, using hair Mn as exposure biomarker. Children aged 7–9 years were examined in Marietta (Ohio), home to the largest operating ferromanganese industry in North America. Both low and high Mn concentrations in both blood and hair were negatively associated with the total IQ scores (Haynes et al. 2015). Other studies were conducted in various geographic locations, with moderate sample size, although usually higher than 200 subjects, and adjusted for several covariates including maternal education and intellectual ability, child age, child gender, and nutritional status (Bouchard et al. 2011; Menezes-Filho et al. 2011; Lin et al. 2013; Chung et al. 2015). A few cross-sectional studies in adolescents reported no significant association with IQ using blood Mn (Lucchini et al. 2012a; Bhang et al. 2013), although one of these reported significant associations with deficits on the Learning Disability Evaluation Scale (LDES) (Bhang et al. 2013). Despite limited sample size, a cross-sectional study of 1–4-year-old Uruguayan children reported also significant inverse associations of hair Mn with cognition and language (Rink et al. 2014). Taken all together, these studies support the conclusion that elevated early-life Mn exposure adversely impacts childhood cognition with particularly consistent associations with the IQ (Sanders et al. 2015).

The WISC battery is also used to assess cognitive effects caused by interaction between Mn and other developmental neurotoxicants such as Pb and As. Interaction between Pb or As and Mn and their co-exposure is associated with neurodevelopmental deficiencies that are more severe than expected based on the effects of exposure to each metal alone (Claus Henn et al. 2012; Rodrigues et al. 2016). Similarly, in utero exposure to high Pb and high Mn was associated with larger deficits in cognition and language development compared to low exposure to both metals or to exposure to high levels of only one metal at a time (Lin et al. 2013). In contrast, no statistical interaction was found between Pb and Mn on IQ in 11- to 14-year-old Italian children (Lucchini et al. 2012a).

The WISC has been used to assess both children's general cognitive abilities (IQ) and more specific cognitive functions through its subtests. A study by Rahman et al. (2016) aimed to evaluate potential adverse effects of elevated exposure to Mn in drinking water (W-Mn) from fetal life to school age in a large cohort of boys and girls during 10 years in Bangladesh. Gender was strongly influential in the models of prenatal exposure to W-Mn, with the different cognitive ability measures and the interaction between gender and W-Mn resulting significant for full IQ scale and subscales of verbal comprehension, working memory, and processing speed. Elevated prenatal W-Mn exposure was positively associated with cognitive function in girls, while boys appeared unaffected. This gender influence is observed in several children studies, although the underlying mechanism is still unclear. Several hypotheses have been suggested such as gender-related differences in epigenetic and/or hormonal factors (Barker et al. 2013; Tarrade et al. 2015) or different kinetics (Berglund et al. 2011; Oulhote et al. 2014). In experimental studies, postnatal exposure to Mn has been shown to alter the levels of monoamines and corticosterone in a sex-dependent manner (Vorhees et al. 2014) and cause more morphological changes in striatal medium spiny neurons in male than in female mice (Madison et al. 2011).

Executive Function

Manganese exposure during childhood can impact the executive function (FE), a set of cognitive processes including attentional and inhibitory control, working memory and cognitive flexibility, reasoning, problem-solving, and planning, necessary for cognitive control of behavior (Diamond 2013). Data from experimental study corroborate epidemiological research in children and suggest that exposure to Mn during neurodevelopment significantly alters dopaminergic synaptic environments in brain nuclei and in fronto-striatal circuits that mediate the control of executive function (Kern et al. 2010; Carvalho et al. 2014).

Children aged 6–12 years showed a significant association between Mn in blood and impaired visual attention, while Mn in hair was related to impaired performance of working memory. High levels of Mn from drinking water can affect inhibitory control (Nascimento et al. 2016). Similarly, in Brazilian children living near a ferromanganese alloy plant, airborne Mn exposure was associated to lower IQ and neuropsychological performance in tasks of inhibition responses, strategic visual formation, and verbal working memory (Carvalho et al. 2014).

All together, these results confirm a negative association between executive function and high Mn exposures reported also in a large body of occupational literature (Bowler et al. 2015).

Memory

Significant associations between Mn exposure and cognitive function have been observed also in the domains of learning and memory skills. A group of 174 Mexican children aged 7–11 years was evaluated with the Children's Auditory Verbal Learning Test (CAVLT). Compared to nonexposed subjects, they showed higher hair and blood Mn ($p < 0.001$) as well as lower scores ($p < 0.001$) for all the CAVLT subscales. Hair Mn was inversely associated with most CAVLT subscales, especially those evaluating long-term memory and learning. Blood Mn showed also a negative but nonsignificant association with the CAVLT scores (Torres-Agustín et al. 2013). This study confirmed the findings by Wright et al. (2006) based on the California Verbal Learning Test-Children (CVLTC) and the Wide Range Assessment of Memory and Learning (WRAML) scales. Children with higher Mn levels in that study recalled fewer words on the learning trials of the CVLTC as well as on both the short delay free recall and long delay free recall trials and fewer elements on the WRAML stories.

Imaging research on nonhuman primates has also shown that in addition to the basal ganglia, Mn affects the frontal cortex (Guilarte 2013), an area associated with strategic encoding, organization, and retrieval of verbal and visual memories (Stuss and Alexander 2007).

Academic Achievement

Although adverse effects of Mn on cognitive function raise concern about potential repercussion on children academic achievement, little scientific evidence is available on this aspect. In a rural area of Bangladesh, a cross-sectional study was conducted in 840 children, to investigate associations between the levels of Mn and As in drinking water and academic achievement in mathematics and languages among elementary school children aged 8–11 years. The annual scores of the study children in languages (Bangla and English) and mathematics were obtained from the academic achievement records of the elementary schools. No significant relation was observed between W-Mn and academic achievement in either language. Neither W-As was significantly related to any of the three academic achievement scores. Diversely, W-Mn at levels above the WHO standard of 400 µg/L was associated with a 6.4% score loss in the mathematics achievement test scores, after adjustment for W-As and other sociodemographic variables. These results suggest that deficits in mathematics may be induced by high concentrations of Mn in drinking water (Khan et al. 2012).

Effects on Motor Functions

Although a high number of studies have historically investigated the relation between Mn exposure and motor impairment in workers and adults, little research has focused on these issues in children. In a recent study of Rodrigues et al. (2016), a sample of Bangladeshi children aged 20–40 months were assessed using a translated, culturally adapted version of the Bayley Scales of Infant and Toddler Development, Third Edition (BSID-III). Age-adjusted z-scores were calculated for the five test's domains (i.e., cognitive, receptive language, expressive language, fine motor, and gross motor). The results indicated that most associations between As, Pb, and Mn exposures and the BSID-III z-scores were linear, with the exception of W-Mn concentrations and fine motor scores, for which an inverse U-shaped curve was observed. The interpretation of an inverse U relationship is that at W-Mn <400 µg/L, Mn is beneficial to fine motor development, whereas at W-Mn >400 µg/L, Mn exposure is detrimental for motor function. These results differ from a previous study conducted on 375 Canadian children (Oulhote et al. 2014) that showed a significant association between the estimated Mn intake from water consumption and decreased performance at motor function tests. In the same study, no significant association was also observed between hair Mn and motor function.

A sample of 55 children, aged 7–9 years, was enrolled to determine the association between Mn and Pb exposure with neuromotor function in children. All measures of Mn exposure (blood, hair, and time-weighted distance from a ferromanganese emission) were significantly associated with poor postural balance. Low-level blood Pb was also negatively associated with balance outcomes (Rugless et al. 2014). In a previous study, adolescents aged 11–14 years were recruited in Val Camonica, a region impacted by ferroalloy plants emissions for a century until 2001 and the reference area of Lake Garda (Italy). Several motor tasks were used including the Luria-Nebraska Motor Battery, finger tapping, visual simple reaction time, pursuit aiming, tremor test, and body sway. Regression models showed a significant impairment of motor coordination (Luria-Nebraska test), hand dexterity (pursuit aiming), and odor identification (Sniffin' task), as associated with soil Mn measured at the participants' house. Tremor intensity was directly associated with blood and hair Mn concentrations (Lucchini et al. 2012b).

However, few prospective studies have looked at the effects of both prenatal and postnatal Mn exposure on child cognitive and motor functions. Recently, a cohort of 248 children living near agricultural fields treated with Mn-containing fungicides in Salinas, California, have been studied longitudinally (Mora et al. 2015). Manganese levels was measured in prenatal and early postnatal dentine of shed teeth and confronted to behavior, cognition, memory, and motor functioning examined at the ages of 7, 9, and/or 10.5 years. Motor functions were assessed using finger tapping, the Pegboard tests, and five subtests of the Luria-Nebraska Motor Battery that have shown sensitivity to Mn exposure (Lucchini et al. 2012b). Results showed that higher prenatal and early postnatal Mn levels in dentine of deciduous teeth were adversely associated with behavioral outcomes in school-aged boys and girls. In contrast, higher Mn levels in prenatal and postnatal dentine were associated with better memory abilities and cognitive and motor abilities in school-aged boys (Mora et al. 2015). Hernández-Bonilla et al. (2011) assessed the association between Mn exposure and motor function in 195 children (100 exposed and 95 not exposed), aged 7–11 years, living in Mexico near a Mn mine. Motor functions were assessed with the Grooved Pegboard, the Finger Tapping, and the Santa Ana Test. Comparing exposed and not-exposed groups, a significant difference emerged in the number of errors on the Grooved Pegboard, where exposed subjects made errors more frequently during the test; no differences were observed between groups in the other two motor tests. An inverse association was observed between MnB and Finger Tapping performance for each hand.

The effects of As and Mn ingestion through drinking water, on children's motor functions, were further studied by Parvez et al. (2011). They investigated the association of W-As and W-Mn with motor function in a population of 304 children (8–11 years) from Bangladesh. They assessed motor functions using the Bruininks-Oseretsky test, generating a summary score (total motor composite, TMC) and four subscales: fine manual control (FMC), manual coordination, body coordination, and strength and agility. Adjusted model found an inverse association between As in blood and three motor scales: TMC, FMC, and BC. No associations were observed between MnB or PbB and motor function.

Behavior

Several studies showed an exposure-response relationship between Mn and neurobehavioral effects, but not conclusive. Most of the studies analyzed the exposure during childhood and fewer studies also during prenatal exposure.

Sanders et al. in an epidemiologic review (2015) about early-life Mn exposure identified seven studies that examined the association between early-life exposure to Mn and children/adolescent behavior. Taken together, these studies provide some evidence of a link between early-life Mn and ADHD (attention deficit hyperactivity disorder), ASD (autism spectrum disorder), and other adverse behavioral outcomes in children.

Attention Deficit Hyperactivity Disorder

ADHD is a neurodevelopmental disorder manifested by symptoms of inattention, hyperactivity, and impulsivity; it affects approximately 3–7% of school-aged children. Its persistence into adulthood may result in an approximately 1–4% prevalence of adult ADHD. Children with ADHD are at higher risk of developing psychiatric comorbidity (Hong et al. 2014; Sharma and Couture 2014). The exact etiology of ADHD is still unknown (Sharma and Couture 2014). The proposed neurotoxic mechanisms of Mn involve striatal dopamine neurotransmission, implicated in the pathophysiology of ADHD (Hong et al. 2014).

Sanders et al. (2015) considered a case-control study in the United Arab Emirates [although with the limits of a small case group (18 cases vs 74 controls) and not adjusted for any confounding variables] that reported increased odds of ADHD with increased blood Mn levels. They highlight also a large cross-sectional study of South Korean children where blood Mn levels were associated with poorer scores of commission on one of the three ADHD tests but with no association with doctor-diagnosed ADHD.

In addition, Mora et al. (2015) found that prenatal and early postnatal Mn levels in dentine of deciduous teeth were adversely associated with behavioral outcomes – namely, maternal reports of, using Conners' Rating Scales, internalizing and externalizing symptoms and hyperactivity problems, in school-aged boys and girls.

In a study by Benko et al. (2010), children with ADHD show significantly higher serum Mn concentrations. A cross-sectional study (Bouchard et al. 2007), using the Revised Conners' Teacher and Parent Rating Scales, demonstrated greater hyperactive and oppositional classroom behavior, associated with higher hair Mn from children, on average, 11 years old.

About remediation for ADHD, a case-cohort study in Brazil found that the treatment of adolescent ADHD with the common medication methylphenidate (Ritalin®) significantly reduced blood Mn levels (Farias et al. 2010). A recent study found that methylphenidate administered following chronic postnatal Mn exposure resulted in improved motor function in rats; however, there was no effect on Mn blood levels (Beaudin et al. 2015). If Mn metabolism is part of the underlying biologic pathway for ADHD, this finding may support the evidence for a biological role of Mn in the ADHD causation. Conversely, if methylphenidate alters Mn metabolism independently from its effect on ADHD, this may represent a source of bias. Further research is needed to replicate and understand this relationship.

Autism Spectrum Disorder

Autism spectrum disorder (ASD) is a neurodevelopmental disorder that impairs social interaction and communication. Currently, the etiology of ASD is still not well understood. A number of studies have indicated that ASD has a genetic factor. Others have suggested that there are combinations of factors that influence the etiology of ASD, including the interaction of genetic predisposition and environmental exposures (Rahbar et al. 2014; Rossignol et al. 2014).

Several studies examined the relationship between ASD and Mn exposure, as measured by air distribution, tooth enamel, hair, urine, and red blood cells, but with conflicting findings (Rahbar et al. 2014). Rossignol et al. (2014) supported the idea of an association between environmental toxicants and ASD. On the other side, they declare that many of the reviewed studies contain significant weaknesses and reveal a need for more high-quality epidemiological studies concerning e-relation between environmental toxicants and ASD. In particular, they highlight for Mn a study conducted with 325 children with ASD vs 22,101 controls. The study reported that perinatal exposure to the highest versus lowest quintile of air pollutants was significantly associated with an increased risk of ASD, including Mn (OR = 1.5; 95%), and an overall measure of metals was significantly associated with ASD suggesting that perinatal exposure to air pollutants may increase risk for ASD. Notably, a stronger association was observed in boys compared with girls for most pollutants, suggesting a sex-specific interaction (Roberts et al. 2013).

Comparing the blood Mn levels between ASD children and healthy Jamaican children, Rahbar et al. (2014) found no significant association between BMn and ASD, suggesting that there is no significant association between Mn exposures and ASD. The authors however underscore that blood Mn cannot be used to assess early exposure at potentially more susceptible time period.

Moreover, a case-control study of children in the USA with ASD found that cases had marginally significantly lower levels of tooth enamel Mn, representing postnatal exposure, compared to controls (Abdullah et al. 2012). These findings should be interpreted cautiously, however, because tooth enamel does not track early-life timing of exposure as well as tooth dentine, due to its longer maturation process (Arora et al. 2012).

The potential correlation between hair toxic metal concentrations and ASD severity was examined by Geier (2012) in a prospective cohort of participants diagnosed with moderate to severe ASD. Only hair Hg concentrations resulted significantly correlated with increased ASD severity, and for Mn, no significant correlations were observed for ASD severity.

Other Behavioral Outcomes

Neurobehavioral toxicities associated with Mn also include other behavioral aspects, often evaluated using checklist and questionnaires.

Sanders et al. (2015) analyze in their review three studies examining scales of adolescent behavior and reporting conflicting results. Firstly a cross-sectional study by Lucchini et al. (2012b) on 11- to 14-year-old Italian adolescents found that blood Mn was not associated with any of the other behavioral Conners' Rating Scale subscales. A second cross-sectional study of 8- to 11-year-olds in Bangladesh reported a significant association between drinking water Mn, but not blood Mn, with internalizing and externalizing behaviors (Khan et al, 2012). The third cross-sectional study of 7- to 12-year-old subjects in Brazil found significantly impaired performance on attention when comparing higher hair Mn tertiles to lower Mn but did not report a significant linear relationship (Carvalho et al. 2014).

Moreover, Rahman et al. (2016) assessed behavior problems in 1265 10-year-old children in rural Bangladesh. Elevated prenatal and early childhood exposure levels to W-Mn appeared to increase the risk of children's behavioral problems at 10 years of age. Behavioral problems were assessed using a specific questionnaire assessing conduct problems, hyperactivity/inattention, emotional symptoms, and peer relationship problems. Early-life W-Mn exposure appeared to adversely affect children's behavior. Results showed that W-Mn at all time points was significantly associated with increased risk of conduct problems. They found a significant interaction between gender and prenatal W-Mn for hyperactivity and between gender and W-Mn at 10 years of age for peer problems. Stratifying the models by gender indicated slightly stronger associations of prenatal W-Mn with conduct problems in boys (statistically significant) than in girls (not significant).

Menezes-Filho et al. (2014) verified externalizing behaviors and attention problems using Child Behavior Checklist (CBCL). For girls, CBCL was significantly associated with higher hair Mn.

Hong et al. (2014) demonstrated a correlation between blood Mn levels and behavioral problems like anxiety, social behavior, and aggression in ADHD.

Khan et al. (2011, 2012) in a cross-sectional study demonstrated a dose-response relationship between blood Mn levels and externalizing behavior problems like disruptive behavior and conduct problems and showed that higher water Mn concentrations are associated with lower achievement scores, IQ, and behavioral scores among children.

Effects on Olfactory Function

Elevated Mn exposure during pre-/early adolescence plays an important role in human neurotoxicity, and it is associated with olfactory function in children and elderly (Aschner et al. 2005; Lucchini et al. 2012b, Zoni and Lucchini 2013; Iannilli et al. 2016).

Inhalation of ultrafine particles represents one of the primary routes for neurotoxicity. Manganese exposure reduces significantly the surviving adult-born cells in the olfactory bulb and markedly inhibits their differentiation into mature neurons, resulting in an overall decreased neurogenesis in this brain's region (Fu et al. 2016). Furthermore Mn, as other metals, is readily transported via olfactory pathways and may access to structures located within the brain (Zoni et al. 2012; Guarneros et al. 2013). Mn bypasses the blood-brain barrier and the homeostatic mechanisms that regulate absorption and excretion to keep Mn levels in the desired range. Through the inhalation route, Mn can reach the prefrontal cortex and the striatum altering monoaminergic signaling pathways, particularly dopaminergic transmission, in these two areas (Ye and Kim 2015).

The dopaminergic system plays a central role in the regulation of motor and olfactory function. Various clinical diseases which manifest in the adult life are known to present with olfactory loss, including Parkinson's disease (PD) and Alzheimer-type dementia (AD), which both present with significant smell loss in more than 70% of patients (Lucchini et al. 2014).

Olfactory deficit is an early sign of PD. Living in a Mn-affected environment area can cause impairment in the olfactory functions that may be potentially considered as an early warning for the onset of late neurodegenerative effects in the older age (Lucchini et al. 2012b).

Few studies explored the relation between olfactory functions and Mn exposure. Iannilli et al. (2016) in a pilot study comparing exposed preadolescent (who have spent their whole life span in contaminated areas) with not exposed controls assessed an fMRI experiment pointing at differences of brain activities. They found a generally lower sensory-odor response, and the decreased activity in the relevant brain olfactory areas suggests that young subjects exposed to Mn exhibit a significative reduction of subjective odor sensitivity and olfactory bulb volume. Moreover Mn exposure induces less activation of the limbic system, suggesting an alteration of brain network linked to odor perception and emotional responses (Iannilli et al. 2016).

Lucchini et al. (2012b) used the Sniffin' Sticks test (Hummel et al. 1997) to assess the response to a standardized odor source in 311 healthy adolescents living near a ferroalloy plant. They found a significant impairment of motor functions and of odor identification associated with soil Mn.

References

Abdullah MM, Ly AR, Goldberg WA, Clarke-Stewart KA, Dudgeon JV, Mull CG, Chan TJ, Kent EE, Mason AZ, Ericson JE. Heavy metal in children's tooth enamel: related to autism and disruptive behaviors? J Autism Dev Disord. 2012;42(6):929–36.

Adriano DC. Trace elements in terrestrial environments. Biogeochemistry, bioavailability and risks of metals. New York: Springer; 2001. 867 p.

Andra SS, Austin C, Arora M. Tooth matrix analysis for biomonitoring of organic chemical exposure: current status, challenges, and opportunities. Environ Res. 2015;142:387–406.

Arora M, et al. Determining fetal manganese exposure from mantle dentine of deciduous teeth. Environ Sci Technol. 2012;46(9):5118–25.

Aschner M, Erikson KM, Dorman DC. Manganese dosimetry: species differences and implications for neurotoxicity. Crit Rev Toxicol. 2005;35(1):1–32.

ATSDR (Agency for Toxic Substances and Disease Registry). Toxicological profile for manganese [ATSDR Tox profile]. Atlanta: U.S. Department of Health and Human Services, Public Health Service; 2012.

Barker D, Barker M, Fleming T, Lampl M. Developmental biology: support mothers to secure future public health. Nature. 2013;504(7479):209–11. Review

Beaudin SA, Strupp BJ, Lasley SM, Fornal CA, Mandal S, Smith DR. Oral methylphenidate alleviates the fine motor dysfunction caused by chronic postnatal manganese exposure in adult rats. Toxicol Sci. 2015;144(2):318–27.

Benko CR, Cordeiro ML, Costa MT, Cunha A, Farias AC, et al. Manganese in children with attention-deficit/hyperactivity disorder: relationship with methylphenidate exposure. J Child Adolesc Psychopharmacol. 2010;20:113.

Berglund M, Lindberg AL, Rahman M, Yunus M, Grandér M, Lönnerdal B, Vahter M. Gender and age differences in mixed metal exposure and urinary excretion. Environ Res. 2011;111(8):1271–9.

Bhang SY, Cho SC, Kim JW, Hong YC, Shin MS, Yoo HJ, Cho IH, Kim Y, Kim BN. Relationship between blood manganese levels and children's attention, cognition, behavior, and academic performance–a nationwide cross-sectional study. Environ Res. 2013;126:9–16. Review

Borgese L, Federici S, Zacco A, Gianoncelli A, Rizzo L, Smith DR, Donna F, Lucchini R, Depero LE, Bontempi E. Metal fractionation in soils and assessment of environmental contamination in Vallecamonica. Italy Environ Sci Pollut Res. 2013;20:5067–75.

Bouchard M, Laforest F, Vandelac L, Bellinger D, Mergler D. Hair manganese and hyperactive behaviors: pilot study of school-age children exposed through tap water. Environ Health Perspect. 2007;115(1):122–7.

Bouchard MF, Sauvé S, Barbeau B, Legrand M, Brodeur MÈ, Bouffard T, Limoges E, Bellinger DC, Mergler D. Intellectual impairment in school-age children exposed to manganese from drinking water. Environ Health Perspect. 2011;119(1):138–43.

Bowler RM, Kornblith ES, Gocheva VV, Colledge MA, Bollweg G, Kim Y, Beseler CL, Wright CW, Adams SW, Lobdell DT. Environmental exposure to manganese in air: associations with cognitive functions. Neurotoxicology. 2015;49:139–48. Review

Carrizales L, Razo I, Tellez-Hernandez JI, Torres-Nerio R, Torres A, Batres LE, Cubillas AC, Diaz-Barriga F. Exposure to arsenic and lead of children living near a Cu-smelter in San Luis Potosi, Mexico: importance of soil contamination for exposure of children. Environ Res. 2006;101:1–10.

Carvalho CF, Menezes-Filho JA, de Matos VP, Bessa JR, Coelho-Santos J, Viana GF, Argollo N, Abreu N. Elevated airborne manganese and low executive function in school-aged children in Brazil. Neurotoxicology. 2014;45:301–8.

Chung SE, Cheong HK, Ha EH, Kim BN, Ha M, Kim Y, Hong YC, Park H, Oh SY. Maternal blood manganese and early neurodevelopment: the mothers and children's environmental health (MOCEH) study. Environ Health Perspect. 2015;123(7):717–22.

Claus Henn B, Ettinger AS, Schwartz J, Téllez-Rojo MM, Lamadrid-Figueroa H, Hernández-Avila M, Schnaas L, Amarasiriwardena C, Bellinger DC, Hu H, Wright RO. Early postnatal blood manganese levels and children's neurodevelopment. Epidemiology. 2010;21(4):433–9.

Claus Henn B, Schnaas L, Ettinger AS, Schwartz J, Lamadrid-Figueroa H, Hernández-Avila M, Amarasiriwardena C, Hu H, Bellinger DC, Wright RO, Téllez-Rojo MM. Associations of early childhood manganese and lead coexposure with neurodevelopment. Environ Health Perspect. 2012;120(1):126–31.

Coetzee DJ, McGovern PM, Rao R, Harnack LJ, Georgieff MK, Stepanov I. Measuring the impact of manganese exposure on children's neurodevelopment: advances and research gaps in biomarker-based approaches. Environ Health. 2016;15(1):91.

DeWitt MR, Chen P, Aschner M. Manganese efflux in Parkinsonism: insights from newly characterized SLC30A10 mutation. Biochem Biophys Res Commun. 2013;432:1–4.

Diamond A. Executive functions. Annu Rev Psychol. 2013;64:135–68.

Eastman RR, Jursa TP, Benedetti C, Lucchini RG, Smith DR. Hair as a biomarker of environmental manganese exposure. Environ Sci Technol. 2013;47(3):1629–37.

Elder A, Gelein R, Silva V, Feikert T, Opanashuk L, Carter J, Potter R, Maynard A, Ito Y, Finkelstein J, Oberdorster G. Translocation of inhaled ultrafine manganese oxide particles to the central nervous system. Environ Health Perspect. 2006;114:1172–8.

Erikson KM, Thompson K, Aschner J, Aschner M. Manganese neurotoxicity: a focus on the neonate. Pharmacol Ther. 2007;113(2):369–77.

Farias AC, et al. Manganese in children with attention-deficit/hyperactivity disorder: relationship with methylphenidate exposure. J Child Adolesc Psychopharmacol. 2010;20(2):113–8.

Ferri R, Donna F, Smith DR, Guazzetti S, Zacco A, Rizzo L, Bontempi E, Zimmerman NJ, Lucchini RG. Heavy metals in soil and salad in the proximity of historical ferroalloy emission. J Environ Prot. 2012;3:374–85.

Ferri R, Hashim D, Smith DR, Guazzetti S, Donna F, Ferretti E, Curatolo M, Moneta C, Beone GM, Lucchini RG. Metal contamination of home garden soils and cultivated vegetables in the province of Brescia, Italy: implications for human exposure. Sci Total Environ. 2015;518–519:507–17.

Food and Agriculture Organization of the United Nations (FAO), Maneb manganese ethylenebisdithiocarbamate. FAP Specifications, FAO Protection Products, Rome, Italy. 1979.

Frisbie SH, Mitchell EJ, Dustin H, Maynard DM, Sarkar B. World Health Organization discontinues its drinking-water guideline for manganese. Environ Health Perspect. 2012;120(6):775–8.

Fu S, Jung W, Gao X, Zeng A, Cholger D, Cannon J, Chen J, Zheng W. Aberrant adult neurogenesis in the Subventricular zone-rostral migratory stream-olfactory bulb system following Subchronic manganese exposure. Toxicol Sci. 2016;150(2):347–68.

Fulk F, Haynes EN, Hilbert TJ, Brown D, Petersen D, Reponen T. Comparison of stationary and personal air sampling with an air dispersion model for children's ambient exposure to manganese. J Expos Sci Environ Epidemiol. 2016;26(5):494–502.

Gonzalez-Merizalde MV, Menezes-Filho JA, Cruz-Erazo CT, Bermeo-Flores SA, Sanchez-Castillo MO, Hernandez-Bonilla D, Mora A. 2016. Manganese and mercury levels in water, sediments, and children living near gold-mining areas of the Nangaritza River basin, Ecuadorian Amazon. Archives Environ Contam Toxicol. 2016;71:171–82.

Guan H, Wang M, Li X, Piao F, Li Q, Xu L, Kitamura F, Yokoyama K. Manganese concentrations in maternal and umbilical cord blood: related to birth size and environmental factors. Eur J Pub Health. 2014;24(1):150–7.

Grandjean P, Landrigan PJ. Neurobehavioural effects of developmental toxicity. Lancet Neurol. 2014;13(3):330–8.

Guarneros M, Ortiz-Romo N, Alcaraz-Zubeldia M, Drucker-Colín R, Hudson R. Nonoccupational environmental exposure to manganese is linked to deficits in peripheral and central olfactory function. Chem Senses. 2013;38(9):783–91.

Guilarte TR. Manganese neurotoxicity: new perspectives from behavioral neuroimaging, and neuropathological studies in humans and non-human primates. Front Aging Neurosci. 2013;5:23.

Gulson B, Mizon K, Taylor A, Korsch M, Davis JM, Louie H, Wu M, Gomez L, Antin L. Pathways of Pb and Mn observed in a 5-year longitudinal investigation in young children and environmental measures from an urban setting. Environ Pollut. 2014;191:38–49.

Gulson B, Mizon K, Taylor A, Korsch M, Stauber J, Davis MJ, Louie H, Wu M, Swan H. Changes in manganese and lead in the environment and young children associated with the introduction of methylcyclopentadienyl manganese tricarbonyl in gasoline—preliminary results. Environ Res. 2006;100:100–14.

Gunier RB, Arora M, Jerrett M, Bradman A, Harley KG, Mora AM, Kogut K, Hubbard A, Austin C, Holland N, Eskenazi B. Manganese in teeth and neurodevelopment in young Mexican-American children. Environ Res. 2015;142:688–95.

Gunier RB, Bradman A, Jerrett M, Smith DR, Harley KG, Austin C, Vedar M, Arora M, Eskenazi B. Determinants of manganese in prenatal dentin of shed teeth from CHAMACOS children living in an agricultural community. Environ Sci Technol. 2013;47:11249–57.

Harris AR, Davidson CI. The role of resuspended soil in lead flows in the California south coast Air Basin. Environ Sci Technol. 2005;39(19):7410–5.

Haynes EN, Heckel P, Ryan P, Roda S, Leung YK, Sebastian K, Succop P. Environmental manganese exposure in residents living near a ferromanganese refinery in Southeast Ohio: a pilot study. Neurotoxicology. 2010;31(5):468–74.

Haynes EN, Ryan P, Chen A, Brown D, Roda S, Kuhnell P, Wittberg D, Terrell M, Reponen T. Assessment of personal exposure to manganese in children living near a ferromanganese refinery. Sci Total Environ. 2012;427–428:19–25.

Haynes EN, Sucharew H, Kuhnell P, Alden J, Barnas M, Wright RO, Parsons PJ, Aldous KM, Praamsma ML, Beidler C, Dietrich KN. Manganese exposure and neurocognitive outcomes in rural school-age children: the communities actively researching exposure study (Ohio, USA). Environ Health Perspect. 2015;123(10):1066–71.

Hong SB, Kim JW, Choi BS, Hong YC, Park EJ, Shin MS, Kim BN, Yoo HJ, Cho IH, Bhang SY, Cho SC. Blood manganese levels in relation to comorbid behavioral and emotional problems in children with attentiondeficit/hyperactivity disorder. Psychiatry Res 2014;220(1–2):418–25.

Hernández-Bonilla D, Schilmann A, Montes S, Rodríguez-Agudelo Y, Rodríguez-Dozal S, Solís-Vivanco R, Ríos C, Riojas-Rodríguez H. Environmental exposure to manganese and motor function of children in Mexico. Neurotoxicology. 2011;32(5):615–21.

Hough RL, Breward N, Young SD, Crout NMJ, Tye AM, Moir AM, Thornton I. Assessing potential risk of heavy metal exposure from consumption of homeproduced vegetables by urban populations. Environ Health Perspect. 2004;112(2):215–21.

Hummel T, Sekinger B, Wolf SR, Pauli E, Kobal G. 'Sniffin' sticks': olfactory performance assessed by the combined testing of odour identification, odor discrimination and olfactory threshold. Chem Senses. 1997;22:39–52.

Iannilli E, Gasparotti R, Hummel T, Zoni S, Benedetti C, Fedrighi C, Ying Tang C, Van Thriel C, Lucchini RG, et al. PLoS One. 2016:11(1).

IOM. Dietary reference intakes for vitamin A, vitamin K, arsenic, boron, chromium, copper, iodine, iron, manganese, molybdenum, nickel, silicon, vanadium and zinc. Washington, DC: Institute of Medicine, Food and Nutrition Board/National Academy Press; 2002. p. 10-1–10-22.

Kern CH, Stanwood GD, Smith DR. Preweaning manganese exposure causes hyperactivity, disinhibition, and spatial learning and memory deficits associated with altered dopamine receptor and transporter levels. Synapse. 2010;64(5):363–78.

Khan K, et al. Manganese exposure from drinking water and children's classroom behavior in Bangladesh. Environ Health Perspect. 2011;119(10):1501–6.

Khan K, Wasserman GA, Liu X, Ahmed E, Parvez F, Slavkovich V, Levy D, Mey J, Van Geen A, Graziano JH, Factor-Litvak P. Manganese exposure from drinking water and children's academic achievement. Neurotoxicology. 2012;33(1):91–7.

Kim SH, Chang KH, Chi JG, Cheong HK, Kim JY, Kim YM, Han MH. Sequential change of MR signal intensity of the brain after manganese administration in rabbits. Correlation with manganese concentration and histopathologic findings. Investig Radiol. 1999;34(6):383–93.

Kim Y, Kim BN, Hong YC, Shin MS, Yoo HJ, Kim JW, Bhang SY, Cho SC. Co-exposure to environmental lead and manganese affects the intelligence of school-aged children. Neurotoxicology. 2009;30(4):564–71.

Laohaudomchok W, Lin X, Herrick RF, Fang SC, Cavallari JM, Christiani DC, Weisskopf MG. Toenail, blood and urine as biomarkers of manganese exposure. J Occup Environ Med. 2011;53(5):506–10.

Lezak MD, Howieson DB, Bigler ED, et al. Neuropsychological assessment. New York: Oxford University Press; 2012.

Lin CC, Chen YC, Su FC, Lin CM, Liao HF, Hwang YH, Hsieh WS, Jeng SF, Su YN, Chen PC. In utero exposure to environmental lead and manganese and neurodevelopment at 2 years of age. Environ Res. 2013;123:52–7.

Lioy PJ, Freeman NCG, Millette JR. Dust: a metric for use in residential and building exposure assessment and source characterization. Environ Health Persp. 2002;110:969–83.

Liu W, Huo X, Liu D, Zeng X, Zhang Y, Xu X. S100β in heavy metal-related child attention-deficit hyperactivity disorder in an informal e-waste recycling area. Neurotoxicology. 2014;45:185–91.

Ljung K, Vahter M. Time to re-evaluate the guideline value for manganese in drinking water? Environ Health Perspect. 2007;115(11):1533–8. Review

Ljung KS, Kippler MJ, Goessler W, Grandér GM, Nermell BM, Vahter ME. Maternal and early life exposure to manganese in rural Bangladesh. Environ Sci Technol. 2009;43(7):2595–601.

Lucas EL, Bertrand P, Guazzetti S, Donna F, Peli M, Jursa TP, Lucchin R, Smith DR. Impact of ferromanganese alloy plants on household dust manganese levels: implications for childhood exposure. Environ Res. 2015;138:279–90.

Lucchini RG, Zoni S, Guazzetti S, Bontempi E, Micheletti S, Broberg K, Parrinello G, Smith DR. Inverse association of intellectual function with very low blood lead but not with manganese exposure in Italian adolescents. Environ Res. 2012a;118:65–71.

Lucchini RG, Guazzetti S, Zoni S, Donna F, Peter S, Zacco A, Salmistraro M, Bontempi E, Zimmerman NJ, Smith DR. Tremor, olfactory and motor changes in Italian adolescents exposed to historical ferro-manganese emission. Neurotoxicology. 2012b;33(4):687–96.

Lucchini RG, Dorman DC, Elder A, Veronesi B, et al. Neurotoxicology. 2012c;33:838–41.

Lucchini RG, Guazzetti S, Zoni S, Benedetti C, Fedrighi C, Peli M, Donna F, Bontempi E, Borgese L, Micheletti S, Ferri R, Marchetti S, Smith DR. Neurofunctional dopaminergic impairment in elderly after lifetime exposure to manganese. Neurotoxicology. 2014;45:309–17.

Madison JL, Wegrzynowicz M, Aschner M, Bowman AB. Gender and manganese exposure interactions on mouse striatal neuron morphology. Neurotoxicology. 2011;32(6):896–906.

Menezes-Filho JA, Bouchard M, Sarcinelli P, Moreira JC. Manganese exposure and the neuropsychological effect on children and adolescents: a review. Rev Panam Salud Publica. 2009a;26(6):541–8.

Menezes-Filho JA, Paes CR, Pontes AM, Moreira JC, Sarcinelli PN, Mergler D. High levels of hair manganese in children living in the vicinity of a ferro—manganese alloy production plant. Neurotoxicology. 2009b;30:1207–13.

Menezes-Filho JA, Novaes Cde O, Moreira JC, Sarcinelli PN, Mergler D. Elevated manganese and cognitive performance in school-aged children and their mothers. Environ Res. 2011;111(1):156–63.

Menezes-Filho JA, de Carvalho-Vivas CF, Viana GF, Ferreira JR, Nunes LS, Mergler D, et al. Elevated manganese exposure and school-aged children's behavior: a gender-stratified analysis. Neurotoxicology. 2014;45:293–300.

Menezes-Filho JA, De Souza KOF, Rodrigues JLG, dos Santos NR, Bandeira MD, Koin NL, Oliveira SSD, Leonor A, Godoy PC, Mergler D. Manganese and lead in dust fall accumulation in elementary schools near a ferromanganese alloy plant. Environ Res. 2016;148:322–9.

Millaleo R, Reyes-Diaz M, Ivanov AG, Mora ML, Alberdi M. Manganese as essential and toxic element for plants: transport, accumulation and resistance mechanisms. J Soil Sci Plant Nutr. 2010;10:470–81.

Miller BL, Cummings JL. The human frontal lobes. New York: The Guilford Press; 2007.

Mora AM, Arora M, Harley KG, Kogut K, Parra K, Hernández-Bonilla D, Gunier RB, Bradman A, Smith DR, Eskenazi B. Prenatal and postnatal manganese teeth levels and neurodevelopment at 7, 9, and 10.5 years in the CHAMACOS cohort. Environ Int. 2015;84:39–54.

Nadaska G, Lesny J, Michalik I, Environmental aspect of manganese chemistry. 2012. http://heja.szif.hu/ENV/ENV-100702-A/env100702a.pdf. Accessed 22 Aug 2016.

Nascimento S, Baierle M, Göethel G, Barth A, Brucker N, Charão M, Sauer E, Gauer B, Arbo MD, Altknecht L, Jager M, Dias AC, de Salles JF, Saint Pierre T, Gioda A, Moresco R, Garcia SC. Associations among environmental exposure to manganese, neuropsychological performance, oxidative damage and kidney biomarkers in children. Environ Res. 2016;147:32–43.

Oulhote Y, Mergler D, Bouchard MF. Sex- and age-differences in blood manganese levels in the U.S. general population: national health and nutrition examination survey 2011–2012. Environ Health. 2014;13:87.

Parvez F, Wasserman GA, Factor-Litvak P, Liu X, Slavkovich V, Siddique AB, Sultana R, Sultana R, Islam T, Levy D, Mey JL, van Geen A, Khan K, Kline J, Ahsan H, Graziano

JH. Arsenic exposure and motor function among children in Bangladesh. Environ Health Perspect. 2011;119(11):1665–70.

Pavilonis BT, Lioy PJ, Guazzetti S, Bostick BC, Donna F, Peli M, Zimmerman NJ, Bertrand P, Lucas EL, Smith DR, Georgopoulos PG, Mi Z, Royce SG, Lucchini RG. Manganese concentrations in soil and settled dust in an area with historic ferroalloy production. J Expos Sci Environ Epidemiol. 2014;1–8

Pearson GF, Greenway GM. Recent developments in manganese speciation. Trac-Trends Anal Chem. 2005;24(9):803–9.

Poulakis E, Theodosi C, Bressi M, Sciare J, Ghersi V, Mihalopoulos N. Airborne mineral components and trace metals in Paris region: spatial and temporal variability. Environ Sci Pollut Res Int. 2015;22(19):14663–72.

Pruvot C, Douay F, Herve F, Waterlot C. Heavy metals in soil, crops and grass as a source of human exposure in the former mining areas. J Soil Sedim. 2006;6:215–20.

Rahbar MH, Samms-Vaughan M, Dickerson AS, Loveland KA, Ardjomand-Hessabi M, Bressler J, Shakespeare-Pellington S, Grove ML, Pearson DA, Boerwinkle E. Blood manganese concentrations in Jamaican children with and without autism spectrum disorders. Environ Health 2014;13:69.

Rahman SM, Kippler M, Tofail F, Bölte S, Hamadani JD, Vahter M. Manganese in drinking water and cognitive abilities and behavior at 10 years of age: a prospective cohort study. Environ Health Perspect. 2016.; [Epub ahead of print]

Rink SM, Ardoino G, Queirolo EI, Cicariello D, Mañay N, Kordas K. Associations between hair manganese levels and cognitive, language, and motor development in preschool children from Montevideo. Uruguay Arch Environ Occup Health. 2014;69(1):46–54.

Riojas-Rodríguez H, Solís-Vivanco R, Schilmann A, Montes S, Rodríguez S, Ríos C, Rodríguez-Agudelo Y. Intellectual function in Mexican children living in a mining area and environmentally exposed to manganese. Environ Health Perspect. 2010;118(10):1465–70.

Roberts AL, Lyall K, Hart JE, Laden F, Just AC, Bobb JF, et al. Perinatal air pollutant exposures and autism spectrum disorder in the children of nurses' health study II participants. Environ Health Perspect. 2013;121:978–84.

Rodrigues EG, Bellinger DC, Valeri L, Hasan MO, Quamruzzaman Q, Golam M, Kile ML, Christiani DC, Wright RO, Mazumdar M. Neurodevelopmental outcomes among 2- to 3-year-old children in Bangladesh with elevated blood lead and exposure to arsenic and manganese in drinking water. Environ Health. 2016;15:44.

Rossignol DA, Genuis SJ, Frye RE. Environmental toxicants and autism spectrum disorders: a systematic review. Transl Psychiatry. 2014;4:e360.

Rugless F, Bhattacharya A, Succop P, Dietrich KN, Cox C, Alden J, Kuhnell P, Barnas M, Wright R, Parsons PJ, Praamsma ML, Palmer CD, Beidler C, Wittberg R, Haynes EN. Childhood exposure to manganese and postural instability in children living near a ferromanganese refinery in southeastern Ohio. Neurotoxicol Teratol. 2014;41:71–9.

Sanders AP, Claus Henn B, Wright RO. Perinatal and childhood exposure to cadmium, manganese, and metal mixtures and effects on cognition and behavior: a review of recent literature. Curr Environ Health Rep. 2015;2(3):284–94. Review

Sharma A, Couture J. A review of the pathophysiology, etiology, and treatment of attention-deficithyperactivity disorder (ADHD). Ann Pharmacother 2014;48(2):209–25.

SIVR, Società Italiana Valori di Riferimento. Valori di riferimento degli elementi di interesse biologico e tossicologico, 2011. http://www.valoridiriferimento.it

Smith D, Gwiazda R, Bowler R, et al. Biomarkers of Mn exposure in humans. Am J Ind Med. 2007;50:801–11.

Stuss DT, Alexander MP. Is there a dysexecutive syndrome? Philos Trans R Soc Lond Ser B Biol Sci. 2007;362(1481):901–15. Review

Takser L, Mergler D, de Grosbois S, Smargiassi A, Lafond J. Blood manganese content at birth and cord serum prolactin levels. Neurotoxicol Teratol. 2004;26(6):811–5.

Tarrade A, Panchenko P, Junien C, Gabory A. Placental contribution to nutritional programming of health and diseases: epigenetics and sexual dimorphism. J Exp Biol. 2015;218(Pt 1):50–8. Review

Thorpe A, Harrison RM. Sources and properties of non-exhaust particulate matter from road traffic: a review. Sci Total Environ. 2008;400(1–3):270–82.

Torres-Agustín R, Rodríguez-Agudelo Y, Schilmann A, Solís-Vivanco R, Montes S, Riojas-Rodríguez H, Cortez-Lugo M, Ríos C. Effect of environmental manganese exposure on verbal learning and memory in Mexican children. Environ Res. 2013;121:39–44.

US Environmental Protection Agency. Health effects support document for manganese. Washington, DC: US EPA, Office of Water; 2002.

US Environmental Protection Agency. User's guide for the AMS/EPA regulatory model AERMOD. Washington, DC: US EPA; 2005. EPA-454/B-03-001

Vorhees CV, Graham DL, Amos-Kroohs RM, Braun AA, Grace CE, Schaefer TL, Skelton MR, Erikson KM, Aschner M, Williams MT. Effects of developmental manganese, stress, and the combination of both on monoamines, growth, and corticosterone. Toxicol Rep. 2014;1:1046–61.

Walsh MP. The global experience with lead in gasoline and the lessons we should apply to the use of MMT. Am J Ind Med. 2007;50(11):853–60.

Wasserman GA, Liu X, Parvez F, Ahsan H, Levy D, Factor-Litvak P, Kline J, van Geen A, Slavkovich V, LoIacono NJ, Cheng Z, Zheng Y, Graziano JH. Water manganese exposure and children's intellectual function in Araihazar. Bangladesh Environ Health Perspect. 2006;114(1):124–9.

Wasserman GA, Liu X, Parvez F, Factor-Litvak P, Ahsan H, Levy D, Kline J, van Geen A, Mey J, Slavkovich V, Siddique AB, Islam T, Graziano JH. Arsenic and manganese exposure and children's intellectual function. Neurotoxicology 2011;32(4):450–7.

WHO (World Health Organization). Guidelines for Drinking-Water Quality. 4th ed. Geneva: WHO; 2011b.

Wright RO, Amarasiriwardena C, Woolf AD, Jim R, Bellinger DC. Neuropsychological correlates of hair arsenic, manganese, and cadmium levels in school-age children residing near a hazardous waste site. Neurotoxicology. 2006;27(2):210–6.

Ye Q, Kim J. Loss of the function reverses impaired recognition memory caused by olfactory manganese exposure in mice. Toxicol Res. 2015;31(1):17–23.

Zheng W, Fu SX, Dydak U, Cowan DM. Biomarkers of manganese intoxication. Neurotoxicology. 2011;32(1):1–8.

Zoni S, Bonetti G, Lucchini R. Olfactory functions at the intersection between environmental exposure to manganese and Parkinsonism. J Trace Elem Med Biol. 2012;26(2–3):179–82.

Zoni S, Lucchini RG. Manganese exposure: cognitive, motor and behavioral effects on children: a review of recent findings. Curr Opin Pediatr. 2013;25(2):255–60.

Zota AR, Schaider LA, Ettinger AS, Wright RO, Shine JP, Spengler JD. Metal sources and exposures in the homes of young children living near a mining impacted superfund site. J Expo Sci Environ Epidemiol. 2011;21:495–505.

Zota AR, Riederer AM, Ettinger AS, Schaider LA, Shine JP, Amarasiriwardena CJ, Wright RO, Spengler JD. Associations between metals in residential environmental media and exposure biomarkers over time in infants living near a mining-impacted site. J Expos Sci Environ Epidemiol. 2016;26(5):510–9.

Inherited Disorders of Manganese Metabolism

Charles E. Zogzas and Somshuvra Mukhopadhyay

Abstract While the neurotoxic effects of manganese were recognized in 1837, the first genetic disorder of manganese metabolism was described only in 2012 when homozygous mutations in *SLC30A10* were reported to cause manganese-induced neurotoxicity. Two other genetic disorders of manganese metabolism have now been described – mutations in *SLC39A14* cause manganese toxicity, while mutations in *SLC39A8* cause manganese and zinc deficiency. Study of rare genetic disorders often provides unique insights into disease pathobiology, and the discoveries of these three inherited disorders of manganese metabolism are already transforming our understanding of manganese homeostasis, detoxification, and neurotoxicity. Here, we review the mechanisms by which mutations in *SLC30A10*, *SLC39A14*, and *SLC39A8* impact manganese homeostasis to cause human disease.

Keywords Manganese • SLC30A10 • SLC39A14 • SLC39A8 • Homeostasis • Transporter

Introduction

Manganese (Mn) is an essential element required for the activities of numerous enzymes (Aschner et al. 2009). In adults, adequate intake is 1.8 mg/day in females and 2.3 mg/day in males (Freeland-Graves et al. 2016). As Mn is essential, its deficiency is linked to a number of health conditions including impaired cognitive function, asthma, osteoporosis, and dyslipidemia (Freeland-Graves et al. 2016). However, the causal role of Mn in these diseases is not clear. In contrast, more information is available about Mn toxicity. When systemic levels of Mn increase, the metal accumulates in the brain, primarily in the basal ganglia, and leads to the onset of a parkinsonian-like movement disorder (Aschner et al. 2009; Olanow 2004; Perl and Olanow 2007). Historically, Mn-induced neurotoxicity was reported in

C.E. Zogzas • S. Mukhopadhyay (✉)
Division of Pharmacology & Toxicology, College of Pharmacy; Institute for Cellular & Molecular Biology; and Institute for Neuroscience, The University of Texas at Austin, 3.510E BME, 107 W. Dean Keeton, Austin, TX 78712, USA
e-mail: som@austin.utexas.edu

© Springer International Publishing AG 2017
M. Aschner and L.G. Costa (eds.), *Neurotoxicity of Metals*, Advances in Neurobiology 18, DOI 10.1007/978-3-319-60189-2_3

individuals exposed to elevated Mn from occupational sources (e.g., welding, manufacture of dry batteries and steel, and mining) (Aschner et al. 2009). Recent studies suggest that Mn-induced neurotoxicity may also occur due to environmental exposure to elevated Mn (Lucchini et al. 2014, 2012). Furthermore, individuals with defective liver function, due to diseases such as cirrhosis, fail to excrete Mn and may develop Mn-induced neurotoxicity in the absence of exposure to elevated Mn (Butterworth 2013). Calculations show that, under physiologic conditions, brain Mn levels are ~5–14 ng Mn/mg protein (corresponding to 20–53 µM Mn) (Bowman and Aschner 2014). In mammalian systems, neurotoxicity occurs when there is ~3-fold elevation in brain Mn levels, which corresponds to 16–42 ng Mn/mg protein or 60–158 µM Mn (Bowman and Aschner 2014). Thus, intracellular levels of Mn need to be maintained within a narrow physiologic range. Earlier studies on Mn homeostasis used yeast as a model organism and led to the discovery of an elegant regulatory system in which the Mn influx transporters, Smf1p and Smf2p, are degraded when intracellular Mn levels increase (Culotta et al. 2005; Jensen et al. 2009). However, in mammalian systems, Mn-induced downregulation of DMT1, the homolog of Smf1 proteins, has not been observed, suggesting that there may be fundamental differences between the regulatory systems that control cellular Mn in yeast and mammals. A major breakthrough in understanding the mechanisms that regulate Mn homeostasis in humans came in 2012 with the discovery that individuals harboring homozygous mutations in *SLC30A10* suffer from familial Mn-induced neurotoxicity (Quadri et al. 2012; Tuschl et al. 2008, 2012). Remarkably, soon after this, two other genetic diseases of Mn metabolism were discovered. In 2016, mutations in *SLC39A14* were also reported to induce Mn neurotoxicity (Tuschl et al. 2016), and in 2015, mutations in *SLC39A8* were identified to cause Mn and zinc (Zn) deficiency (Boycott et al. 2015; Park et al. 2015). *SLC30A10*, *SLC39A14*, and *SLC39A8* all code for Mn transporters and induce disease by directly altering cellular and tissue Mn (and Zn, in case of *SLC39A8*) levels. Below, we describe the mechanisms by which loss-of-function mutations in these genes impact Mn metabolism in greater detail.

SLC30A10

The first detailed clinical study of a patient later shown to harbor homozygous mutations in *SLC30A10* was published in 2008 (Tuschl et al. 2008). In this report, the authors described findings from a 12-year-old female patient who was born to consanguineous parents and developed difficulty in walking and in conducting fine movements with her hands. Clinical analyses revealed that the patient had a characteristic "cock-walk" gait and signs of dystonia along with ~10-fold increase in blood Mn (Tuschl et al. 2008). Magnetic resonance imaging provided evidence of Mn deposition in the basal ganglia, anterior pituitary, and the dentate nucleus and white matter of the cerebellum (Tuschl et al. 2008). Liver biopsy revealed that the patient had cirrhosis and that liver Mn levels were elevated (Tuschl et al. 2008). The patient also had polycythemia and hyperbilirubinemia (Tuschl et al. 2008). Importantly, there was no history of exposure to elevated Mn from environmental sources, and levels of other essential metals in plasma [copper (Cu) and Zn] were normal (Tuschl et al. 2008). These findings raised the possibility that the patient may have a defect in Mn metabolism, perhaps of genetic origin, which led to Mn retention and subsequent neurotoxicity.

In 2012, analyses from additional patients who presented with the above-described clinical picture were published (Quadri et al. 2012; Tuschl et al. 2012). Whole-genome homozygosity mapping and exome sequencing revealed that affected patients carried homozygous mutations in the *SLC30A10* gene (Quadri et al. 2012; Tuschl et al. 2012). The disease exhibited autosomal recessive inheritance, and unaffected siblings and parents were heterozygous for mutations in *SLC30A10* (Quadri et al. 2012; Tuschl et al. 2012). Interestingly, the clinical presentation described in 2008 and 2012 recapitulated findings from a prior case report published in 2000 (Gospe et al. 2000). This patient underwent follow-up examinations and was included in the 2012 studies, and genetic analyses identified homozygous mutations in *SLC30A10* (Lechpammer et al. 2014; Tuschl et al. 2012). He recently died and findings obtained from a complete autopsy were published in 2014 (Lechpammer et al. 2014). Features observed included hepatomegaly and micronodular cirrhosis with portal hypertension in the liver and severe neuronal loss in the globus pallidus (Lechpammer et al. 2014). There was a 16-fold increase in Mn levels in the basal ganglia and a 9-fold increase in the liver, but levels of Zn and iron (Fe) were normal in the brain (there was a 2–3-fold increase in liver Zn and Cu, but it was likely due to cirrhosis and compromised hepatic function) (Lechpammer et al. 2014). Neuronal loss in the globus pallidus is consistent with changes seen in humans patients exposed to elevated Mn from occupational sources (Olanow 2004; Perl and Olanow 2007).

The clinical and genetic studies described above suggest that mutations in SLC30A10 affect Mn metabolism in a manner that leads to Mn retention, likely due to decreased biliary excretion of Mn, and the observed neurotoxicity develops as a consequence of secondary Mn accumulation in the brain. The hepatic injury seen in affected patients is an important aspect of the disease, may be life-threatening, and can be explained by the observed deposition of Mn in the liver (Quadri et al. 2012; Tuschl et al. 2012). Polycythemia, another hallmark of the disease, may be a direct consequence of increased Mn levels as Mn and other transition metals, such as cobalt and nickel, have the ability to mimic the effects of hypoxia on erythropoietin gene expression by stabilizing hypoxia-inducible factor 1α (Ebert and Bunn 1999).

Some insights into the molecular mechanisms that lead to Mn retention when SLC30A10 function is compromised are now available. SLC30A10 belongs to the SLC30 family of metal transporters, which has ten members, SLC30A1–SLC30A10 (Huang and Tepaamorndech 2013; Kambe et al. 2015; Kolaj-Robin et al. 2015). SLC30A1–SLC30A8 transport Zn from the cytosol to the cell exterior or into the lumen of cellular organelles (i.e., mediate Zn efflux; the classification of SLC30A9 as a transporter is likely incorrect, and recent evidence suggests that it functions as a nuclear receptor coactivator) (Huang and Tepaamorndech 2013; Kambe et al.

2015; Kolaj-Robin et al. 2015). While SLC30A10 was initially thought to act as a Zn efflux transporter (Bosomworth et al. 2012), the fact that patients harboring mutations in this gene had elevated Mn levels suggested that it may mediate efflux of Mn, instead of Zn, and that disease-causing mutations may interfere with its Mn efflux function. To test this hypothesis, we performed a set of mechanistic experiments in cell and neuronal culture and in *C. elegans* (Leyva-Illades et al. 2014). Localization assays revealed that $SLC30A10_{WT}$ trafficked to the cell surface; in contrast, disease-causing mutants tested were trapped in the endoplasmic reticulum (Leyva-Illades et al. 2014). Mn measurement assays in cell culture revealed that $SLC30A10_{WT}$, but not a disease-causing mutant, reduced intracellular Mn levels. We performed a pulse-chase assay and confirmed that the reduction in intracellular Mn on $SLC30A10_{WT}$ expression was due to an increase in Mn efflux and not a reduction in Mn influx (Leyva-Illades et al. 2014). Additional studies revealed that expression of $SLC30A10_{WT}$, but not a disease-causing mutant, protected HeLa and GABAergic AF5 cells and primary midbrain neurons against Mn toxicity (Leyva-Illades et al. 2014). In contrast, siRNA-mediated knockdown of SLC30A10 in GABAergic AF5 cells led to Mn accumulation and heightened sensitivity to Mn toxicity (Lcyva-Illades et al. 2014). In *C. elegans*, expression of $SLC30A10_{WT}$ protected dopaminergic neurons against Mn-induced neurodegeneration, rescued a Mn-induced locomotor defect, and enhanced viability when worms were exposed to high Mn; these effects were not evident when a disease-causing mutant was expressed (Leyva-Illades et al. 2014). These results suggest that SLC30A10 functions as a cell surface-localized Mn efflux transporter that enhances manganese efflux and protects against Mn toxicity. Mutations that induce human disease block the Mn efflux activity of the transporter, leading to increased Mn accumulation within cells.

The fact that SLC30A10 mediates Mn efflux implies that there is an important difference between its substrate specificity and that of other SLC30 family transporters, which mediate Zn efflux. Additionally, available evidence suggests that SLC30A10 lacks observable Zn efflux activity in cells and organisms. To summarize here, in cell culture, expression of $SLC30A10_{WT}$ reduced intracellular Mn and protected against Mn toxicity, but did not reduce Zn levels or alter viability in response to Zn toxicity (Leyva-Illades et al. 2014; Zogzas et al. 2016). Similarly, in *C. elegans*, expression of $SLC30A10_{WT}$ protected against Mn, but not Zn, toxicity (Chen et al. 2015; Leyva-Illades et al. 2014). Finally, in humans, loss-of-function mutations in SLC30A10 increased Mn levels in the liver, blood and brain, but increases in plasma or brain Zn levels have not been reported (as described earlier, an increase in liver Zn was reported in one patient, but this was likely due to hepatic decompensation) (Lechpammer et al. 2014; Quadri et al. 2012; Tuschl et al. 2008, 2012). The mechanisms underlying the difference in metal specificity of SLC30A10 and other SLC30 proteins are largely unknown, but two recent papers have begun to provide some understanding (Nishito et al. 2016; Zogzas et al. 2016). We generated a predicted structure of SLC30A10, based on the solved crystal structure of the related bacterial Zn transporter YiiP, and performed structure-function assays (Zogzas et al. 2016). YiiP has a transmembrane domain with six membrane spanning

Fig. 1 Comparison of the solved crystal structure of YiiP with the predicted structure of SLC30A10. (**a**) Crystal structure of YiiP (Protein Data Bank code 3H90) is depicted in the cartoon format in *gray*; residues corresponding to Site A are shown as *red* sticks; Zn ion is shown as *yellow* sphere. (**b**) Predicted structure of SLC30A10 is shown in the cartoon format in *green*; residues corresponding to Site A of YiiP, which form the putative Site A of SLC30A10, are shown as *red* sticks. (**c**) Predicted structure of SLC30A10 is shown in the cartoon format in *green*; residues in the transmembrane segments that are required for Mn efflux activity are shown as *red* sticks. Note that not all required residues are depicted here and that residue H244 is not required by itself, but acts cooperatively with N127. Further details are provided in our ref. (Zogzas et al. 2016)

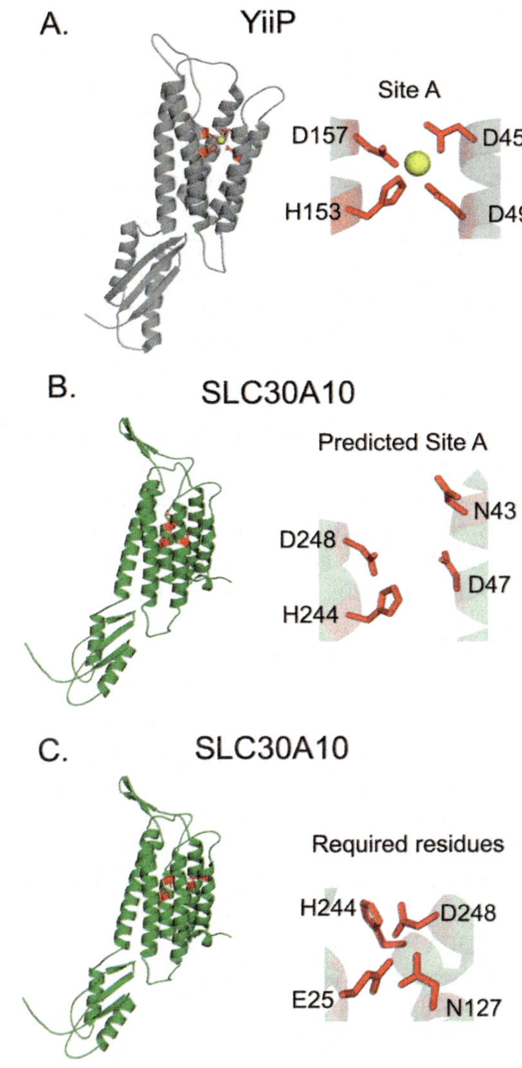

segments and a C-terminal domain (Lu et al. 2009, Lu and Fu 2007; Fig. 1a). In YiiP, side chains of amino acid residues Asp-45 and Asp-49 in the second and His-153 and Asp-157 in the fifth transmembrane segments form a metal coordination site, referred to as Site A, which coordinates Zn and is required for transport (Lu et al. 2009, Lu and Fu 2007; Fig. 1a). Prior studies show that, in other transporters of this superfamily (named cation diffusion facilitator), including in other SLC30 family proteins studied, residues that correspond to Site A of YiiP are crucial for metal coordination, specificity, and transport (Hoch et al. 2012; Huang and Tepaamorndech 2013; Kambe et al. 2015; Kolaj-Robin et al. 2015; Lu et al. 2009; Lu and Fu 2007; Martin and Giedroc 2016; Montanini et al. 2007; Ohana et al.

2009; Shusterman et al. 2014). Sequence analyses revealed that, in SLC30A10, residues corresponding to Site A of YiiP are Asn-43, Asp-47, His-244, and Asp-248 (Zogzas et al. 2016; Fig. 1b). The presence of Asn-43 was interesting because asparagine residues have a higher propensity to coordinate with Mn than with Zn (Dokmanic et al. 2008). Indeed, in Zn-transporting SLC30 proteins, this asparagine residue is replaced with histidine (Zogzas et al. 2016), which has a higher propensity to coordinate with Zn than asparagine (Dokmanic et al. 2008). Based on this, we hypothesized that perhaps a single amino acid change, from histidine in Zn-transporting SLC30 proteins to asparagine in SLC30A10, conferred Mn-transport capability to SLC30A10. To test this idea, first, we analyzed the obtained predicted structure of SLC30A10. Intriguingly, we discovered that, in SLC30A10, the side chain of Asn-43 pointed away from the putative ion binding pocket, located in the space between the second and fifth transmembrane domains, reducing the likelihood that it was involved in ion coordination and transport (Zogzas et al. 2016; Fig. 1b). Consistent with this, Mn transport assays in cells transfected with SLC30A10$_{WT}$ or mutants revealed that, among putative Site A residues, only Asp-248 was required for transport and side chains of Asn-43 and Asp-47 were not required (Zogzas et al. 2016; Fig. 1b and c). Instead, side chains of Glu-25 and Asn-127, located in the first and fourth transmembrane segment respectively and facing Asp-248, were required (Zogzas et al. 2016; Fig. 1c). His-244, a putative Site A residue, was not required by itself, but acted cooperatively with Asn-127 (Zogzas et al. 2016; Fig. 1c). Thus, in SLC30A10, only one of four residues that form its putative Site A is required for transport activity, suggesting that the mechanism of ion coordination in this transporter may be substantially different from that of Zn-transporting SLC30 proteins. Further biochemical and structural assays are now required to elucidate the nature of the difference in ion coordination and transport between SLC30A10 and other SLC30 proteins.

In the above discussion, it is important to note that, so far, our structure-function assays have been performed in HeLa cells, which do not express endogenous SLC30A10 and are amenable to genetic manipulation. However, substantial differences exist between immortal cell lines and neuronal and hepatic systems, where SLC30A10 functions under physiologic conditions. Transport kinetics of SLC30A10 mutants may differ between cell types due to changes in intracellular localization or differences in available interacting partners. The relevance of cell type-specific effects is underlined by findings in another recent paper on SLC30A10 function in which the authors expressed human SLC30A10$_{WT}$ or mutants in a chicken cell line (Nishito et al. 2016). They discovered that, in the avian system, SLC30A10 was largely trapped in the Golgi [the wild-type protein traffics to the plasma membrane in mammalian systems (Leyva-Illades et al. 2014; Zogzas et al. 2016)] and the side chain of the Asn-43 residue was required to protect against Mn-induced cell death (Nishito et al. 2016). Thus, in addition to biochemical and structural studies in minimal systems, it will be important to validate transport activities of SLC30A10 mutants in physiologically relevant primary neuronal and hepatocyte cultures.

How can the loss of the Mn efflux function of SLC30A10 at the cellular level lead to the phenotype observed in patients? SLC30A10 is expressed in the liver

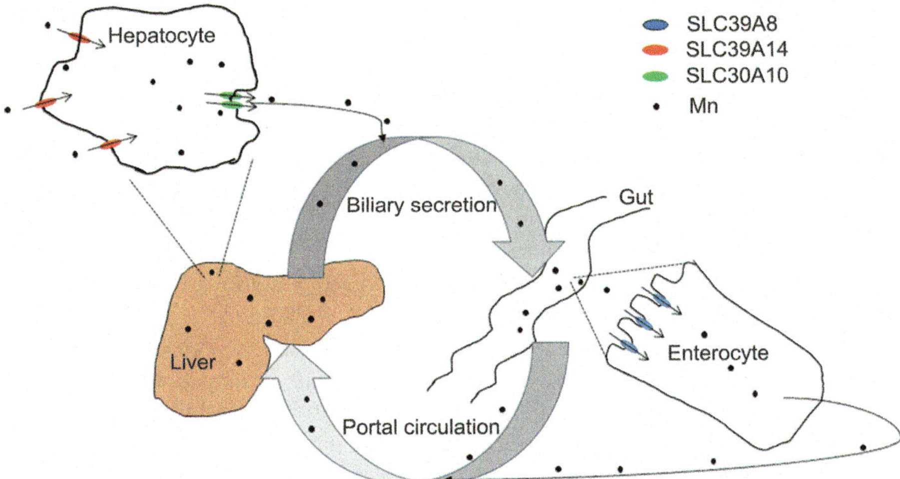

Fig. 2 Proposed model for the role of SLC39A8, SLC39A14, and SLC30A10 in regulating Mn homeostasis and detoxification. SLC39A8 may localize to the apical domain of enterocytes and mediate Mn (and Zn) influx into enterocytes. Metal ions would then be exported from enterocytes into blood by other transporters. SLC39A14 may localize to the basolateral aspect of hepatocytes and mediate influx of Mn from blood into hepatocytes. Finally, SLC30A10 may localize to the canalicular domain of hepatocytes and mediate efflux of Mn into bile. Consistent with the clinical data, this model predicts that loss of function of SLC39A8 should reduce Mn levels in blood; loss of function of SLC39A14 should increase Mn in blood, but not in the liver; and loss of function of SLC30A10 should increase Mn in the blood and liver

(Lechpammer et al. 2014; Quadri et al. 2012). Presumably, the efflux activity in the liver mediates biliary Mn excretion (Fig. 2). Loss of SLC30A10 function should decrease Mn excretion, and the retained Mn should accumulate in the liver leading to hepatic injury. Decreased Mn excretion should also lead to Mn accumulation in the blood and brain. Additionally, SLC30A10 is expressed in neurons of the basal ganglia, including in the globus pallidus (Lechpammer et al. 2014; Quadri et al. 2012). Loss of SLC30A10 function in the basal ganglia may further enhance Mn accumulation and induced damage in regional neurons. This sequence of events may culminate with the development of Mn-induced neurotoxicity. The above model remains to be experimentally tested, but is consistent with the available clinical data. To obtain more concrete evidence, we and others are generating global and tissue-specific SLC30A10 knockout mice to determine whether neurotoxicity is due to loss of efflux activity in the liver, brain, or both organs.

SLC39A14

In 2016, mutations in *SLC39A14* were reported to lead to the onset of Mn-induced neurotoxicity in humans (Tuschl et al. 2016). SLC39A14 belongs to the SLC39 family of metal transporters (Jeong and Eide 2013). While most members of this family mediate Zn influx into cells, SLC39A14 is known to mediate influx of Mn, Fe, and cadmium (Cd), in addition to Zn (Jeong and Eide 2013; Liuzzi et al. 2006; Girijashanker et al. 2008; Jenkitkasemwong et al. 2012; Pinilla-Tenas et al. 2011; Taylor et al. 2005). Patients harboring homozygous mutations in this gene were born to consanguineous parents (Tuschl et al. 2016). Clinical signs were evident early in life and included loss of developmental milestones, progressive dystonia, and bulbar dysfunction. Around age 10 years, patients developed severe, generalized dystonia that was resistant to treatment, spasticity, limb contractures, and scoliosis and lost the ability to move about by themselves. Some patients showed features of parkinsonism, such as hypomimia, tremor, and bradykinesia (Tuschl et al. 2016). Levels of Mn in blood were ~3–25-fold greater than normal; in contrast, blood levels of Fe, Zn, and Cd in tested patients were normal (Tuschl et al. 2016). Magnetic resonance imaging showed evidence of Mn deposition in the globus pallidus, striatum (lesser than the pallidus), and anterior pituitary and extensive involvement of the white matter (cerebellum, spinal cord, and dorsal pons). In some patients, evidence of cerebral and cerebellar atrophy was present (Tuschl et al. 2016). Importantly, however, there was no evidence of Mn deposition in the liver (Tuschl et al. 2016), suggesting that the transport activity of SLC39A14 may be required for import of Mn into hepatocytes. Consistent with the absence of Mn deposition in the liver, affected patients did not develop liver disease seen in individuals carrying SLC30A10 mutations (Tuschl et al. 2016). Polycythemia was also absent (Tuschl et al. 2016). Autopsy findings were available from one individual and revealed severe neuronal loss in the globus pallidus and the dentate nucleus of the cerebellum (Tuschl et al. 2016). Neurons in the caudate, putamen, thalamus, and cerebral cortex were largely preserved (Tuschl et al. 2016).

What are the molecular mechanisms that lead to Mn retention and neurotoxicity in patients harboring *SLC39A14* mutations? In HEK-293 cells, SLC39A14$_{WT}$, as well as disease-causing mutants tested, appeared to traffic to the cell surface (Tuschl et al. 2016). However, in this system, Mn influx was greater in cells expressing SLC39A14$_{WT}$ compared to those expressing disease-causing mutants, suggesting that the mutations inhibited the Mn transport activity of the protein (Tuschl et al. 2016). The mechanism by which loss-of-function mutations in a Mn importer lead to Mn toxicity may rely on the localization of the transporter in cells and tissues. SLC39A14 is expressed in the liver as well as in neurons, including in the globus pallidus (Tuschl et al. 2016). SLC39A14 may function to transport Mn from blood into hepatocytes, while SLC30A10 may function to transport Mn from within hepatocytes to bile (Fig. 2; also see Tuschl et al. 2016). In such a scenario, the influx transporter SLC39A14 and efflux transporter SLC30A10 would function synergistically to mediate biliary Mn excretion (Fig. 2). This model remains to be experimentally tested, but is consistent with available data. The model predicts that loss of SLC30A10 function should increase liver, blood, and brain Mn, while loss of SLC39A14 function should increase Mn in blood, but not in the liver. These features are seen in patients who harbor loss-of-function mutations of the respective genes (Lechpammer et al. 2014; Quadri et al. 2012; Tuschl et al. 2008, 2012, 2016). Notably, patients carrying SLC39A14 mutations accumulate Mn in the brain (Tuschl et al. 2016). Similar findings were reported when SLC39A14 was depleted in zebrafish (Tuschl et al. 2016). These results imply that while the transport activity of SLC39A14 may be crucial for hepatic Mn influx, it is not required for the uptake of Mn into neuronal cells. Overall, the neurotoxicity seen in patients harboring SLC39A14 mutations is probably a consequence of a decrease in the biliary excretion of Mn, which leads to Mn accumulation in the brain and subsequent neuronal injury. Finally, the reason why patients with mutations in SLC30A10, but not SLC39A14, develop polycythemia remains to be clarified. As described earlier, a possible mechanism for polycythemia in patients with SLC30A10 mutations is Mn-induced chemical hypoxia followed by increased erythropoietin production. It may be that SLC39A14 is required for Mn influx into erythropoietin-producing cells, and when this transporter is mutated, the chemical hypoxia leading to erythropoietin production does not occur.

SLC39A8

In 2015, mutations in *SLC39A8* were reported to cause an inherited disorder of Mn and Zn deficiency (Boycott et al. 2015; Park et al. 2015). SLC39A8 belongs to the SLC39 family of transporters, similar to SLC39A14, and is known to mediate influx of Zn, Mn, Fe, Cd, and cobalt (Jenkitkasemwong et al. 2012; Jeong and Eide 2013; Wang et al. 2012). One of the reports described clinical findings from a German child born to unrelated parents who was initially referred at 4 months of age (Park et al. 2015). The patient presented with infantile spasms, dwarfism, cranial asymmetry, and hearing loss. Radiology demonstrated craniosynostosis of the coronary and lambdoid sutures with asymmetrical brain atrophy; the cerebellum was normal (Park et al. 2015). The second report identified six children from the Hutterite ethno-religious group, which is genetically isolated due to sociocultural practices, and two children of Egyptian descent born to consanguineous parents (Boycott et al. 2015). Patients from the Hutterite group presented with profound intellectual disability, developmental delay, hypotonia, strabismus, and cerebellar atrophy. Signs were evident soon after birth and head control was achieved only in early childhood. Additional features included short stature, osteopenia, recurrent infections, and, in most cases, an inability to walk (Boycott et al. 2015). The Egyptian children were siblings who presented at age 2 and 8 years with intellectual disability, developmental delay, hypotonia, and strabismus. One of the siblings had myoclonic seizures (Boycott et al. 2015). Genetic analyses revealed that affected patients carried missense mutations in both copies of *SLC39A8* (Boycott et al. 2015; Park et al. 2015). The Hutterite and Egyptian patients had homozygous mutations that changed glycine at amino acid 38 to arginine; the German patient carried the above glycine-to-arginine mutation and an isoleucine-to-asparagine mutation at amino acid 340 (Boycott et al. 2015; Park et al. 2015). In all cases tested, parents were heterozygous for these mutations (Boycott et al. 2015; Park et al. 2015). Importantly, plasma Mn levels were below detectable limits in the German patient, while serum Zn levels were normal (Park et al. 2015). However, in the Hutterite and Egyptian groups, blood or erythrocyte Mn levels were low only in four out of seven patients (Mn values were not available for one patient) and within the normal range in the other three; plasma or serum Zn values were mildly decreased in five patients (Boycott et al. 2015). The reason for the difference in metal ion levels in the patients is unknown. It may be reflective of the difference in the mutations present. Immunoblot analyses from fibroblasts isolated from a control subject and a Hutterite patient who was homozygous for the glycine-to-arginine mutation demonstrated comparable SLC39A8 protein levels (Boycott et al. 2015), suggesting that this mutation, by itself, did not affect protein expression. One possibility is that the isoleucine-to-asparagine mutation at amino acid 340 may mislocalize, degrade, and/or profoundly inhibit Mn transport activity of SLC39A14, and this may be the underlying reason that Mn levels are undetectable in the patient harboring this mutation. Differences in Zn levels in patients reported in the two papers may also be related to differential effects of the mutations. It may be that when SLC39A8 function is completely abolished, activities of other Zn transporters are altered to compensate; this compensatory effect may not be evident when SLC39A8 activity is only partially inhibited. Further understanding of the cellular changes leading to the disease phenotype requires in-depth knowledge of the effects of the disease-causing mutations on SLC39A8 transport activity, which is not yet available. Biochemical and cell biological approaches used to study functional consequences of SLC30A10 and SLC39A14 mutations should provide necessary insights.

While the mechanisms by which mutations in SLC39A8 lead to human disease are not yet clear, changes in glycosylation may play a role. The German patient described above exhibited a defect in *N*-linked glycosylation, with a primary problem in galactosylation (Park et al. 2015). The phenotypic presentation of this patient had similarities to that seen in a congenital disorder of glycosylation when *SLC35A2*, a UDP-galactose transporter, is mutated (Park et al. 2015). SLC35A2 imports UDP-galactose from the cytosol into the Golgi (Ng et al. 2013). Within the Golgi, galactose is transferred to acceptor proteins by galactosyltransferase enzymes (Hennet 2002), several of which require Mn for activity (Wagner and Cynkin 1971; Schachter et al. 1971). Conceivably, when intracellular Mn is low, activities of galactosyltransferases may be inhibited, leading to the observed glycosylation defect. In support of this idea, prior studies have demonstrated that glycosylation defects occur when SPCA1, the Golgi localized P-type pump that transports Mn from the cytosol into the Golgi, is depleted (Ramos-Castaneda et al. 2005). Thus, it is not surprising that there are phenotypic similarities between patients with *SLC35A2* and *SLC39A8* mutations. Indeed, after the discovery of the glycosylation defect, Park et al. went on to screen for *SLC39A8* mutations in patients who had impaired glycosylation of unknown origin and identified another patient with mutations in the *SLC39A8* gene; Mn levels in blood were below the detectable limit in this patient (Park et al. 2015). However, the extent to which a defect in glycosylation contributes to the development of the disease and the severity of the phenotype in all patients who have mutations in SLC39A8 is not clear. Patients who were homozygous for the glycine-to-arginine mutation did not have a severe Mn deficiency and, compared to patients without detectable Mn in blood or plasma, exhibited milder defects in glycosylation; yet, they had extensive neurologic damage (Boycott et al. 2015; Park et al. 2015). These results imply that while there may be a relationship between Mn levels and glycosylation efficiency, the contribution of the glycosylation defect to disease development and severity needs further assessment. The relative contribution of Mn versus Zn deficiency to disease pathobiology is also unclear and needs to be elucidated. Overall, mutations in SLC39A8 lead to severe neurological disease that is related to Mn and Zn deficiency (Fig. 2) and a defect in *N*-linked glycosylation. Future studies need to determine the effect of disease-causing mutations on transporter activity and to elucidate the mechanisms by which altered transport activity changes metal levels; affects cellular processes, such as glycosylation; and induces disease.

Concluding Perspectives

The discoveries of the above-described genetic diseases are poised to revolutionize our understanding of Mn homeostasis, detoxification, and induced toxicity at the cellular and organismal level. All three disorders were identified within the last 5 years; therefore, detailed understanding of disease pathobiology cannot be expected. However, rapid progress is anticipated as multiple laboratories are now working on elucidating the mechanisms by which these proteins mediate Mn transport and regulate Mn homeostasis and detoxification, by which the transporters themselves are regulated in response to changing Mn levels and by which changes in transporter activities alter brain Mn to induce disease. While these genetic diseases are rare, studying these disorders is expected to improve our understanding of Mn biology in general and the mechanisms of Mn-induced neurotoxicity in particular. Indeed, *SLC39A14* knockout mice are available (Aydemir et al. 2012), and *SLC30A10* knockout mice are being generated. It is likely that these genetic models will make invaluable contributions to the neurotoxicity field. It also noteworthy that recent population-based studies identified a common single nucleotide polymorphism in *SLC30A10* that was associated with increased Mn levels in blood, altered neurological function, and decreased *SLC30A10* expression (Wahlberg et al. 2016). Similarly, single nucleotide polymorphisms in *SLC39A8* and *SLC39A14* have been associated with increased Cd levels in humans (Rentschler et al. 2014). These discoveries raise the intriguing possibility that single nucleotide polymorphisms in these genes may alter the risk for the development of toxicity from Mn and other metals in the general population. Finally, increasing Mn efflux may be an effective strategy for protection against or treatment of Mn-induced neurotoxicity (Leyva-Illades et al. 2014; Mukhopadhyay and Linstedt 2011). As SLC30A10, SLC39A14, and SLC39A8 appear to be the primary transporters responsible for maintaining homeostatic control of Mn in humans, understanding the function and regulation of these proteins may directly augment the ability to generate therapeutically viable and effective efflux enhancing drugs.

Acknowledgments Supported by R00-ES020844 and R01-ES024812 from NIH/NIEHS (to S. M.).

References

Aschner M, Erikson KM, Herrero Hernandez E, Tjalkens R. Manganese and its role in Parkinson's disease: from transport to neuropathology. NeuroMolecular Med. 2009;11:252–66.

Aydemir TB, Sitren HS, Cousins RJ. The zinc transporter Zip14 influences c-met phosphory-lation and hepatocyte proliferation during liver regeneration in mice. Gastroenterology. 2012;142:1536–46.

Bosomworth HJ, Thornton JK, Coneyworth LJ, Ford D, Valentine RA. Efflux function, tissue-specific expression and intracellular trafficking of the Zn transporter ZnT10 indicate roles in adult Zn homeostasis. Metallomics. 2012;4:771–9.

Bowman AB, Aschner M. Considerations on manganese (Mn) treatments for in vitro studies. Neurotoxicology. 2014;41:141–2.

Boycott KM, Beaulieu CL, Kernohan KD, Gebril OH, Mhanni A, Chudley AE, Redl D, Qin W, Hampson S, Kury S, Tetreault M, Puffenberger EG, Scott JN, Bezieau S, Reis A, Uebe S, Schumacher J, Hegele RA, Mcleod DR, Galvez-Peralta M, Majewski J, Ramaekers VT, Care4Rare Canada Consortium, Nebert DW, Innes AM, Parboosingh JS, Abou Jamra R. Autosomal-Recessive Intellectual Disability with Cerebellar Atrophy Syndrome Caused by Mutation of the Manganese and Zinc Transporter Gene SLC39A8. *Am J Hum Genet.* 2015;97:886–93.

Butterworth RF. Parkinsonism in cirrhosis: pathogenesis and current therapeutic options. Metab Brain Dis. 2013;28:261–7.

Chen P, Bowman AB, Mukhopadhyay S, Aschner M. SLC30A10: A novel manganese transporter. WormBook. 2015;4:e1042648.

Culotta VC, Yang M, Hall MD. Manganese transport and trafficking: lessons learned from *Saccharomyces cerevisiae*. Eukaryot Cell. 2005;4:1159–65.

Dokmanic I, Sikic M, Tomic S. Metals in proteins: correlation between the metal-ion type, coor-dination number and the amino-acid residues involved in the coordination. Acta Crystallogr D Biol Crystallogr. 2008;64:257–63.

Ebert BL, Bunn HF. Regulation of the erythropoietin gene. Blood. 1999;94:1864–77.

Freeland-Graves JH, Mousa TY, Kim S. International variability in diet and requirements of man-ganese: causes and consequences. J Trace Elem Med Biol. 2016;

Girijashanker K, He L, Soleimani M, Reed JM, Li H, Liu Z, Wang B, Dalton TP, Nebert DW. Slc39a14 gene encodes ZIP14, a metal/bicarbonate symporter: similarities to the ZIP8 transporter. Mol Pharmacol. 2008;73:1413–23.

Gospe SM Jr, Caruso RD, Clegg MS, Keen CL, Pimstone NR, Ducore JM, Gettner SS, Kreutzer RA. Paraparesis, hypermanganesaemia, and polycythaemia: a novel presentation of cirrhosis. Arch Dis Child. 2000;83:439–42.

Hennet T. The galactosyltransferase family. Cell Mol Life Sci. 2002;59:1081–95.

Hoch E, Lin W, Chai J, Hershfinkel M, Fu D, Sekler I. Histidine pairing at the metal transport site of mammalian ZnT transporters controls Zn2+ over Cd2+ selectivity. Proc Natl Acad Sci U S A. 2012;109:7202–7.

Huang L, Tepaamorndech S. The SLC30 family of zinc transporters - a review of current understanding of their biological and pathophysiological roles. Mol Asp Med. 2013;34:548–60.

Jenkitkasemwong S, Wang CY, Mackenzie B, Knutson MD. Physiologic implications of metal-ion transport by ZIP14 and ZIP8. Biometals. 2012;25:643–55.

Jensen LT, Carroll MC, Hall MD, Harvey CJ, Beese SE, Culotta VC. Down-regulation of a manganese transporter in the face of metal toxicity. Mol Biol Cell. 2009;20:2810–9.

Jeong J, Eide DJ. The SLC39 family of zinc transporters. Mol Asp Med. 2013;34:612–9.

Kambe T, Tsuji T, Hashimoto A, Itsumura N. The physiological, biochemical, and molecular roles of zinc transporters in zinc homeostasis and metabolism. Physiol Rev. 2015;95:749–84.

Kolaj-Robin O, Russell D, Hayes KA, Pembroke JT, Soulimane T. Cation diffusion facilitator family: structure and function. FEBS Lett. 2015;589:1283–95.

Lechpammer M, Clegg MS, Muzar Z, Huebner PA, Jin LW, Gospe SM Jr. Pathology of inherited manganese transporter deficiency. Ann Neurol. 2014;75:608–12.

Leyva-Illades D, Chen P, Zogzas CE, Hutchens S, Mercado JM, Swaim CD, Morrisett RA, Bowman AB, Aschner M, Mukhopadhyay S. SLC30A10 is a cell surface-localized manganese efflux transporter, and parkinsonism-causing mutations block its intracellular trafficking and efflux activity. J Neurosci. 2014;34:14079–95.

Liuzzi JP, Aydemir F, Nam H, Knutson MD, Cousins RJ. Zip14 (Slc39a14) mediates non-transferrin-bound iron uptake into cells. Proc Natl Acad Sci U S A. 2006;103:13612–7.

Lu M, Fu D. Structure of the zinc transporter YiiP. Science. 2007;317:1746–8.

Lu M, Chai J, Fu D. Structural basis for autoregulation of the zinc transporter YiiP. Nat Struct Mol Biol. 2009;16:1063–7.

Lucchini RG, Guazzetti S, Zoni S, Donna F, Peter S, Zacco A, Salmistraro M, Bontempi E, Zimmerman NJ, Smith DR. Tremor, olfactory and motor changes in Italian adolescents exposed to historical ferro-manganese emission. Neurotoxicology. 2012;33:687–96.

Lucchini RG, Guazzetti S, Zoni S, Benedetti C, Fedrighi C, Peli M, Donna F, Bontempi E, Borgese L, Micheletti S, Ferri R, Marchetti S, Smith DR. Neurofunctional dopaminergic impairment in elderly after lifetime exposure to manganese. Neurotoxicology. 2014;45:309–17.

Martin JE, Giedroc DP. Functional determinants of metal ion transport and selectivity in paralogous cation diffusion facilitator transporters CzcD and MntE in Streptococcus Pneumoniae. J Bacteriol. 2016;198:1066–76.

Montanini B, Blaudez D, Jeandroz S, Sanders D, Chalot M. Phylogenetic and functional analysis of the cation diffusion facilitator (CDF) family: improved signature and prediction of substrate specificity. BMC Genomics. 2007;8:107.

Mukhopadhyay S, Linstedt AD. Identification of a gain-of-function mutation in a Golgi P-type ATPase that enhances Mn2+ efflux and protects against toxicity. Proc Natl Acad Sci U S A. 2011;108:858–63.

Ng BG, Buckingham KJ, Raymond K, Kircher M, Turner EH, He M, Smith JD, Eroshkin A, Szybowska M, Losfeld ME, Chong JX, Kozenko M, Li C, Patterson MC, Gilbert RD, Nickerson DA, Shendure J, Bamshad MJ, University of Washington Center for Mendelian Genomics, Freeze HH. Mosaicism of the UDP-galactose transporter SLC35A2 causes a congenital disorder of glycosylation. Am J Hum Genet. 2013;92:632–6.

Nishito Y, Tsuji N, Fujishiro H, Takeda TA, Yamazaki T, Teranishi F, Okazaki F, Matsunaga A, Tuschl K, Rao R, Kono S, Miyajima H, Narita H, Himeno S, Kambe T. Direct comparison of manganese detoxification/efflux proteins and molecular characterization of ZnT10 protein as a manganese transporter. J Biol Chem. 2016;291:14773–87.

Ohana E, Hoch E, Keasar C, Kambe T, Yifrach O, Hershfinkel M, Sekler I. Identification of the Zn2+ binding site and mode of operation of a mammalian Zn2+ transporter. J Biol Chem. 2009;284:17677–86.

Olanow CW. Manganese-induced parkinsonism and Parkinson's disease. Ann N Y Acad Sci. 2004;1012:209–23.

Park JH, Hogrebe M, Gruneberg M, Duchesne I, von der Heiden AL, Reunert J, Schlingmann KP, Boycott KM, Beaulieu CL, Mhanni AA, Innes AM, Hortnagel K, Biskup S, Gleixner EM, Kurlemann G, Fiedler B, Omran H, Rutsch F, Wada Y, Tsiakas K, Santer R, Nebert DW, Rust S, Marquardt T. SLC39A8 deficiency: a disorder of manganese transport and glycosylation. Am J Hum Genet. 2015;97:894–903.

Perl DP, Olanow CW. The neuropathology of manganese-induced parkinsonism. J Neuropathol Exp Neurol. 2007;66:675–82.

Pinilla-Tenas JJ, Sparkman BK, Shawki A, Illing AC, Mitchell CJ, Zhao N, Liuzzi JP, Cousins RJ, Knutson MD, Mackenzie B. Zip14 is a complex broad-scope metal-ion transporter whose functional properties support roles in the cellular uptake of zinc and nontransferrin-bound iron. Am J Physiol Cell Physiol. 2011;301:C862–71.

Quadri M, Federico A, Zhao T, Breedveld GJ, Battisti C, Delnooz C, Severijnen LA, di Toro Mammarella L, Mignarri A, Monti L, Sanna A, Lu P, Punzo F, Cossu G, Willemsen R, Rasi F, Oostra BA, van de Warrenburg BP, Bonifati V. Mutations in SLC30A10 cause parkinsonism and dystonia with hypermanganesemia, polycythemia, and chronic liver disease. Am J Hum Genet. 2012;90:467–77.

Ramos-Castaneda J, Park YN, Liu M, Hauser K, Rudolph H, Shull GE, Jonkman MF, Mori K, Ikeda S, Ogawa H, Arvan P. Deficiency of ATP2C1, a Golgi ion pump, induces secretory pathway defects in endoplasmic reticulum (ER)-associated degradation and sensitivity to ER stress. J Biol Chem. 2005;280:9467–73.

Rentschler G, Kippler M, Axmon A, Raqib R, Skerfving S, Vahter M, Broberg K. Cadmium concentrations in human blood and urine are associated with polymorphisms in zinc transporter genes. Metallomics. 2014;6:885–91.

Schachter H, Mcguire EJ, Roseman S. Sialic acids. 13. A uridine diphosphate D-galactose: mucin galactosyltransferase from porcine submaxillary gland. J Biol Chem. 1971;246:5321–8.

Shusterman E, Beharier O, Shiri L, Zarivach R, Etzion Y, Campbell CR, Lee IH, Okabayashi K, Dinudom A, Cook DI, Katz A, Moran A. ZnT-1 extrudes zinc from mammalian cells functioning as a Zn(2+)/H(+) exchanger. Metallomics. 2014;6:1656–63.

Taylor KM, Morgan HE, Johnson A, Nicholson RI. Structure-function analysis of a novel member of the LIV-1 subfamily of zinc transporters, ZIP14. FEBS Lett. 2005;579:427–32.

Tuschl K, Mills PB, Parsons H, Malone M, Fowler D, Bitner-Glindzicz M, Clayton PT. Hepatic cirrhosis, dystonia, polycythaemia and hypermanganesaemia--a new metabolic disorder. J Inherit Metab Dis. 2008;31:151–63.

Tuschl K, Clayton PT, Gospe SM Jr, Gulab S, Ibrahim S, Singhi P, Aulakh R, Ribeiro RT, Barsottini OG, Zaki MS, del Rosario ML, Dyack S, Price V, Rideout A, Gordon K, Wevers RA, Chong WK, Mills PB. Syndrome of hepatic cirrhosis, dystonia, polycythemia, and hypermanganesemia caused by mutations in SLC30A10, a manganese transporter in man. Am J Hum Genet. 2012;90:457–66.

Tuschl K, Meyer E, Valdivia LE, Zhao N, Dadswell C, Abdul-Sada A, Hung CY, Simpson MA, Chong WK, Jacques TS, Woltjer RL, Eaton S, Gregory A, Sanford L, Kara E, Houlden H, Cuno SM, Prokisch H, Valletta L, Tiranti V, Younis R, Maher ER, Spencer J, Straatman-Iwanowska A, Gissen P, Selim LA, Pintos-Morell G, Coroleu-Lletget W, Mohammad SS, Yoganathan S, Dale RC, Thomas M, Rihel J, Bodamer OA, Enns CA, Hayflick SJ, Clayton PT, Mills PB, Kurian MA, Wilson SW. Mutations in SLC39A14 disrupt manganese homeostasis and cause childhood-onset parkinsonism-dystonia. Nat Commun. 2016;7:11601.

Wagner RR, Cynkin MA. Glycoprotein metabolism: a UDP-galactose-glycoprotein galactosyltransferase of rat serum. Biochem Biophys Res Commun. 1971;45:57–62.

Wahlberg K, Kippler M, Alhamdow A, Rahman SM, Smith DR, Vahter M, Lucchini RG, Broberg K. Common polymorphisms in the solute carrier SLC30A10 are associated with blood manganese and neurological function. Toxicol Sci. 2016;149:473–83.

Wang CY, Jenkitkasemwong S, Duarte S, Sparkman BK, Shawki A, Mackenzie B, Knutson MD. ZIP8 is an iron and zinc transporter whose cell-surface expression is up-regulated by cellular iron loading. J Biol Chem. 2012;287:34032–43.
Zogzas CE, Aschner M, Mukhopadhyay S. Structural elements in the transmembrane and cytoplasmic domains of the metal transporter SLC30A10 are required for its manganese efflux activity. J Biol Chem. 2016;291:15940–57.

Part II
Neurodegenerative Disorders

Chemical Speciation of Selenium and Mercury as Determinant of Their Neurotoxicity

C.S. Oliveira, B.C. Piccoli, M. Aschner, and J.B.T. Rocha

Abstract The antagonism of mercury toxicity by selenium has been well documented. Mercury is a toxic metal, widespread in the environment. The main target organs (kidneys, lungs, or brain) of mercury vary depending on its chemical forms (inorganic or organic). Selenium is a semimetal essential to mammalian life as part of the amino acid selenocysteine, which is required to the synthesis of the selenoproteins. This chapter has the aim of disclosing the role of selenide or hydrogen selenide (Se^{-2} or HSe^{-}) as central metabolite of selenium and as an important antidote of the electrophilic mercury forms (particularly, Hg^{2+} and MeHg). Emphasis will be centered on the neurotoxicity of electrophile forms of mercury and selenium. The controversial participation of electrophile mercury and selenium forms in the development of some neurodegenerative disease will be briefly presented. The potential pharmacological use of organoseleno compounds (Ebselen and diphenyl diselenide) in the treatment of mercury poisoning will be considered. The central role of thiol (−SH) and selenol (−SeH) groups as the generic targets of electrophile mercury forms and the need of new *in silico* tools to guide the future biological researches will be commented.

Keywords Selenide • Selenoproteins • Selenocysteine • Cysteine • Neurotoxicity

C.S. Oliveira • B.C. Piccoli
Programa de Pós-graduação em Ciências Biológicas: Bioquímica Toxicológica, Centro de Ciências Naturais e Exatas, Universidade Federal de Santa Maria, Santa Maria, RS, Brazil

M. Aschner
Department of Molecular Pharmacology, Albert Einstein College of Medicine, Bronx, New York, NY, USA

J.B.T. Rocha (✉)
Programa de Pós-graduação em Ciências Biológicas: Bioquímica Toxicológica, Centro de Ciências Naturais e Exatas, Universidade Federal de Santa Maria, Santa Maria, RS, Brazil

Departamento de Bioquímica e Biologia Molecular, Centro de Ciências Naturais e Exatas, Universidade Federal de Santa Maria, Santa Maria, RS, Brazil
e-mail: jbtrocha@gmail.com

© Springer International Publishing AG 2017
M. Aschner and L.G. Costa (eds.), *Neurotoxicity of Metals*, Advances in Neurobiology 18, DOI 10.1007/978-3-319-60189-2_4

Introduction

Selenium is an element located in the group 16 (formerly 6A) of the periodic table, and it is in the same family of oxygen, sulfur, tellurium, and polonium (Housecroft and Sharpe 2012). From the physiological point of view, selenium can partially imitate the chemistry of sulfur, particularly when present in the form of the selenol group ($-SeH$). The -SeH is analog to the thiol group ($-SH$) found in the amino acids selenocysteine (Sec) and cysteine (Cys), respectively (Fig. 1) (Rocha et al. 2017). Selenium is sometimes erroneously classified in the same category of the toxic metals such as Hg, Cd, and Pb, among others (Frost 1972). However, from the chemical point of view, selenium is a semimetal (Housecroft and Sharpe 2012).

Selenium is an essential element to the life maintenance, but not for all kind of organisms. For instance, several prokaryotes and almost all the animals studied require selenium in some redox reactions catalyzed by selenoproteins (Drosophila 12 genomes Consortium 2007; Chapple and Guigó 2008; Lobanov et al. 2008). High plants and fungi can have selenium in low- and high-molecular (selenium-containing proteins) mass molecules, but these molecules do not have a defined physiological role (Lobanov et al. 2009). The main and, possibly, the only role of selenium in cell physiology is associated with its incorporation in the organic moiety of the amino acid Sec. This amino acid is found in specific selenoproteins, where its –SeH group participates in important redox reactions (Heverly-Coulson and Boyd 2010; Nauser et al. 2012; Hatfield et al. 2014; Labunskyy et al. 2014). The incorporation of selenium into the phosphoester of serinephosphate, which forms the selenocysteinyl residue, is complex and involves obligatorily its metabolism to selenide or hydrogen selenide (Se^{2-} or HSe^- and H_2Se). Thus, though we can ingest different forms of selenium (either inorganic or organic), all the "physiologically active" selenium in mammals will have to be metabolized to selenide.

In contrast to its analogous sulfide or hydrogen sulfide (S^{2-} or HS^- and H_2S; Fig. 1), which have been demonstrated to be an endogenous gaseous transmitter (Wang 2012; Yang et al. 2008), up to now no definite physiological role have been attribute to selenide.

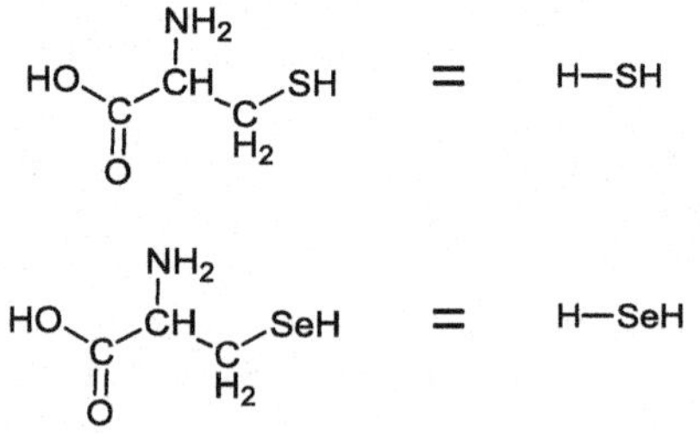

Fig. 1 Hydrogen sulfide (H_2S) can be considered an equivalent of the thiol group (R-SH) of thiol-containing organo-molecules (for instance, cysteine). In analogy, hydrogen selenide (H_2Se) can be considered a nonorganic equivalent of the selenol ($-SeH$) group

For adult humans, the dietary recommended daily intake of selenium is about 1 µg/kg of body weight (Metanis et al. 1995; NAS 2000). The deficiency of dietary selenium has been considered an important factor in diseases such as the Keshan disease (endemic cardiomyopathy) and Kashin-Beck disease (degenerative osteoarthropathy) (Navarro-Alarcon and López-Martínez 2000; Chen 2012). In contrast, intentional, accidental, or chronic intake of selenium has been associated with problems in nail and hair structure, in the gastrointestinal function or neurological damage (Vinceti et al. 2001, 2009; Aldosary et al. 2012). Recent studies have been indicating that selenium supplementation can increase the probability of developing diabetes type 2 (Ogawa-Wong et al. 2016), amyotrophic lateral sclerosis (ALS), and certain types of cancer (WHO 2003; Vinceti et al. 2010, 2012).

The modern use of selenium by the man can be divided in three categories: (1) as a nutritional supplement, which can be achieved either by ingesting selenium-containing formulations or enriched-selenium crops (Fagan et al. 2015; Malagoli et al. 2015), (2) in organic synthesis as intermediate or to produce bioactive molecules (Nogueira and Rocha 2011), and (3) as electronic component (for instance, as semiconductor in quantum-DOTS). The use of selenium in DOTS may have environmental and toxicological significance, because the salts of selenium found in DOTS have toxic metals in their composition, for instance, cadmium, mercury, lead, and bismuth (Peng et al. 2000; Ellingson et al. 2005; Khan and Wang 2009).

The objective of this chapter is to furnish general information about the physiological role of selenium in vertebrate cells and the importance of selenide as intermediate of selenium metabolism. The fate of selenium in the environment will be briefly presented to indicate how this element is incorporated both in a nonspecific way (plants) and in a specific way (humans) in organic molecules. The interaction of selenium and mercury will be discussed, because selenide is one of the most important "antidote of electrophilic forms of mercury." The neuroprotective effects of selenide, particularly in relation to neurotoxicity of mercury, will be discussed with more detail, because selenide and selenol can be considered the strongest coordinating forms of inorganic and organic mercury compounds. The neurotoxicity of selenide and selenium will be discussed, despite the available data about this subject is still scant (Vinceti et al. 2013, 2014).

Selenium in the Environment and in the Living Organisms

The fate of selenium in the biosphere is presented in this chapter to give to the readers an idea on how selenium is incorporated by living organisms. The environmental selenium levels is determined by: (1) the natural presence of selenium in the environment (which is mainly determined by the occurrence of different chemical forms of selenium in a particular soil or in aquatic sediments) and (2) the anthropogenic release of selenium (for instance, selenium mobilized from the combustion of coal at electric powers or by its use in crop fertilization) (Chapman 1999; Fordyce 2007). Irrigation of seleniferous soils has produced subsurface drainage of high-selenium

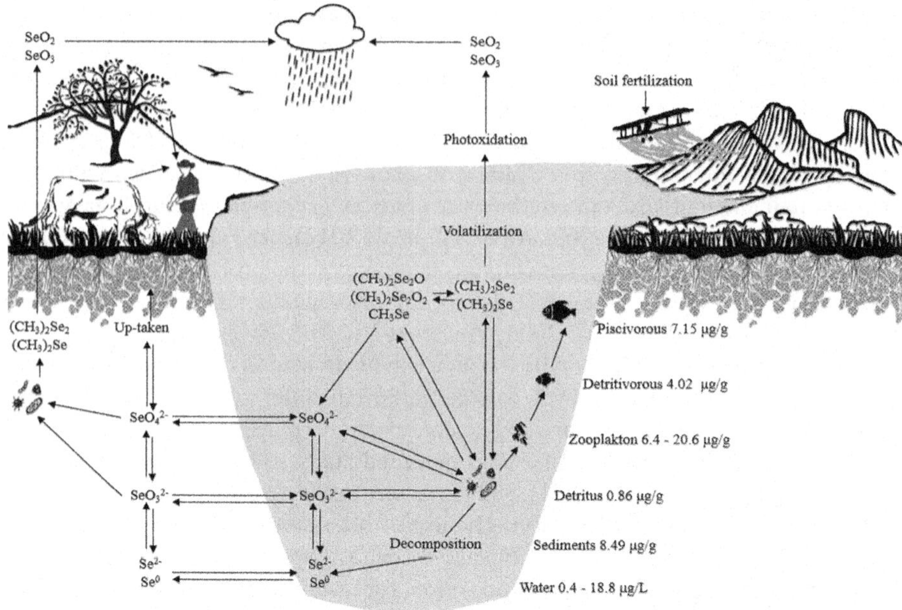

Fig. 2 Cycle of selenium in the environment. Selenium is naturally found in terrestrial (soil) or aquatic ecosystems in different chemical forms. The most biologically important chemical forms of selenium are selenite (Se^{4+}) and selenate (Se^{6+}). Anthropogenic sources can also increase the levels of selenium in a given ecosystem, for example, soil Se-enrichment or drainage of selenium via irrigation of seleniferous regions. Once in the environment, the Se^0 can be either oxidized to selenite and selenate or reduced to selenide by microorganisms; it is important to note that these reactions are reversible. Selenide can be incorporated into organic compounds, for instance, selenium-containing proteins and other small molecules in plants and selenoproteins in animals. Studies have demonstrated the biomagnification of selenium in the aquatic food chain; here the selenium levels were based in studies of Lemly (1996) and Barwick and Maher (2003). Thus, the main source of selenium to humans is fish from the top of the food chain and plants containing selenium (e.g., the Brazil nuts)

concentrations, contaminating wetlands and poisoning fish and migratory birds at several locations in the Western United States (Lemly 1996). A schematic representation of the biogeochemical cycle of selenium can be found in Fig. 2.

Selenium is widely distributed in the earth's crust; however, the content of selenium in soils from different regions can vary considerably (Rosenfeld and Beath 1964; Dumont et al. 2006). For instance, in China the soil selenium concentration varies from low (0.2 mg/kg) to extremely high levels (10–40 mg/kg) (Dumont et al. 2006). Several factors in the soil, including the physical, biological, and chemical properties, interfere in the bioavailability of selenium to plants (Rosenfeld and Beath 1964).

A critical factor for selenium absorption by plants is its valence number or chemical form (Fig. 3). Selenate can be found in alkaline soils, where it is soluble and easily available to plants. Indeed, the selenate competes with its sulfur analog,

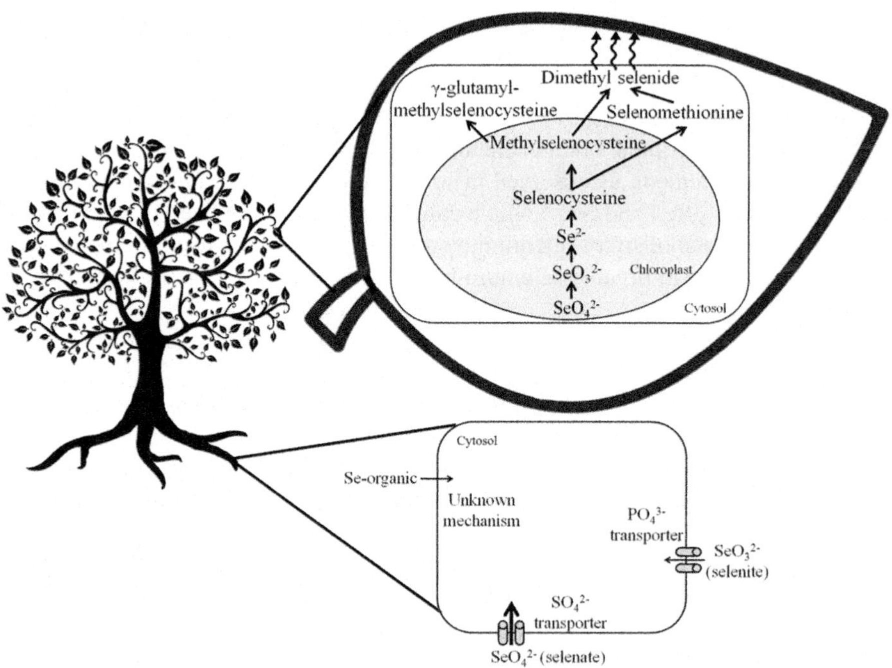

Fig. 3 Fate of selenium in the higher plants. Briefly, the uptake of selenium by root plasma membrane depends on its valence number, for instance, the selenate form enters in the root cells by the sulfate transporters. Moreover, the selenite form enters in the cells by the phosphate transporters. Organic forms of selenium can also be found in the soil, but the mechanism by which they enter into the cell is unknown. In the chloroplast, selenate is reduced to selenide, which can nonspecifically be incorporated into the o-acetylserine molecule forming the amino acid Sec instead of Cys. Thus, this organic selenium molecule can be metabolized to selenomethionine, methylselenocysteine, and γ-glutamyl-methylselenocysteine (mainly in the selenium accumulator plants) or to the volatile molecules: dimethyl selenide and dimethyl diselenide

sulfate, for uptake by root plasma membrane of plants (Dumont et al. 2006). On the other hand, selenite can be absorbed at least partly by phosphate transporters (Zhang et al. 2014; Winkel et al. 2015). Once inside the root cells, inorganic selenium is transported to the shoot part of the plant, specifically to the leaves, where it is metabolized into organic selenocompounds, for instance, Sec, selenomethionine, methylselenocysteine, and γ-glutamyl-methylselenocysteine. Moreover, some of the organo-selenocompounds are metabolized to the volatile dimethyl diselenide and/or to dimethyl selenide (Chapman 1999; Dumont et al. 2006; Winkel et al. 2015). Although it is not the main subject of the present chapter, it is important to highlight that, different from animals, in plants, the synthesis of Sec is nonspecific. In plant cells, the enzyme cysteine synthase can incorporate a selenium atom into an o-acetylserine molecule forming Sec instead of Cys (Metanis et al. 1995). Consequently, the concentration of selenium in organic moieties in plants will be influenced by the availability of selenium in the environment. Indeed, the first description of organic compounds containing selenium with amino acid-like structure was made in 1940 (Horn and Jones 1940), and the first clear demonstration of a selenoamino acid, i.e., methylselenocysteine in selenium accumulator plants was made in 1960 (Trelease et al. 1960).

Selenium has a complex biogeochemistry in the aquatic environment (see Fig. 2). In aquatic environment, as observed in soil, Se can be found in different oxidation states (Se^{4+}, Se^{6+}, Se^0, and Se^{2-}) which can be reduced or oxidized by the microorganisms. The metabolism of selenium by microorganisms can result in the incorporation of selenium in organic selenium molecules, which can have a high mobility in the food chain.

Environmental surveys have demonstrated the biomagnification of selenium in the aquatic environment (Lemly 1996; Barwick and Maher 2003). Of particular environmental and toxicological importance, anthropogenic-released selenium can result in its biomagnification, which can be dangerous to the animals located at the top of food chain (Lemly 1996; Hamilton 2004). Though the toxic effects of selenium bioaccumulated in the food chain have not been reported for humans, the consumption of fish from moderately selenium contaminated ecosystems can hypothetically result in acute or chronic exposure to toxic levels of selenium.

Inside living cells, selenium can be found in different organic chemical forms. Selenomethionine and methylselenocysteine are important organic forms of selenium that can be absorbed and metabolized by the animals. Regarding to mammalian cells, the only physiologically significant organic selenium compound is the amino acid Sec that is found incorporated in selenoproteins. There is no free pool of Sec, because it is much more reactive and unstable than Cys (Huber and Criddle 1967), and in the presence of oxygen, it is rapidly oxidized to selenocystine (Nogueira and Rocha 2010).

The general fate of selenium in the body of mammalians is depicted in Fig. 4. The incorporation of inorganic or organic selenium into the Sec is complex. In short, the organic forms of selenium, for instance, selenomethionine, selenocystine, Sec, methylselenocysteine, or cationic inorganic forms of selenium have to be metabolized to selenide. The selenide is then metabolized to selenophosphate that is subsequently incorporated into Sec, and the Sec is incorporated in selenoproteins.

The metabolism of Sec released after the degradation of selenoproteins can generate selenide in a reaction catalyzed by β-selenocysteine lyase (β-Sec lyase) or cysteine desulfurase (Esaki et al. 1982). The release of selenide from Sec is important to recycle selenium in mammalian cells. Selenomethionine can be metabolized by β-lyase, forming methylselenol/methylselenolate (CH_3SeH/CH_3Se^-). The methylselenol intermediate can be demethylated enzymatically forming selenide. Although a small amount of free Sec can be formed in liver of rats (Esaki et al. 1982), this possibly is not metabolically important. Although hydrogen selenide can be potentially toxic to mammals (Alderman and Bergin 1986), there is no clear indication that selenide generated metabolically can be toxic.

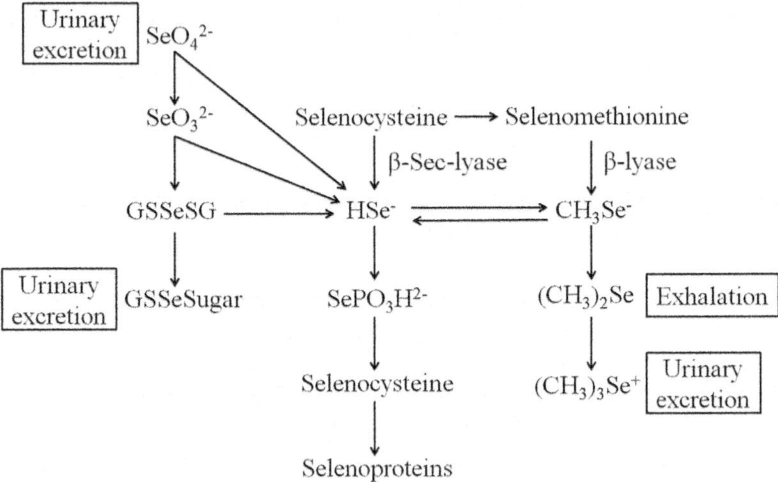

Fig. 4 General fate of selenium inside the mammalian cells

Selenium: Neurotoxic or Neuroprotector?

The neurotoxicity of selenium has not been investigated in detail. Indeed, different forms of selenium have been reported to induce changes in neurochemical markers in the brain of experimental animals; however, the link between these changes and the gross neurotoxic manifestation found both in rodents and humans is uncertain. Some morphological studies have demonstrated the neurotoxicity of selenium to motor neurons (particularly Se^{4+}) in pigs exposed to high doses of inorganic and organic forms of selenium (for review, see Vinceti et al. 2014). In view of the quite distinct neurotoxicity of inorganic electrophilic selenium forms, when compared with organic forms, Vinceti et al. (2014) have shrewdly pointed out the inappropriateness of the generic term "Se neurotoxicity." The authors have indicated the necessity of specifying the chemical form of the selenium.

The superior toxicity of Se^{4+} followed by Se^{6+} in relation to other forms of selenium was observed by the earlier investigators of selenium toxicity (for review, see Rocha et al. 2017). It is worth mentioning the electrophilic character of these two forms and the capacity of them to catalytically oxidize –SH groups of biomolecules. Organoselenium compounds can also catalyze the oxidation of SH-containing proteins, but selenite has a greater potency than organic forms (Nogueira et al. 2004; Rocha et al. 2012, 2017). Thus, in addition of promoting the production of reactive oxygen species (ROS), electrophile forms of selenium can oxidize –SH groups of important proteins, which may contribute to their neurotoxicity (Nogueira et al. 2004; Nogueira and Rocha 2011; Rocha et al. 2017). However, our knowledge about the interaction of specific forms of selenium with relevant thiol-containing targets under complex in vivo conditions is very limited.

As cited above, the chemical form of selenium can influence considerably the neurotoxicity of selenium. The injection of aliphatic selenium to rats and mice has been reported to cause severe neurotoxicity (convulsions) and lethality, depending on the aliphatic chain size (Nogueira et al. 2004).

The neurotoxicity caused by organo-diselenides in rodents has been attributed to several phenomena (for review, see Nogueira and Rocha 2011). However, the identification of the primary molecular target(s) of different forms of selenium are still incipient. The interaction of diselenides and electrophilic selenium cations (Se^{4+} and Se^{6+}) with SH-containing targets has been demonstrated to occur in vitro (Nogueira and Rocha 2011), but the phenomena and mechanisms that operate in vivo are elusive. It has been postulated that selenium compounds can promote the overproduction of oxidative stress and the oxidation of specific SH-containing proteins (Nogueira and Rocha 2011; Rocha et al. 2012, 2017). However, the connection between these alterations and the neurotoxicity of selenium has not been proved yet.

Although the experimental exposure to different forms of selenium can cause neurotoxicity in mammals, the points of evidence that high-selenium ingestion can cause primary neurotoxicity in humans are still scarce (Vinceti et al. 2014). As cited above, there are only a few numbers of studies demonstrating the neurotoxicity of selenium. The reasons of low interest in the potential neurotoxicity of selenium may be related to the physiological antioxidant role of selenium as a part of selenoproteins.

Neurodegeneration and neurotoxicity of different agents are associated with oxidative stress, and it has been proved that selenium confers neuroprotection in different experimental in vitro and in vivo models of neurotoxicity (Imam et al. 1999; Glaser et al. 2010, 2013; Heath et al. 2010; Erken et al. 2014). However, inorganic and organic selenium compounds can have potential toxicity both in experimental animals and in humans (Vinceti et al. 2013, 2014).

Of particular toxicological significance, Vinceti et al. (2013) have demonstrated the importance of the chemical form of selenium found in the cerebrospinal fluid of patients with ALS. Vinceti et al. (2013) pointed out that the relative risk of ALS increased with increasing selenite in the cerebrospinal fluid; however, selenate levels did not change the relative risk of having ALS. In contrast, the levels of selenium bound to selenoprotein P (SepP1) indicated a decreased in the relative risk of ALS. Taken together, the results from this pioneering study of Vinceti and collaborators highlighted the importance of selenium speciation in the cerebrospinal fluid as potential predictor of neurodegenerative disease development. However, even having a small sample size, this study is noteworthy once it highlights the neurotoxic potential of the electrophilic selenite and the neuroprotective potential of SepP1.

Recently, some studies with small number of patients have indicated that selenium levels can vary in autism spectrum disorder (ASD). In one of the studies, the hair selenium levels were increased in children with communication disorder (CD) and ASD (Skalny et al. 2016a). But the levels of selenium were slightly decreased or not altered in the serum or hair of ASD children (Lakshmi Priya and Geetha

2011; Blaurock-Busch et al. 2012; De Palma et al. 2012; Tabatadze et al. 2015; Skalny et al. 2016b, c). In short, the association of selenium and ASD is far from being elucidated, and more robust studies will be needed to solve this question.

From Selenide to Selenoproteins

The discovery of the 21st proteinogenic amino acid Sec generated great progress in the knowledge about the potential benefits of selenium for human health. The discovery of selenocysteinyl residue was first made in bacteria in 1976 (for review, see Hatfield et al. 2014). Two years later, a selenocysteinyl residue was described in the active center of hepatic glutathione peroxidase (GPx) of rats (Forstrom et al. 1978).

The pathway of Sec synthesis is complex and involves first the combination of selenide with a phosphate (forming selenophosphate). The selenophosphate then reacts with the phosphoester of serine bound to the Sec tRNA (designated tRNA[Ser] [Sec]) (Lee et al. 1989). The critical experiments demonstrating that the carbon backbone of selenocysteine was derived from serine were made by Sunde and Evenson (1987). Thus, the synthesis of Sec takes place in its tRNA[Ser]Sec.

The first demonstration that a specific tRNA could be aminoacylated by radioactive-labeled selenite was made by Hawkes et al. (1982). Hawkes and Tappel (1983) also demonstrated that the selenocysteinyl-tRNA (formed from labeled selenite) supplied the selenocysteinyl residue found in the active center of hepatic GPx.

There are two isoforms of tRNA[Ser]Sec, which differ by the methyl group in the uracil nucleotide at position 34 (Um34). The methylation of this nucleotide occurs during the maturation process of the tRNA[Ser]Sec, which is modified by the selenium status (Hatfield et al. 1991; Howard et al. 2013). In the case of selenium deficiency, the levels of the tRNA[Ser]Sec with non-methylated uracil (designated 5-methoxycarbonylmethyluridine or mcm^{5}U) are increased, while the tRNA[Ser]Sec with methylated uracil are decreased (designated 5-methoxycarbonylmethyl-20-O-methyluridine or mcm^{5}Um) in the liver, kidney, heart, and muscles (Diamond et al. 1993). In the presence of high levels of mcm^{5}Um (i.e., when selenium is available), the expression of stress-related selenoproteins (e.g., GPx1) is favored in the same tissues cited above. In contrast, in the presence of high levels of mcm^{5}U (i.e., selenium deficiency), the expression of housekeeping selenoproteins (e.g., thioredoxin reductase 1, thioredoxin reductase 3, and selenoprotein W) is favored (Diamond et al. 1993; Carlson et al. 2005; Howard et al. 2013).

Another particularity of Sec incorporation into selenoproteins is its decoding by the UGA codon, which normally is a stop codon (for a comprehensive review, see Labunskyy et al. 2014). This is possible because there is a structural factor in the mRNA that allows the interpretation of the UGA codon not as the end of the protein but as the place of Sec incorporation. The structural element found in the 3'non-translated portion of the mRNA is termed Sec insertion sequence (SECIS). There

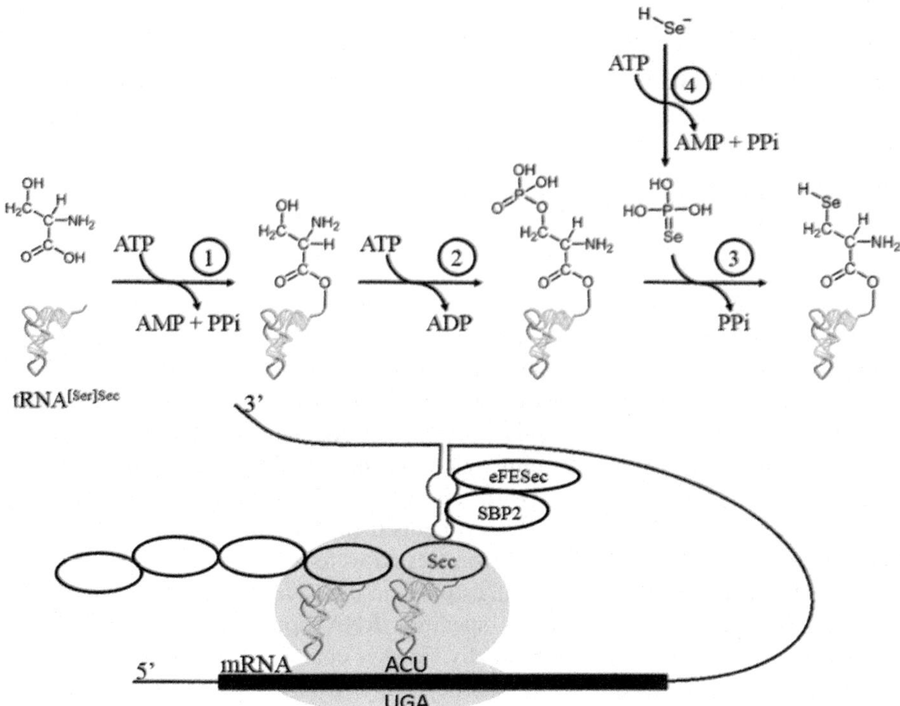

Fig. 5 The synthesis of Sec starts with the aminoacylation of tRNA[Ser]Sec with serine (Ser), which is catalyzed by seryl-tRNA synthetase (SerS (1)) in the presence of ATP. Then, the -OH group of serine is phosphorylated by the phosphoseryl-tRNA kinase (PSTK (2)) forming phosphoseryl (PO_4^3 Ser). Concomitantly, selenophosphate synthetase 2 (SPS2 (4)) catalyzes the reaction of selenide with ATP, forming monoselenophosphate. The incorporation of a selenium atom in the serine backbone is catalyzed by phosphoseryl-tRNA selenium transferase enzyme (SEPSecS (3)) resulting in the synthesis of Sec residue covalently bound to the tRNA[Ser]Sec. Once the selenocysteine is synthetized, it can be incorporated into selenoproteins, for instance, the tRNA[Ser]Sec interacts with selenocysteine insertion sequence (SECIS), SECIS-binding protein (SBP2), and Sec-specific elongation factor (eEFSec) which permits the decoding of the UGA codon as a Sec

are at least two other macromolecular required elements in eukaryotes for correct reading of Sec-UGA codon: (1) the SECIS-binding protein (SBP2) and (2) Sec-specific elongation factor (eEFSec), which interact with the SECIS of the mRNA coding for selenoproteins. The SBP2 is believed to stabilize the interaction between the SECIS region with the ribosome. The binding of SECIS to the ribosome induces a conformational change in the organelle structure, facilitating the interaction with the eEFSec. All these interactions and conformational changes allow the association of the tRNA[Ser]Sec with the UGA coding the selenocysteinyl residue in selenoproteins (Fig. 5) (Hatfield et al. 2006, 2014).

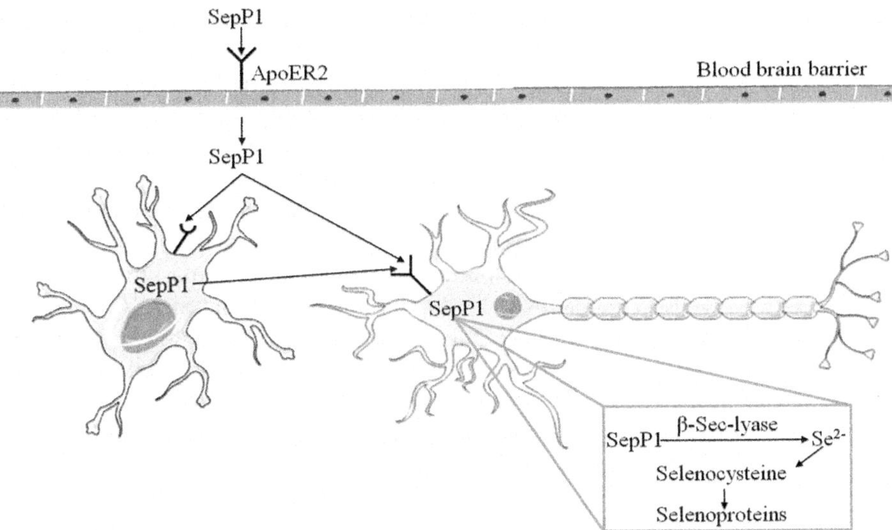

Fig. 6 Fate of selenium in the brain. The majority of the selenium enters into the brain in the form of selenoprotein P1 (SepP1, which contains around ten selenocysteinyl residues).Via the interaction with ApoER2 receptor, SepP1 crosses the blood-brain barrier. Once in the brain, SepP1 can enter in the neuronal cells by the interaction with the ApoER2 receptor. During a selenium deficiency, the SepP1 present in the astrocyte can be transported to the neurons. Then it can be degraded, and by the action of β-Sec-lyase, the selenide can be utilized in the synthesis of metabolically active selenocysteinyl residue bound to the tRNA[Ser]Sec

Selenoproteins in the Brain

The human genome has 25 selenoproteins, and about half of them have well characterized biochemical functions (Hatfield et al. 2014; Pillai et al. 2014; Cardoso et al. 2015). The selenoproteins are vital to mammalian life, and the brain requires constant levels of selenium to maintenance of its functions, e.g., to the synthesis of selenoproteins (Burk et al. 2014; Cardoso et al. 2015). In fact, during a selenium deficiency, the brain levels of selenium do not change appreciably, which contrast with a drastic decrease in the liver and kidney selenium content (Trapp and Millam 1975; Burk et al. 1991). Accordingly, hepatic and renal selenium deficiency is associated with a dramatic decrease in the activity of cytosolic GPx (Maquat 2001). The decrease in GPx1 synthesis is modulated by the levels of Um34 (Li et al. 1990; Diamond et al. 1993), and in the presence of low selenium, the available selenium is directed to the syntheses of hierarchically more important selenoproteins.

SepP 1, a Sec-rich C terminal domain selenoprotein, supplies the organs with selenium, especially the brain. A study from Byrns et al. (2014) demonstrated that the depletion of SepP1 decreases the brain selenium levels and the animals became more susceptible to exhibit seizures. The neurotoxic phenotype was ameliorated by dietary supplementation with selenium. Thus, SepP1 is critical, but not the only mechanism involved in the selenium transport into the brain. In the blood-brain barrier, SepP1 is recognized by the ApoER2 receptor and then this complex is internalized (Burk et al. 2014). Neurons also have the ApoER2 receptor which facilitated the entry of the SepP1 into the cell, where it can be degraded releasing Sec. The amino acid can be metabolized to selenide that can be used in the synthesis of more critical selenoproteins (Fig. 6) (Burk et al. 2014). Interestingly, during a starvation of selenium, the astrocyte may synthetize SepP1 to supply the neurons with selenium (Steinbrenner and Sies 2013).

Selenoproteins are extremely important to the brain, particularly by catalyzing critical redox reactions (Lobanov et al. 2009; Steinbrenner and Sies 2013; Hatfield et al. 2014; Labunskyy et al. 2014). Impairments in the selenoprotein functioning cause disturbances in the redox balance, which can result in the overproduction of ROS. The disproportionate ROS production causes damage to brain macromolecules, which increase the incidence of several disease related to oxidative stress (Hatfield et al. 2014; Hassan et al. 2015; Table 1).

Mercury in the Environment and in the Living Organisms

Mercury is an ubiquitous metal found naturally in the environment; however with the processes of urbanization and industrialization, the mercury levels have been increased in the environment (Fig. 7) (Muntean et al. 2014). Although several countries have decreased the mercury emission (Zhang et al. 2016), the levels of mercury are still a health concern in the entire planet. In fact, mercury has no biochemical or physiological function in living organisms, and even exposure to low levels of mercury forms is of toxicological significance (Clarkson 2002; Farina et al. 2011a; Brandão et al. 2015).

In the past, the man used mercury (in inorganic or organic forms) for different purposes (Clarkson 1997, 2002). Nowadays, ionic inorganic mercury can be found in whitening skin creams, which is used in several countries (Chan 2011). Although the cutaneous absorption of cationic forms of mercury is low, the exposure to mercury via whitening skin creams has been associated with serious neurotoxicological effects in humans (Benz et al. 2011; Peregrino et al. 2011; Gbetoh and Amyot 2016).

Hg^0 is still used as part of medical equipment, as a catalyst in chloride and caustic soda factories and as part of dental amalgams (Horowitz et al. 2014). In Brazil and in other developing countries, Hg^0 was intensely used in artisanal gold mining (Kristensen et al. 2014; Marques et al. 2015). Although the gold mining decreased in Brazil some decades ago, the use of Hg^0 for mining is still a serious toxicological concern to humans and to the environment (Branco et al. 2007). It has been estimated that artisanal gold mining is an important source of Hg release into to the atmosphere (UNEP 2013). In fact, in areas near to the artisanal mining, the mercury in fish exceeds the safe limits for human consumption (0.5 µg/g) (WHO 2007).

In the ancient China, mercury (in the form of HgS) was used in the composition of a red ink and as component of herbal medicines (Wu et al. 2016). The use of HgS

Table 1 Selenoprotein in the brain

Selenoprotein	Selenoprotein in the brain	Experimental model
Glutathione peroxidase 1 (Gpx1)	*Depletion*: increased the susceptibility to ischemic and toxic insults (Crack et al. 2006)	Mice
	Distribution in the brain: abundantly found in microglia and low in neurons (Power and Blumbergs 2009)	Human
	Presence: suppressed free radical generation and protected dopaminergic neurons against 6-hydroxydopamine (Gardaneh et al. 2011)	Cell culture
Glutathione peroxidase 4 (Gpx4)	*Distribution in the brain*: found in neurons of cerebral cortex, hippocampus, and cerebellum and low in glial cells (Savaskan et al. 2007)	Rats
	In neurodegenerative disease: increased relative to the cell density of surviving nigral cells (Bellinger et al. 2011)	Human Parkinson's brain
Thioredoxin reductase 1 (TRx1)	*Overexpression*: extended the life span (Takagi et al. 1999)	Mice
	In neurodegenerative disease: decreased the expression in the substantia nigra pars compacta (Liu et al. 2013)	Mice
Selenophosphate synthetase 2 (SPS2)	*Presence*: detected in brain (Kim and Stadtman 1995).	Rat
Selenoprotein P (SepP1)	*Deficiency*: severe alterations in synaptic transmission, short-term plasticity, and long-term potentiation in hippocampus (Peters et al. 2006)	Mice
	Function: maintenance of selenium in brain (Nakayama et al. 2007)	Mice
	Knockout: severe neurological dysfunction, neurodegeneration, and audiogenic seizures (Byrns et al. 2014)	Mice
	Knockout: neurological damage was exacerbated in male when compared to female (Raman et al. 2012)	Mice
	Uptake: depends on the apolipoprotein receptor 2 (ApoER2) (Burk et al. 2014)	Mice
	In neurodegenerative disease: increased in brain of Alzheimer's disease (Rueli et al. 2015)	Human
Selenoprotein W (SelW)	*Distribution in the brain*: widespread in neurons (Raman et al. 2013)	Mice
15 kDa selenoprotein (Sep 15)	*Knockout:* normal brain morphology (Kasaikina et al. 2011)	Mice
Selenoprotein M (SelM)	*Deletion:* absence of deficits in motor coordination and cognitive function (Pitts et al. 2013)	Mice
	Se deficiency: reduction of the expression (Huang et al. 2016)	Chicken

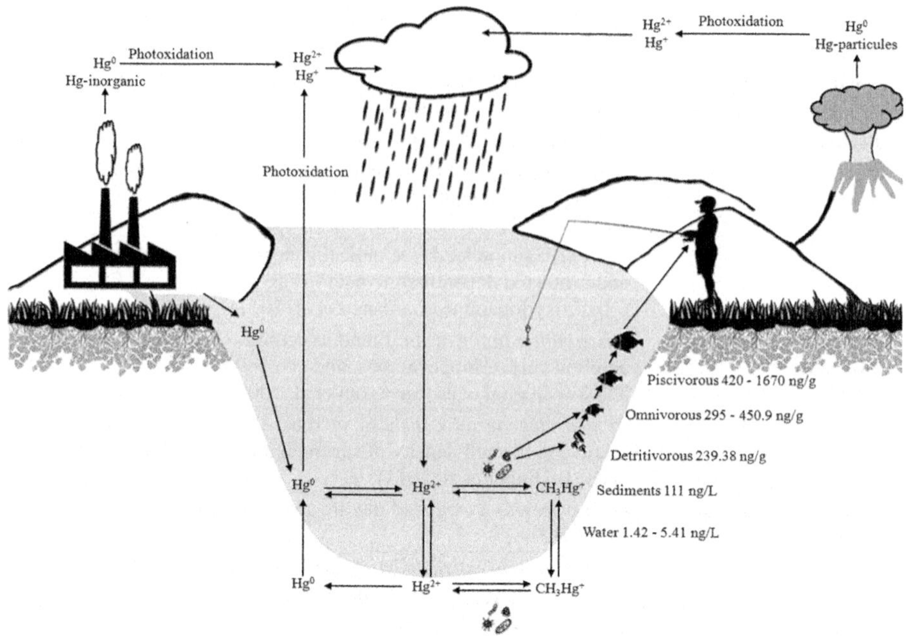

Fig. 7 Cycle of mercury in the environment. Briefly, mercury is released in the environment by natural (volcanism or erosion) or anthropogenic sources (for instance, gold mining, waste of factories). In the environment, the chemical species of mercury can be methylated or oxidized by the microorganisms. Although these reactions are reversible, the concentration of organic mercury is higher in aquatic biotic ecosystem than the concentration of inorganic species in the abiotic ecosystem. Accordingly, the biomagnification of MeHg in the food chain (the range was based in Bowles et al. (2001) and Barbosa et al. (2003)) can be 1-6 orders of magnitude over the levels of mercury found in the water and sediments

in traditional medicine is still practiced in China and in other countries, regardless of the unpredictable toxicological effects of HgS (Yu et al. 2015; Dong et al. 2016). Based in the toxicological and chemical point of view, it is noteworthy the biased comparisons of the lower toxicity of HgS (or traditional formulations containing it) with $HgCl_2$ and MeHg. In view of the HgS low solubility, it is obvious that HgS will have lower bioavailability and, consequently, lower toxicity than $HgCl_2$. However, by no means, this indicates that HgS or any combination containing it is nontoxic.

Although inorganic mercury (Hg^0, Hg^+, or Hg^{2+}) is a toxic agent that can threaten the human health, the exposure to these forms of mercury is expected to decrease in the next decades, and possibly the exposure will be limited to those occupationally exposed to them or using traditional medicine formulations containing cinnabar (Ye et al. 2016).

The interest in the toxicity of mercury increased considerably after the outbreak of Minamata Bay. The use of Hg^{2+} as a catalyst in the synthesis of acetaldehyde resulted in a heavy contamination of Minamata Bay via mercury biomagnification. The demonstration that MeHg found in fish from Minamata Bay was the causative

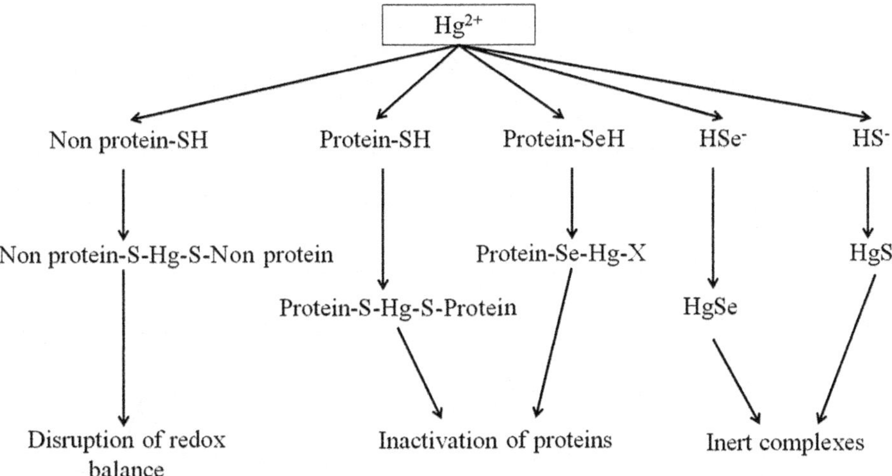

Fig. 8 Hg^{2+} generic molecular targets. The most abundant targets of mercury are the SH-containing molecules, but Hg^{2+} can also bind to selenol-containing targets. Hg^{2+} can coordinate with two SH-containing molecules. In the case of –SeH, Hg^{2+} possibly coordinates with the –SeH and with one SH-containing molecule. X = selenium or sulfur

factor of Minamata disease and that MeHg was metabolically formed from Hg^{2+} took a long time. The exposure of adults and developing humans to MeHg resulted in catastrophic cases of neurotoxicity (Ekino et al. 2007; Tsuda et al. 2009) and alarmed the industrialized societies to the dangerous of the industrial waste release.

In the last decades, investigators have been speculating that exposure to very low levels of mercury during critical phases of the brain development could be associated with an increased incidence of neuropathologies or cognitive disabilities, particularly with ASD (Mutter et al. 2005; Bjørklund et al. 2016; Kern et al. 2016; Farina et al. 2017) and Alzheimer's disease (Mutter et al. 2010; Farina et al. 2017). However, the clinical and epidemiological points of evidence supporting a specific role for mercury in such type of disorders are still doubtful (Kern et al. 2015; Skalny et al. 2016a).

After the Minamata Bay outbreak, search for detailed mechanistic information about on how mercury causes its toxicity became intense. Although the toxicity of electrophilic mercury forms (particularly Hg^{2+} and MeHg) differs at molecular level, the basic process of their toxicological effects can be explained in chemical terms. Hg^{2+} and MeHg behave as strong and soft electrophiles, and, consequently, in the living systems, they will have high affinity for soft nucleophile centers (Figs. 8 and 9) (Farina et al. 2011a,b, 2017).

In mammals, we have two important physiological soft nucleophile centers, i.e., the –SH and the –SeH groups. The –SH group is found in thousands of proteins and in low-molecular-mass compounds (for instance, Cys and glutathione), whereas the –SeH group is found only in 20–40 types of selenoproteins of vertebrate cells (Hatfield et al. 2014; Rocha et al. 2017). Thus, the much higher affinity of mercury

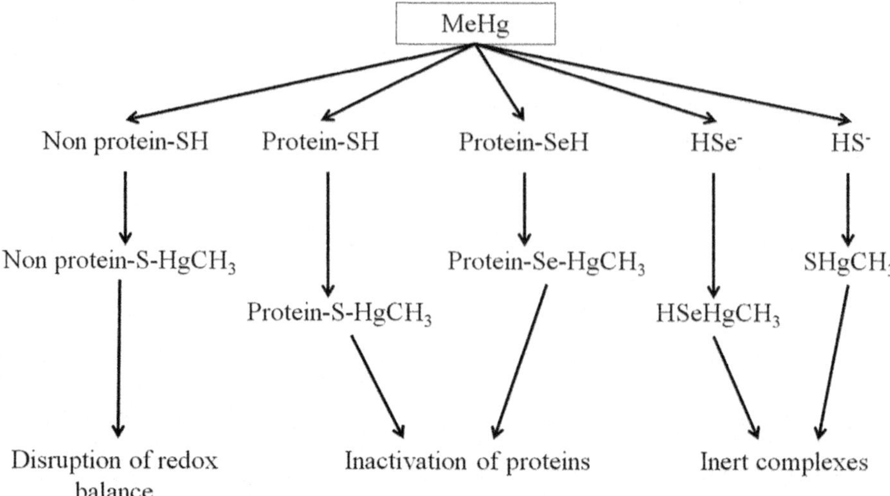

Fig. 9 MeHg possible molecular targets. The most abundant targets of mercury are the thiol-containing molecules, but Hg^{2+} can also bind to selenol-containing targets. MeHg can coordinate with one thiol or selenol group. X = selenium or sulfur

forms for the –SeH group of selenoproteins (Sugiura et al. 1976; Falnoga and Tušek-Žnidarič 2007; Farina et al. 2011a, b) over the –SH group, and the much higher concentration of non-protein-SH and protein-SH over the protein–SeH makes the study about the distribution of mercurials extremely complex in living cells. Similarly, the definition of the primary molecular targets of mercurials in living organisms is very complex. The targeting of –SH- or –SeH-containing proteins are certainly involved in the toxicity of mercurials (Figs. 8, 9, and 10), but distributional factors will influence profoundly the targeting of different organs by Hg^{2+} and MeHg (Clarkson 1997; Zalups 2000; Bridges and Zalups 2016).

For instance, the binding and the strong coordination of Hg^{2+} to two cysteine molecules will direct the Hg^{2+} to the kidney (Zalups 2000; Bridges and Zalups 2016), whereas the coordination of MeHg to one cysteine facilitates its entry into the brain by molecular mimicry (Aschner 1989; Bridges and Zalups 2016; Clarkson et al. 2007).

Inorganic Mercury Neurotoxicity

The inorganic forms of Hg can be divided in two types: (1) Hg^0 (elemental mercury), number of valence 0, and (2) Hg^+ and Hg^{2+} (ionic and electrophile mercury forms) (Clarkson et al. 2007).

Elemental mercury is liquid, but because of its high vapor pressure, it can be found as vapor at room temperature (Magos and Clarkson 2006). Studies about

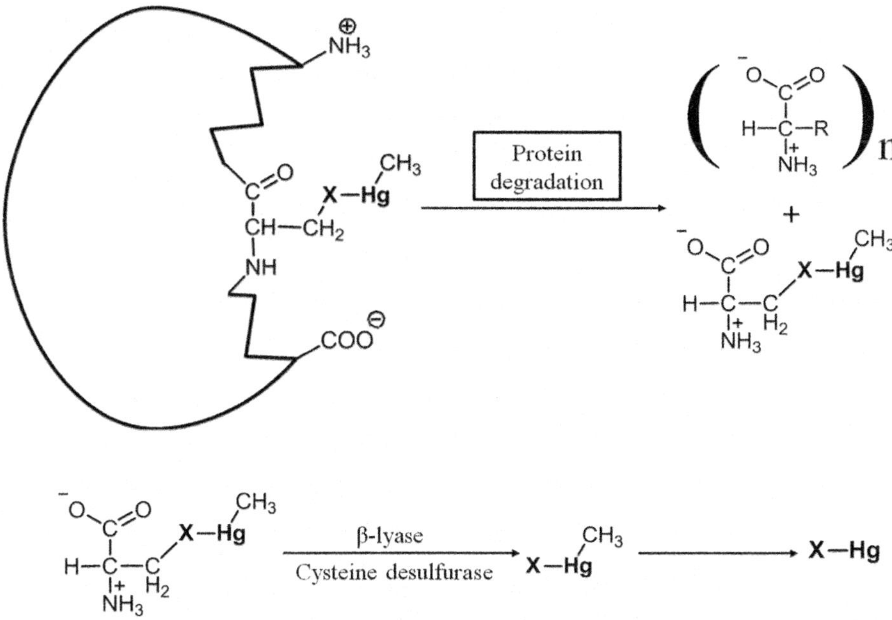

Fig. 10 MeHg interaction with –SH- or SeH-containing proteins. MeHg as a strong and soft electrophile has a great affinity for the soft nucleophiles –SH or –SeH groups. In the proteins, these nucleophiles are found in the amino acids Cys and Sec. Thus, MeHg interaction with these groups can inactivate the protein function. After the protein degradation, the enzymes β-lyase or cysteine desulfurase catalyze the release of selenide or sulfide group bound to the Hg atom (i.e., as the complexes HgS or HgSe). The bioavailability of the HgSe is much lower than that of HgS, and the HgSe is possibly inert toxicologically. However, the understanding about the inertness of HgSe is limited (X = selenium or sulfur)

elemental mercury toxicity demonstrated that when the exposure is acute, the respiratory system is primarily affected, since the exposure is through inhalation (Rowens et al. 1991). However, in the case of chronic exposure to low levels of Hg vapor, the toxic effects are generalized and can affect the nervous, hepatic, and renal systems (Oz et al. 2012).

After the inhalation, part of the Hg^0 can be oxidized to Hg^{2+} by the catalase in the erythrocytes (Hursh et al. 1988); however, the residence time of the Hg^0 in the circulation is sufficiently long to allow the diffusion of Hg^0 into the organs. In fact, studies have demonstrated the accumulation of Hg in the brain of different mammal species after exposure to Hg^0 through inhalation (Berlin et al. 1969; Warfvinge 2000). The exposition to Hg^0 can be associated with a variety of neurological symptoms, such as loss of memory, erythrism, mood alteration, sleep disturbances (Hilt et al. 2009), tremor, and postural instability (Kern et al. 2014), and some of them can be observed even 30 years after the cessation of exposure (Letz et al. 2000).

As briefly mentioned above, Hg^0 can be converted to Hg^{2+}, and the cationic mercury will target the kidney (Zalups 2000; Peixoto and Pereira 2007; Oliveira et al.

2015, 2016). The alterations in the kidney can contribute to the cardiovascular toxicity (Kim et al. 2014) and neurotoxicity of Hg^{2+} (Peixoto et al. 2007; Moraes-Silva et al. 2014). However, the potential contribution of non-neural changes to the neurotoxicity of inorganic mercury has not been investigated in detail.

The accumulation of Hg^{2+} in the brain either derived from the oxidation of Hg^0 or from MeHg has been speculated to be associated with neurodegenerative diseases development, for instance, ALS (Pamphlett and Kum Jew 2013), Alzheimer's disease (Mutter et al. 2010), or autism (Curtis et al. 2011). However, the causal relationship between mercury exposure and neurodegenerative diseases is still a matter of discussion.

Organic Mercury Neurotoxicity

MeHg is by far the most studied neurotoxic form of mercury. There are several reviews and studies about the mechanisms of neurotoxicity caused by MeHg (Aschner 1989; Atchison and Hare 1984; Aschner et al. 2007; Ceccatelli et al. 2010; Farina et al. 2011a,b, 2017; Ishihara et al. 2016; Ruszkiewicz et al. 2016). Consequently, here we will not do a comprehensive review about the neurotoxicity of MeHg. Although the primary targets of MeHg have not been identified, persuasive points of evidence have indicated that glutamatergic system (particularly the NMDA receptor over activation), intracellular Ca^{2+} dysregulation, mitochondrial toxicity, and oxidative stress are important phenomena involved in MeHg neurotoxicity (Atchison and Hare 1994; Aschner et al. 2007; Ishihara et al. 2016).

As explained above, the reactive electrophile forms of mercury (particularly Hg^{2+} and MeHg) are not found freely in the biological systems. Due to their strong affinity for soft nucleophile centers, they are coordinated to –SH- or SeH-containing molecules. Furthermore, the binding of MeHg to low-molecular-mass thiols (e.g., Cys) has been proposed as an important step in the entrance of MeHg through the blood-brain barrier as a mimic of the amino acid methionine. The amino acid carrier System L (LAT1 and LAT2) transports MeHg bound to Cys (MeHg-S-Cys) into the brain (Aschner 1989; Bridges and Zalups 2016).

Recently, an in vitro study using primary porcine choroid plexus epithelial cells as a model of the blood cerebrospinal fluid barrier demonstrated that MeHg effectively crosses out of the cerebrospinal fluid side to the blood side (Lohren et al. 2015) which may point to a new MeHg excretion route in the brain. However, as we mentioned above, the MeHg is not found free in the cells; thus, more studies using the MeHg forms present in the cell, for instance, MeHg-S-Cys, are necessary to determine the importance of this Hg-efflux system.

Although MeHg is the predominant chemical form of Hg found in the brain, studies have demonstrated the accumulation of Hg^{2+} in the brain after MeHg exposure. Consequently, part of the neurotoxicity of MeHg could be mediated by Hg^{2+} accumulation (Yamamoto et al. 1986; Ishitobi et al. 2010).

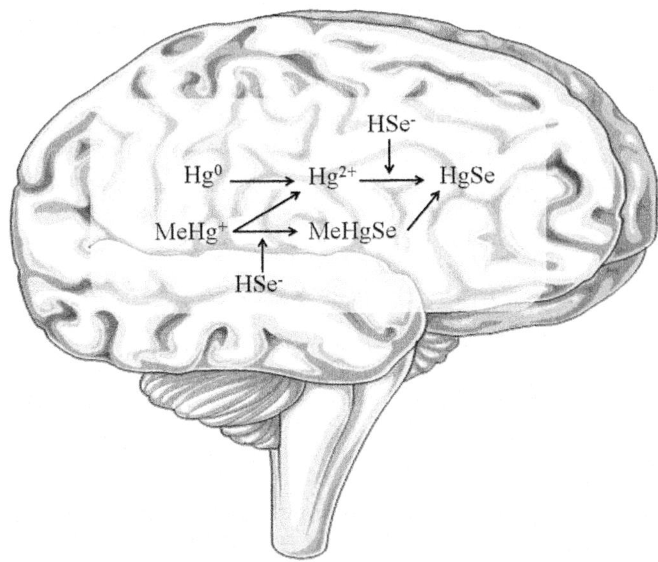

Fig. 11 Mercury interaction with selenium in the brain. Inside the brain, Hg^0 can be oxidized to Hg^{2+}, which can bind to selenide forming an inert complex (HgSe). MeHg can be demethylated releasing HgSe by a mechanism that may involve the binding to the selenocysteinyl residue found in selenoproteins. The release of selenocysteinyl residue after the intracellular digestion of selenoproteins and the metabolism of Sec can form the HgSe complex. Moreover, MeHg may interact directly with selenide, which can facilitate the breakage of C-Hg bound, releasing the HgSe complex. The HgSe complex may be retained into the brain without causing toxic effects

The break of the C-Hg bond occurs very slowly, but the coordination of MeHg with SeH-containing amino acids can accelerate the process (Figs. 10 and 11) (Asaduzzaman and Schreckenbach 2011). Similarly, the binding of MeHg to selenide can facilitate the breakage of the C-Hg bond in vitro (Iwata et al. 1982).

Selenium and Mercury: Brain Antagonistic Interactions

Selenium as Antagonist of Inorganic Mercury Neurotoxicity

The data about the effects of selenium against the neurotoxicity induced by elemental mercury are rare. A study from Suzuki et al. (1986) with workers from a thermometer manufacturing indicated a negative correlation between urinary selenium and mercury levels. On the other hand, Alexander et al. (1983) observed an increase in urinary selenium levels in chloralkali plant workers. It has been postulated that only high levels of Hg^0 exposure could change the elimination of selenium (Hongo et al. 1985). One important aspect about the interaction of Hg^0 and selenide that has not been tackled in the literature is the potential formation of HgSe by brain cells.

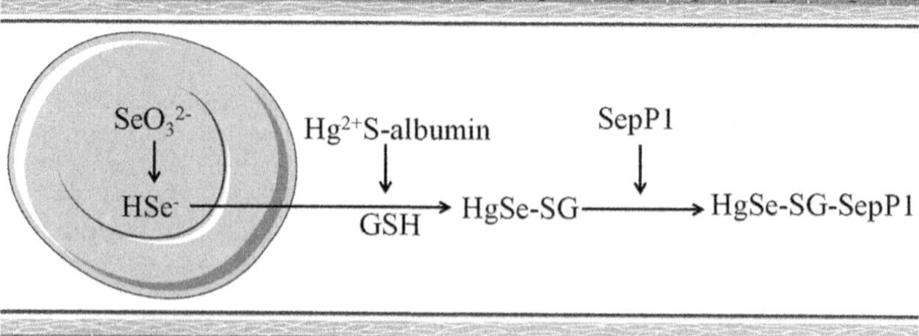

Fig. 12 Inorganic mercury and inorganic selenium interactions within the blood stream. Selenite is oxidized to selenide in the erythrocytes. In the plasma, selenide plus reduced glutathione (GSH) may interact with Hg^{2+}-albumin complexes, forming HgSe-SG complex, which binds to SepP1, diverging Hg^{2+} from the brain and kidneys

In contrast to Hg^0, several studies demonstrated the antagonistic effect of inorganic selenium against the toxic effects of Hg^{2+} (El-Demerdash 2001; Karaboduk et al. 2015; Uzunhisarcikli et al. 2015). The first study was published by Parizek and Ostadalova (1967), which demonstrated an increase in rat survival rate when co-exposed to inorganic mercury and sodium selenite. The pre-exposure to sodium selenite causes changes in the mercury distribution in the body, decreasing the kidney and brain mercury content and increasing selenium in blood (Orct et al. 2015). The conceivable explanation involves the reduction of selenite to selenide inside the erythrocytes and the formation of complexes with Hg^{2+} (HgSe) in the plasma (Chmielnicka et al. 1979; Naganuma et al. 1984) (Fig. 12). This complex formed in the blood stream has low mobility to the organs (brain and kidney, among others). It has been suggested that the HgSe complex binds to the SepP1 and is transported to the liver, where it is excreted via bile to the feces (Khan and Wang 2009).

The organic selenium molecules (for example, Ebselen and diphenyl diselenide (PhSe)$_2$) selenium compounds from yeast and selenomethionine, have been tested against the toxicity of ionic mercury. Studies have demonstrated that pre-exposure to (PhSe)$_2$ protects against alterations caused by Hg^{2+} exposure (Nogueira and Rocha 2010; Fiuza Tda et al. 2015). In sharp contrast, the administration of (PhSe)$_2$ 30 minutes after HgCl$_2$ was ineffective (Nogueira and Rocha 2010) or even potentiated the mercury toxic effects (Brandão et al. 2011; de Freitas et al. 2012). Interestingly, when a chlorine atom was added to (PhSe)$_2$ molecule, 4,4-dichloro-diphenyl diselenide (ClPhSe)$_2$, the toxicity of Hg^{2+} was diminished (de Freitas et al. 2012). Taken together, the results indicate that the interaction of (PhSe)$_2$ and its analog (ClPhSe)$_2$ with Hg^{2+} is rather complex and will depend on the time and type of exposure (pre- or post-treatment), metabolism, and distribution of the organoselenium compounds. The formation of complexes, for instance (PhSe-Hg-SeR or PhSe-Hg-SR), may facilitate the uptake of mercury by the target organs (kidney and liver). The accumulation of mercury complexes in the kidney can reach pathological levels, potentiating the toxicity of Hg^{2+}.

Of particular therapeutic significance for remediating the toxicity of mercury, Li et al. (2012) demonstrated that inhabitants of extensive mercury mining areas in China supplemented with 100 µg of organic selenium (selenium-enriched yeast) daily presented an increase in the excretion of mercury in the urine and a decrease in markers of oxidative stress (urinary malondialdehyde and 8-hydroxy-2-deoxyguanosine), when compared to the placebo group. The study was performed during 90 days, and the beneficial effects of organic selenium was detected from 30 to 90 days after starting the supplementation. Authors have also determined the selenium in urine, but the speciation of selenium and mercury was not determined.

Selenium as an Antagonist of the Neurotoxicity Induced by Organic Mercury

Inorganic selenium decreases the neurotoxicity caused by MeHg even increasing the brain mercury retention (Magos and Webb 1977; Glynn et al. 1993; Newland et al. 2006). The break of Hg-C bond after the formation of HSeHg-CH$_3$ complex is thought to be involved, and after the breakage, the mercury is trapped in the inert complex HgSe. Accordingly, Korbas et al. (2010) showed that the major part of the mercury deposited in human brain of individuals intoxicated with MeHg was in the form of HgSe complex(es). Recently, an elegant study demonstrated the deposition of HgSe as nanometric particles in the brain and liver of pilot whales. The authors also demonstrated a positive correlation between the particle size and the whales' age (Gajdosechova et al. 2016).

The major health concern about the HgSe complex formation is the potential depletion of selenium, which can disrupt the synthesis of seleno-antioxidant enzymes (Usuki et al. 2011), resulting in overproduction of ROS. Accordingly, several studies demonstrated MeHg neurotoxicity associated with the overproduction of ROS (Aschner et al. 2007; Kirkpatrick et al. 2015; Feng et al. 2016). Interestingly, selenium deficiency was showed to be associated with increased MeHg neurotoxicity in vivo (Ralston and Raymond 2010) and in vitro (Kim et al. 2005).

Different types of organoselenium compounds have been used to counteract the neurotoxicity of MeHg both in vitro and in vivo. For instance, Ebselen and (PhSe)$_2$ blunted the oxidative stress and the decrease in the glutamate uptake caused by MeHg administration in vivo (Farina et al. 2003a,b). Similarly, in vitro, Ebselen blunted the neurochemical modifications produced by MeHg in different brain preparations (Moretto et al. 2004, 2005; Roos et al. 2009).

(PhSe)$_2$, which is a simple aromatic diselenide, decreased the neurotoxicity of MeHg in vivo and decreased considerably the deposition of mercury in the brain (cerebrum and cerebellum), liver, and kidneys of mice (de Freitas et al. 2009; Glaser et al. 2013, 2014). (PhSe)$_2$ was also reported to decrease the toxicity of MeHg in vitro (Moretto et al. 2005; Roos et al. 2009). However, (PhSe)$_2$ did not reduce the deposition of mercury in rats exposed to MeHg (Dalla Corte et al. 2013).

Conclusion

The toxicity of mercury and selenium is rather complex, and our knowledge about them is still incipient. Mercury is a nonphysiological element and, in the electrophilic state (Hg^{2+} and MeHg, among others), will target any available –SH or –SeH group. Thus, the toxicity of mercury compounds will be determined by its interaction with abundant –SH groups (for instance, Cys or glutathione) and with specific –SH groups found in proteins with higher nucleophilicity and accessibility than Cys and glutathione. The interaction of electrophile mercury-containing forms with the –SeH group of selenoproteins is expected to be favored over the –SH-containing proteins, because electrophile mercury has a much higher affinity for –SeH than for –SH groups. However, the abundance of –SeH-containing proteins in the cell is much lower than that of –SH-containing molecules. Thus, the interaction of neurotoxic forms of mercury (e.g., MeHg) with its targets will depend on a variety of complex factors. Our knowledge about the primary molecular targets that trigger the cascade of events involved in the cellular toxicity of MeHg is still elusive. We know that MeHg can cause oxidative stress and disruption of glutamatergic neurotransmission. However, we know little about the proteins that are targeted first and, most importantly, almost nothing about the proteins that are chemically modified by MeHg and participate in the primary processes that will culminate in the final neuropathological insult. For instance, some –SH- or –SeH-containing proteins have been demonstrated to be inhibited by MeHg either in vitro or in vivo. But we cannot establish a causal relationship between the inhibitory effects and the final pathological events.

The problems briefly cited here can also be applied to selenium, because selenium has different chemical states, and this element can be either potentially a strong nucleophile (for instance, in the form of selenide or –SeH) or an electrophile (in the form of Se^{4+} or Se^{6+}). In the electrophile forms, selenium can also oxidize –SH-containing molecules. Consequently, Se^{4+} or Se^{6+} can hit some of the targets of electrophilic mercury compounds. The overlapping of some targets or secondary processes of low concentrations of mercury and high concentrations of selenium (e.g., SH-containing proteins and induction of oxidative stress) can explain partially some of the neurotoxic effects of compounds containing these elements. But, as cited for mercury, our knowledge about the additive or synergic toxic effects of chemicals is still very limited. Effective advances in the field of the molecular toxicology of reactive chemicals (for instance, mercury and selenium compounds) will depend on the development of new *in silico* tools that will be able to predict the interaction of different chemical agents with their critical targets under complex chemical media. The new *in silico* methods will have to predict also the complex cascade of events that will follow the hitting of the presupposed primary target proteins. Consequently, our better understanding about the neurotoxicity of different chemical forms of mercury and selenium will depend on the development of new and complex computational methodologies.

References

Alderman LC, Bergin JJ. Hydrogen selenide poisoning: an illustrative case with review of the literature. Arch Environ Health. 1986;41:354–8.

Aldosary BM, Sutter ME, Schwartz M, Morgan BW. Case series of selenium toxicity from a nutritional supplement. Clin Toxicol. 2012;50:57–64.

Alexander J, Thomassen Y, Aaseth J. Increased urinary excretion of selenium among workers exposed to elemental mercury vapor. J Appl Toxicol. 1983;3:143–5.

Asaduzzaman AM, Schreckenbach G. Degradation mechanism of methyl selenoamino acid complexes: a computational study. Inorg Chem. 2011;50:2366–72.

Aschner M. Brain, kidney and liver ^{203}Hg-methyl mercury uptake in the rat: relationship to the neutral amino acid carrier. Basic Clinic Pharmacol Toxicol. 1989;65:17–20.

Aschner M, Syversen T, Souza DO, Rocha JB, Farina M. Involvement of glutamate and reactive oxygen species in methylmercury neurotoxicity. Braz J Med Biol Res. 2007;40:285–91.

Atchison WD, Hare MF. Mechanisms of methylmercury-induced neurotoxicity. FASEB J. 1984;8:622–9.

Barbosa AC, de Souza J, Dórea JG, Jardim WF, Fadini OS. Mercury Biomagnification in a tropical black water, Rio Negro, Brazil. Arch Environ Contam Toxicol. 2003;45:235–46.

Barwick M, Maher W. Biotransference and biomagnification of selenium copper, cadmium, zinc, arsenic and lead in a temperate seagrass ecosystem from Lake Macquarie Estuary, NSW, Australia. Mar Environ Res. 2003;56:471–502.

Bellinger FP, Bellinger MT, Seale LA, Takemoto AS, Raman AV, Miki T, Manning-Boğ AB, Berry MJ, White LR, Ross GW. Glutathione peroxidase 4 is associated with neuromelanin in substantia nigra and dystrophic axons in putamen of Parkinson's brain. Mol Neurodegener. 2011;21:8.

Benz MR, Lee SH, Kellner L, Döhlemann C, Berweck S. Hyperintense lesions in brain MRI after exposure to a mercuric chloride-containing skin whitening cream. Eur J Pediatr. 2011;170:747–50.

Berlin M, Fazackerley J, Nordberg G, Kand M. The uptake of mercury in the brains of mammals exposed to mercury vapor and to mercuric salts. Arch Environ Health. 1969;18:719–29.

Bjørklund G, Aaseth J, Ajsuvakova OP, Nikonorov AA, Skalny AV, Skalnaya MG, Tinkov AA. Molecular interaction between mercury and selenium in neurotoxicity. Coord Chem Rev. 2016; doi:10.1016/j.ccr.2016.10.009.

Blaurock-Busch E, Amin OR, Dessoki HH, Rabah T. Toxic metals and essential elements in hair and severity of symptoms among children with autism. Maedica (Buchar). 2012;7:38–48.

Bowles KC, Apte SC, Maher WA, Kawei M, Smith R. Bioaccumulation and biomagnification of mercury in Lake Murray, Papua New Guinea. Can J Fish Aquat Sci. 2001;58:888–97.

Branco V, Caito S, Farina M, Rocha JBT, Aschner M, Carvalho C. Biomarkers of mercury toxicity: Past, present, and future trends. J Toxicol Environ Health Part B. 2007; doi:10.1080/1093740 4.2017.1289834.

Brandão F, Cappello T, Raimundo J, Santos MA, Maisano M, Mauceri A, Pacheco M, Pereira P. Unravelling the mechanisms of mercury hepatotoxicity in wild fish (*Liza aurata*) through a triad approach: bioaccumulation, metabolomic profiles and oxidative stress. Metallomics. 2015;7:1352–63.

Brandão R, Moresco RN, Bellé LP, Leite MR, de Freitas ML, Bianchini A, Nogueira CW. Diphenyl diselenide potentiates nephrotoxicity induced by mercuric chloride in mice. J Appl Toxicol. 2011;31:773–82.

Bridges CC, Zalups RK. Mechanisms involved in the transport of mercuric ions in target tissues. Arch Toxicol. 2016; doi:10.1007/s00204-016-1803-y.

Byrns CN, Pitts MW, Gilman CA, Hashimoto AC, Berry MJ. Mice lacking selenoprotein P and selenocysteine lyase exhibit severe neurological dysfunction, neurodegeneration, and audiogenic seizures. J Biol Chem. 2014;289:9662–74.

Burk RF, Hill KE, Read R, Bellew T. Response of rat selenoprotein P to selenium administration and fate of its selenium. Am J Phys. 1991;261:26–30.

Burk RF, Hill KE, Motley AK, Winfrey VP, Kurokawa S, Mitchell SL, Wanqi Zhang W. Selenoprotein P and apolipoprotein E receptor-2 interact at the blood-brain barrier and also within the brain to maintain an essential selenium pool that protects against neurodegeneration. FASEB J. 2014;

Cardoso BR, Roberts BR, Bush AI, Hare DJ. Selenium, selenoproteins and neurodegenerative diseases. Metallomics. 2015;7:1213–28.

Carlson BA, Xu MX, Gladyshev VN, Hatfield DL. Um34 in selenocysteine tRNA is required for the expression of stress-related selenoproteins in mammals. Top Curr Genet. 2005;12:431–8.

Ceccatelli S, Daré E, Moors M. Methylmercury-induced neurotoxicity and apoptosis. Chem Biol Interact. 2010;188:301–8.

Chan TY. Inorganic mercury poisoning associated with skin-lightening cosmetic products. Clin Toxicol. 2011;49:886–91.

Chapman PM. Selenium – a potential time bomb or just another contaminant. Hum Ecol Risk Assessm. 1999;5:1123–38.

Chapple CE, Guigó R. Relaxation of selective constraints causes independent selenoprotein extinction in insect genomes. PLoS One. 2008;4(7) doi:10.1371/journal.pone.0002968.

Chen J. An original discovery: selenium deficiency and Keshan disease (an endemic heart disease). Asia Pac J Clin Nutr. 2012;21:320–6.

Chmielnicka J, Komsta-Szumska E, Jedrychowski R. Organ and subcellular distribution of mercury in rats as dependent on the time of exposure to sodium selenite. Environ Res. 1979;20:80–6.

Clarkson TW. The toxicology of mercury. Crit Rev Clin Lab Sci. 1997;34:369–403.

Clarkson TW. The three modern faces of mercury. Environ Health Perspect. 2002;110:11–23.

Clarkson TW, Vyas JB, Ballatori N. Mechanisms of mercury disposition in the body. Am J Ind Med. 2007;50:757–64.

Crack PJ, Cimdins K, Ali U, Hertzog PJ, Iannello RC. Lack of glutathione peroxidase-1 exacerbates Abeta-mediated neurotoxicity in cortical neurons. J Neural Transm. 2006;113:645–57.

Curtis JT, Chen Y, Buck DJ, Davis RL. Chronic inorganic mercury exposure induces sex-specific changes in central TNF expression: importance in autism? Neurosci Lett. 2011;504:40–4.

Dalla Corte CL, Wagner C, Sudati JH, Comparsi B, Leite GO, Busanello A, Soares FAA, Aschner M, Rocha JBT. Effects of diphenyl Diselenide on methylmercury toxicity in rats. BioMed Res Intern. 2013; doi:10.1155/2013/983821.

de Freitas AS, Funck VR, Rotta Mdos S, Bohrer D, Mörschbächer V, Puntel RL, Nogueira CW, Farina M, Aschner M, Rocha JB. Diphenyl diselenide, a simple organoselenium compound, decreases methylmercury-induced cerebral, hepatic and renal oxidative stress and mercury deposition in adult mice. Brain Res Bull. 2009;79:77–84.

de Freitas ML, da Silva AR, Roman SS, Brandão R. Effects of 4,4′-dichloro-diphenyl diselenide (ClPhSe)2 on toxicity induced by mercuric chloride in mice: a comparative study with diphenyl diselenide (PhSe)2. Environ Toxicol Pharmacol. 2012;34:985–94.

De Palma G, Catalani S, Franco A, Brighenti M, Apostoli P. Lack of correlation between metallic elements analyzed in hair by ICP-MS and autism. J Autism Dev Disord. 2012;42:342–53.

Diamond AM, Choin IS, Grain PF, Hashizumell T, Pomerantzll SC, Cruz R, Steer CJ, Hill KE, Burk RF, McCloskey HDL. Dietary selenium affects methylation of the wobble nucleoside in the anticodon of Selenocysteine tRNA[Ser]Sec. J Biol Chem. 1993;268:14215–23.

Dong W, Liu J, Wei L, Jingfeng Y, Chernick M, Hinton DE. Developmental toxicity from exposure to various forms of mercury compounds in medaka fish (*Oryzias latipes*) embryos. Peer J. 2016;23:2282.

Drosophila 12 Genomes Consortium. Evolution of genes and genomes on the drosophila phylogeny. Nature. 2007;450:203–18.

Dumont E, Vanhaecke F, Cornelis R. Selenium speciation from food source to metabolites: a critical review. Anal Bioanal Chem. 2006;385:1304–23.

Ekino S, Susa M, Ninomiya T, Imamura K, Kitamura T. Minamata disease revisited: an update on the acute and chronic manifestations of methyl mercury poisoning. J Neurol Sci. 2007;262:131–44.

El-Demerdash FM. Effects of selenium and mercury on the enzymatic activities and lipid peroxidation in brain, liver, and blood of rats. J Environ Sci Health B. 2001;36:489–99.

Ellingson RJ, Beard MC, Johnson JC, Yu P, Micic OI, Nozik AJ, Shabaev A, Efros AL. Highly efficient multiple exciton generation in colloidal PbSe and PbS quantum dots. Nano Lett. 2005;5:865–71.

Erken HA, Koç ER, Yazıcı H, Yay A, Önder GÖ, Sarıcı SF. Selenium partially prevents cisplatin-induced neurotoxicity: a preliminary study. Neurotoxicology. 2014;42:71–5.

Esaki N, Nakamura T, Tanaka H, Soda K. Selenocysteine lyase, a novel enzyme that specifically acts on selenocysteine. Mammalian distribution, purification, and properties of pig liver enzyme. J Biol Chem. 1982;257:4386–91.

Fagan S, Owens R, Ward P, Connolly C, Doyle S, Murphy R. Biochemical comparison of commercial selenium yeast preparations. Biol Trace Elem Res. 2015;166:245–59.

Falnoga I, Tušek-Žnidarič M. Selenium–mercury interactions in man and animals. Biol Trace Elem Res. 2007;119:212–20.

Farina M, Frizzo ME, Soares FA, Schwalm FD, Dietrich MO, Zeni G, Rocha JBT, Souza DO. Ebselen protects against methylmercury-induced inhibition of glutamate uptake by cortical slices from adult mice. Toxicol Lett. 2003a;144:351–7.

Farina M, Dahm KC, Schwalm FD, Brusque AM, Frizzo ME, Zeni G, Souza DO, Rocha JBT. Methylmercury increases glutamate release from brain synaptosomes and glutamate uptake by cortical slices from suckling rat pups: modulatory effect of ebselen. Toxicol Sci. 2003b;73:135–40.

Farina M, Aschner M, Rocha JBT. Oxidative stress in MeHg-induced neurotoxicity. Toxicol Applied Pharmacol. 2011a;256:405–17.

Farina M, Rocha JBT, Aschner M. Mechanisms of methylmercury-induced neurotoxicity: evidence from experimental studies. Life Sci. 2011b;89:555–63.

Farina M, Aschner M, Rocha JBT. The catecholaminergic neurotransmitter system in methylmercuryinduced neurotoxicity

Feng S, Xu Z, Wang F, Yang T, Liu W, Deng Y, Xu B. Sulforaphane prevents methylmercury-induced oxidative damage and excitotoxicity through activation of the Nrf2-ARE pathway. Mol Neurobiol. 2016;7:1–17.

Fiuza Tda L, Oliveira CS, da Costa M, Oliveira VA, Zeni G, Pereira ME. Effectiveness of (PhSe)2 in protect against the HgCl2 toxicity. J Trace Elem Med Biol. 2015;29:255–62.

Fordyce F. Selenium geochemistry and health. Ambio. 2007;36:94–7.

Forstrom JW, Zakowski JJ, Tappel AL. Identification of the catalytic site of rat liver glutathione peroxidase as selenocysteine. Biochemistry. 1978;27:2639–44.

Frost DV. The two faces of selenium – can selenophobia be cured? Crit Rev Toxicol. 1972;

Gajdosechova Z, Lawan MM, Urgast DS, Raab A, Scheckel KG, Lombi E, Kopittke PM, Loeschner K, Larsen EH, Woods G, Brownlow A, Read FL, Feldmann J, Krupp EM. In vivo formation of natural HgSe nanoparticles in the liver and brain of pilot whales. Sci Rep. 2016; doi:10.1038/srep34361.

Gardaneh M, Gholami M, Maghsoudi N. Synergy between glutathione peroxidase-1 and astrocytic growth factors suppresses free radical generation and protects dopaminergic neurons against 6-hydroxydopamine. Rejuvenation Res. 2011;14:195–204.

Gbetoh MH, Amyot M. Mercury, hydroquinone and clobetasol propionate in skin lightening products in West Africa and Canada. Environ Res. 2016;150:403–10.

Glaser V, Nazari EM, Müller YM, Feksa L, Wannmacher CM, Rocha JB, De Bem AF, Farina M, Latini A. Effects of inorganic selenium administration in methylmercury-induced neurotoxicity in mouse cerebral cortex. Int J Dev Neurosci. 2010;28:631–7.

Glaser V, Moritz B, Schmitz A, Dafré AL, Nazari EM, Rauh Müller YM, Feksa L, Straliottoa MR, de Bem AF, Farina M, da Rocha JB, Latini A. Protective effects of diphenyl diselenide in a mouse model of brain toxicity. Chem Biol Interact. 2013;206:18–26.

Glaser V, Martins Rde P, Vieira AJ, Oliveira Ede M, Straliotto MR, Mukdsi JH, Torres AI, de Bem AF, Farina M, da Rocha JB, De Paul AL, Latini A. Diphenyl diselenide administration enhances cortical mitochondrial number and activity by increasing hemeoxygenase type 1 content in a methylmercury-induced neurotoxicity mouse model. Mol Cell Biochem. 2014;390:1–8.

Glynn AW, Ilback N-G, Brabencova D, Carlsson L, Enqvist E-C, Netzel E, Oskarsson A. Influence of sodium selenite on ^{203}Hg absorption, distribution, and elimination in male mice exposed to methyl^{203}Hg. Biol Trace Elem Res. 1993;39:97–107.

Hamilton SJ. Review of selenium toxicity in the aquatic food chain. Sci Total Environ. 2004;326:1–31.

Hassan W, Oliveira CS, Noreen H, Kamdem JP, Nogueira CW, Rocha JBT. Organoselenium compounds as potential neuroprotective therapeutic agents. Curr Org Chem. 2015;20:218–31.

Hawkes WC, Tappel AL. In vitro synthesis of glutathione peroxidase from selenite. Translational incorporation of selenocysteine. Biochim Biophys Acta. 1983;739:225–34.

Hawkes WC, Lyons DE, Tappel AL. Identification of a selenocysteine-specific aminoacyl transfer RNA from rat liver. Biochim Biophys Acta. 1982;31:183–91.

Hatfield DL, Lee BJ, Diamond AM. Selenium induces changes in the selenocysteine tRNA[Ser]sec population in mammalian cells. Nucleic Acids Res. 1991;19:939–43.

Hatfield DL, Carlson BA, Xu XM, Mix H, Gladyshev VN. Selenocysteine incorporation machinery and the role of selenoproteins in development and health progress nucleic acid. Res Mol Biol. 2006;81:97–142.

Hatfield DL, Tsuji PA, Carlson BA, Gladyshev VN. Selenium and selenocysteine: roles in cancer, health, and development. Trends Biochem Sci. 2014;39:112–20.

Heath JC, Banna KM, Reed MN, Pesek EF, Cole N, Li J, Newland MC. Dietary selenium protects against selected signs of aging and methylmercury exposure. Neurotoxicology. 2010;31:169–79.

Heverly-Coulson GS, Boyd RJ. Reduction of hydrogen peroxide by glutathione peroxidase mimics: reaction mechanism and energetics. J Phys Chem A. 2010;114:1996–2000.

Hilt B, Svendsen K, Syversen T, Aas O, Qvenild T, Sletvold H, Melø I. Occurrence of cognitive symptoms in dental assistants with previous occupational exposure to metallic mercury. Neurotoxicology. 2009;30:1202–6.

Hongo T, Suzuki T, Himeno S, Watanabe C, Satoh H, Shimada Y. Does mercury vapor exposure increase urinary selenium excretion? Ind Health. 1985;23:163–5.

Horn MJ, Jones DB. Isolation from *Astragalus pectinatus* of a crystalline amino acid complex containing selenium and sulfur. J Biol Chem. 1940;139:649–60.

Horowitz HM, Jacob DJ, Amos HM, Streets DG, Sunderland EM. Historical mercury releases from commercial products: global environmental implications. Environ Sci Technol. 2014;48:10242–50.

Housecroft C, Sharpe AG. Inorganic chemistry. 4th ed. Harlow: Pearson Education Limited; 2012. chapter 15

Howard MT, Carlson BA, Anderson CB, Hatfield DL. Translational redefinition of UGA codons is regulated by selenium availability. J Biol Chem. 2013;2:122–8.

Huang JQ, Ren FZ, Jiang YY, Lei X. Characterization of Selenoprotein M and its response to selenium deficiency in chicken brain. Biol Trace Elem Res. 2016;170:449–58.

Huber RE, Criddle RS. Comparison of the chemical properties of selenocysteine and selenocystine with their sulfur analogs. Arch Biochem Biophys. 1967;122:164–73.

Hursh JB, Sichak SP, Clarkson TW. In vitro oxidation of mercury by the blood. Pharmacol Toxicol. 1988;63:26–273.

Imam SZ, Newport GD, Islam F, Slikker W, Ali SF. Selenium, an antioxidant, protects against methamphetamine-induced dopaminergic neurotoxicity. Brain Res. 1999;818:575–8.

Ishihara Y, Tsuji M, Kawamoto T, Yamazaki T (2016) Involvement of reactive oxygen species derived from mitochondria in neuronal injury elicited by methylmercury. J Clin Biochem Nutr 16-19.

Ishitobi H, Stern S, Thurston SW, Zareba G, Langdon M, Gelein R, Weiss B. Organic and inorganic mercury in neonatal rat brain after prenatal exposure to methylmercury and mercury vapor. Environ Health Perspec. 2010;118:242–8.

Iwata H, Masukawa T, Kito H, Hayashi M. Degradation of methylmercury by selenium. Life Sci. 1982;31:859–66.

Karaboduk H, Uzunhisarcikli M, Kalender Y. Protective effects of sodium selenite and vitamin e on mercuric chloride-induced cardiotoxicity in male rats. Braz Arch Biol Technol. 2015;58:229–38.

Kasaikina MV, Fomenko DE, Labunskyy VM, Lachke SA, Qiu W, Moncaster JA, Zhang J, Wojnarowicz MW Jr, Natarajan SK, Malinouski M, Schweizer U, Tsuji PA, Carlson BA, Maas RL, Lou MF, Goldstein LE, Hatfield DL, Gladyshev VN. Roles of the 15-kDa selenoprotein (Sep15) in redox homeostasis and cataract development revealed by the analysis of Sep 15 knockout mice. J Biol Chem. 2011;286:33203–12.

Kern JK, Geier DA, Bjørklund G, King PG, Homme KG, Haley BE, Sykes LK, Geier MR. Evidence supporting a link between dental amalgams and chronic illness, fatigue, depression, anxiety, and suicide. Neuro Endocrinol Lett. 2014;35:535–52.

Kern JK, Geier DA, Deth RC, Sykes LK, Hooker BS, Love JM, Bjørklund G, Chaigneau CG, Haley BE, Geier MR. Systematic assessment of research on Autism Spectrum Disorder and mercury reveals conflicts of interest and the need for transparency in autism research. Sci Eng Ethics. 2015; doi:10.1007/s11948-015-9713-6.

Kern JK, Geier DA, Sykes LK, Haley BE, Geier MR. The relationship between mercury and autism: a comprehensive review and discussion. J Trace Elem Med Biol. 2016;37:8–24.

Khan MAK, Wang F. Mercury-selenium compounds and their toxicological significance: toward a molecular understanding of the mercury-selenium antagonism. Environ Toxicol Chem. 2009;28:1567–77.

Kim IY, Stadtman TC. Selenophosphate synthetase: Detection in extracts of rat tissues by immunoblot assay and partial purification of the enzyme from the archaean Methanococcus vannielii (mammalian selenophosphate synthetase). Proc Natl Acad Sci USA. 1995;92:7710–3.

Kim YJ, Chai YG, Ryu JC. Selenoprotein W as molecular target of methylmercury in human neuronal cells is down-regulated by GSH depletion. Biochem Biophys Res Commun. 2005;20:1095–10200.

Kim YN, Kim YA, Yang AR, Lee BH. Relationship between blood mercury level and risk of cardiovascular diseases: results from the fourth Korea National Health and nutrition examination survey (KNHANES IV) 2008-2009. Prev Nutr Food Sci. 2014;19:333–42.

Kirkpatrick M, Benoit J, Everett W, Gibson J, Rist M, Fredette N. The effects of methylmercury exposure on behavior and biomarkers of oxidative stress in adult mice. NeuroToxicol. 2015;50:170–8.

Korbas M, O'Donoghue JL, Watson GE, Pickering IJ, Singh SP, Myers G, Clarkson TW, George GN. The chemical nature of mercury in human brain following poisoning or environmental exposure. ACS Chem Neurosci. 2010;1:810–8.

Kristensen AKB, Thomsen JF, Mikkelsen S. A review of mercury exposure among artisanal small-scale gold miners in developing countries. Int Arch Occup Environ Health. 2014;87:579–90.

Labunskyy VM, Hatfield DL, Gladyshev VN. Selenoproteins: molecular pathways and physiological roles. Physiol Rev. 2014;94:739–77.

Lakshmi Priya MD, Geetha A. Level of trace elements (copper, zinc, magnesium and selenium) and toxic elements (lead and mercury) in the hair and nail of children with autism. Biol Trace Elem Res. 2011;142:148–58.

Lee BJ, Worland PJ, Davis JN, Stadtman TC, Hatfield DL. Identification of a selenocysteinyl-tRNA[ser] in mammalian cells that recognizes the nonsense codon, UGA. J Biol Chem. 1989;264:9724–7.

Lemly AD. Assessing the toxic threat of selenium to fish and aquatic birds. Environ Monit Assess. 1996;43:19–35.

Letz R, Gerr F, Cragle D, Green RC, Watkins J, Fidler AT. Residual neurologic deficits 30 years after occupational exposure to elemental mercury. Neurotoxicology. 2000;21:459–74.

Li N, Reddy PS, Thyagaraju K, Reddy AP, Hsu BL, Scholz RW, Tu C-P D, Reddy CC. Elevation of rat liver mRNA for selenium-dependent glutathione peroxidase by selenium deficiency. J Biol Chem. 1990;265:108–13.

Li YF, Dong Z, Chen C, Li B, Gao Y, Qu L, Wang T, Fu X, Zhao Y, Chai Z. Organic selenium supplementation increases mercury excretion and decreases oxidative damage in long-term mercury-exposed residents from Wanshan, China. Environ Sci Technol. 2012;46:11313–8.

Liu ZH, Jing YH, Yin J, Mu JY, Yao TT, Gao LP. Downregulation of thioredoxin reductase 1 expression in the substantia nigra pars compacta of Parkinson's disease mice. Neural Regener Res. 2013;8:3275–83.

Lobanov AV, Hatfield DL, Gladyshev VN. Selenoproteinless animals: selenophosphate synthetase SPS1 functions in a pathway unrelated to selenocysteine biosynthesis. Protein Sci. 2008;17:176–82.

Lobanov AV, Hatfield DL, Gladyshev VN. Eukaryotic selenoproteins and selenoproteomes. Biochim at Biophys Acta. 2009;1790:1424–8.

Lohren H, Bornhorst J, Gallab H, Schwerdtle T. The blood–cerebrospinal fluid barrier – first evidence for an active transport of organic mercury compounds out of the brain. Metallomics. 2015;7:1420–30.

Magos L, Clarkson TW. Overview of the clinical toxicity of mercury. Ann Clin Biochem. 2006;43:257–68.

Magos L, Webb M. The effect of selenium on the brain uptake of methylmercury. Arch Toxicol. 1977;38:201–7.

Malagoli M, Schiavon M, Dall'Acqua S, Pilon-Smits EA. Effects of selenium biofortification on crop nutritional quality. Front Plant Sci. 2015;21:280.

Maquat LE. Evidence that selenium deficiency results in the cytoplasmic decay of GPx1 mRNA dependent on pre-mRNA splicing proteins bound to the mRNA exon-exon junction. Biofactors. 2001;14:37–42.

Marques RC, Bernardi JVE, Abreu L, Dórea JG. Neurodevelopment outcomes in children exposed to organic mercury from multiple sources in a tin-ore mine environment in Brazil. Arch Environ Contam Toxicol. 2015;68:432–41.

Metanis N, Beld J, Hilvert D. Chapter 19: The chemistry of selenocysteine. In: Patai S, editor. The chemistry of organic selenium and tellurium compounds, vol. 3. New York: Wiley; 1995.

Moraes-Silva L, Siqueira LF, Oliveira VA, Oliveira CS, Ineu RP, Pedroso TF, Fonseca MM, Pereira ME. Preventive effect of $CuCl_2$ on behavioral alterations and mercury accumulation in central nervous system induced by $HgCl_2$ in newborn rats. J Biochem Mol Toxicol. 2014;28:328–35.

Moretto MB, Franco J, Posser T, Nogueira CW, Zeni G, Rocha JBT. Ebselen protects Ca^{2+} influx blockage but does not protect glutamate uptake inhibition caused by Hg^{2+}. Neurochem Res. 2004;29:1801–6.

Moretto MB, Funchal C, Santos AQ, Gottfried C, Boff B, Zeni G, Pessoa-Pureur R, Souza D, Wofchuk S, Rocha JBT. Ebselen protects glutamate uptake inhibition caused by methyl mercury but does not by Hg^{2+}. Toxicology. 2005;214:57–66.

Muntean M, Janssens-Maenhout G, Song S, Selin NE, Olivier JGJ, Guizzardi D, Maas R, Dentener F. Trend analysis from 1970 to 2008 and model evaluation of EDGARv4 global gridded anthropogenic mercury emissions. Sci Total Environ. 2014;494–495:337–50.

Mutter J, Naumann J, Schneider R, Walach H, Haley B. Mercury and autism: accelerating evidence. Neuroendocrinol Lett. 2005;26:439–6.

Mutter J, Curth A, Naumann J, Deth R, Walach H. Does inorganic mercury play a role in Alzheimer's disease? A systematic review and an integrated molecular mechanism. J Alzheimers Dis. 2010;22:357–74.

Nakayama A, Hill KE, Austin LM, Motley AK, Burk RF. All regions of mouse brain are dependent on selenoprotein P for maintenance of selenium. J Nutr. 2007;137:690–3.

Naganuma A, Ishii Y, Imura N. Effect of administration sequence of mercuric chloride and sodium selenite on their fates and toxicities in mice. Ecotoxicol Environ Saf. 1984;8:572–80.

NAS. Dietary reference intakes for vitamin C, vitamin E, selenium, and carotenoids. A report of the Panel on Dietary Antioxidants and Related Compounds, Subcommittees on Upper

Reference Levels of Nutrients and Interpretation and Uses of Dietary Reference Intakes, and the Standing Committee on the Scientific Evaluation of Dietary Reference Intakes. Washington, DC: National Academy of Sciences, Institute of Medicine, Food and Nutrition Board; 2000.

Nauser T, Steinmann D, Koppenol WH. Why do proteins use selenocysteine instead of cysteine? Amino Acids. 2012;42:39–44.

Navarro-Alarcon M, López-Martínez MC. Essentiality of selenium in the human body: relationship with different diseases. Sci Total Environ. 2000;249:347–71.

Newland MC, Reed MN, LeBlanc A, Donlin W. Brain and blood mercury and selenium after chronic and developmental exposure to methylmercury. Neurotoxicology. 2006;27:710–20.

Nogueira CW, Zeni G, Rocha JB. Organoselenium and organotellurium compounds: toxicology and pharmacology. Chem Rev. 2004;104:6255–86.

Nogueira CW, Rocha JBT. Diphenyl diselenide: a Janus faced compound. J Braz Chem Soc. 2010;21:2055–71.

Nogueira CW, Rocha JBT. Toxicology and pharmacology of selenium: emphasis on synthetic organoselenium compounds. Arch Toxicol. 2011;85:1313–59.

Orct T, Lazarus M, Ljubojević M, Sekovanić A, Sabolić I, Blanuša M. Metallothionein, essential elements and lipid peroxidation in mercury-exposed suckling rats pretreated with selenium. Biometals. 2015;28:701–12.

Ogawa-Wong AN, Mj B, Seale LA. Selenium and metabolic disorders: an emphasis on type 2 diabetes risk. Forum Nutr. 2016;8:1–19.

Oliveira CS, Joshee L, Zalups RK, Bridges CC. Compensatory renal hypertrophy and the handling of an acute nephrotoxicant in a model of aging. Exp Gerontol. 2016;75:16–23.

Oliveira CS, Joshee L, Zalups RK, Pereira ME, Bridges CC. Disposition of inorganic mercury in pregnant rats and their offspring. Toxicology. 2015;335:62–71.

Oz SG, Tozlu M, Yalcin SS, Sozen T, Guven GS. Mercury vapor inhalation and poisoning of a family. Inhal Toxicol. 2012;24:652–8.

Pamphlett R, Kum Jew S. Uptake of inorganic mercury by human locus ceruleus and cortico-motor neurons: implications for amyotrophic lateral sclerosis. Acta Neuropathol Commun. 2013;9:1–13.

Parizek J, Ostadalova I. The protective effect of small amounts of selenite in sublimate intoxication. Experientia. 1967;23:142–3.

Peixoto NC, Pereira ME. Effectiveness of $ZnCl_2$ in protecting against nephrotoxicity induced by $HgCl_2$ in newborn rats. Ecotoxicol Environ Saf. 2007;66:441–6.

Peixoto NC, Roza T, Morsch VM, Pereira ME. Behavioral alterations induced by $HgCl_2$ depend on the postnatal period of exposure. Int J Dev Neurosci. 2007;25:39–46.

Peng X, Manna L, Yang W, Wickham J, Scher E, Kadavanich A, Alivisatos AP. Shape control of CdSe nanocrystals. Nature. 2000;404:59–61.

Peregrino CP, Moreno MV, Miranda SV, Rubio AD, Leal LO. Mercury levels in locally manufactured Mexican skin-lightening creams. Int J Environ Res Public Health. 2011;8:2516–23.

Peters MM, Hill KE, Burk RF, Weeber EJ. Altered hippocampus synaptic function in selenoprotein P deficient mice. Mol Neurodegener. 2006;1:1–13.

Pillai R, Uyehara-Lock JH, Bellinger FP. Selenium and selenoprotein function in brain disorders. IUBMB Life. 2014;66:229–39.

Pitts MW, Reeves MA, Hashimoto AC, Ogawa A, Kremer P, Seale LA, Berry MJ. Deletion of selenoprotein M leads to obesity without cognitive deficits. J Biol Chem. 2013;288:26121–34.

Power JH, Blumbergs PC. Cellular glutathione peroxidase in human brain: cellular distribution, and its potential role in the degradation of Lewy bodies in Parkinson's disease and dementia with Lewy bodies. Acta Neuropathol. 2009;117:63–73.

Ralston NV, Raymond LJ. Dietary selenium's protective effects against methylmercury toxicity. Toxicology. 2010;278:112–23.

Raman AV, Pitts MW, Seyedali A, Hashimoto AC, Bellinger FP, Berry MJ. Selenoprotein W expression and regulation in mouse brain and neurons. Brain Behavior. 2013;3:562–74.

Raman AV, Pitts MW, Seyedali A, Hashimoto AC, Seale LA, Bellinger FP, Berry MJ. Absence of selenoprotein P but not selenocysteine lyase results in severe neurological dysfunction. Genes, Brain and Behav. 2012;11:601–13.

Rocha JBT, Piccoli BC, Oliveira CS. Biological and chemical interest in selenium: a brief historical account. ARKIVOC. 2017; doi:10.3998/ark.5550190.p009.784.

Rocha JBT, Saraiva RA, Garcia SC, Gravina FS, Nogueira CW. Aminolevulinate dehydratase (δ-ALA-D) as marker protein of intoxication with metals and other pro-oxidant situations. Toxicol Res. 2012;1:85–102.

Roos DH, Puntel RL, Santos MM, Souza DO, Farina M, Nogueira CW, Aschner M, Burger ME, Barbosa NB, Rocha JB. Guanosine and synthetic organoselenium compounds modulate methylmercury-induced oxidative stress in rat brain cortical slices: involvement of oxidative stress and glutamatergic system. Toxicol In Vitro. 2009;23:302–7.

Rosenfeld I, Beath OA. Selenium: geobotany, biochemistry, toxicity, and nutrition. Chapter 3. New York: Academic Press INC; 1964.

Rowens B, Guerrero-Betancourt D, Gottlieb CA, Boyes RJ, Eichenhorn MS. Respiratory failure and death following acute inhalation of mercury vapor. A clinical and histologic perspective. Chest J. 1991;99:185–90.

Rueli RHLH, Parubrub AC, Dewing AST, Hashimoto AC, Bellinger MT, Weeber EJ, Uyehara-Lock JH, White LR, Berry MJ, Bellinger FP. Increased selenoprotein P in choroid plexus and cerebrospinal fluid in Alzheimer's disease brain. J Alzheimer's Disease. 2015;44:379–83.

Ruszkiewicz JA, Bowman AB, Farina M, Rocha JB, Aschner M. Sex-and structure-specific differences in antioxidant responses to methylmercury during early development. Neurotoxicology. 2016;56:118–26.

Savaskan NE, Borchert A, Bräuer AU, Kuhn H. Role for glutathione peroxidase-4 in brain development and neuronal apoptosis: specific induction of enzyme expression in reactive astrocytes following brain injury. Free Radic Biol Med. 2007;15:191–201.

Skalny AV, Simashkova NV, Klyushnik TP, Grabeklis AR, Radysh IV, Skalnaya MG, Nikonorov AA, Tinkov AA. Assessment of serum trace elements and electrolytes in children with childhood and atypical autism. J Trace Elem Med Biol. 2016a; doi:10.1016/j.jtemb.2016.09.009.

Skalny AV, Simashkova NV, Klyushnik TP, Grabeklis AR, Radysh IV, Skalnaya MG, Tinkov AA. Analysis of hair trace elements in children with autism Spectrum disorders and communication disorders. Biol Trace Elem Res. 2016b; doi:10.1007/s12011-016-0878-x.

Skalny AV, Simashkova NV, Klyushnik TP, Grabeklis AR, Bjørklund G, Skalnaya MG, Nikonorov AA, Tinkov AA. Hair toxic and essential trace elements in children with autism spectrum disorder. Metab Brain Dis. 2016c; doi:10.1007/s11011-016-9899-6.

Sugiura Y, Hojo Y, Tamai Y, Tanaka H. Letter: selenium protection against mercury toxicity. Binding of methylmercury by the selenohydryl-containing ligand. J Am Chem Soc. 1976;98:2339–41.

Sunde RA, Evenson JK. Serine incorporation into the selenocysteine moiety of glutathione peroxidase. J Biol Chem. 1987;15:933–7.

Steinbrenner H, Sies H. Selenium homeostasis and antioxidant selenoproteins in brain: implications for disorders in the central nervous system. Arch Biochem Biophys. 2013;536:152–7.

Suzuki T, Himeno S, Hongo T, Watanabe C, Satoh H. Mercury-selenium interaction in workers exposed to elemental mercury vapor. J Appl Toxicol. 1986;6:149–53.

Tabatadze T, Zhorzholiani L, Kherkheulidze M, Kandelaki E, Ivanashvili T. Hair heavy metal and essential trace element concentration in children with autism spectrum disorder. Georgian Med News. 2015;248:77–82.

Takagi Y, Mitsui A, Nishiyama A, Nozaki K, Sono H, Gon Y, Hashimoto N, Yodo J. Overexpression of thioredoxin in transgenic mice attenuates focal ischemic brain damage. Proc Natl Acad Sci U S A. 1999;96:4131–6.

Trapp GA, Millam J. The distribution of 75Se in brains of selenium-deficient rats. J Neurochem. 1975;24:593–5.

Trelease SF, Di Somma AA, Jacobs AL. Seleno-amino acid found in *Astragalus bisulcatus*. Science. 1960;132:618.

Tsuda T, Yorifuji T, Takaob S, Miyai M, Babazono A. Minamata disease: catastrophic poisoning due to a failed public health response. J Public Health Policy. 2009;30:54–67.

United Nations Environment Programme (UNEP). Global Mercury Assessment. Sources, emissions, releases and environmental transport. UNEP: Geneva, Switzerland; 2013. p. 2013.

Uzunhisarcikli M, Aslanturk A, Kalender S, Apaydin FG, Bas H. Mercuric chloride induced hepatotoxic and hematologic changes in rats: The protective effects of sodium selenite and vitamin E. Toxicol Ind Health. 2015:0748233715572561.

Usuki F, Yamashita A, Fujimura M. Post-transcriptional defects of antioxidant selenoenzymes cause oxidative stress under methylmercury exposure. J Biol Chem. 2011;286:6641–9.

Vinceti M, Wei ET, Malagoli C, Bergomi M, Vivoli G. Adverse health effects of selenium in humans. Rev Environ Health. 2001;16:233–51.

Vinceti M, Maraldi T, Bergomi M, Malagoli C. Risk of chronic low-dose selenium overexposure in humans: insights from epidemiology and biochemistry. Rev Environ Health. 2009;24:231–48.

Vinceti M, Bonvicini F, Rothman KJ, Vescovi L, Wang F. The relation between amyotrophic lateral sclerosis and inorganic selenium in drinking water: a population-based case–control study. Environ Health. 2010;9:77.

Vinceti M, Crespi CM, Malagoli C, Bottecchi I, Ferrari A, Sieri S, Krogh V, Alber D, Bergomi M, Seidenari S, Pellacani G. A case–control study of the risk of cutaneous melanoma associated with three selenium exposure indicators. Tumori. 2012;98:287–95.

Vinceti M, Solovyev N, Mandrioli J, Crespi CM, Bonvivini F, Arcolin E, Georgoulopoulou E, Michalke B. Cerebrospinal fluido f newly diagnosed amyotrophic lateral sclerosis patients exhibits abnormal levels of selenium species including elevated selenite. Neurotoxicology. 2013;38:25–32.

Vinceti M, Mandrioli J, Borella P, Michalke B, Tsatsakis A, Finkelstein Y. Selenium neurotoxicity in humans: bridging laboratory and epidemiologic studies. Toxicol Lett. 2014;230:295–303.

Wang R. Physiological implications of hydrogen sulfide: a whiff exploration that blossomed. Physiol Rev. 2012;92:791–896.

Warfvinge K. Mercury distribution in the neonatal and adult cerebellum after mercury vapor exposure of pregnant squirrel monkeys. Environ Res. 2000;83:93–101.

Winkel LHE, Vriens B, Jones GD, Schneider LS, Pilon-Smits E, Bañuelos GS. Selenium cycling across soil-plant-atmosphere interfaces: a critical review. Forum Nutr. 2015;7:4199–239.

WHO (World Health Organization) (2003) Selenium in drinking-water: Background document for development of WHO guidelines for drinking-water quality.

WHO (World Health Organization) (2007) Exposure to mercury: a major public health concern. Environmental Health Criteria. Geneva: World Health Organization.

Wu Y, Guo X, Wang W, Chen X, Zhao Z, Xia X, Yang Y. Red pigments and Boraginaceae leaves in mortuary ritual of late Neolithic China: a case study of Shengedaliang site. Microsc Res Tech. 2016; doi:10.1002/jemt.22791.

Yamamoto R, Suzuki T, Satoh H, Kawais K. Generation and dose as modifying factors of inorganic mercury accumulation in brain, liver, and kidneys of rats fed methylmercury. Environ Res. 1986;41:309–18.

Yang G, Wu L, Jiang B, Yang W, Qi J, Cao K, Meng Q, Mustafa AK, Mu W, Zhang S, Snyder SH, Wang R. H$_2$S as a physiologic vasorelaxant: hypertension in mice with deletion of cystathionine γ-lyase. Science. 2008;322:587–90.

Ye BJ, Kim BG, Jeon MJ, Kim SY, Kim HC, Jang TW, Chae HJ, Choi WJ, Ha MN, Hong YS. Evaluation of mercury exposure level, clinical diagnosis and treatment for mercury intoxication. Ann Occup Environ Med. 2016; doi:10.1186/s40557-015-0086-8.

Yu WH, Zhang N, Qi JF, Sun C, Wang YH, Lin M. Arsenic and mercury containing traditional chinese medicine (Realgar and cinnabar) strongly inhibit organic anion transporters, Oat1 and Oat3, in vivo in mice. Biomed Res Int. 2015; doi:10.1155/2015/863971.

Zalups RK. Molecular interactions with mercury in the kidney. Pharmacol Rev. 2000;52:113–43.

Zhang L, Hu B, Li W, Che R, Deng K, Li H, Yu F, Ling H, Li Y, Chu C. OsPT2, a phosphate transporter, is involved in the active uptake of selenite in rice. New Phytol. 2014;201:1183–91.

Zhang Y, Jacob DJ, Horowitz HM, Chen L, Amos HM, Krabbenhoft DP, Sleim F, Louis VLS, Sunderland EM. Observed decrease in atmospheric mercury explained by global decline in anthropogenic emissions. Proc Natl Acad Sci USA. 2016;113:526–31.

Metals and Paraoxonases

Lucio G. Costa, Toby B. Cole, Jacqueline M. Garrick, Judit Marsillach, and Clement E. Furlong

Abstract The paraoxonases (PONs) are a three-gene family which includes PON1, PON2, and PON3. PON1 and PON3 are synthesized primarily in the liver and a portion is secreted in the plasma, where they are associated with high-density lipoproteins (HDLs), while PON2 is an intracellular enzyme, expressed in most tissues and organs, including the brain. PON1 received its name from its ability to hydrolyze paraoxon, the active metabolite of the organophosphorus (OP) insecticide parathion, and also more efficiently hydrolyzes the active metabolites of several other OPs. PON2 and PON3 do not have OP-esterase activity, but all PONs are lactonases and are capable of hydrolyzing a variety of lactones, including certain drugs, endogenous compounds, and quorum-sensing signals of pathogenic bacteria. In addition, all PONs exert potent antioxidant effects. PONs play important roles in cardiovascular diseases and other oxidative stress-related diseases, modulate susceptibility to infection, and may provide neuroprotection (PON2). Hence, significant attention

L.G. Costa (✉)
Department of Environmental and Occupational Health Sciences, University of Washington, 4225 Roosevelt Way NE, Suite 100, Seattle, WA 98105, USA

Department of Medicine & Surgery, University of Parma, Parma, Italy
e-mail: lgcosta@uw.edu

T.B. Cole
Department of Environmental and Occupational Health Sciences, University of Washington, 4225 Roosevelt Way NE, Suite 100, Seattle, WA 98105, USA

Center on Human Development and Disability, University of Washington, 4225 Roosevelt Way NE, Suite 100, Seattle, WA 98105, USA

J.M. Garrick
Department of Environmental and Occupational Health Sciences, University of Washington, 4225 Roosevelt Way NE, Suite 100, Seattle, WA 98105, USA

J. Marsillach
Department of Medicine (Division of Medical Genetics), University of Washington, 4225 Roosevelt Way NE, Suite 100, Seattle, WA 98105, USA

C.E. Furlong
Department of Medicine (Division of Medical Genetics), University of Washington, 4225 Roosevelt Way NE, Suite 100, Seattle, WA 98105, USA

Department of Genome Sciences, University of Washington, 4225 Roosevelt Way NE, Suite 100, Seattle, WA 98105, USA

© Springer International Publishing AG 2017
M. Aschner and L.G. Costa (eds.), *Neurotoxicity of Metals*, Advances in Neurobiology 18, DOI 10.1007/978-3-319-60189-2_5

has been devoted to their modulation by a variety of dietary, pharmacological, life-style, or environmental factors. A number of metals have been shown in in vitro, animal, and human studies to mostly negatively modulate expression of PONs, particularly PON1, the most studied in this regard. In addition, different levels of expression of PONs may affect susceptibility to toxicity and neurotoxicity of metals due to their aforementioned antioxidant properties.

Keywords Paraoxonases • Metals • Lead • Mercury • Cadmium • Manganese • Oxidative stress

Abbreviations

Ag	Silver
Al	Aluminum
As	Arsenic
Ba	Barium
Cd	Cadmium
Ce	Cesium
Co	Cobalt
Cr	Chromium
Cu	Copper
Fe	Iron
Gd	Gadolinium
HDL	High-density lipoprotein
Hg	Mercury
L	Leucine
La	Lanthanum
LDL	Low-density lipoprotein
M	Methionine
MeHg	Methylmercury
Mn	Manganese
Ni	Nickel
OP	Organophosphate
Pb	Lead
PCR	Polymerase chain reaction
PON	Paraoxonase
ppb	Parts per billion
ppm	Parts per million
Q	Glutamine
R	Arginine
Sm	Samarium
Y	Yttrium
Zn	Zinc

Introduction

The paraoxonases (PONs) are a three-gene family which includes PON1, PON2, and PON3, all clustered in tandem on the long arm of human chromosome 7 (7q21.22). PON1 and PON3 are synthesized primarily in the liver and a portion is secreted in the plasma, where they are associated with high-density lipoproteins (HDLs); low levels of PON1 and PON3 may be expressed in a number of other tissues, primarily in epithelia (Primo-Parmo et al. 1996; Marsillach et al. 2008). In contrast, PON2 is an intracellular enzyme, expressed in most tissues and organs, including the brain. PON1 received its name from its ability to hydrolyze paraoxon, the active metabolite of the organophosphorus (OP) insecticide parathion, which is its first and most studied substrate. PON1 more efficiently hydrolyzes the active metabolites of several other OP insecticides (e.g., chlorpyrifos oxon, diazoxon) and less efficiently nerve agents such as sarin and soman (Costa et al. 2003, 2013a). PON2 and PON3 do not have OP-esterase activity, but all PONs are lactonases and are capable of hydrolyzing a variety of lactones, including certain drugs (bioactivating some, e.g., the antibacterial prodrug prulifloxacin, or inactivating others, e.g., glucocorticoids), endogenous compounds (e.g., lactone metabolites of arachidonic acid), and N-acyl homoserine lactones, which are quorum-sensing signals of pathogenic bacteria (Draganov et al. 2005; Teiber et al. 2008). All PONs have potent antioxidant effects: PON1 and PON3 protect low-density lipoproteins (LDLs) (Mackness et al. 1991), as well as HDL from oxidation (Aviram et al. 1998; reviewed in Costa et al. 2003), while PON2 exerts intracellular antioxidant effects (Costa et al. 2014). PON1 is the most studied of the PONs, because of its important roles in modulating susceptibility to OP neurotoxicity and in cardiovascular disease and other diseases (Costa and Furlong 2002; Costa et al. 2003; Furlong et al. 2010). PON2 has received more attention recently, and novel important roles in the central nervous system and in tumor cells are emerging (Costa et al. 2014; Witte et al. 2011). PON3 is the least studied of the three PONs, but there is evidence that it plays important roles in cardiovascular disease, in susceptibility to infection, and in tumor cells (Shih et al. 2007; Schweikert et al. 2012a, b; Marsillach et al. 2015).

PON1

Human Polymorphisms of PON1 and Definition of PON1 Status

Earlier observations on the polymorphic distribution of serum paraoxonase activity in human populations led to the purification, cloning, and sequencing of human (and rabbit) PON1, as well as in the molecular characterization of its polymorphisms (Furlong et al. 1993; Humbert et al. 1993). Of the two polymorphisms observed in the PON1 coding sequence (Q192R and L55M), the former significantly affects the catalytic efficiency of PON1 for some substrates (Humbert et al. 1993).

The $PON1_{R192}$ allozyme hydrolyzes paraoxon or chlorpyrifos oxon more readily than $PON1_{Q192}$, while the opposite is true in the case of sarin or soman (Davies et al. 1996). In the case of diazoxon, both PON1 alloforms hydrolyze this compound with the same efficiency, and both alloforms are able to provide in vivo protection against exposure (Li et al. 2000). Lactones are hydrolyzed preferentially by either $PON1_{R192}$ or $PON1_{Q192}$, depending on their structure (Draganov et al. 2005). For example, $PON1_{R192}$ is more efficient at hydrolyzing homocysteine thiolactone (HCL), while gamma-valerolactone and 2-coumaranone are more rapidly hydrolyzed by $PON1_{Q192}$. However, it is important to note that hydrolysis of HCL by PON1 is orders of magnitude less efficient than by bleomycin hydrolase and especially by biphenyl hydrolase-like protein (Marsillach et al. 2014). Furthermore, $PON1_{Q192}$ has also a higher efficiency in protecting against LDL oxidation than the $PON1_{R192}$ allozyme (Mackness et al. 1998).

The L/M polymorphism at position 55 does not appear to affect catalytic activity, but has been associated with plasma PON1 protein levels, with $PON1_{M55}$ being associated with low plasma PON1 activity (Mackness et al. 1998). However, this appears to result primarily from linkage disequilibrium with the low-efficiency –108T allele of the −108 promoter region polymorphism (Brophy et al. 2002). Of the several additional polymorphisms found in the noncoding region of the PON1 gene, one of the most significant is this polymorphism at position −108, with the −108C allele providing levels of PON1 about twice as high on average as those seen with the –108T allele (Brophy et al. 2001).

Most studies investigating the association of PON1 with various diseases have examined nucleotide polymorphisms (mainly Q192R, L55M, C-108T) with PCR-based assays. A functional genomic activity analysis, however, provides a much more informative approach, as measurement of an individual's PON1 function (serum activity) takes into account all polymorphisms and other factors that might affect PON1 activity or expression. This is accomplished through the use of high-throughput enzyme assays involving two PON1 substrates (traditionally diazoxon and paraoxon at high salt concentration, but more recently the nontoxic phenyl acetate at high salt, and 4-(chloromethyl)phenyl acetate at low salt; Richter and Furlong 1999; Richter et al. 2008, 2009). Both the earlier assay with the two OP substrates and the new assays using the safer non-OP substrates provide a clear separation of the three $PON1_{192}$ functional genotypes (QQ, QR, RR), as well as information on enzyme activity within each genotype (Richter and Furlong 1999). This approach, which provides a functional assessment of the plasma $PON1_{192}$ alloforms, including information on the plasma level of PON1 for each individual, has been referred to as the determination of PON1 "status" for an individual (Richter and Furlong 1999). In a given population, plasma PON1 activity can vary up to 40–50-fold, and differences in PON1 protein levels up to 13–15-fold are also present within a single $PON1_{192}$ genotype in adults (Richter and Furlong 1999). The use of PON1 substrates that are not affected by the Q192R polymorphism (e.g., phenyl acetate hydrolysis at low salt to measure arylesterase activity) provides a surrogate measure of PON1 plasma protein level as does direct analysis of PON1 protein concentration (e.g., by ELISA or mass spectrometry). In contrast, given that PON1 activity is strongly determined by enzyme genotype, assays using paraoxon as a substrate would provide equivocal results, if each group is not matched for genotype, since each $PON1_{192}$ alloform hydrolyzes paraoxon with different efficiencies. A good example of analyzing individuals within each $PON1_{192}$ functional genotype is provided in the study of PON1 status and stroke (Jarvik et al. 2000).

The importance of PON1 status in modulating susceptibility to the acute toxicity of a number of OP insecticides has been shown by several studies (Shih et al. 1998; Li et al. 2000; Cole et al. 2005). Studies with transgenic animal models have shown that PON1-deficient mice are highly susceptible to the toxicity of specific OPs (Shih et al. 1998; Li et al. 2000). Depending on the OP, PON1 levels alone (as in the case of diazoxon) or $PON1_{192}$ functional genotype as well as activity level (as in the case of chlorpyrifos oxon) may determine the degree of protection against a specific OP (Li et al. 2000). Alterations in circulating PON1 levels have been found in a variety of diseases involving oxidative stress, including cardiovascular disease, diabetes, Alzheimer's disease, chronic renal failure, and chronic liver impairment (Costa and Furlong 2002; Costa et al. 2003; Marsillach et al. 2007a, b; Furlong et al. 2010; Androutsopoulos et al. 2011). Studies investigating the role of PON1 in cardiovascular disease have provided evidence that PON1 status (encompassing genotype and activity levels) is a much better predictor of disease than PON1 genotype alone (Mackness et al. 2001; Jarvik et al. 2003).

Modulation of PON1 Activity and Expression

Given the role of PON1 in protecting against toxic pesticide exposures and cardiovascular disease, and its decreased activity levels in a number of pathological conditions, it is not surprising that particular attention has been devoted to factors that may positively modulate PON1, i.e., increase its activity or expression (reviewed in Costa et al. 2005, 2011; Camps et al. 2009). While a major determinant of PON1 activity is represented by genetic polymorphisms, age also plays an important role, as PON1 activity is very low before birth and gradually increases during the first year or two of life in humans (Cole et al. 2003). PON1 activity may also decline with aging, possibly because of the development of oxidative stress conditions (reviewed in Costa et al. 2005). An influence of gender has also been suggested, with female mice displaying higher PON1 activity (reviewed in Costa et al. 2005). Several studies investigating modulation of PON1 have involved pharmaceutical drugs, particularly lipid-lowering compounds such as statins and fibrates, as well as other drugs (reviewed in Costa et al. 2005, 2011). As PON1 is easily inactivated by exogenous or endogenous oxidants, several strategies to increase PON1 have focused on the administration of dietary antioxidants such as vitamin C (ascorbic acid), vitamin E (alpha-tocopherol), and several dietary polyphenols, particularly quercetin and pomegranate juice and extract, which contain several polyphenolic compounds such as punicalagin, gallic acid, and ellagic acid (reviewed in Costa et al. 2011). Dietary lipids (e.g., olive oil or omega-3 fatty acids) and moderate doses of alcohol also increase PON1 activity and expression (reviewed in Costa et al. 2005, 2011).

While most attention has been devoted to identifying pharmacological or dietary factors that may increase PON1 activity, other factors that, in contrast, may negatively impact PON1 should also be considered, as they may increase susceptibility to diseases and/or toxic effects. High alcohol consumption, smoking, and consumption of certain high-fat diets have been shown to decrease PON1 expression (reviewed in Costa et al. 2005, 2011). Several studies have shown that metals can also negatively modulate PON1 (see section "Interactions of metals with PONs").

PON2

PON2 as an Intracellular Antioxidant Enzyme

PON2, a PON isozyme less studied than PON1, is nevertheless emerging as an important defense system toward oxidative stress and inflammation. In contrast to PON1 and PON3, PON2 is a ubiquitously expressed intracellular enzyme, but is not present in plasma (Mochizuki et al. 1998; Ng et al. 2001; Marsillach et al. 2008; Giordano et al. 2011). In peripheral tissues, PON2 is considered important in modulating sensitivity to bacterial infections because of its high acyl-HSL hydrolytic activity, and also plays a significant role in atherosclerosis (Ng et al. 2006), and in antagonizing oxidative and inflammatory processes that may affect mucosal integrity in the gastrointestinal tract (Levy et al. 2007). Subcellular distribution studies have shown that PON2 is localized primarily in the mitochondria, endoplasmic reticulum, and perinuclear region (Devarajan et al. 2011; Giordano et al. 2011), a major source of free radical-related oxidative stress. More recently it was reported that PON2 translocates its catalytic domain to the outside of the cell under certain conditions of oxidative stress, to protect membrane lipids from oxidation (Hagmann et al. 2014).

Two common coding region polymorphisms in strong disequilibrium (A147G and S311C) have been found in human PON2 (Primo-Parmo et al. 1996; Mochizuki et al. 1998). The PON2 S311C polymorphism has been shown to affect lactonase activity, but does not appear to influence antioxidant activity of PON2 (Altenhofer et al. 2010). PON2 mRNA and protein have been found in the central nervous system (CNS) of several species including mouse, rat, nonhuman primate, and human (Costa et al. 2014). In mouse brain, the highest levels of PON2 are in the dopaminergic regions (substantia nigra, striatum, nucleus accumbens), with lower levels in other brain areas (Giordano et al. 2011). In every brain region, as well as in peripheral tissues, PON2 levels are higher in female mice than in male mice. PON2 exerts a protective effect toward oxidative stress, for example, the cytotoxicity of the oxidants hydrogen peroxide (H_2O_2) and 2,3-dimethoxy-1,4-naphthoquinone (DMNQ) is much greater in brain cells from PON2 knockout mice (Giordano et al. 2011). The different levels of expression of PON2 protein between male and female mice are also reflected in a differential susceptibility to neurotoxicity (Giordano et al. 2011, 2013). While the apparent anti-apoptotic properties of PON2 may underlie neuroprotection, the same characteristic in cancer cells makes them more resistant to chemotherapy-induced apoptosis (Witte et al. 2011; Krüger et al. 2015).

Modulation of PON2

The higher levels of PON2 in tissues from female mice appear to be related to a positive modulatory effect by estrogens, as suggested by various lines of evidence (Giordano et al. 2013). For example, 17-beta estradiol increases the levels of PON2 in striatal astrocytes from male mice; the effect is due to transcriptional activation of the PON2 gene and appears to be mediated by activation of estrogen receptor alpha (Giordano et al. 2013). In addition, PON2 levels (protein and mRNA) in ovariectomized female mice are significantly reduced in brain regions and in the liver, approaching the levels found in male mice (Giordano et al. 2013).

Activation of dopamine D2 receptors in the kidney positively modulates PON2 expression through activation of NADPH oxidase, leading to a decrease in ROS production (Yang et al. 2012). In the CNS, the highest levels of dopamine D2 receptors are found in the same areas (e.g., striatum, nucleus accumbens, substantia nigra) that also have the highest level of PON2 expression (Giordano et al. 2011). If a similar mechanism as observed in kidneys also occurs in the CNS, the loss of dopamine associated with Parkinson's disease would lead to decreased PON2 levels, thus fostering a spiral of events further aggravating neurodegeneration. The functional consequences of a higher expression of PON2 in females may have important ramifications. For example, oxidative stress plays a highly relevant role in the etiopathology of Parkinson's disease, whose incidence is 90% higher in males (Surmeier et al. 2011; Wirdefeld et al. 2011). Furthermore, as PON2 is expressed in most tissues and levels appear to be higher in females in each tissue examined (Giordano et al. 2011), the reported higher sensitivity of males to oxidative stress in the heart, to atherosclerosis, and to infections may all be related to a differential expression of PON2 (Klein 2000; Kardys et al. 2007; Wang et al. 2010).

In contrast to PON1 and PON3, PON2 expression is increased by oxidative stress (Rosenblat et al. 2003). Additionally, arachidonic acid, unesterified cholesterol, the licorice phytoestrogen glabridin, extracts of yerba mate (*Ilex paraguariensis*), and the hypocholesterolemic drug atorvastatin also upregulate PON2 expression in various cell types (Rosenblat et al. 2004; Fernandes et al. 2012; Yehuda et al. 2016). A recent study found that quercetin increases PON2 protein expression in the brain, thereby providing neuroprotection (Costa et al. 2013b). In contrast to studies on PON1 regulation, no studies on negative modulation of PON2, other than by metals (see section "Interactions of metals with PONs"), have been identified.

PON3: Activity, Polymorphisms, Physiological Functions, and Modulation

PON3 is synthesized mainly by the liver and is found in circulation in HDLs (Reddy et al. 2001) and intracellularly in endoplasmic reticulum (Rothem et al. 2007) and mitochondria (Schweikert et al. 2012a). In mice, PON3 is undetectable in serum or HDL (Ng et al. 2007), but its protein expression has been identified in multiple tissues (Marsillach et al. 2008). PON3 is the least characterized of the PON family of enzymes. It does not hydrolyze OPs, but possesses lipo-lactonase and N-acyl homoserine lactone activities (Draganov et al. 2005). Compared to PON1, PON3 has a higher catalytic activity for statin lactones (such as lovastatin), which are commonly used to monitor PON3 activity (Draganov et al. 2005).

There are only few studies on polymorphisms in the PON3 gene (reviewed in Furlong et al. 2016). Two missense mutations (S311T, G324D) in exons III, IV, and IX of *PON3* were identified in healthy subjects from Southern Italy (Campo et al. 2004), and later in children with diagnosed inflammatory bowel disease (Sanchez et al. 2006), but no relationship between the *PON3* genetic variants and disease was observed.

Human PON3 concentration in serum is about two orders of magnitude lower than PON1 (Aragones et al. 2011). However, recombinant rabbit PON3 seems to be more potent than recombinant rabbit PON1 in protecting LDL from copper-induced oxidative modifications in vitro (Draganov et al. 2000). PON3 and PON2 protect murine macrophages against oxidative damage, with cellular PON3 activity being decreased under oxidative stress (Rosenblat et al. 2003). In vivo, mice overexpressing PON3 are more resistant to atherosclerosis and obesity (Shih et al. 2007; Ng et al. 2007). Interestingly, these effects were only seen in male mice although a protective role of PON3 in obesity has also been reported in female mice with the PON3 gene knocked out (Shih et al. 2015). As previously reported for PON1, human serum PON3 concentration significantly increases in some disease states such as chronic liver disease, coronary and peripheral artery disease, and HIV infection (Garcia-Heredia et al. 2011; Rull et al. 2012; Aragones et al. 2012), while another study has recently reported a significant decrease in PON3 in HDL from patients with autoimmune disease (type 1 diabetes or systemic lupus erythematosus) and subclinical atherosclerosis (Marsillach et al. 2015).

Despite PON3's beneficial role in protecting against a variety of oxidative stress-related diseases, an unexpected finding is a role for PON3 (and PON2), in cancer where PON3 is upregulated (PON3 being much more overexpressed in cancer cells than PON2) and protects tumor cells against mitochondrial superoxide-mediated apoptosis. Also similarly to PON2, PON3 has an important role in the defense against *P. aeruginosa* virulence (Schweikert et al. 2012b).

With the exception of being negatively affected by oxidative stress (Rosenblat et al. 2003), no other information, except for the interaction with metals (see section "Interactions of metals with PONs"), is available on positive or negative modulation of PON3.

Interactions of Metals with PONs

Metals are often defined by their physical properties of the element in the solid state (e.g., high electrical and thermal conductivity, mechanical ductility), but their toxicological relevance is linked to their ability to lose one or more electrons to form cations (Tokar et al. 2013). In addition, metals often exhibit variable oxidation states. Over 75% of elements in the periodic table are regarded as metals or metalloids. Metals are found naturally in the Earth's crust, and their level varies across the continents; they are redistributed naturally in the environment by geologic and biologic cycles. However, human intervention can shorten the residence of metals in ore, may form new compounds, and may increase worldwide distribution. Due to their wide and early use, the toxicity of metals has been known for centuries. Initially, concerns were primarily related to acute effects, though later metal toxicology has shifted to more subtle, chronic, low-dose effects. Some metals (e.g., lithium, bismuth, platinum compounds) can also have beneficial effects and are used as pharmaceutical drugs. Exposure of humans to metals can occur through multiple pathways (mostly inhalation and oral) and in a variety of settings. Occupational exposure to a number of metals (e.g., manganese, lead, cadmium, mercury) is quite common. Exposure to metals through food (e.g., methylmercury in fish) or water (e.g., arsenic, lead, manganese) is also common.

Various metal ions have been studied in relationship to their interactions with PONs. Most studies have examined interactions of metals with PON1, while very few have also included PON2 and PON3. Several in vitro studies have been carried out, but only a handful of in vivo studies in experimental animals are available. A few human studies investigating associations between serum PON1 and blood metal levels in various populations are also available. Below, we review the current information on the interactions of metals with PONs.

In Vitro Effects of Metals on PONs

A number of studies have examined the in vitro effects of metals on PONs, all with one exception on PON1. The first indication that PON1 activity may be affected by metals can be found in an abstract from the 1950s in which Aldridge (1951) reported that E 600 (paraoxon) hydrolase (later identified as PON1) was "inhibited by mercury, copper and nickel (10^{-6} M)". A few years later, Erdös et al. (1959, 1960) investigated the effects of several metals on blood arylesterase activity (identifiable as PON1 by calcium dependence and inhibition by EDTA). In addition to heavy metals (e.g., cadmium, lead, mercury), these investigators also tested rare earth metals such as cerium, samarium, gadolinium, lanthanum, as well as yttrium. The latter metals were actually more potent than other metals in inhibiting arylesterase activity, as the IC_{50} values were in the sub-micromolar range (Erdös et al. 1960; Table 1).

L.G. Costa et al.

Table 1 In vitro effects of metals on PON1 activity

Study/metal	1	2	3	4	5	6	7	8	9	10
Ag	0.7									
Al								>1000/>1000		
Ba	60	5580/290	930							
Cd	0.3			3.3		150/250		0.05/0.08	730	2840/980
Ce	0.04									
Co	3.0	2330/90		64	80	800/4200				2050/1810
Cr							1991			
Cu	8.0	170/850	63	178	317	50/300			290	747/530
Fe							3960	100/200		
Gd	0.04									
Hg	0.7	520/320	21	4.7	4.0	120/1100		0.2/0.5	490	1810/2840
La	0.06	1116/280	310	85						
Mn	10	3740/170		151	199	300/700				5140/1980
Ni	3.0			21.3		1000/1200			2000	3390/4220
Pb	2.0						838	0.2/1.0		
Sm	0.2									
Zn	1.0	1060/130	920	6.2			7410	0.05/0.1		0.08/0.2
Y	0.2									

Shown are values of IC_{50} (μM). Studies are (1) Erdös et al. 1960; (2) Gil et al. 1994 [plasma/liver]; (3) Gonzalvo et al. 1997; (4) Debord et al. 2003; (5) Pla et al. 2007; (6) Gencer and Arslan 2009 [R/Q]; (7) Ekinci and Beydemir 2010; (8) Cole et al. 2002; Figs. 1 and 2 [R/Q]; (9) Sayin et al. 2012; (10) Erol et al. 2013

Gil et al. (1994) reported that rat plasma paraoxonase activity was inhibited by a number of metals with IC_{50} values ranging from 170 μM (copper) to 5580 μM (barium) (Table 1). Kinetic studies indicated that inhibition was either competitive (Cu, La, Zn, Co) or noncompetitive (Hg, Mn, Ba). Results were somewhat different when the same metals were tested on the paraoxonase activity of the rat hepatic microsomal fraction. In particular, cobalt was the most potent metal, followed by zinc and manganese (Table 1). All metals appeared to be more potent (by up to 25-fold) in inhibiting liver paraoxonase activity compared to plasma, with the exception of copper, for which the opposite was found. In addition, kinetic characteristics of inhibition in the liver were different from plasma (Gil et al. 1994). In a follow-up study, Gonzalvo et al. (1997) studied the ability of a number of metals to inhibit paraoxon hydrolysis in human liver microsomes (Table 1). They found that mercury was the most potent inhibitor ($IC_{50} = 21$ μM), followed by copper and lanthanum.

Debord et al. (2003) reported that several metals inhibited arylesterase activity of human serum, with copper being the most potent (Table 1). The paraoxonase activity of PON1 purified from rat liver was inhibited by various heavy metals, including mercury, manganese, copper, and cobalt (Pla et al. 2007; Table 1). Similarly, paraoxonase activity of purified human PON1 (no indication of genotype) was found to be inhibited by metals, and lead was the most potent in this regard (Ekinci and Beydemir 2010; Table 1). Kinetic studies indicated that inhibition was of the competitive type for lead and iron and noncompetitive for chromium and zinc. Sayin et al. (2012) and Erol et al. (2013) describe the purification of PON1 from blood of sharks and of two breeds of sheep (merino and kivircik) and the effect of various metals on paraoxonase activity (Table 1). Of the metals tested, cadmium and copper appeared to be the most potent. Kinetic analysis of the type of inhibition provided results similar to those of Gil et al. (1994); for example, inhibition by copper and cobalt was of the competitive type, while that of mercury was noncompetitive. A study by Sukketsiri et al. (2013) examined the effects of lead acetate on PON1 arylesterase activity in HepG2 (human hepatoma) cells. In contrast to other studies, at concentrations of up to 100 μg/ml (263.6 μM), lead had no effect on PON1 arylesterase activity.

Gencer and Arslan (2009) were the first to investigate the relative effects of metals on $PON1_{R192}$ and $PON1_{Q192}$ allozymes. For all metals tested (Cd, Co, Cu, Hg, Mn, Ni), the $PON1_{R192}$ allozyme displayed a higher sensitivity to inhibition, ranging from less than two- to ninefold. In our laboratory, we tested the ability of cadmium, mercury, iron, zinc, lead, and aluminum (all chloride salts) to inhibit plasma PON1 activity. Metals were incubated with PON1 for various lengths of time (15 min, 4 h, 24 h), followed by measurement of PON1 arylesterase activity. Initial experiments used human PON1 purified from individuals expressing either the $PON1_{R192}$ or $PON1_{Q192}$ alloform (Fig. 1). All metals tested inhibited PON1 arylesterase activity to some degree. Inhibition was similar at the three incubation times tested (15 min, 4 h, and 24 h), and results for 4 h are shown in Fig. 1. Cadmium, mercury, lead, and zinc were potent inhibitors of PON1, with nearly complete inhibition at 0.75 μM and significant inhibition in some cases at less than 0.1 μM. For these four metals,

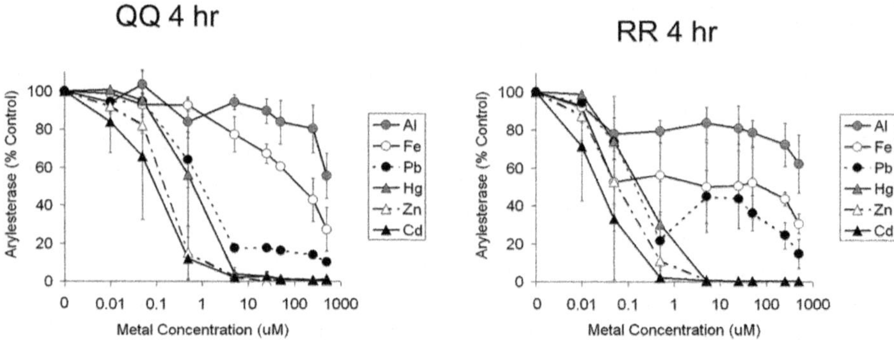

Fig. 1 Effect of various metals on the activity of purified human $PON1_{Q192}$ and $PON1_{R192}$ after 4 h incubation. Results are expressed as means (± SE) with n = 4

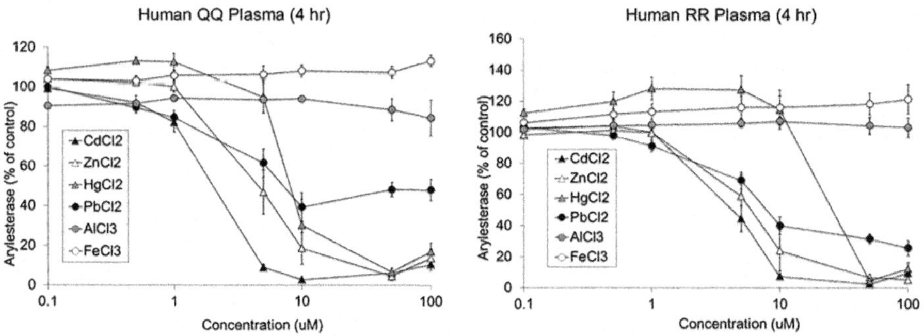

Fig. 2 Effects of various metals on arylesterase activity in plasma from individual homozygotes for the 192Q or 192R alleles of PON1, after 4 h incubation. Results are expressed as means (± SE) with n = 4

the $PON1_{R192}$ alloform was more susceptible to inhibition by cadmium than the $PON1_{Q192}$ alloform (Table 1; Fig. 1), in agreement with the findings of Gencer and Arslan (2009). Iron also inhibited PON1, with significant inhibition at 0.75 µM; inhibition was mainly observed with the $PON1_{R192}$ alloform, while the $PON1_{Q192}$ alloform was relatively resistant to inhibition. Aluminum was the weakest inhibitor of PON1, with <50% inhibition in both PON1 allozymes.

In a second series of experiments, the ability of the same metals to inhibit PON1 arylesterase activity was measured in plasma of individuals homozygous for the $PON1_{R192}$ or the $PON1_{Q192}$ allele. Results of these experiments for the 4 h time point are shown in Fig. 2, and similar results were obtained for the shorter (15 min) and longer (24 h) incubation times. In general, the concentration-response curves for metal inhibition of PON1 were shifted to the right compared to those obtained with purified PON1, suggesting that factors present in plasma provide some protection against metal inhibition of PON1 (compare Fig. 1 and Fig. 2). For example, cadmium

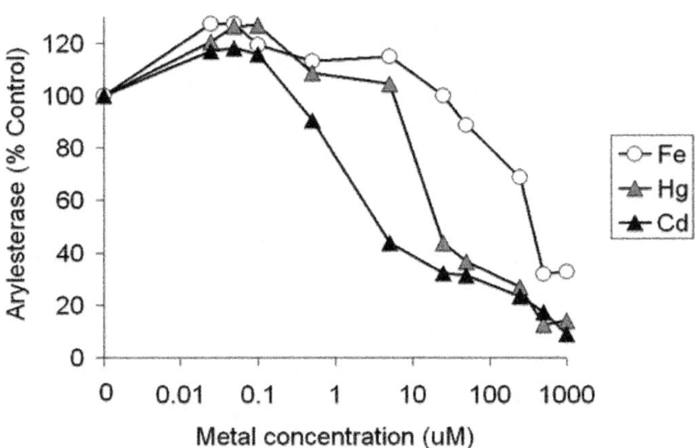

Fig. 3 Effects of cadmium, mercury, and iron (all chlorides) on arylesterase activity in mouse plasma after 4 h incubation. Results are expressed as means (± SE) with n = 4

was still the most potent inhibitor of PON1, but its potency was decreased by >10-fold compared to the purified enzyme. Pb, Zn, and Hg had an intermediate potency, while Al and Fe were devoid of any inhibitory activity (Fig. 2). For all metals, the differences in inhibitory potency between the RR and the QQ genotype were minimal (Fig. 2).

The ability of three metals (Cd, Hg, and Fe) to inhibit serum PON1 arylesterase activity in vitro was also assessed in mouse plasma. As shown in Fig. 3, cadmium was the most potent inhibitor of PON1, followed by mercury and by iron. No significant differences were found when comparing results obtained after 4 h incubation (Fig. 3) with 15 min or 24 h incubations (not shown).

In contrast to PON1, information on the possible effects of metals on PON2 and/ or PON3 is scarce. In HepG2 human hepatoma cells, lead acetate significantly inhibited lactonase activity (dihydrocoumarin hydrolysis) at concentrations as low as 0.13 μM (Sukketsiri et al. 2013). Inhibition appeared to increase with the length of incubation (4–72 h), particularly at the lower concentrations. However, lactonase activity was attributed solely to PON2, while all PONs are expressed in the liver and all have lactonase activity. Levels of PON2 protein were not affected by lead, but PON2 mRNA levels were increased, though not in a concentration-dependent manner. Interestingly, addition of calcium (1 mM) prevented the inhibitory effect of lead on PON2 activity. With regard to PON3, there is only one study by Pla et al. (2007) who purified this enzyme from rat liver. Various metals inhibited PON3 lactonase activity (dihydrocoumarin hydrolysis) including (IC_{50}, μM) mercury (2), copper (36), manganese (318), and cobalt (1898). EDTA also inhibited PON3 activity, confirming that activity of this enzyme, like PON1, is also calcium dependent.

Effects of Metals on PONs: Animal Studies

Very few studies have investigated in experimental animals the effects of in vivo exposure to metals on PONs, and all, to our knowledge, have focused on PON1. Tas et al. (2006) examined the effects of treatment of rats rendered diabetic with strep-tozotocin with vanadyl sulfate (a vanadium salt) on serum paraoxonase and aryles-terase activities. Vanadyl sulfate had no effect on PON1 activity in control rats; however, it was capable of partially reversing the decrease in paraoxonase and arylesterase activities induced by streptozotocin, in virtue, according to the authors, of its antioxidant properties (Tas et al. 2006). This represents the only animal study in which a metal derivative was found to "increase" PON1 activity.

Rats given cadmium chloride in drinking water at the levels of 15 or 100 ppm for 2 months had serum levels of Cd of ~6 and ~15 ppb, respectively, compared to ~0.5 ppb in controls (Ferramola et al. 2012). At the highest dose, Cd caused an increased in oxidative stress in serum and a 40% decrease of serum PON1 paraox-onase activity. In another study, female C57Bl/6 mice were given cadmium chloride in drinking water (5 mg/L) for 1 month (Ramambason et al. 2016), yielding levels of Cd in the liver of 0.33 μg/g (Thijssen et al. 2007). A significant 30% decrease in liver PON1 arylesterase activity was found in Cd-treated animals (Ramambason et al. 2016). In contrast, in a study in which cadmium chloride was given by i.p. injections for 2 weeks at doses of 0.1 to 0.5 mg/kg, no changes in serum PON1 activity were found (Cole et al. 2002). Ebabe Elle et al. (2013) fed rats a standard diet supple-mented with 500 mg/kg silver nanoparticles for 81 days. Silver caused oxidative stress and inflammation in the liver and decreased plasma PON1 paraoxonase activ-ity by 15%. In contrast, administration of aluminum by intraperitoneal injections to Wistar rats for 2 weeks did not alter PON1 activity in plasma (Maghraoui et al. 2014).

As Hg was a potent in vitro inhibitor of human and mouse PON1 arylesterase activity (Figs. 1, 2 and 3), we examined the effects of Hg exposure in vivo on plasma and liver PON1 activity. Male C57/B6 mice were exposed by subcutaneous (s.c.) injections of methylmercury hydroxide (10, 20, or 30 μmol/kg/d, equivalent to 2.33, 4.65, and 6.98 mg/kg/d) for 14 days. Unfortunately, Hg tissue levels were not mea-sured in this study, though based on other similar studies they are expected to be in the sub-micromolar to low nanomolar range. Somewhat unexpectedly, neither plasma nor liver PON1 activity (diazoxonase and paraoxonase) was decreased by treatment (not shown). To investigate the potential effect of iron on PON1 activity, we took advantage of an ongoing Fe overload study in female $apoE^{-/-}$ mice 6–8 weeks of age. Groups of mice were fed a low-iron (0.02% Fe) or a high-iron (2%) diet for 12 weeks. Serum nonheme iron and liver iron levels were determined together with diazoxonase and paraoxonase activity in plasma. Figure 4a shows that serum Fe levels increased significantly in mice fed the high-iron diet, and liver iron levels increased to a much greater extent. Activity of diazoxonase in plasma was decreased only by a nonsignificant 12%, while plasma paraoxonase activity was decreased by ~20% (Fig. 4b and c). In this case the minimal effect of Fe on plasma PON1 activity in vivo is not surprising, given the limited effectiveness of this metal

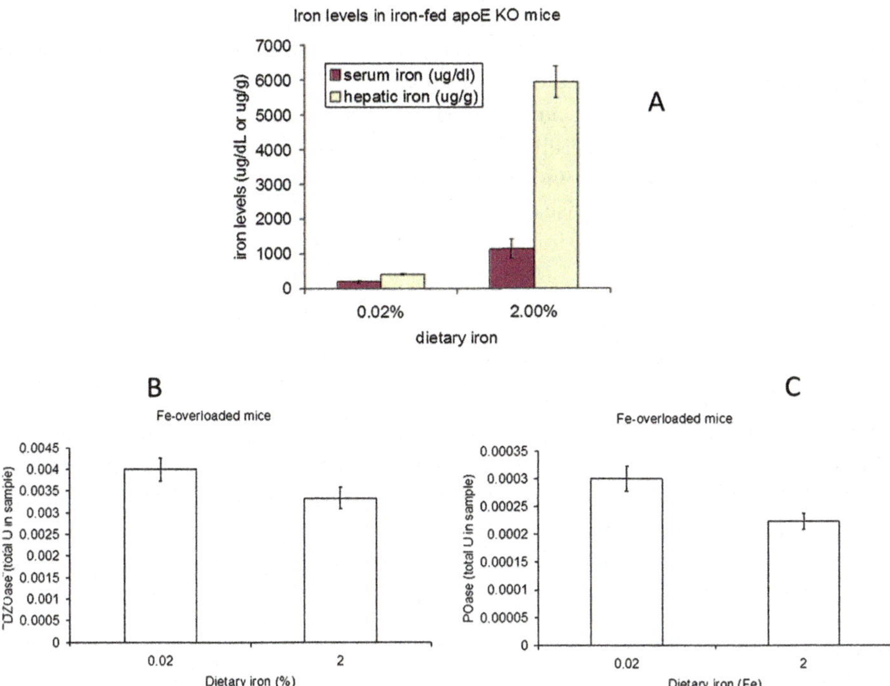

Fig. 4 Effect of dietary iron overload on plasma PON1 activity in apoE knockout mice. (**a**) Levels of Fe in serum and liver. (**b**) Diazoxonase activity in plasma; (**c**) paraoxonase activity in plasma. Results represent the mean (± SE) of four mice

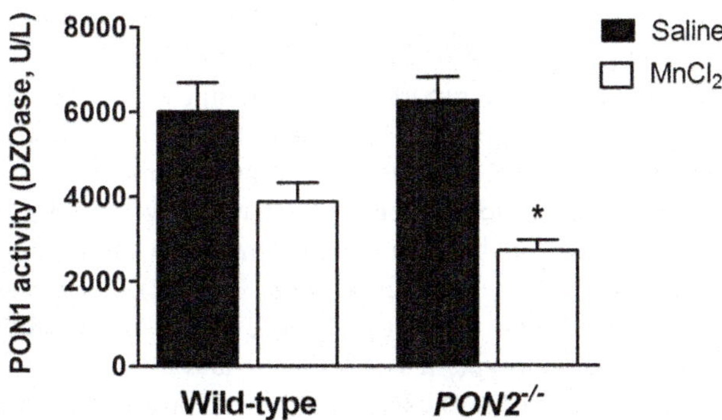

Fig. 5 Effect of manganese chloride on serum PON1 activity (diazoxonase) in male wild-type and PON2$^{-/-}$ mice. Results represent the mean (± SE) of four mice

in inhibiting PON1 in human and mouse plasma (Figs. 2 and 3). In an additional preliminary experiment, male wild-type and PON2 knockout (PON2$^{-/-}$) mice were given three doses of manganese (MnCl$_2$, 100 mg/kg, s.c.) and sacrificed 1 week later. As shown in Fig. 5, Mn decreased serum PON1 activity (measured as diazoxonase), and its effects were more pronounced in $PON2^{-/-}$ mice, suggesting that PON2 may protect PON1 from oxidative stress related to metal exposure (Marsillach et al., unpublished observations).

Overall, it is evident that animal studies investigating the effects of metals on PONs are limited and almost all focus on PON1, with little attention paid so far to PON2 and PON3. Given the widespread exposure to metals and the relevance of all three PONs in a variety of diseases, further animal studies of the effects of metals on PONs activity and expression are certainly warranted.

Metals and PONs: Human Studies

A number of studies have examined the association between blood metal levels and PON1 activity and/or expression in humans. While the in vitro studies described in section "In vitro effects of metals on PONs" were presented in chronological order of publication, as numerous metals were tested in each study, human studies are described below by grouping them for each metal investigated.

Lead

Li et al. (2006) examined the associations between blood Pb levels and PON1 activity (measured as paraoxonase, arylesterase, and diazoxonase activities) in about 600 workers in Taiwan. Workers were divided into three groups on the basis of blood Pb levels (μg/dL): ≤ 10, $10 \leq 40$, and ≥ 40. There was a small (10–13%) but significant decrease in PON1 activity with increasing blood Pb concentrations. The three most relevant PON1 polymorphisms (Q192R, L55M, and C-108T) were also determined and found to be similar to those reported for the Chinese population. The strongest inverse association between Pb and PON1 was found in $PON1_{R192}$ homozygotes (RR), while the results in heterozygotes (QR) and QQ homozygotes were not statistically significant (Li et al. 2006). Levels of Pb in blood were in the low micromolar range, a concentration that had been shown in some studies to inhibit PON1 activity in vitro (Erdös et al. 1960; Cole et al. 2002; Figs. 1 and 2). In agreement with these in vivo findings, Pb had been found to be more potent in inhibiting arylesterase activity of the $PON1_{R192}$ genotype in vitro (Figs. 1 and 2).

A study by Pollack et al. (2014) examined the association between blood Pb and PON1 activity in a group of 250 women, and their findings are in agreement with those of Li et al. (2006) in that a decreased PON1 activity was associated with Pb, but only in individuals homozygous for the $PON1_{R192}$ allele. Levels of blood Pb in this cohort were very low, about 1 μg/dL (Pollack et al. 2014). Further support for an in vivo effect of Pb on PON1 activity has been provided by another study which investigated the association between blood Pb and PON1 in a group of 100 workers of a lead battery factory (Kamal et al. 2011). Blood Pb levels in all workers averaged 45.7 μg/dL (vs. 12.5 μg/dL in controls), and PON1 was decreased by an average of

60%. When stratifying workers based on blood Pb levels (< 40, 40–59, ≥ 60 µg/dL), PON1 activity was decreased by 36, 63, and 69%, respectively. These investigators also reported that PON1 activity was lowest in the Pb-exposed workers homozygous for the $PON1_{R192}$ allele, thus substantiating previous findings in humans (Li et al. 2006; Pollack et al. 2014) and in vitro (Figs. 1 and 2). An additional study has reported associations between blood Pb levels and PON1 activity. In a group of Pb-exposed earthenware factory workers in Thailand (n = 65; mean blood Pb level 31.4 µg/dL), PON1 arylesterase activity was decreased by 24% (Permpongpaiboon et al. 2011). However, PON1 paraoxonase activity did not differ between Pb-exposed workers and controls; this may be related to a differential $PON1_{192}$ genotype distribution between Pb-exposed workers and controls, though this was not determined. An increase in blood parameters of oxidative stress was also found in this study, and the authors attributed the decrease in PON1 arylesterase activity to oxidative stress, which is known to negatively affect PON1 (Nguyen and Sok 2003).

Arsenic

A single study examined the influence of arsenic (As) exposure on PON1 activity in 196 residents from an arseniasis-endemic area in Southwestern Taiwan (Li et al. 2009). However, consumption of As-contaminated well water had ceased for several years, and indeed only urinary excretion of inorganic As was higher in the endemic group. Overall, plasma PON1 activity was similar between controls and the endemic group and was actually higher in $PON1_{Q192}$ homozygotes with As exposure compared to controls. However, high As exposure together with low PON1 activity increased the risk for developing atherosclerosis by 5.7-fold (Li et al. 2009). When analyzing $PON1$ (Q192R and C-108T) and $PON2$ (A148G, C311S) polymorphism distribution in the control and As-exposed populations, some differences were found, whose significance is unclear. Hernandez et al. (2009) also examined the association between urinary levels of As and plasma PON1 activity in a population of healthy individuals (n = 536). They found no associations between As levels and PON1 activity, though carriers of the $PON1_{R192}$ alloform had higher levels of urinary As (Hernandez et al. 2009).

Methylmercury and Selenium

Various human studies have also explored possible association between exposure to methylmercury (MeHg) and PON1 activity. A study in Nunavik, Canada, of 896 Inuit adults found a significant inverse correlation between plasma Hg levels and PON1 activity (Ayotte et al. 2011). With increasing blood Hg levels (from <30 to >100 nmol/L; geometric mean = 53.2 nmol/L), PON1 activity decreased to a maximum of 14%; however, there was no association with any specific $PON1_{192}$ genotype. Interestingly, as found in other studies (Cayir et al. 2014; Laird et al. 2015), blood concentrations of selenium (Se) appeared to counteract the effect of Hg on

PON1. Selenium is present in the active site of several enzymes, many of which are involved in modulation of oxidative stress (e.g., thioredoxin reductase, glutathione peroxidase), and this may explain its "protective" effect toward PON1. The study by Ayotte et al. (2011) was utilized by Ginsberg et al. (2014) to mechanistically address the issue of a possible association of MeHg exposure with cardiovascular disease suggested by some, but not all epidemiological studies (Virtanen et al. 2007; Mozaffarian et al. 2011). Ginsberg et al. (2014) calculated that a dose of 0.3 mg/kg/day of MeHg would cause an average 6.1% decrease in PON1 level, and this would increase the risk of cardiovascular disease by 9.7%.

In another Canadian population (Cree people in Eastern James Bay) with lower MeHg exposure (blood Hg geometric mean = 16.7 nmol/L; n = 369), no association between blood Hg levels and PON1 activity was found (Drescher et al. 2014). However, a negative correlation was found between Hg and PON1 in carriers of the (rare) TT allele at position -108 (the low PON1 activity allele). This polymorphism disrupts a recognition site for Sp1, a zinc-finger transcription factor whose DNA-binding activity can be inhibited by Hg ions through interaction with Cys_2His_2 zinc-binding domains (Rodgers et al. 2001; Deakin et al. 2003). People with the -108T allele, who have a compromised interaction between Sp1 and the promoter sequence ($PON1_{-108T}$), may thus be more susceptible to further disruption by Hg (Drescher et al. 2014).

Additional information on potential associations between Hg and PON1 were provided by the studies of Hernandez et al. (2009), Pollack et al. (2014), and Laird et al. (2015). In contrast to most other studies, Hernandez et al. (2009) found a positive association between urinary levels of Hg and PON1 in a population of healthy subjects in Spain, i.e., higher PON1 activity with increasing blood Hg levels. On the other hand, Pollack et al. (2014) found that women exposed to Hg and homozygous for the $PON1_{R192}$ genotype had 23% lower PON1 activity in plasma. Finally, a recent study examined PON1 activity in plasma in relationship to blood levels of Hg, Pb, Cd, and Se in a population of over 2000 Inuit in Canada (Laird et al. 2015). PON1 activity was measured with a commercially available kit which utilizes 7-diethylphospho-6,8-difluor-methylumbelliferyl as a substrate, and no assessment of PON1 genotype was done. As expected, PON1 activity in the population varied by 27-fold. No correlations were found between Pb and PON1, in contrast to previous studies (Li et al. 2006; Hernandez et al. 2009). Somewhat in agreement with the findings of Hernandez et al. (2009), a positive correlation between blood Hg and PON1 activity was found. Blood Se levels were also positively associated with PON1 activity, as also reported by others (Ayotte et al. 2011). Similarly, Cayir et al. (2014) found that in obese children from Turkey (who already have ~40% lower serum PON1 activity than normal weight children) blood Se levels are positively associated with plasma PON1 activity.

Cadmium

Three studies examined the associations between cadmium (Cd) and PON1. In the recent one by Laird et al. (2015), Cd was the only metal negatively associated with PON1 activity after adjustment for a number of co-variables. This result confirmed previous findings by Hernandez et al. (2009) and Pollack et al. (2014). In the former study, blood Cd was negatively associated with PON1 paraoxonase, arylesterase, and diazoxonase activities (Hernandez et al. 2009), while Pollack et al. (2014) found a similar association but only in $PON1_{R/R192}$ and $PON1_{Q/R192}$ individuals.

Other Metals

Limited or no information exists with regard to other metals. Levels of blood copper were negatively associated with PON1 activity in obese children (Cayir et al. 2014). No association was found between manganese blood levels and PON1 in two studies (Hernandez et al. 2009; Cayir et al. 2014). Finally, with regard to zinc, one study reported no association with PON1 (Cayir et al. 2014), while another found a decreased PON1 activity (Hernandez et al. 2009), in agreement with in vitro results.

Overall, human studies indicate, for the most part, an association between metal exposure and decreased PON1 activity, especially in individuals homozygous for the $PON1_{R192}$ allele. However, most studies are incomplete and lack important information on exposure to metals and to other potential confounding factors (e.g., smoking, alcohol, drugs), duration of exposure, blood levels of metals, PON1 genotype, accurate PON1 activity measurements, and levels of PON1 protein.

Potential Mechanisms of PON1 Modulation by Metals

Metals may reduce PON1 activity and/or expression by generating oxidative stress, by directly interacting with the enzyme, by interfering with its transcription/translation, or by combinations of these mechanisms. Most metals cause oxidative stress (Jaishankar et al. 2014; Matovic et al. 2015; Valko et al. 2016), and PON1 is known to be inactivated by oxidative stress (Nguyen and Sok 2003). The same is true for PON3 (Rosenblat et al. 2003), while PON2 expression is increased by oxidative stress (Aviram and Rosenblat 2004). Certain antioxidants may increase PON1 activity by preventing its oxidative inactivation (Aviram et al. 1999), and this may explain the results observed with selenium. A direct interaction between metals and the PON1 protein is also likely, as suggested by the findings of the in vitro studies detailed above. Some metals (e.g., zinc, nickel) may bind to histidine (His) in positions 115, 134, 155, and 243, which are essential for PON1 activity (Josse et al. 2002). Other metals (e.g., mercury, lead) may bind to free sulfhydryl groups on the enzyme. PON1 has three cysteine (Cys) residues (in positions 42, 284, and 353), with a disulfide bond between Cys-42 and Cys-353, while Cys-284 is a free thiol (Josse et al. 2002). The disulfide-linked Cys-42 and Cys-353 are essential for PON1 esterase activity, while Cys-284 is not; however, Cys-284 is close to the active site for catalytic activity of PON1, and its covalent modification interferes with PON1 activity (Sorenson et al. 1995). Most importantly, the free Cys-284 is essential for PON1 to be protective against LDL oxidation (Aviram et al. 1998; Josse et al. 2002).

However, this Cys is buried, and thus it is unclear if it may represent a target for some metals (Harel et al. 2004; Hernandez et al. 2009; Laird et al. 2015).

Calcium-binding sites on PON1 are also most relevant. The crystal structure for a recombinant PON1 has indicated that PON1 is a six-bladed β-propeller, which in the central tunnel contains two calcium ions, one of which is essential for enzyme activity and the other for stability of the protein (Harel et al. 2004). Removal of calcium ions from PON1 by the chelating agents EDTA or EGTA leads to inhibition of its esterase activity, but not of its ability to protect against LDL oxidation (Kuo and La Du 1998; Aviram et al. 1998). Lead is known to mimic calcium, causing stimulation or inhibition of calcium-dependent enzymes depending on its concentration (Simons 1993). This may explain the observed decrease in PON1 activity attributed to Pb exposure in several studies (Li et al. 2006; Kamal et al. 2011; Pollack et al. 2014). However, alternative mechanisms have also been proposed to explain the effect of Pb on PON1. For example, Pb may interfere with copper utilization, and Cu deficiency has been suggested to decrease PON1 activity (Klevay 2006; Laird et al. 2015). However, this hypothesis appears unlikely, as in vitro studies have consistently shown that copper is a relatively potent inhibitor of PON1 activity (Table 1).

The GGCGGG consensus sequence in the binding site for transcription factor Sp1 in the 5′ regulatory region of PON1 has been shown to be the site of the C-108T mutation which affects PON1 expression (Brophy et al. 2001; Deakin et al. 2003). This site has been shown to be the target for the positive modulation of PON1 by statins and by low alcohol consumption (reviewed in Costa et al. 2005, 2011) and has been suggested as a possible target of mercury (Drescher et al. 2014). As more information on the interactions of metals with PONs emerges, further mechanistic investigations are also warranted.

Summary and Conclusions

Humans are exposed to a several different metals and organometallic compounds in a variety of settings. Occupational exposure to certain metals (e.g., manganese, lead, cadmium, mercury) is quite common. In addition, exposure of large populations to metals through the diet (e.g., methylmercury in fish) or through contaminated drinking water (e.g., arsenic, manganese) also occurs. It is known that metals can exert a number of adverse health effects involving multiple mechanisms, with oxidative stress being a major one (Jaishankar et al. 2014; Matovic et al. 2015; Valko et al. 2016).

The three PONs exert significant roles as antioxidant and anti-inflammatory proteins, and several studies indicate their involvement in a variety of diseases. Evidence summarized in this chapter from available in vitro, animal, and human studies indicates that metals can modulate PON activity and expression, and this in turn may relate to some of their adverse health effects. The majority of studies have focused so far on the most studied of the PONs, PON1; however, as knowledge on the other

PONs increases, so should information on the interactions of metals with PON2 and PON3. Given the important role of PON1 in cardiovascular disease (reviewed in Costa and Furlong 2002; Furlong et al. 2008), effects of metals on this enzyme activity and/or expression should be further investigated, also considering the existence of possible differential effects linked to PON1 polymorphisms. Similar considerations also apply to PON2 and PON3, which also play a role in cardiovascular disease. Two other areas that deserve investigation are related to the role of PONs in modulating bacterial infection or in protecting tumor cells (PON2 and PON3 in particular) and the possibility that metal exposure may alter important homeostatic mechanisms by affecting their activity. Furthermore, in the case of PON2, a negative modulation of this enzyme by metals may affect neuroprotective mechanisms, and in the case of PON3, it may induce obesity and related metabolic syndromes.

In addition to the multiple potential effects of metals on PONs and the possible involvement of these in adverse effects of metals, another toxicological aspect involving the reverse effect should also be considered. The level of expression of PONs, determined by genetic background or by other factors, may be relevant in modulating metal toxicity. For example, results shown in Fig. 5 show that the effects of Mn on PON1 are more pronounced in $PON2^{-/-}$ mice, suggesting that levels of PON2 (e.g., in males vs. females) may affect Mn toxicity. Low level of PON2 may increase susceptibility to neurotoxic metals, and low levels of any of the three PONs may increase susceptibility to metal-induced cardiovascular effects and microbial infections.

Acknowledgments Studies by the authors were supported in part by grants from NIEHS (P42ES004696, P30ES007033). We thank Dr. Wan-Fen Li and Ms. Rebecca Richter for their help with some of the experiments shown in Figs. 1, 2, and 3 and Dr. Renee LeBoeuf for providing the tissues of experiments shown in Fig. 4.

References

Aldridge WN. The enzymic hydrolysis of diethyl p-nitrophenyl phosphate (E 600). Biochem J. 1951;49:i.

Altenhofer S, Witte I, Teiber JF, Wilgenbus P, Pautz A, Li H, Daiber A, Witan H, Clement AM, Forstermann U, Horke S. One enzyme, two functions. PON2 prevents mitochondrial superoxide formation and apoptosis independent from its lactonase activity. J Biol Chem. 2010;285:24398–403.

Androutsopoulos VP, Kanavouras K, Tsatsakis AM. Role of paraoxonase 1 (PON1) in organophosphate metabolism: implications in neurodegenerative diseases. Toxicol Appl Pharmacol. 2011;256:418–24.

Aragones G, Guardiola M, Barreda M, Marsillach J, Beltran-Debon R, Rull A, Mackness B, Mackness M, Joven J, Simo JM, Camps J. Measurement of serum PON-3 concentration: method evaluation, reference values, and influence of genotypes in a population-based study. J Lipid Res. 2011;52:1055–61.

Aragones G, Garcia-Heredia A, Guardiola M, Rull A, Beltran-Debon R, Marsillach J, Alonso-Villaverde C, Mackness B, Mackness M, Pedro-Botet J, Pardo-Reche P, Joven J, Camps

J. Serum paraoxonase-3 concentration in HIV-infected patients. Evidence for a protective role against oxidation. J Lipid Res. 2012;53:168–74.

Aviram M, Rosenblat M. Paraoxonases 1, 2, and 3, oxidative stress, and macrophage foam cell formation during atherosclerosis development. Free Rad Biol Med. 2004;37:1304–16.

Aviram M, Billecke S, Sorenson R, Bisgaier CL, Newton RS, Rosenblat M, et al. Paraoxonase active site required for protection against LDL oxidation involved its free sulfhydryl group and is different than that required for its arylesterase/paraoxonase activities: selective action of human paraoxonase allozymes Q and R. Arterioscler Thromb Vasc Biol. 1998;18:1617–24.

Aviram M, Rosenblat M, Billecke S, Erogul J, Sorenson R, Bisgaier CL, et al. Human serum paraoxonase (PON1) is inactivated by oxidized low density lipoprotein and preserved by antioxidants. Free Rad Biol Med. 1999;26:892–904.

Ayotte P, Carriesr A, Oullet N, Boiteau V, Abdous B, Laouan Sidi EA, Chateau-Degat ML, Dewailly E. Relation between methylmercury exposure and plasma paraoxonase activity in Inuit adults from Nunavik. Environ Health Perspect. 2011;119:1077–83.

Brophy VH, Jampsa RL, Clendenning JB, McKinstry LA, Furlong CE. Effects of 5′ regulatory – region polymorphisms on paraoxonase gene (PON1) expression. Am J Hum Genet. 2001;68:1428–36.

Brophy VH, Jarvik GP, Furlong CE. PON1 polymorphisms. In: Costa LG, Furlong CE, editors. Paraoxonase (PON1) in health and disease: basic and clinical aspects. Norwell: Kluwer Academic Publishers; 2002. p. 53–77.

Campo S, Sardo AM, Campo GM, Avenoso A, Castaldo M, D'Ascola A, Giunta E, Calatroni A, Saitta A. Identification of paraoxonase 3 gene (PON3) missense mutations in a population of southern Italy. Mutat Res. 2004;546:75–80.

Camps J, Marsillach J, Joven J. Pharmacological and lifestyle factors modulating serum paraoxonase-1 activity. Mini Rev Med Chem. 2009;9:911–20.

Cayir Y, Cayir A, Turan MI, Kurt N, Kara M, Lalaoglu E, Ciftel M, Yildirim A. Antioxidant status in blood of obese children: the relation between trace elements, paraoxonase, and arylesterase values. Biol Trace El Res 2014; 160: 155–60.

Cole TB, Li WF, Richter RJ, Furlong CE, Costa LG. Inhibition of paraoxonase (PON1) by heavy metals. Toxicol Sci. 2002;66(Suppl. 1):312.

Cole TB, Jampsa RL, Walter BJ, Arndt TL, Richter RJ, Shih DM, Tward A, Lusis AJ, Jack RM, Costa LG, Furlong CE. Expression of human paraoxonase during development. Pharmacogenetics. 2003;13:1–8.

Cole TB, Walter J, Shih DM, Tward AD, Lusis AJ, Timchalk C, Richter RJ, Costa LG, Furlong CE. Toxicity of chlorpyrifos oxon in a transgenic mouse model of the human paraoxonase (PON1) Q192R polymorphism. Pharmacogenet Genom. 2005;15:589–98.

Costa LG, Furlong CE, editors. Paraoxonase (PON1) in health and disease: basic and clinical aspects. Boston, MA: Kluwer Academic Publishers; 2002. p. 216.

Costa LG, Cole TB, Jarvik GP, Furlong CE. Functional genomics of the paraoxonase (PON1) polymorphisms: effects on pesticide sensitivity, cardiovascular disease, and drug metabolism. Annu Rev Med. 2003;54:371–92.

Costa LG, Vitalone A, Cole TB, Furlong CE. Modulation of paraoxonase (PON1) activity. Biochem Pharmacol. 2005;69:541–50.

Costa LG, Giordano G, Furlong CE. Pharmacological and dietary modulators of paraoxonase 1 (PON1) activity and expression: the hunt goes on. Biochem Pharmacol. 2011;81:337–44.

Costa LG, Giordano G, Cole TB, Marsillach J, Furlong CE. Paraoxonase 1 (PON1) as a genetic determinant of susceptibility to organophosphate toxicity. Toxicology. 2013a;307:115–22.

Costa LG, Tait L, de Laat R, Dao K, Giordano G, Pellacani C, Cole TB, Furlong CE. Modulation of paraoxonase 2 (PON2) in mouse brain by the polyphenol quercetin: a mechanism of neuroprotection? Neurochem Res. 2013b;38:1809–18.

Costa LG, de Laat R, Dao K, Pellacani C, Cole TB, Furlong CE. Paraoxonase-2 (PON2) in brain and its potential role in neuroprotection. Neurotoxicology. 2014;43:3–9.

Davies H, Richter RJ, Kiefer M, Broomfield C, Sowalla J, Furlong CE. The human serum paraoxonase polymorphism is reversed with diazoxon, soman and sarin. Nature Genet. 1996;14:334–6.

Deakin S, Leviev I, Brulhart-Meynet MC, James RW. Paraoxonase-1 promoter haplotypes and serum paraoxonase: a predominant role for polymorphic position −107, implicating the Sp1 transcription factor. Biochem J. 2003;372:643–9.

Debord J, Bollinger JC, Merle L, Dantoine T. Inhibition of human serum arylesterase by metal chlorides. J Inorg Biochem. 2003;94:1–4.

Devarajan A, Bourquard N, Hama S, Navab M, Grijalva VR, Morvardi S, Clarke C, Vergnes L, Reue K, Teiber JF, Reddy ST. Paraoxonase 2 deficiency alters mitochondrial function and exacerbates the development of atherosclerosis. Antiox Redox Signal. 2011;14:341–51.

Draganov DI, Stetson PI, Watson CE, Billecke SS, La Du BN. Rabbit serum paraoxonase 3 (PON3) is a high density lipoprotein-associated lactonase and protects low density lipoprotein against oxidation. J Biol Chem. 2000;275:33435–42.

Draganov DI, Teiber JF, Speelman A, Osawa Y, Sunahara R, La Du BN. Human paraoxonases (PON1, PON2 and PON3) are lactonases with overlapping and distinct substrate specificities. J Lipid Res. 2005;46:1239–47.

Drescher O, Dewailly E, Diorio C, Oullet N, Laouan Sidi EA, Abdous B, Valera B, Ayotte P. Methylmercury exposure, PON1 gene variants and serum paraoxonase activity in eastern James Bay Cree adults. J Exp Sci Environ Epidemiol. 2014;24:608–14.

Ebabe Elle R, Gaillet S, Vide J, Romain C, Lauret C, Rugani N, Cristol JP, Rouanet JM. Dietary exposure to silver nanoparticles in Sprague-Dawley rats: effects on oxidative stress and inflammation. Food Chem Toxicol. 2013;60:297–301.

Ekinci D, Beydemir S. Purification of PON1 from human serum and assessment of enzyme kinetics against metal toxicity. Biol Trace Elem Res. 2010;135:112–20.

Erdös EG, Debay CR, Westerman MP. Activation and inhibition of the arylesterase of human serum. Nature. 1959;184:430–1.

Erdös EG, Debay CR, Westerman MP. Arylesterase in blood: effect of calcium and inhibitors. Bichem Pharmacol. 1960;5:173–86.

Erol K, Gencer N, Arslan M, Arslan O. Purification, characterization, and investigation of in vitro inhibition by metals of paraoxonase from different sheep breeds. Artif Cells Nanomed Biotechnol. 2013;41:125–30.

Fernandes ES, Machado MO, Becker AM, de Andrade F, Maraschin M, da Silva EL. Yerba mate (Ilex paraguariensis) enhances the gene modulation activity of paraoxonase-2: in vitro and in vivo studies. Nutrition. 2012;28:1157–64.

Ferramola ML, Perez Diaz MFF, Honore SM, Sanchez SS, Anton RI, Anzulovich AC, Gimenez MS. Cadmium-induced oxidative stress and histological damage in the myocardium. Effects of a soy- based diet. Toxicol Appl Pharmacol. 2012;265:380–9.

Furlong CE, Costa LG, Hassett C, Richter RJ, Sundstrom JA, Adler DA, Disteche CM, Omiecinski CJ, Chapline C, Crebb JW, Humbert R. Human and rabbit paraoxonases: purification, cloning, sequencing, mapping and role of polymorphism in organophosphate detoxification. Chem Biol Inter. 1993;87:35–48.

Furlong CE, Richter RJ, Li WF, Brophy VH, Carlson C, Meider M, Nickerson D, Costa LG, Ranchalis J, Lusis AJ, Shih DM, Tward A, Jarvik GP. The functional consequences of polymorphisms in the human PON1 gene. In: The paraoxonases: their role in disease, development, and xenobiotic metabolism. (Mackness, B, Mackness M, Aviram M, Paragh G, eds.), Springer, Dordrecht, 2008; pp. 267–281.

Furlong CE, Suzuki SM, Stevens RC, Marsillach J, Richter RJ, Jarvik GP, Checkoway H, Samii A, Costa LG, Griffith A, Roberts JW, Yearout D, Zabetian CP. Human PON1, a biomarker of disease and exposure. Chem Biol Interact. 2010;187:355–61.

Furlong CE, Marsillach J, Jarvik JP, Costa LG. Paraoxonases −1, −2, and −3: what are their functions? Chem Biol Inter. 2016.; (in press)

Garcia-Heredia A, Marsillach J, Aragones G, Guardiola M, Rull A, Beltran-Debon R, Folch A, Mackness B, Mackness M, Pedro-Botet J, Joven J, Camps J. Serum paraoxonase-3 concentration is associated with the severity of hepatic impairment in patients with chronic liver disease. Clin Biochem. 2011;44:1320–4.

Gencer N, Arslan O. Purification human PON1Q192 and PON1R192 isoenzymes by hydrophobic interaction chromatography and investigation of the inhibition by metals. J Chromatogr B. 2009;877:134–40.

Gil F, Gonzalvo MC, Hernandez AF, Villanueva E, Pla A. Differences in the kinetic properties effect of calcium and sensitivity to inhibitors of paraoxon hydrolase activity in rat plasma and microsomal fractions from rat liver. Biochem Pharmacol. 1994;48:1559–68.

Ginsberg G, Sonawane B, Nath R, Lewandowski P. Methylmercury-induced inhibition of paraoxonase-1 (PON1) – implications for cardiovascular risk. J Toxicol Environ Health A. 2014;77:1004–23.

Giordano G, Cole TB, Furlong CE, Costa LG. Paraoxonase 2 (PON2) in the mouse central nervous system: a neuroprotective role? Toxicol Appl Pharmacol. 2011;256:369–78.

Giordano G, Tait L, Furlong CE, Cole TB, Kavanagh TJ, Costa LG. Gender differences in brain susceptibility to oxidative stress are mediated by levels of paraoxonase-2 (PON2) expression. Free Rad Biol Med. 2013;58:98–108.

Gonzalvo MC, Gil F, Hernandez AF, Villanueva E, Pla A. Inhibition of paraoxonase activity in human liver microsomes by exposure to EDTA, metals and mercurials. Chem Biol Interact. 1997;105:169–79.

Hagmann H, Kuczkowski A, Ruehl M, Lamkemeyer T, Brodesser S, Horke S, Dryer S, Schermer B, Benzing T, Brinkkoetter PT. Breaking the chain at the membrane: paraoxonase 2 counteracts lipid peroxidation at the plasma membrane. FASEB J. 2014;28:1769–79.

Harel M, Aharoni A, Gaidukov L, Brumshtein B, Khersonsky O, Meged R, et al. Structure and evolution of the serum paraoxonase family of detoxifying and ant-atherosclerotic enzymes. Nat Struct Mol Biol. 2004;11:412–9.

Hernandez AF, Gil F, Leno E, Lopez O, Rodrigo L, Pla A. Interaction between human serum esterases and environmental metal compounds. Neurotoxicology. 2009;30:628–35.

Humbert R, Adler DA, Disteche CM, Omiecinski CJ, Furlong CE. The molecular basis of the human serum paraoxonase polymorphisms. Nature Genet. 1993;3:73–6.

Jaishankar M, Tseten T, Anbalagan N, Mathew BB, Beeregowda KN. Toxicity, mechanism and health effects of some heavy metals. Interdiscip Toxicol. 2014;7:60–72.

Jarvik GP, Rozek LS, Brophy VH, Hatsukami TS, Richter RJ, Schellenberg GD, Furlong CE. Paraoxonase (PON1) phenotype is a better predictor of vascular disease than is PON1(192) or PON1(55) genotype. Arterioscl Thromb Vasc Biol. 2000;20:2441–7.

Jarvik GP, Hatsukami TS, Carlson C, Richter RJ, Jampsa R, Brophy VH, Margolin S, Rieder M, Nickerson D, Schellenberg GD, Heagerty PJ, Furlong CE. Paraoxonase activity, but not haplotype utilizing the linkage equilibrium structure, predicts vascular disease. Arterioscl Thromb Vasc Biol. 2003;23:1465–71.

Josse D, Bartels C, Lockridge O, Masson P. PON1 structure. In: Paraoxonase (PON1) in health and disease: basic and clinical aspects (Costa LG, Furlong CE, eds.), Kluwer Academic Publishers, Norwell, 2002; pp. 27–52.

Kamal M, Fathy MM, Taher E, Hasan M, Tolba M. Assessment of the role of paraoxonase gene polymorphism (Q192R) and paraoxonase activity in the susceptibility to atherosclerosis among lead exposed workers. Ann Saudi Med. 2011;31:481–7.

Kardys I, Vliegenthart R, Oudkerk M, Hofman A, Witteman JCM. The female advantage in cardiovascular disease: do vascular beds contribute equally? Am J Epidemiol. 2007;166:403–12.

Klein SL. The effects of hormones on sex differences in infection: from genes to behavior. Neurosci Biobehav Rev. 2000;24:627–38.

Klevay LM. How dietary deficiency, genes and a toxin can cooperate to produce arteriosclerosis and ischemic heart disease. Cell Mol Biol. 2006;52:11–5.

Krüger M, Pabst AM, Al-Nawas B, Horke S, Moergel M. Paraoxonase 2 (PON2) protects oral squamous cell cancer cells against irradiation-induced apoptosis. J Cancer Res Oncol. 2015;141:1757–66.

Kuo CL, La Du BN. Calcium binding by human and rabbit serum paraoxonases. Structural stability and enzymatic activity. Drug Metab Dispos. 1998;26:653–60.

Laird BD, Gongharov AB, Ayotte P, Chan HM. Relationship between the esterase paraoxonase-1 (PON1) and metal concentrations in whole blood of Inuit in Canada. Chemosphere. 2015;120:479–85.

Levy E, Trudel K, Bendayan M, Seidman E, Delvin E, Elchebly M, Lavoie JC, Precourt LP, Amre D, Sinnett D. Biological role, protein expression, subcellular localization, and oxidative stress response of paraoxonase 2 in the intestine of human and rats. Am J Physiol Gastrointest Liver Physiol. 2007;293:G1252–61.

Li WF, Costa LG, Richter RJ, Hagen T, Shih DM, Tward A, Lusis AJ, Furlong CE. Catalytic efficiency determines the in vivo efficacy of PON1 for detoxifying organophosphates. Pharmacogenetics. 2000;10:767–79.

Li WF, Pan MH, Chung MC, Ho CK, Chuang HY. Lead exposure is associated with decreased serum paraoxonase 1 (PON1) activity and genotypes. Environ Health Perspect. 2006;114:1233–6.

Li WF, Sun CW, Chang TJ, Chang KH, Chen CJ, Wang SL. Risk of carotid atherosclerosis is associated with low serum paraoxonase (PON1) activity among arsenic exposed residents of southwestern Taiwan. Toxicol Appl Pharmacol. 2009;236:246–53.

Mackness MI, Arrol S, Durrington PN. Paraoxonase prevents accumulation of lipoperoxides in low-density lipoprotein. FEBS Lett. 1991;286:152–4.

Mackness B, Mackness MI, Arrol T, Turkie W, Durrington PN. Effect of the human serum paraoxonase 55 and 192 genetic polymorphisms on the protection by high density lipoprotein against low density lipoprotein oxidative modifications. FEBS Lett. 1998;423:57–60.

Mackness B, Davies GK, Turkie W, Lee E, Roberts DH, Hill E, Roberts C, Durrington PN, Mackness MI. Paraoxonase status in coronary heart disease: are activity and concentration more important than genotype? Arterioscler Thromb Vasc Biol. 2001;21:1451–7.

Maghraoui S, Clichici S, Ayadi A. Login C, Moldovan R, Daicoviciu D, Decea N, Muresan a, Tekaya L. oxidative stress in blood and testicle of rat following intraperitoneal administration of aluminum and indium. Acta Physiol. Ther Hung. 2014;101:47–58.

Marsillach J, Martínez-Vea A, Marcas L, Mackness B, Mackness M, Ferré N, Joven J, Camps J. Administration of exogenous erythropoietin beta affects lipid peroxidation and serum paraoxonase-1 activity and concentration in predialysis patients with chronic renal disease and anaemia. Clin Exp Pharmacol Physiol. 2007a;34:347–9.

Marsillach J, Ferré N, Vila MC, Lligoña A, Mackness B, Mackness M, Deulofeu R, Solá R, Parés A, Pedro-Botet J, Joven J, Caballeria J, Camps J. Serum paraoxonase-1 in chronic alcoholics: relationship with liver disease. Clin Biochem. 2007b;40:645–50.

Marsillach J, Mackness B, Mackness M, Riu F, Beltran R, Joven J, Camps J. Immunohistochemical analysis of paraoxonases-1, 2 and 3 expression in normal mouse tissues. Free Rad Biol Med. 2008;45:146–57.

Marsillach J, Suzuki SM, Richter RJ, McDonald MG, Rademacher PM, MacCoss MJ, Hsieh EJ, Rettie AE, Furlong CE. Human valacyclovir hydrolase/biphenyl hydrolase-like protein is a highly efficient homocysteine thiolactonase. PLoS One. 2014;9:e110054.

Marsillach J, Becker JO, Vaisar T, Hahn BH, Brunzell JD, Furlong CE, de Boer IH, McMahon MA, Hoofnagle AN, DCCT/EDIC Research Group. Paraoxonase-3 is depleted from the high-density lipoproteins of autoimmune disease patients with subclinical atherosclerosis. J Proteome Res. 2015;14:2046–54.

Matovic V, Buha A, Dukic-Cosic D, Bulat Z. Insight into the oxidative stress induced by lead and/or cadmium in blood, liver and kidneys. Food Chem Toxicol. 2015;78:130–40.

Mochizuki H, Scherer SW, Xi T, Nickle DC, Majer M, Huizenga JJ, Tsui LC, Prochazka M. Human PON2 gene at 7q21.3: cloning, multiple mRNA forms, and missense polymorphisms in the coding sequence. Gene. 1998;213:149–57.

Mozaffarian D, Shi P, Morris JS, Spiegelman D, Granjean P, Siscovick DS, Willett WC, Rimm EB. Mercury exposure and risk of cardiovascular disease in two U.S. cohorts. New Engl J Med. 2011;364:1116–24.

Ng CJ, Wadleigh DJ, Gangopadhyyay A, Hama S, Grijalva VR, Navab M, Fogelman AM, Redd ST. Paraoxonase-2 is a ubiquitously expressed protein with antioxidant properties and is capable of preventing cell-mediated oxidative modification of low density lipoprotein. J Biol Chem. 2001;276:44444–9.

Ng CJ, Bourquard N, Grijalva V, Hama S, Shih DM, Navab M, Fogelman AM, Lusis AJ, Young S, Reddy ST. Paraoxonase-2 deficiency aggravates atherosclerosis in mice despite lower apolipoprotein-B-containing lipoproteins. Antiatherogenic role for paraoxonase-2. J Biol Chem. 2006;281:29491–500.

Ng CJ, Bourquard N, Hama SY, Shih D, Grijalva VR, Navab M, Fogelman AM, Reddy ST. Adenovirus-mediated expression of human paraoxonase 3 protects against the progression of atherosclerosis in apolipoprotein E-deficient mice. Arterioscler Thromb Vasc Biol. 2007;27:1368–74.

Nguyen SD, Sok DE. Oxidative inactivation of paraoxonase 1, an antioxidant protein and its effect on antioxidant action. Free Rad Res. 2003;37:1319–30.

Permpongpaiboon T, Nagila A, Pidetcha P, Tuangmungsakulchai K, Tantrarongroj S, Porntadavity S. Decreased paraoxonase 1 activity and increased oxidative stress in low-lead exposed workers. Hum Exp Toxicol. 2011;30:1196–203.

Pla A, Rodrigo L, Hernandez AF, Gil F, Lopez O. Effect of metal ions and calcium on purified PON1 and PON3 from rat liver. Chem Biol Interact. 2007;167:63–70.

Pollack AZ, Sjaarda L, Ahrens KA, Mumford SL, Browne RW, Wactawski-Wende J, Schisterman EF. Association of cadmium, lead and mercury with paraoxonase 1 activity in women. PLoS One 2014; e92152.

Primo-Parmo SL, Sorenson RC, Teiber J, La Du BN. The human serum paraoxonase/arylesterase gene (PON1) is one member of a multigene family. Genomics. 1996;33:498–507.

Ramambason C, Moroy G, Daubigney F, Paul JL, Janel N. Effect of cadmium administration in hyperhomocysteinemic mice due to cystathionine beta synthase deficiency. Exp Toxicol Pathol. 2016.; (in press)

Reddy ST, Wadleigh DJ, Grijalva V, Ng C, Hama S, Gangopadhyay A, Shih DM, Lusis AJ, Navab M, Fogelman AM. Human paraoxonase-3 is an HDL-associated enzyme with biological activity similar to paraoxonase-1 protein but is not regulated by oxidized lipids Arterioscler. Thromb Vasc Biol. 2001;21:542–7.

Richter RJ, Furlong CE. Determination of paraoxonase (PON1) status requires more than genotyping. Pharmacogenetics. 1999;9:745–53.

Richter RJ, Jarvik GP, Furlong CE. Determination of paraoxonase 1 (PON1) status without the use of toxic organophosphate substrates. Circ Cardiovasc Genet. 2008;1:147–52.

Richter RJ, Jarvik JP, Furlong CE. Paraoxonase (PON1) status and substrate hydrolysis. Toxicol Appl Pharmacol. 2009;235:1–9.

Rodgers JS, Hocker JR, Hanas RJ, Nwosu EC, Hanas JS. Mercuric ion inhibition of eukaryotic transcription factor binding to DNA. Biochem Pharmacol. 2001;61:1543–50.

Rosenblat M, Draganov D, Watson CE, Bisgaier CL, La Du BN, Aviram M. Mouse macrophage paraoxonase-2 activity is increased whereas cellular paraoxonase 3 activity is decreased under oxidative stress. Arterioscler Thromb Vasc Biol. 2003;23:468–74.

Rosenblat M, Hayek T, Hussein K, Aviram M. Decreased macrophage paraoxonase 2 expression in patients with hypercholesterolemia is the results of their increases cellular cholesterol content: effect of atorvastatin therapy. Arterioscler Thromb Vasc Biol. 2004;24:175–80.

Rothem L, Hartman C, Dahan A, Lachter J, Eliakim R, Shamir R. Paraoxonases are associated with intestinal inflammatory diseases and intracellularly localized to the endoplasmic reticulum. Free Radic Biol Med. 2007;43:730–9.

Rull A, Garcia R, Fernandez-Sender L, Garcia-Heredia A, Aragones G, Beltran-Debon R, Marsillach J, Alegret JM, Martin-Paredero V, Mackness B, Mackness M, Joven J, Camps J. Serum paraoxonase-3 concentration is associated with insulin sensitivity in peripheral artery disease and with inflammation in coronary artery disease. Atherosclerosis. 2012;220:545–51.

Sanchez R, Levy E, Seidman E, Amre D, Costea F, Sinnett D. Paraoxonase 1, 2 and 3 DNA variants and susceptibility to childhood inflammatory bowel disease. Gut. 2006;55:1820–1.

Sayin D, Cakir DT, Gencer N, Arslan O. Effects of some metals on paraoxonase activity from shark *Scyliorhinus canicula*. J Enz Inhib Med Chem. 2012;27:595–8.

Schweikert EM, Devarajan A, Witte I, Wilgenbus P, Amort J, Forstermann U, Shabazian A, Grijalva V, Shih DM, Farias-Eisner R, Teiber JF, Reddy ST, Horke S. PON3 is upregulated in cancer tissues and protects against mitochondrial superoxide-mediated cell death. Cell Death Differ. 2012a;19:1549–60.

Schweikert EM, Amort J, Wilgenbus P, Forstermann U, Teiber JF, Horke S. Paraoxonases-2 and -3 are important defense enzymes against *Pseudomonas aeruginosa* virulence factors due to their anti-oxidative and anti-inflammatory properties. J Lipids. 2012b:art. 352857.

Shih DM, Gu L, Xia YR, Navab M, Li WF, Hama S, Castellani LW, Furlong CE, Costa LG, Fogelman AM, Lusis AJ. Mice lacking serum paraoxonase are susceptible to organophosphate toxicity and atherosclerosis. Nature. 1998;394:284–7.

Shih DM, Xia YR, Wang XP, Wang SS, Bourquard N, Fogelman AM, Lusis AJ, Reddy ST. Decreased obesity and atherosclerosis in human paraoxonase 3 transgenic mice. Circul Res. 2007;100:1200–7.

Shih DM, Yu JM, Vergnes L, Dali-Youcef N, Champion MD, Devarajan A, Zhang P, Castellani LW, Brindley DN, Jamey C, Auwerx J, Reddy ST, Ford DA, Reue K, Lusis AJ. PON3 knockout mice are susceptible to obesity, gallstone formation, and atherosclerosis. FASEB J. 2015;29:1185–97.

Simons TJB. Lead-calcium interactions in cellular lead toxicity. Neurotoxicology. 1993;14:77–86.

Sorenson RC, Primo-Parmo SL, Kuo CL, Adkins S, Lockridge O, La Du BN. Reconsideration of the catalytic center and mechanism of mammalian paraoxonase/arylesterase. Proc Natl Acad Sci U S A. 1995;92:7187–91.

Sukketsiri W, Porntadavidity S, Phivthong-ngam L, Lawanprasert S. Lead inhibits paraoxonase 2 but not paraoxonase 1 activity in human hepatoma HepG2 cells. J Appl Toxicol. 2013;33:631–7.

Surmeier DJ, Guzman JN, Sanchez-Padilla J, Goldberg JA. The origins of oxidant stress in Parkinson's disease and therapeutic strategies. Antioxid Redox Signal. 2011;14:1289–301.

Tas S, Srandol E, Ziyanok-Ayvalik S, Ocak N, Serdar Z, Dirican M. Vanadyl sulfate treatment improves oxidative stress and increases serum paraoxonase activity in streptozotocin-induced diabetic rats. Nutr Res. 2006;26:670–6.

Teiber JF, Horke S, Haines DC, Chowdhary PK, Xiao J, Kramer GL, Haley RW, Draganov DI. Dominant role of paraoxonases in inactivation of the Pseudomonas Aeruginosa quorum-sensing signal N-(3-oxododecanoyl)-L-homoserine lactone. Infect Immun. 2008;76:2512–9.

Thijssen S, Maringwa J, Faes C, Lambrichts I, Van Kerkhove E. Chronic exposure of mice to environmentally relevant, low doses of cadmium leads to early renal damage, not predicted by blood or urine cadmium levels. Toxicology. 2007;229:145–56.

Tokar EJ, Boyd WA, Freedman JH, Waalkes MP. Toxic effects of metals. In: Casarett and Doull's toxicology: the basic science of poisons (Klaassen CD, Ed.), McGraw Hill, New York, 2013; pp. 981–1030.

Valko M, Jomova K, Rhodes CJ, Kuča K, Musilek K. Redox- and non-redox-metal-induced formation of free radicals and their role in human disease. Arch Toxicol. 2016;90:1–37.

Virtanen JK, Rissanen TH, Voutilanen S, Tuomainen TP. Mercury as a risk factor for cardiovascular diseases. J Nutr Biochem. 2007;18:75–85.

Wang F, He Q, Sun Y, Dai X, Yang XP. Female adult mouse cardiomyocytes are protected against oxidative stress. Hypertension. 2010;55:1172–8.

Wirdefeld K, Adami HO, Cole P, Trichopoulos D, Mandel J. Epidemiology and etiology of Parkinson's disease: a review of the evidence. Eur J Epidemiol. 2011;26:S1–S58.

Witte I, Altenhöfer S, Wilgenbus P, Amort J, Clement AM, Pautz A, Li H, Förstermann U, Horke S. Beyond reduction of atherosclerosis: PON2 provides apoptosis resistance and stabilizes tumor cells. Cell Death Dis. 2011;2:e212.

Yang Y, Zhang Y, Cuevas-Gonzalez S, Villar VA, Escano C, Asico L, Yu P, Grandy DK, Felder RA, Armando I, Jose PA. Paraoxonase 2 decreases renal reactive oxygen species production, lowers blood pressure, and mediates dopamine D2 receptor-induced inhibition of NADPH oxidase. Free Rad Biol Med. 2012;53:437–46.

Yehuda I, Madar Z, Leikin-Frenkel A, Szuchman-Sapir A, Magzal F, Markma G, Tamir S. Glabridin, an isoflavan from licorice root, upregulates paraoxonase 2 expression under hyperglycemia and protects it from oxidation. Mol Nutr Food Res. 2016;60:287–99.

Manganese and the Insulin-IGF Signaling Network in Huntington's Disease and Other Neurodegenerative Disorders

Miles R. Bryan and Aaron B. Bowman

Abstract Huntington's disease (HD) is an autosomal dominant neurodegenerative disease resulting in motor impairment and death in patients. Recently, several studies have demonstrated insulin or insulin-like growth factor (IGF) treatment in models of HD, resulting in potent amelioration of HD phenotypes via modulation of the PI3K/AKT/mTOR pathways. Administration of IGF and insulin can rescue microtubule transport, metabolic function, and autophagy defects, resulting in clearance of Huntingtin (HTT) aggregates, restoration of mitochondrial function, amelioration of motor abnormalities, and enhanced survival. Manganese (Mn) is an essential metal to all biological systems but, in excess, can be toxic. Interestingly, several studies have revealed the insulin-mimetic effects of Mn—demonstrating Mn can activate several of the same metabolic kinases and increase peripheral and neuronal insulin and IGF-1 levels in rodent models. Separate studies have shown mouse and human striatal neuroprogenitor cell (NPC) models exhibit a deficit in cellular Mn uptake, indicative of a Mn deficiency. Furthermore, evidence from the literature reveals a striking overlap between cellular consequences of *Mn deficiency* (i.e., impaired function of Mn-dependent enzymes) and known HD endophenotypes including excitotoxicity, increased reactive oxygen species (ROS) accumulation, and decreased mitochondrial function. *Here we review published evidence support-*

M.R. Bryan (✉)
Department of Neurology, Vanderbilt University Medical Center, Nashville, TN 37232, USA

Vanderbilt Brain Institute, Vanderbilt University Medical Center, Nashville, TN 37232, USA

Department of Pediatrics, Vanderbilt University Medical Center, Nashville, TN 37232, USA
e-mail: miles.r.bryan@vanderbilt.edu

A.B. Bowman
Vanderbilt Center in Molecular Toxicology, Vanderbilt University Medical Center, Nashville, TN 37232, USA

Department of Neurology, Vanderbilt University Medical Center, Nashville, TN 37232, USA

Vanderbilt Brain Institute, Vanderbilt University Medical Center, Nashville, TN 37232, USA

Department of Pediatrics, Vanderbilt University Medical Center, Nashville, TN 37232, USA
e-mail: aaron.b.bowman@vanderbilt.edu

© Springer International Publishing AG 2017
M. Aschner and L.G. Costa (eds.), *Neurotoxicity of Metals*, Advances in Neurobiology 18, DOI 10.1007/978-3-319-60189-2_6

ing a hypothesis that (1) *the potent effect of IGF or insulin treatment on HD models,* (2) *the insulin-mimetic effects of Mn, and* (3) *the newly discovered Mn-dependent perturbations in HD may all be functionally related.* Together, this review will present the intriguing possibility that intricate regulatory cross-talk exists between Mn biology and/or toxicology and the insulin/IGF signaling pathways which may be deeply connected to HD pathology and, perhaps, other neurodegenerative diseases (NDDs) and other neuropathological conditions.

Keywords Neuroprogenitor cell (NPC) • Autophagy • Mitochondria • Cargo recognition • Dysregulation

Introduction

Between 1 and 3 out of 100,000 individuals are diagnosed with Huntington's disease (HD) in the USA. However, given the autosomal dominant etiology and near 100% penetrance of HD, generations of families are devastated by this disease. HD is caused by an expanded trinucleotide CAG repeat in the HTT gene. If these repeats surpass 35–40 repeats, there is a near 100% chance that the patient will present with Huntington's disease at some point in their lifetime (usually in middle-late adulthood). While the *HTT* gene was discovered in 1993, there is still no cure for HD though several drugs have been used to treat symptoms (i.e., tetrabenazine for chorea). Furthermore, researchers still do not fully understand (1) the exact function(s) of wild-type HTT is in the human brain or (2) how mutant HTT (HTT >35 CAG repeats) causes neurotoxicity and HD. Two of the posited causes for HD are (1) mitochondrial dysfunction (2) autophagic dysfunction and aggregate accumulation. Recently, a series of studies have shown that insulin/insulin-like growth factor (IGF) treatment in HD models can ameliorate both of these pathogenic mechanisms.

Manganese (Mn) has only been recently implicated in HD, and studies have suggested that a Mn deficiency may underlie some of HD pathology (Tidball et al. 2015a; Kwakye et al. 2011; Williams et al. 2010a; b; Stansfield et al. 2014). Interestingly, Mn can modulate insulin/IGF homeostasis, shown to be essential for mitochondrial function, and able to stimulate neuroprotective pathways associated with the activation of autophagy, namely, insulin/IGF signaling (IIS). *This review explores the functional intersection of these two modifiers of HD—(a) Mn biology and (b) insulin/IGF signaling (IIS)—both have been shown to regulate autophagy and mitochondrial health/function.* Here we will review a role for Mn and IGF joint dysregulation in HD pathology and briefly explore some of the implications of this co-regulation in the context of other neurodegenerative diseases and conditions.

While Huntington's disease will be discussed in detail, other neurodegenerative diseases (NDDs) will also be referenced when studies provide mechanistic understanding of the roles of Mn and IGF/insulin given the shared cellular pathologies between NDDs and HD (i.e., aggregate accumulation, reactive oxygen species, mitochondrial dysfunction). It is plausible that the mechanisms of these NDD pathologies might be quite similar to HD.

IIS Signaling and Its Role in the Brain

Insulin and insulin growth factor (IGF) are homologous growth hormones that classically regulate cellular metabolism. Their role in peripheral tissues has been well elucidated. However, only more recently has their role in brain health and development been studied. In the brain, IIS is necessary for synaptic maintenance and activity, neurogenesis, neurite outgrowth, neuronal survival, mitochondrial function and maintenance, as well as upper-level processes including memory and feeding behavior, and thus dysregulation in neurotrophic support has long been proposed as a mechanism of neurodegenerative diseases (Greenwood and Fleshner 2008; Trejo et al. 2007; Xing et al. 2007, 2006, 2010; Liou et al. 2003; Ozdinler and Macklis 2006; Skeberdis et al. 2001; Sosa et al. 2006; Liu et al. 2009; Dentremont et al. 1999; O'Kusky et al. 2000; Hurtado-Chong et al. 2009; Oishi et al. 2009; Chiu et al. 2008; Ciucci et al. 2007; Hodge et al. 2004; D'Ercole et al. 1996, 2002; Root et al. 2011; Jiu et al. 2010; Marks et al. 2009; Haj-ali et al. 2009; Dhamoon et al. 2009; Zhao et al. 1999; Deijen et al. 1998). Insulin and IGF are mainly produced in the pancreas and liver, respectively, and transported to the brain from the periphery through the blood-brain barrier. Alternatively, IGF and insulin can enter the brain through CSF in the choroid plexus. IGF is also produced locally in all brain regions. Upon binding with their respective ligands, IGF receptors (IGFR) and insulin receptors (IR) undergo autophosphorylation at three tyrosine residues required for activation. Subsequently, the IR kinase domain phosphorylates IR substrates (IRSs) which act as secondary messengers, impinging upon a variety of cell signaling pathways including PI3K/AKT, mTOR, and MAPK/ERK, to exert their biological effects (e.g., energy metabolism, cell stress responses) (Fernandez and Torres-Alemán 2012). However, individual receptors can heterodimerize forming hybrid IGF/insulin receptors which can bind either insulin or IGF and activate both the PI3K/AKT and MAPK/ERK pathways. S6, a downstream target of mTOR, acts as negative feedback, phosphorylating and inactivating IRSs. Six IGF binding proteins exist (IGFBPs) and act to regulate IGF-R binding and modulate signaling. IGFBPs show a selective expression pattern, being in distinct portions of the brain, where they presumably act on specific IIS signaling within anatomical subsets of neurons. These proteins have a higher affinity for IGF than do IGF receptors, allowing tight control of IGF bioavailability. The regulation of neuronal IGFBPs is still quite unknown, but evidence suggests specific mechanisms for each protein including control by epigenetic markers and neuronal activity of specific cell types (Baxter 2014; Clemmons et al. 1995).

Most kinases in humans are either magnesium (Mg) or Mn dependent. Though most are Mg dependent, several are preferentially activated by Mn including ATM and mTOR (Chan et al. 2000; Sato et al. 2009). Furthermore, insulin and IGF receptors have been shown to be Mn-dependent (Morrison et al. 1989, Xu et al. 1995) While little research has been done to explore the role of Mn as a signaling molecule, its inherent role in kinase activation suggests that Mn is essential for cell signaling. Several other proteins are also activated by Mn including Arg, MRE11, Mn-SOD, glutamine synthetase, pyruvate decarboxylase, protein phosphatase 1, and many integrin-related proteins (Horning et al. 2015; Paull and Gellert 1998; Trujillo et al. 1998; D'Antonio et al. 2012; Kanyo et al. 1996; Neulen et al. 2007; Woźniak-Celmer et al. 2001; Maydan et al. 2010; Wedler and Ley 1994; Dormond et al. 2004). Interestingly Mn has been shown to activate several of the same pathways as IGF/insulin including AKT, mTOR, ERK/MAPK, and even the insulin/IGF receptor itself—all of which have been found to be neuroprotective in HD (Dormond et al. 2004; Bae et al. 2006; Cordova et al. 2012a, 2013; Crittenden and Filipov 2011; Exil et al. 2014; Jang 2009; Dearth et al. 2014; Srivastava et al. 2016; Zhang et al. 2013; Hiney et al. 2011).

Mn and Insulin/IGF Homeostasis

Mn toxicity has been linked to neuronal cell death and neurodegenerative conditions for several decades—namely, Parkinson's disease (PD) and manganism. Though recent studies have yielded greater understanding of the toxic effect of Mn on neuronal function, very little is known about basic, neuronal Mn homeostasis. While brain imaging studies have revealed where Mn accumulates within the brain, there is disagreement on what sub-compartment(s) Mn primarily accumulates within a neural cell. The field is in some contention as some studies suggest mitochondria, while others suggest within the nucleus (Gunter et al. 2009; Kalia et al. 2008; Morello et al. 2008). Surprisingly few studies have examined whether Mn primarily accumulates in neurons vs glial cells. Lastly, there is poor understanding of how Mn is transported within a cell, primarily due to the high promiscuity of Mn transporters for other metal ions (Horning et al. 2015; Chen et al. 2015). Muddying this understanding, at present there is only one transporter which seems specific for Mn, SCL30A10, an efflux transporter. Interestingly, mutations in this transporter lead to Mn accumulation in vitro and in vivo and have been linked to increased brain Mn and PD in patients (Chen et al. 2015; DeWitt et al. 2013; Leyva-Illades et al. 2014; Quadri et al. 2012). The answers to these basic questions could offer invaluable understanding of Mn biology in the context of both diseased and healthy brains.

Evidence of a role for Mn-dependent regulation of IIS has been steadily amassing since the 1980s. Baly and colleagues showed Mn deficiency caused glucose intolerance and reduced insulin production in rats, mimicking diabetic-like phenotypes (Baly 1984; Baly et al. 1986, 1988, 1990). In addition, rats fed a Mn-deficient diet exhibited reduced pancreatic insulin output following a glucose stimulus. Furthermore, they and others found Mn to be an insulin mimetic, promoting insulin excretion and activating insulin-related metabolic kinases (Baly 1984; Baly et al. 1986, 1988, 1990; Keen et al. 1984; Subasinghe et al. 1985). Around this same time, another study showed that Mn-deficient rats exhibited decreased circulating IGF-1 and insulin and increased IGFBP3—potentially suggesting Mn might regulate circulating IGF-1 levels via modulating IGFBP3 activity (Clegg et al. 1998). Later, Lee and colleagues reported that Mn supplementation could protect against diet-induced diabetes in mice via increased insulin excretion, amelioration of glucose intolerance, and increased expression of Mn superoxide dismutase (MnSOD), a Mn-dependent antioxidant enzyme in mitochondria (Lee et al. 2013). These results were consistent with reports that diabetic patients were responsive to oral Mn treatment as well as reports of reduced blood Mn in diabetic patients (Ekmekcioglu et al. 2001; Koh et al. 2014; Rubenstein et al. 1962). Concurrently, other groups established that Mn deficiency was associated with reductions of IGF-1 in serum and Mn supplementation could increase IGR-R and IGF-1 expression in the hypothalamus of rats (Dearth et al. 2014; Srivastava et al. 2011, 2013, 2016; Hiney et al. 2011; Clegg et al. 1998; Lee et al. 2006, 2007). *However, the mechanisms by which Mn increases IGF-1 and insulin levels remain unknown.* Together, these findings suggest a functional link between Mn and the regulation of IGF-1/insulin levels in both peripheral tissues and brain. While such studies clearly link Mn to diabetes and hypothalamic/pubertal development, the role of this potent regulatory mechanism has never been studied in the context of a neuronal disease or manganese toxicity.

HD Pathobiology

HD is an autosomal dominant neurodegenerative disease which results in hyperkinetic movements, behavioral changes in cognition and mood, and ultimately death. An expanded trinucleotide (CAG) repeat in the *Huntingtin* gene (*HTT*) resulting in a mutant HTT protein (mHTT) causes HD. Higher CAG repeats are correlated with increased disease severity and younger age of onset though both are highly variable even between patients with similar repeat size (Bates et al. 2014; Landles and Bates 2004). Usually, the disease manifests in adulthood (though juvenile cases do occur) and gives rise to a combination of motor, cognitive, and psychiatric symptoms which ultimately result in death. A hallmark symptom of HD is chorea, uncontrolled hyperkinetic movements, which has been associated with mHTT-dependent cell death within the striatum. Degeneration in other brain regions (cortex, hypothalamus) usually follows, contributing to the variability in symptoms. As HTT is ubiquitously expressed, the basis for the selective neurotoxicity of mHTT for striatal medium spiny neurons (MSNs) and a handful of other neuronal subpopulations remains a mystery (Landles and Bates 2004; Bates et al. 2015; Kumar et al. 2015; Cattaneo et al. 2005; Gusella and MacDonald 2009).

Mn Dysregulation in HD

Mn dysregulation has only recently been implicated in HD. In normal brains, Mn accumulation is enriched in the basal ganglia—the part of the brain which most severely degenerates in HD—suggesting Mn serves an important role in this brain region (Morello et al. 2008; Prohaska 1987; Larsen et al. 1979). Recently a set of studies revealed a Mn-transport deficit, *indicative of a brain-specific Mn deficiency*, in an HD immortalized striatal neuroprogenitor cell line (STHdhQ111/Q111), in HD hiPSC-derived striatal NPCs cells, and also in the striata of YAC128Q mouse model of HD (Tidball et al. 2015a; Williams et al. 2010b). The mechanism of this Mn-transport deficit has been difficult to resolve as so little is known about Mn subcellular transport. Analysis of Mn homeostasis is complicated by the high promiscuity of proposed Mn transporters for other essential metals (Horning et al. 2015; Chen et al. 2015; Takeda 2003; Tidball et al. 2015b).

However, Mn is known to activate several of the signaling pathways dysregulated in HD including ATM/p53 and AKT/mTOR (Tidball et al. 2015a, b; Dearth et al. 2014; Srivastava et al. 2016; Cordova et al. 2012b; Guilartc 2010a). STHdh Q111/Q111 and hiPSC-derived striatal neuroprogenitor HD cell models exhibit decreased net Mn uptake leading to diminished ATM activation, a Mn-responsive kinase upstream of p53 and other cellular stress response proteins (Tidball et al. 2015a). Similar to ATM/p53, Mn robustly activates AKT and mTOR, both of which are neuroprotective in HD (Lee et al. 2014; Humbert et al. 2002; Blázquez et al. 2015; Gines et al. 2003; Humbert and Saudou 2003; Lopes et al. 2014; Ribeiro et al. 2014; Saavedra et al. 2009). AKT activation can increase HTT-Ser421 phosphorylation, shown to facilitate axonal transport, restoring mitochondrial and autophagic function in HD models (Humbert et al. 2002; Humbert and Saudou 2003; Lopes et al. 2014; Ribeiro et al. 2014; Zala et al. 2008; Naia et al. 2015, 2016; Gauthier et al. 2004). In contrast, Guilarte and colleagues reported decreased HTT-Ser421 phosphorylation by Mn in YAC128 mouse cortical and hippocampal primary cultures, though striatal levels were not assessed (Williams et al. 2010b; Stansfield et al. 2014). Lastly, reinstatement of aberrant mTOR activity in HD models restores autophagic function, enhances aggregate clearance, and increases MSN health, though some reports have shown mTOR inhibition to be neuroprotective in HD (Lee et al. 2014; Pryor et al. 2014; Sarkar et al. 2008).

IIS Dysregulation in HD

Recently, several groups observed impaired IIS in HD. Paradoxically, reduced IGF-1 expression has been detected in patient caudate tissue and skin fibroblasts as well as other nonhuman HD models, while increased IGF-1 has been found peripherally in HD, and this has been correlated with cognitive decline (Lopes et al. 2014; Pouladi et al. 2010a; Saleh et al. 2010). Previous studies have shown mutant HTT disrupts intracellular transport and secretion of insulin, while others have shown Mn can act as a potent insulin-mimetic *in vivo* (Subasinghe et al. 1985). Additionally, several groups reported robust neuroprotective effects of IGF-1 treatment in HD cell and mouse models via increased (1) AKT/ERK signaling, (2) IRS2/VPS34 (Class III PI3K) signaling, and (3) HTT Ser421 phosphorylation. Upregulation of these pathways increased autophagic function and aggregate clearance and ameliorated mitochondrial dysfunction (Lopes et al. 2014; Ribeiro et al. 2014; Naia et al. 2015, 2016; Yamamoto et al. 2006; Duarte et al. 2011; Alexi et al. 1999; Metzler et al. 2010; Warby et al. 2009). Administration of IGF and insulin can also rescue microtubule transport, amelioration of motor abnormalities, MSN health, and enhanced survival in cell and rodent models. IGF-1 is also neuroprotective in models of other NDDs (Quesada et al. 2008; Allodi et al. 2016; Bassil et al. 2014; Gasparini and Xu 2003; Aleman 2012; Homolak et al. 2015; Bernhard et al. 2016; Reger et al. 2007).

Autophagy Deficits in HD, Potential Links to Mn and IIS

The inability to clear toxic mHTT aggregates may be a principle mechanism of HD-related cell death though there is contention about which form(s) and fragment(s) are truly toxic and which are a compensatory/protective reaction to cellular toxicity (Truant et al. 2008; Arrasate and Finkbeiner 2011; Bano et al. 2011; Lim and Yue 2015). Autophagy, a process by which cells degrade complex organelles and proteins to base nutrients, is also the primary process in clearing mHTT aggregates (Arrasate and Finkbeiner 2011; Lim and Yue 2015; Cortes and Spada 2014; Cuervo and Zhang 2015; Martin et al. 2014; Ravikumar and Rubinsztein 2006; Sarkar and Rubinsztein 2008; Williams et al. 2008). HTT acts as a scaffold for autophagy, and this activity is altered or impaired by mHTT, potentially exacerbating pathogenesis (Martin et al. 2014; Rui Y-NN et al. 2015; Gelman et al. 2015; Ochaba et al. 2014; Saudou and Humbert 2016). In HD, autophagic impairment causes failure of cargo recognition and lysosomal function resulting in the accumulation of cellular waste and protein aggregates (Martinez-Vicente et al. 2010). This may trigger a feed-forward pathogenic loop with ever-increasing mHTT levels further impairing clearance (Martin et al. 2014).

IGF treatment incurs robust amelioration of autophagy defects in HD models. Rothman and colleagues observed that IGF-1 upregulates autophagy via an IRS2/VPS34-dependent mechanism in HD cells, resulting in a marked increase in aggregate clearance. This is an AKT/mTOR-independent process, though both AKT and mTOR are activated by IGF-1(Yamamoto et al. 2006). Additionally, other groups have shown that upregulation of mTOR in HD models increases autophagy and aggregate clearance, rescuing HD-related phenotypes, even though mTOR canonically acts as a negative regulator of autophagy by inhibiting ULK1(Lee et al. 2014; Sasazawa et al. 2015). Interestingly, published studies indicate Mn both increases and decreases autophagy in neuronal systems in a biphasic, time-dependent manner (Zhang et al. 2013; Gorojod et al. 2015). Given this regulation of autophagy by Mn and Mn-responsive pathways, it seems plausible that correcting Mn homeostasis in HD models may ameliorate aspects of autophagic dysfunction. To date, however, there have been only a handful of studies exploring the role of Mn in autophagy, and the majority have been done in the context of Mn toxicity, instead of Mn essentiality (Zhang et al. 2013; Zhang et al. 2016). Given clear ties of Mn biology to pathways upstream of autophagy, future studies should interrogate the role of Mn in autophagy during normal neuronal function, in addition to disease states. In particular, we need to establish whether Mn plays a role in basal autophagy or only in the context of Mn toxicity.

Mitochondrial Pathology in HD, Possible Links to Mn and IIS

Mitochondrial dysfunction is another mechanism by which mHTT may cause selective neurodegeneration in HD. Mitochondrial dysfunction may contribute to neurodegenerative diseases (NDDs) for several reasons: (1) High mitochondrial respiration is needed to accommodate high ATP usage in neurons; (2) mitochondria, out of all organelles, produce the highest amount of intracellular reactive oxygen species (ROS); (3) mitochondria are a critical regulator of cell death, a common feature of most NDDs; (4) mitophagy (mitochondrial selective autophagy) is often defective in NDD; and (5) perturbations in various metabolic processes, indicative of mitochondrial dysfunction, are often associated with NDD (Filosto et al. 2011; Johri and Beal 2012; Martin 2012). In HD, specifically, overt metabolic effects such as rapid weight changes and defects in glucose homeostasis have been observed in HD patients and models (Saleh et al. 2009; Farrer 1985; Hurlbert et al. 1999; Lalić et al. 2008; Podolsky et al. 1972; Pouladi et al. 2010b; Mochel et al. 2007; Goodman et al. 2008; Gaba et al. 2005; Josefsen et al. 2007; Oláh et al. 2008). Also, WT HTT has been shown essential for mitochondrial health (Ismailoglu et al. 2014). To this end, several basic studies and clinical trials have investigated metabolic targets as potential therapeutics for HD including creatine and coenzyme Q10 but have found little success (Koroshetz et al. 1997; Schilling et al. 2001; Andreassen et al. 2001; Ferrante et al. 2000; Ferrante et al. 2002; Tabrizi et al. 2003; Verbessem et al. 2003).

Several landmark studies demonstrate IGF-1 restores mitochondrial health in HD models (Ribeiro et al. 2014; Naia et al. 2015; Naia et al. 2016). Given the IIS-mimetic effects of Mn, correcting Mn homeostasis may ameliorate some facets of mitochondrial dysfunction in HD. This hypothesis is consistent with established roles for Mn in mitochondria: (1) Mn accumulates in mitochondria more so than other organelles supporting a functional need in this organelle; (2) Mn has antioxidant functions via the Mn-dependent, mitochondrial enzyme, MnSOD; and (3) Mn is essential for the function of at least two gluconeogenesis enyzmes (Horning et al. 2015; Gunter et al. 2009; Kalia et al. 2008; Chen et al. 2015; Tidball et al. 2015b). Rego and colleagues have reported a series of studies providing a mechanistic understanding of how IGF is capable of such robust amelioration of HD symptoms (Naia et al. 2015, Naia et al. 2016; Johri and Beal 2012; Koroshetz et al. 1997;

Ferreira et al. 2010, Ferreira et al. 2011; Damiano et al. 2010; Kim et al. 2010; Reddy et al. 2009; Martin et al. 2007; Milakovic and Johnson 2005; Weydt et al. 2006; Lou et al. 2016; Gouarné et al. 2013). They found HD models exhibit reduced ATP/ADP ratio, decreased O2 consumption, increased mitochondrial ROS and fragmentation, aberrant lactate/pyruvate levels, and decreased mitochondrial membrane potential—all of which indicates mitochondrial dysfunction. Each of these was shown to be ameliorated by IGF treatment via upregulation of PI3K/AKT signaling in cellular and mouse models of HD.

IIS Signaling and Mn in Other NDDs

Abnormal levels of IGF/insulin and decreased IIS signaling (namely, reduced AKT signaling) have been observed in all neurodegenerative diseases including PD, Alzheimer's disease (AD), amyotrophic lateral sclerosis (ALS), multiple sclerosis (MS), spinocerebellar ataxias (SCA), and other NDD-like conditions such as ataxia telangiectasia (AT). In the case for many models of these diseases, IGF or insulin has been successfully used to ameliorate pathologies in vitro and in vivo, and they have been used or targeted in clinical trials (Bassil et al. 2014; Nagano et al. 2013; Sorenson et al. 2008; Hölscher 2014). Unfortunately, these clinical trials have reported little success. One possible reason for this is control of IGF-1 bioavailability by IGFBPs. This could be overcome by using a modified IGF-1-like peptide which is unable to bind IGFBPs (Rauskolb et al. 2016). Furthermore, although many studies have shown perturbation in metal ion homeostasis in these diseases, few have explored a more specific role for Mn dysregulation. Recent studies elucidating Mn or IGF/insulin dysregulation in NDDs will be reviewed next, emphasizing developments in recent years.

PD and IIS/Mn

PD is a neurodegenerative disorder resulting in bradykinesia and motor rigidity affecting an estimated 10 million people worldwide. Symptoms of the disease mostly occur in late adulthood as a threshold of dopaminergic neurons in the substantia nigra degenerate. Unlike HD, there is no clear genetic predisposition for most cases of PD, though mutations in some genes are correlated to increased risk for PD. Given this and the late onset of the disease, many studies have focused on environmental modifiers of the disease (Sulzer 2007). PD has long been associated with perturbations in metal ion homeostasis—particularly iron (Fe) and Mn. Mn toxicity causes parkinsonian-like symptoms and a disease state known as manganism, but most agree that its pathology is different from that seen in PD. This is mainly because neurodegeneration in PD occurs primarily in the dopaminergic neurons of the substantia nigra, while Mn toxicity manifests within the globus pallidus.

Furthermore, at least some patients with Mn-induced parkinsonism do not produce Lewy bodies and can be unresponsive to levodopa treatment (Guilarte and Gonzales 2015; Cersosimo and Koller 2005; Kwakye et al. 2015). While these two diseases may be distinct, several lines of evidence support a role for Mn dysregulation in PD. Chronic exposure to Mn is associated with increased risk for PD. Also, Mn toxicity has been linked to reduced tyrosine hydroxylase and dopamine levels and DAT cell surface expression, but reports regarding impaired neurotransmission and viability in dopaminergic neurons have been inconsistent (Horning et al. 2015; Kwakye et al. 2015; Bowman et al. 2011; Aschner et al. 2009; Guilarte 2010b; Zhang et al. 2011). Mn toxicity has also been associated with increased alpha-synuclein buildup, but it is unclear if this response is neuroprotective or enhances neurodegeneration (Tong et al. 2009; Harischandra et al. 2015; Peres et al. 2016).

IGF has been studied in the context of PD as well. Previous studies have revealed neuroprotective effects of IGF in PD models and associated with increased dopaminergic survival in the substantia nigra (Quesada et al. 2008; Tong et al. 2009; Ayadi et al. 2016; Offen et al. 2001; Krishnamurthi et al. 2004). However, the majority of recent studies mainly focus on plasma IGF-1 levels as a biomarker for PD progression. Several groups published studies suggesting IGF-1 levels were increased in the sera of PD patients compared to control (Ma et al. 2015; Godau et al. 2010). Furthermore, studies revealed that increased plasma IGF-1 levels correlate with cognitive decline and motor symptoms (Ma et al. 2015; Picillo et al. 2013). While these studies have great utility as a clinical tool and seem to be quite sensitive, they have added minimal mechanistic insight as to if or why IGF-1/insulin and related signaling may be dysregulated on pathogenic consequences. Thus, continued basic and mechanistic experiments to understanding of IGF's role in PD are needed to resolve inconsistencies and provide detail.

AKT has received considerable attention in the PD field via its neuroprotective roles in the brain. Aside from reduced p-AKT levels found in postmortem PD brains, several studies have linked increased AKT and IIS signaling to reduced dopaminergic cell death, reduced alpha-synuclein toxicity, and complex interactions with PD-related proteins including PARKIN, PINK1, and DJ1(Quesada et al. 2008; Ayadi et al. 2016; Gong et al. 2012; Nakaso et al. 2008; Timmons et al. 2009; Xiromerisiou et al. 2008; Xu et al. 2014).

AD and IIS/Mn

AD results primarily from the degeneration of hippocampal neurons which leads to severe cognitive defects in late adulthood. Disease is defined by two hallmark pathological features, neurofibrillary tangles (hyperphosphorylated tau) and amyloid beta plaques, two aggregates which incur neurotoxic stress. Heavy metals have also been associated with AD and its aggregate pathology, though few studies have examined Mn levels or dysregulation (Bandmann et al. 2015; Dieter et al. 2005). However, two recent studies investigated plasma Mn levels in AD and reported opposing results. Dehua and colleagues reported elevated Mn levels which were correlated with increased amyloid beta expression and reduced cognition, while Bush et al. reported reduced Mn levels in sera but no difference in patient erythrocytes (Hare et al. 2016; Chui et al. 2013).

AD may have the most significant ties to IGF dysregulation of all NDDs. AD has been heavily correlated to diabetic status, and mechanistic understanding of metabolic dysfunction in AD has led to it being referred to as "type 3 diabetes," a form of diabetes that specifically affects the brain (Suzanne and Wands 2008). In recent years, studies have focused primarily on the effects of IGF/insulin on amyloid beta accumulation and the use of IGF-1 levels as a biomarker for disease risk and progression. Two studies in 2009 reported that reduced IGF signaling protects against AB accumulation, potentially by acting on the plaques themselves, condensing them to less toxic forms (Schubert et al. 2004; Cohen et al. 2010; Cohen et al. 2009; Freude et al. 2009). These were contrary to a flurry of studies in the early mid-2000s revealing IGF resistance and ameliorative effects by IGF treatment on AB accumulation and cognitive function (Gasparini and Xu 2003; Moloney et al. 2010; Neill 2013; Torres-Aleman 2007; Craft and Watson 2004; de la Monte and Wands 2005; Rivera et al. 2005; Steen et al. 2005; Vidal et al. 2016). A few years later, insulin resistance and reduced IIS signaling were found in postmortem AD brain tissue, and soon after that, lower serum IGF-1 levels were correlated to an increased risk for AD and dementia, while higher levels were associated with greater brain volume (Vidal et al. 2016; Liu et al. 2011). Interestingly, increased IGF has been reported in CSF of patients (Åberg et al. 2015; Johansson et al. 2013). Thus, even though conflicting results have been reported, these studies reveal that AD is deeply tied to IGF biology.

Contrary to PD, excessive AKT signaling has been observed in AD. Several studies have reduced or inhibited IIS signaling and observed delays in symptoms and reduced AB pathology (Neill 2013; Griffin et al. 2005; Rickle et al. 2004). These results, of course, are contrary to aforementioned studies utilizing IGF treatment in AD models. Such conflicting results may be explained by an initial hyperactivation of IIS signaling which eventually desensitizes the pathway. In this way, both IIS inhibition early and IIS treatment late in disease progression may result in ameliorative effects. However, further research will have to be done across disease progression to see if this is indeed the case.

ALS and IIS/Mn

ALS is a neurodegenerative disease which affects more than 12,000 people in the USA. Disease onset is more variable than other diseases and can often occur in younger people. The cause of ALS is unknown, but pathology is attributed to loss of motor neurons in the brain and spinal cord resulting in loss of voluntary muscle control, and, in late stage, patients are unable to move or breathe without ventilator support. ALS has also been associated with metal ion dysregulation. Again, few studies focused on Mn levels, but a few studies have reported increased Mn in CSF and plasma, while the other reports no change in Mn but significant increases in copper and zinc and a reduction in selenium (Roos et al. 2012; Peters et al. 2016; Miyata et al. 1983; Nagata et al. 1985).

IGF dysregulation and insulin resistance have been reported in ALS (Reyes et al. 1984; Adem et al. 1994; Bilic et al. 2006). These data led to a few in vivo studies using IGF-1 treatment in ALS models. While subcutaneous injection into the periphery with IGF-1 was largely found to be ineffective, direct intrathecal injections directly into the CSF resulted in some decrease in motor atrophy (Nagano et al. 2013; Nagano et al. 2005). Given these results, a few clinical trials have been attempted in ALS but have found little success (Sorenson et al. 2008; Saccà et al. 2012; Borasio et al. 1998). One reason may be that these treatments are given peripherally instead of intrathecally (Rauskolb et al. 2016). More recently, IGF2 has been found to be neuroprotective in ALS models (Allodi et al. 2016).

Autophagy in Other Neurodegenerative Diseases

Autophagy has been linked to every neurodegenerative disease—namely—because most NDDs develop aggregate pathology which is often processed by autophagy. Autophagy is activated as a protective process in order to maintain healthy homeostasis of the cell, but hyperactivation can result in autophagy-mediated cell death. Thus, interactions between aggregates and autophagy play a precarious role in NDDs (Harris and Rubinsztein 2011). Recent studies have begun to explore the effects of metal toxicity on autophagy as well (Zhang et al. 2013; Zhang et al. 2016). In PD, autophagy has primarily been investigated in the context of mitophagy (mitochondrial specific autophagy). PD has been linked to mitochondrial toxicity and dysfunction which incurs mitophagy in an attempt to remove unhealthy mitochondria from the neurons to reestablish cellular integrity. PARKIN and PINK1, two proteins associated with familial forms of PD, are essential members of the mitophagy process (Chinta et al. 2010; Deas et al. 2010; Geisler et al. 2010; Michiorri et al. 2010; Narendra et al. 2009; Vives-Bauza and Przedborski 2010). In AD, autophagy is known to regulate both the secretion and degradation of AB which adds increased complexity to its role in disease pathology. Several studies have revealed increased autophagosome accumulation in AD models, but these results have been inconsistent across disease progression (Wolfe et al. 2013; Spilman et al. 2010; Yang et al. 2011; Yu et al. 2005). Recently, ALS studies have revealed that two ALS-associated proteins, TDP-43 and SOD1, are often dysregulated in ALS patients and models (Gomes et al. 2010; Wang et al. 2010; Gal et al. 2009). Interestingly, mutations in these proteins (among several other observed ALS mutation-associated proteins) cause aberrant autophagic processing in neuronal and spinal cord neurons (Chen et al. 2012). Further studies are needed to elucidate mechanistic understanding of these complex relationships to determine whether dysregulated autophagy is a pathogenic mechanism or compensatory "rescue" response. Future investigation must interrogate autophagic flux rather than commonly used end-point measurements as the directionality, and capacity of autophagy is necessary for further understanding and therapeutics. The connections that have been drawn between autophagy and Mn or IGF/insulin warrant continued exploration, but studies should consider potential co-regulation of Mn and IGF/insulin on autophagy processes and dysregulation.

Manganese Toxicity and IGF

Little investigation has been done to examine the role of IGF in manganese toxicity. Tong and colleagues found Mn toxicity caused reduced ATP and insulin/IGF receptor expression. Additionally, as mentioned before, Hiney and colleagues have been revealing a role for Mn-induced toxicity in hypothalamic development via IGF/mTOR-related pathways (Hiney et al. 2011; Dearth et al. 2014; Srivastava et al. 2011; Srivastava et al. 2016; Lee et al. 2006; Lee et al. 2007). It is likely that Mn toxicity in other brain regions is regulated in a similar manner. Given that Mn accumulates in the brain primarily in the basal ganglia, not the hypothalamus, it seems likely that IGF/Mn interaction may play even more crucial roles in other brain regions, particularly in aged model systems. Thus, future studies on Mn toxicity and IGF could be informative on developmental toxicity, chronic environmental exposures, and overall brain health.

The Co-regulation of ATM, Mn, and Insulin/IGF

Interestingly ATM, a Mn-activated kinase, has been linked to both IGF/insulin and Mn signaling. Previous studies have shown that Mn-induced p53 activity is regulated by ATM. Furthermore, this Mn-induced activity is blunted in HD due to lack of bioavailable Mn (Tidball et al. 2015a). Separately, low levels of the IGF-1 receptor and loss of IGF-1 sensitivity have been observed in ataxia telangiectasia (AT), the disease resulting from loss of function mutations in ATM, and in loss-of-function ATM models (Luo et al. 2014; Peretz et al. 2001; Miles et al. 2007; Zhou et al. 2007). Additionally, studies have shown patients with AT have significantly decreased IGF-1 levels (Ehlayel et al. 2014; Kieslich et al. 2010; Nissenkorn et al. 2016; Schubert et al. 2005). Additionally, others have shown ATM is essential for IGF and IGF-R transcription by phosphorylating and relieving transcription factors and complexes including p53 from their respective promotors, allowing for transcription (Luo et al. 2014; Peretz et al. 2001; Goetz et al. 2011; Shahrabani-Gargir et al. 2004; Bhat et al. 2001; Ching et al. 2009). Concurrently, downregulation of IGF-R results in increased radiosensitivity and decreased ATM protein levels (mRNA was unchanged) revealing a potential circular regulation between ATM and IGF-R (Peretz et al. 2001; Chitnis et al. 2014; Valenciano et al. 2014). Also, given that ATM is required for full activation of AKT, it seems likely that the connections between ATM, Mn, and IGF carry some biological relevance in the context of Mn/IGF co-regulation in NDD (Halaby et al. 2008). Mn could act as an initiating signaling molecule within this cascade where Mn activates ATM/p53 which results in increased IGF/IGF-R transcription and subsequent activation of the PI3K/AKT pathway. This hypothetical, albeit plausible, interaction could explain how Mn deficiency in HD might contribute to decreased IIS (AKT/mTOR) and Mn-induced ATM/p53 signaling.

IIS Signaling, Mn, and Cancer

Given the striking parallels and potential co-regulation between Mn and IIS and the pronounced and well-studied roles of IIS in cancer progression, one must wonder if there is role for Mn/IIS co-regulation in cancer etiology. As a pro-growth signaling pathway, IIS is often highly upregulated in cancers particularly during tumor progression (Vara et al. 2004). However, most findings suggest Mn is not significantly carcinogenic, even to exposed workers. In fact, Mn deficiency leads to a higher risk. A plethora of studies, namely, clinical examination of Mn levels in cancer patients, support the role for Mn deficiency in cancer via reduced MnSOD activity and enhanced ROS accumulation in various cancer types (Shen et al. 2015; Behrend et al. 2005; Ho et al. 2001; Hu et al. 2005; Weydert et al. 2006). Of note, Mn has been shown to be essential for the activation of ATM and MRE11, two DNA-damage repair proteins, and able to increase phosphorylation of p53, the most-well studied tumor suppressor gene, which exerts control on cell cycle supporting a role for Mn deficiency in cancer. In fact, many cancers contain mutations in these same proteins. Somewhat paradoxically, HD is associated with reduced Mn bioavailability and reduced risk for cancers (Sørensen et al. 1999). Accumulating data, studies, and clinical trials support a hypothesis that perturbations in IIS and metal ion homeostasis separately contribute to both NDDs and cancer in somewhat opposite fashions, while a dearth of investigation exists to study their potential co-regulation in either disease.

Conclusions

The roles for IGF and Mn separately in HD are far from being fully elucidated. However, the sizeable overlap between their homeostasis and downstream effects supports a need to consider their co-regulation in the context of diseased and healthy states. Neuroprotective cell signaling (i.e., AKT, mTOR, ERK/MAPK), mitochondrial health, and autophagic function have been implicated in all NDDs repeatedly by multiple groups. Past and present research has revealed an essential role for IIS in coordinating these cellular processes. However, little attention has been given to

Mn role even though distinct lines of evidence substantiate its essentiality in these very same processes and even the upstream regulation of insulin/IGF. There is not enough evidence one way or another to draw a clear conclusion whether Mn may be at the heart of IIS dysregulation in NDDs, but there is certainly enough to warrant serious consideration of its role as a contributing factor.

It is still unclear how Mn is exerting its effects on IGF/insulin levels and signaling. Is Mn acting at the levels of transcription, translation, or posttranslationally? The intriguing possibility that Mn might regulate IGF and IGF-R transcription through ATM/p53 is one that merits further study as it may have implications in not only NDDs but cancer and diabetes as well. Furthermore, given the widespread transcriptional targets of p53, Mn could be widely essential for the transcription of various proteins. Mn could also be exerting its control posttranscriptionally—potentially at the blood-brain barrier or via interactions with IGF binding proteins. Clegg and colleagues reported that Mn deficiency resulted in increased IGF-BP3 which they suspected might reduce IGF bioavailability (Clegg et al. 1998). However, little investigation has been done to follow up on these findings or explore Mn's role on other IGF-BPs which could offer a clear mechanism of Mn's regulation of IGF.

We discussed here many examples of overlap between HD etiology, IGF/insulin biology, and Mn homeostasis. While these connections have been more fully elucidated in HD, the inherent overlap between NDD pathology suggests similar roles for Mn and IGF/insulin in other NDDs. However, as reviewed here, there is preliminary evidence that these NDDs often exhibit different trends in Mn and/or insulin/IGF homeostasis—for example, PD is associated with increased Mn, while HD is associated with Mn-deficiency. However, these observations lead to the following additional questions: (1) Are we exploring IGF/insulin and Mn dysregulation at the "right" times during disease progression? (2) Are we inspecting the levels of Mn or IGF/insulin in the correct tissues? (3) Is this dysregulation truly a contributor to disease pathology or simply a downstream effect of a higher mechanism? If IGF/insulin and/or Mn are truly dysregulated in NDDs, one would imagine that there are defined stages of disease progression when specific defects can be observed. Mn or insulin/IGF could be affected early on in the disease prior to symptoms, during early symptom manifestation, or during late-stage progression once significant brain atrophy has occurred (or across the entirety of disease progression). Furthermore, it is likely that this dysregulation may differ in not only magnitude but directionality between each stage of the disease as molecular signaling attempts to compensate or desensitize. While serum and plasma levels offer a potential biomarker of brain Mn dysregulation, further studies must examine how these levels correlate to what is seen in actual brain tissue. Studies have found that changes in IGF by age, sex, diet, BMI, and secondary disease status can cause immense variability between patients, particularly in peripheral samples (Bernhard et al. 2016). Several heavy metals are reported to accumulate in the brain with age and can differ by similar confounds suggesting peripheral Mn may also be an inappropriate measurement for brain Mn. Furthermore, regulation of IGF/insulin and Mn across the blood-brain barrier has been somewhat elucidated, but strict regulation of these molecules is needed to establish brain integrity suggesting that they might be very different from what is seen in serum/plasma or even CSF. Confirming consistencies between serum, plasma, blood, CSF and the brain should be done in rodent models across disease progression to validate IGF/insulin and Mn biomarkers—substantiating their use in clinical studies. For other NDDs, a higher mechanistic understanding of IGF/insulin and Mn biology should be explored at the molecular and cellular levels, similar to what has been done in the HD field. Lastly, given the extended time it takes prior to NDD manifestation, one must ask whether observed defects in IGF/insulin or Mn are either a cause of the disease or instead a consequence of the neurodegeneration. This is a difficult question to answer given the inherent difficulty in working with aged models—namely, mouse models—which often do not fully recapitulate the pathology observed in humans.

Currently, available methods and technology make it quite difficult to truly investigate these questions in a high-throughput manner. Highly sensitive biomarkers for Mn and IGF/insulin levels in the brain are likely required to observe changes across disease progression which are currently unavailable. The high variability and contradictory data of IGF/insulin levels in serum/plasma compared to brain suggest these are not always appropriate measurements for brain levels. While existing techniques can quantify levels of Mn in tissues or cells (ICP-MS, graphite furnace, cellular fura-2 Mn extraction assay (CFMEA)) as well as techniques that allow a cellular/subcellar resolution of Mn localization (X-ray absorption near edge structure (XANES)), high costs and complexities related to maintaining in vivo patterns have limited the understanding of Mn brain homeostasis (Kwakye et al. 2011). Thus, creative approaches will be necessary to answer the outstanding questions.

References

Åberg D, Johansson P, Isgaard J, Wallin A, Johansson J-O, Andreasson U, Blennow K, Zetterberg H, Åberg DN, Svensson J. Increased cerebrospinal fluid level of insulin-like growth factor-II in male patients with Alzheimer's disease. Journal of Alzheimer's disease : JAD. 2015;48(3):637–46.

Adem A, Ekblom J, Gillberg PG, Jossan SS, Höög A, Winblad B, Aquilonius SM, Wang LH, Sara V. Insulin-like growth factor-1 receptors in human spinal cord: changes in amyotrophic lateral sclerosis. J Neural Transm. 1994;97(1):73–84.

Aleman I. Insulin-like growth factor-1 and central neurodegenerative diseases. Endocrinol Metab Clin N Am. 2012;41(2):395–408.

Alexi T, Hughes PE, van Roon-Mom WM, Faull RL, Williams CE, Clark RG, Gluckman PD. The IGF-I amino-terminal tripeptide glycine-proline-glutamate (GPE) is neuroprotective to striatum in the quinolinic acid lesion animal model of Huntington's disease. Exp Neurol. 1999;159(1):84–97.

Allodi I, Comley L, Nichterwitz S, Nizzardo M, Simone C, Benitez JA, Cao M, Corti S, Hedlund E. Differential neuronal vulnerability identifies IGF-2 as a protective factor in ALS. Sci Rep. 2016;6:25960.

Andreassen OA, Dedeoglu A, Ferrante RJ, Jenkins BG, Ferrante KL, Thomas M, Friedlich A, Browne SE, Schilling G, Borchelt DR, Hersch SM, Ross CA, Beal MF. Creatine increase survival and delays motor symptoms in a transgenic animal model of Huntington's disease. Neurobiol Dis. 2001;8(3):479–91.

Arrasate M, Finkbeiner S. Protein aggregates in Huntington's disease. Exp Neurol. 2011;238(1):1–11.

Aschner M, Erikson KM, Hernández E, Hernández E, Tjalkens R. Manganese and its role in Parkinson's disease: from transport to neuropathology. NeuroMolecular Med. 2009;11(4):252–66.

Ayadi AE, Zigmond MJ, Smith AD. IGF-1 protects dopamine neurons against oxidative stress: association with changes in phosphokinases. Exp Brain Res. 2016;234(7):1863–73.

Bae J-H, Jang B-C, Suh S-I, Ha E, Baik H, Kim S-S, Lee M-y, Shin D-H. Manganese induces inducible nitric oxide synthase (iNOS) expression via activation of both MAP kinase and PI3K/ Akt pathways in BV2 microglial cells. Neurosci Lett. 2006;398(1–2):151–4.

Baly DL (1984). Effect of manganese deficiency on insulin secretion and carbohydrate Heomostasis in rats. *JNutrition*.

Baly DL, Keen CL, Hurley LS. Effects of manganese deficiency on pyruvate carboxylase and phosphoenolpyruvate carboxykinase activity and carbohydrate homeostasis in adult rats. Biol Trace Elem Res. 1986;11(1):201–12.

Baly DL, Lee I, Doshi R. Mechanism of decreased insulinogenesis in manganese-deficient rats. Decreased insulin mRNA levels. FEBS Lett. 1988;239(1):55–8.

Baly DL, Schneiderman JS, Garcia-Welsh AL. Effect of manganese deficiency on insulin binding, glucose transport and metabolism in rat adipocytes. J Nutr. 1990;120(9):1075–9.

Bandmann O, Weiss K, Kaler SG. Wilson's disease and other neurological copper disorders. The Lancet Neurology. 2015;14(1):103–13.

Bano D, Zanetti F, Mende Y, Nicotera P. Neurodegenerative processes in Huntington's disease. Cell Death Dis. 2011;2(11)

Bassil F, Fernagut P-O, Bezard E, Meissner WG. Insulin, IGF-1 and GLP-1 signaling in neurodegenerative disorders: targets for disease modification? Prog Neurobiol. 2014;118:1–18.

Bates G, Tabrizi S and Jones L (2014). Huntington's disease. Huntington's disease.

Bates GP, Dorsey R, Gusella JF, Hayden MR, Kay C, Leavitt BR, Nance M, Ross CA, Scahill RI, Wetzel R, Wild EJ and Tabrizi SJ (2015). Huntington disease. *Nature reviews Disease primers*. 1: 15005.

Baxter RC. IGF binding proteins in cancer: mechanistic and clinical insights. Nat Rev Cancer. 2014;14(5):329–41.

Behrend L, Mohr A, Dick T, Zwacka RM. Manganese superoxide dismutase induces p53-dependent senescence in colorectal cancer cells. Mol Cell Biol. 2005;25(17):7758–69.

Bernhard FP, Heinzel S, Binder G, Weber K, Apel A, Roeben B, Deuschle C, Maechtel M, Heger T, Nussbaum S, Gasser T, Maetzler W, Berg D. Insulin-like growth factor 1 (IGF-1) in Parkinson's disease: potential as trait-, progression- and prediction marker and confounding factors. PLoS One. 2016;11(3)

Bhat MA, Rios JC, Lu Y, Garcia-Fresco GP, Ching W, Martin M, Li J, Einheber S, Chesler M, Rosenbluth J, Salzer JL, Bellen HJ. Axon-glia interactions and the domain Organization of Myelinated Axons Requires Neurexin IV/Caspr/Paranodin. Neuron. 2001;30(2):369–83.

Bilic E, Bilic E, Rudan I, Kusec V, Zurak N, Delimar D, Zagar M. Comparison of the growth hormone, IGF-1 and insulin in cerebrospinal fluid and serum between patients with motor neuron disease and healthy controls. Eur J Neurol. 2006;13(12):1340–5.

Blázquez C, Chiarlone A, Bellocchio L, Resel E, Pruunsild P, García-Rincón D, Sendtner M, Timmusk T, Lutz B, Galve-Roperh I, Guzmán M. The CB1 cannabinoid receptor signals striatal neuroprotection via a PI3K/Akt/mTORC1/BDNF pathway. Cell Death Differ. 2015;22(10):1618–29.

Borasio GD, Robberecht W, Leigh PN, Emile J, Guiloff RJ, Jerusalem F, Silani V, Vos PE, Wokke JH, Dobbins T. A placebo-controlled trial of insulin-like growth factor-I in amyotrophic lateral sclerosis. European ALS/IGF-I study group. Neurology. 1998;51(2):583–6.

Bowman AB, Kwakye GF, Hernández E, Aschner M. Role of manganese in neurodegenerative diseases. J Trace Elem Med Biol. 2011;25(4):191–203.

Cattaneo E, Zuccato C, Tartari M. Normal huntingtin function: an alternative approach to Huntington's disease. Nat Rev Neurosci. 2005;6(12):919–30.

Cersosimo MG, Koller WC. The diagnosis of manganese-induced parkinsonism. Neurotoxicology. 2005;27(3):340–6.

Chan DW, Son SC, Block W, Ye R, Khanna KK, Wold MS, Douglas P, Goodarzi AA, Pelley J, Taya Y, Lavin MF, Lees-Miller SP. Purification and characterization of ATM from human placenta. A manganese-dependent, wortmannin-sensitive serine/threonine protein kinase. J Biol Chem. 2000;275(11):7803–10.

Chen S, Zhang X, Song L, Le W. Autophagy dysregulation in amyotrophic lateral sclerosis. Brain Pathol. 2012;22(1):110–6.

Chen P, Chakraborty S, Mukhopadhyay S, Lee E, Paoliello MM, Bowman AB, Aschner M. Manganese homeostasis in the nervous system. J Neurochem. 2015;134(4):601–10.

Ching J, Luebbert SH, Zhang Z, Marupudi N, Banerjee S, Hurd R, Collins IV, Roy L, Ralston L and Fisher JS (2009). Ataxia telangiectasia mutated (ATM) is required in insulin-like growth factor-1 (IGF-1) signaling through the PI3K/Akt pathway. FASEB J. 23.

Chinta SJ, Mallajosyula JK, Rane A, Andersen JK. Mitochondrial α-synuclein accumulation impairs complex I function in dopaminergic neurons and results in increased mitophagy in vivo. Neurosci Lett. 2010;486(3):235–9.

Chitnis MM, Lodhia KA, Aleksic T, Gao S, Protheroe AS, Macaulay VM. IGF-1R inhibition enhances radiosensitivity and delays double-strand break repair by both non-homologous end-joining and homologous recombination. Oncogene. 2014;33(45):5262–73.

Chiu S-L, Chen C-M, Cline HT. Insulin receptor signaling regulates synapse number, dendritic plasticity, and circuit function in vivo. Neuron. 2008;58(5):708–19.

Chui D, Yang H, Wang H, Tuo JI, Yu J, Zhang S, Chen Z, Xiao W. The dishomeostasis of metal ions plays an important role for the cognitive impartment. Mol Neurodegener. 2013;8(S1):1–1.

Ciucci F, Putignano E, Baroncelli L, Landi S, Berardi N, Maffei L. Insulin-like growth factor 1 (IGF-1) mediates the effects of enriched environment (EE) on visual cortical development. PLoS One. 2007;2(5)

Clegg MS, Donovan SM, Monaco MH, Baly DL, Ensunsa JL, Keen CL. The influence of manganese deficiency on serum IGF-1 and IGF binding proteins in the male rat. Proc Soc Exp Biol Med Soc Exper Biol Med NY. 1998;219(1):41–7.

Clemmons DR, Busby WH, Arai T, Nam TJ, Clarke JB, Jones JI, Ankrapp DK. Role of insulin-like growth factor binding proteins in the control of IGF actions. Prog Growth Factor Res. 1995;6(2–4):357–66.

Cohen E, Paulsson JF, Blinder P, Burstyn-Cohen T, Du D, Estepa G, Adame A, Pham HM, Holzenberger M, Kelly JW, Masliah E, Dillin A. Reduced IGF-1 signaling delays age-associated Proteotoxicity in mice. Cell. 2009;139(6):1157–69.

Cohen E, Du D, Joyce D, Kapernick EA, Volovik Y, Kelly JW, Dillin A. Temporal requirements of insulin/IGF-1 signaling for proteotoxicity protection. Aging Cell. 2010;9(2):126–34.

Cordova FM, Aguiar AS, Peres TV, Lopes MW, Gonçalves FM, Remor AP, Lopes SC, Pilati C, Latini AS, Prediger RD, Erikson KM, Aschner M, Leal RB. In vivo manganese exposure modulates Erk, Akt and Darpp-32 in the striatum of developing rats, and impairs their motor function. PLoS One. 2012;7(3)

Cordova FM, Aguiar AS, Peres TV, Lopes MW, Gonçalves FM, Pedro DZ, Lopes SC, Pilati C, Prediger RD, Farina M, Erikson KM, Aschner M, Leal RB. Manganese-exposed developing rats display motor deficits and striatal oxidative stress that are reversed by Trolox. Arch Toxicol. 2013;87(7):1231–44.

Cortes CJ, Spada AR. The many faces of autophagy dysfunction in Huntington's disease: from mechanism to therapy. Drug Discov Today. 2014;19(7):963–71.

Craft S, Watson GS. Insulin and neurodegenerative disease: shared and specific mechanisms. The Lancet Neurology. 2004;3(3):169–78.

Crittenden PL, Filipov NM. Manganese modulation of MAPK pathways: effects on upstream mitogen activated protein kinase kinases and mitogen activated kinase phosphatase-1 in microglial cells. Journal of applied toxicology : JAT. 2011;31(1):1–10.

Cuervo A, Zhang S. Selective autophagy and huntingtin: learning from disease. Cell Cycle. 2015;14(11)

D'Antonio EL, Hai Y, Christianson DW. Structure and function of non-native metal clusters in human arginase I. Biochemistry. 2012;51(42):8399–409.

D'Ercole AJ, Ye P, Calikoglu AS, Gutierrez-Ospina G. The role of the insulin-like growth factors in the central nervous system. Mol Neurobiol. 1996;13(3):227–55.

D'Ercole JA, Ye P, O'Kusky JR. Mutant mouse models of insulin-like growth factor actions in the central nervous system. Neuropeptides. 2002;36(2–3):209–20.

Damiano M, Galvan L, Déglon N, Brouillet E. Mitochondria in Huntington's disease. Biochim Biophys Acta (BBA) - Mol Basis Dis. 2010;1802(1):52–61.

Dearth RK, Hiney JK, Srivastava VK, Hamilton AM, Dees WL. Prepubertal exposure to elevated manganese results in estradiol regulated mammary gland ductal differentiation and hyperplasia in female rats. Exp Biol Med. 2014;239(7):871–82.

Deas E, Wood NW, Plun-Favreau H. Mitophagy and Parkinson's disease: the PINK1-parkin link. Biochim Biophys Acta. 2010;1813(4):623–33.

Deijen JB, de Boer H, van der Veen EA. Cognitive changes during growth hormone replacement in adult men. Psychoneuroendocrinology. 1998;23(1):45–55.

Dentremont KD, Ye P, D'Ercole AJ, O'Kusky JR. Increased insulin-like growth factor-I (IGF-I) expression during early postnatal development differentially increases neuron number and growth in medullary nuclei of the mouse. Brain Res Dev Brain Res. 1999;114(1):135–41.

DeWitt MR, Chen P, Aschner M. Manganese efflux in parkinsonism: insights from newly characterized SLC30A10 mutations. Biochem Biophys Res Commun. 2013;432(1):1–4.

Dhamoon MS, Noble JM, Craft S. Intranasal insulin improves cognition and modulates -amyloid in early ad. Neurology. 2009;72(3):292–4.

Dieter HH, Bayer TA, Multhaup G. Environmental copper and manganese in the pathophysiology of neurologic diseases (Alzheimer's disease and Manganism). Acta Hydrochim Hydrobiol. 2005;33(1):72–8.

Dormond O, Ponsonnet L, Hasmim M, Foletti A, Rüegg C. Manganese-induced integrin affinity maturation promotes recruitment of alpha V beta 3 integrin to focal adhesions in endothelial cells: evidence for a role of phosphatidylinositol 3-kinase and Src. Thromb Haemost. 2004;92(1):151–61.

Duarte AI, Petit GH, Ranganathan S, Li JY, Oliveira CR, Brundin P, Björkqvist M, Rego AC. IGF-1 protects against diabetic features in an in vivo model of Huntington's disease. Exp Neurol. 2011;231(2):314–9.

Ehlayel M, Soliman A, Sanctis V. Linear growth and endocrine function in children with ataxia telangiectasia. Ind J Endocrinol Metabol. 2014;18(7):93–6.

Ekmekcioglu C, Prohaska C, Pomazal K, Steffan I, Schernthaner G, Marktl W. Concentrations of seven trace elements in different hematological matrices in patients with type 2 diabetes as compared to healthy controls. Biol Trace Elem Res. 2001;79(3):205–19.

Exil V, Ping L, Yu Y, Chakraborty S, Caito SW, Wells KS, Karki P, Lee E, Aschner M. Activation of MAPK and FoxO by manganese (Mn) in rat neonatal primary astrocyte cultures. PLoS One. 2014;9(5)

Farrer LA. Diabetes mellitus in Huntington disease. Clin Genet. 1985;27(1):62–7.

Fernandez AM, Torres-Alemán I. The many faces of insulin-like peptide signalling in the brain. Nat Rev Neurosci. 2012;13(4):225–39.

Ferrante RJ, Andreassen OA, Jenkins BG, Dedeoglu A, Kuemmerle S, Kubilus JK, Kaddurah-Daouk R, Hersch SM, Beal MF. Neuroprotective effects of creatine in a transgenic mouse model of Huntington's disease. J Neurosci Off J Soc Neurosci. 2000;20(12):4389–97.

Ferrante RJ, Andreassen OA, Dedeoglu A, Ferrante KL, Jenkins BG, Hersch SM, Beal FM. Therapeutic effects of coenzyme Q10 and remacemide in transgenic mouse models of Huntington's disease. J Neurosci Off J Soc Neurosci. 2002;22(5):1592–9.

Ferreira IL, Nascimento MV, Ribeiro M, Almeida S, Cardoso SM, Grazina M, Pratas J, Santos M, Januário C, Oliveira CR and Rego CA (2010). Mitochondrial-dependent apoptosis in Huntington's disease human cybrids. Exp Neurol 222 (2): 243–255.

Ferreira LI, Cunha-Oliveira T, Nascimento MV, Ribeiro M, Proença TM, Januário C, Oliveira CR, Rego CA. Bioenergetic dysfunction in Huntington's disease human cybrids. Exp Neurol. 2011;231(1):127–34.

Filosto M, Scarpelli M, Cotelli M, Vielmi V, Todeschini A, Gregorelli V, Tonin P, Tomelleri G, Padovani A. The role of mitochondria in neurodegenerative diseases. J Neurol. 2011;258(10):1763–74.

Freude S, Schilbach K, Schubert M. The role of IGF-1 receptor and insulin receptor signaling for the pathogenesis of Alzheimer's disease: from model organisms to human disease. Curr Alzheimer Res. 2009;6(3):213–23.

Gaba AM, Zhang K, Marder K, Moskowitz CB, Werner P, Boozer CN. Energy balance in early-stage Huntington disease. Am J Clin Nutr. 2005;81(6):1335–41.

Gal J, Ström AL, Kwinter DM, Kilty R, Zhang J, Shi P, Fu W, Wooten MW, Zhu H. Sequestosome 1/p62 links familial ALS mutant SOD1 to LC3 via an ubiquitin-independent mechanism. J Neurochem. 2009;111(4):1062–73.

Gasparini L, Xu H. Potential roles of insulin and IGF-1 in Alzheimer's disease. Trends Neurosci. 2003;26(8):404–6.

Gauthier LR, Charrin BC, Borrell-Pagès M, Dompierre JP, Rangone H, Cordelières FP, Mey J, MacDonald ME, Leßmann V, Humbert S, Saudou F. Huntingtin controls neurotrophic support and survival of neurons by enhancing BDNF vesicular transport along microtubules. Cell. 2004;118(1):127–38.

Geisler S, Holmström KM, Skujat D, Fiesel FC, Rothfuss OC, Kahle PJ, Springer W. PINK1/Parkin-mediated mitophagy is dependent on VDAC1 and p62/SQSTM1. Nat Cell Biol. 2010;12(2):119–31.

Gelman A, Rawet-Slobodkin M, Elazar Z. Huntingtin facilitates selective autophagy. Nat Cell Biol. 2015;17(3):214–5.

Gines S, Ivanova E, Seong I-S, Saura CA, MacDonald ME. Enhanced Akt signaling is an early pro-survival response that reflects N-methyl-D-aspartate receptor activation in Huntington's disease knock-in striatal cells. J Biol Chem. 2003;278(50):50514–22.

Godau J, Herfurth M, Kattner B, Gasser T, Berg D. Increased serum insulin-like growth factor 1 in early idiopathic Parkinson's disease. J Neurol Neurosurg Psychiatry. 2010;81(5):536–8.

Goetz EM, Shankar B, Zou Y, Morales JC, Luo X, Araki S, Bachoo R, Mayo LD, Boothman DA. ATM-dependent IGF-1 induction regulates secretory clusterin expression after DNA damage and in genetic instability. Oncogene. 2011;30(35):3745–54.

Gomes C, Escrevente C, Costa J. Mutant superoxide dismutase 1 overexpression in NSC-34 cells: effect of trehalose on aggregation, TDP-43 localization and levels of co-expressed glycoproteins. Neurosci Lett. 2010;475(3):145–9.

Gong L, Zhang QL, Zhang N, Hua WY, Huang YX, Di PW, Huang T, Xu XS, Liu CF, Hu LF, Luo WF. Neuroprotection by urate on 6-OHDA-lesioned rat model of Parkinson's disease: linking to Akt/GSK3β signaling pathway. J Neurochem. 2012;123(5):876–85.

Goodman A, Murgatroyd PR, Medina-Gomez G, Wood NI, Finer N, Vidal-Puig AJ, Morton JA, Barker RA. The metabolic profile of early Huntington's disease- a combined human and transgenic mouse study. Exp Neurol. 2008;210(2):691–8.

Gorojod RM, Alaimo A, Porte Alcon S, Pomilio C, Saravia F, Kotler ML. The autophagic- lysosomal pathway determines the fate of glial cells under manganese- induced oxidative stress conditions. Free Radic Biol Med. 2015;87:237–51.

Gouarné C, Tardif G, Tracz J, Latyszenok V, Michaud M, Clemens L, Yu-Taeger L, Nguyen H, Bordet T, Pruss RM. Early deficits in glycolysis are specific to striatal neurons from a rat model of Huntington disease. PLoS One. 2013;8(11)

Greenwood BN, Fleshner M. Exercise, learned helplessness, and the stress-resistant brain. NeuroMolecular Med. 2008;10(2):81–98.

Griffin RJ, Moloney A, Kelliher M, Johnston JA, Ravid R, Dockery P, O'Connor R, O'Neill C. Activation of Akt/PKB, increased phosphorylation of Akt substrates and loss and altered distribution of Akt and PTEN are features of Alzheimer's disease pathology. J Neurochem. 2005;93(1):105–17.

Guilarte TR. APLP1, Alzheimer's-like pathology and neurodegeneration in the frontal cortex of manganese-exposed non-human primates. Neurotoxicology. 2010a;31(5):572–4.

Guilarte TR. Manganese and Parkinson's disease: a critical review and new findings. Environ Health Perspect. 2010b;118(8):1071–80.

Guilarte TR, Gonzales KK. Manganese-induced parkinsonism is not idiopathic Parkinson's disease: environmental and genetic evidence. Toxicol Sci. 2015;146(2):204–12.

Gunter TE, Gavin CE, Gunter KK. The case for manganese interaction with mitochondria. Neurotoxicology. 2009;30(4):727–9.

Gusella JF, MacDonald ME. Huntington's disease: the case for genetic modifiers. Genome Med. 2009;1(8):80.

Haj-ali V, Mohaddes G, Babri SH. Intracerebroventricular insulin improves spatial learning and memory in male Wistar rats. Behav Neurosci. 2009;123(6):1309.

Halaby M-J, Hibma JC, He J, Yang D-Q. ATM protein kinase mediates full activation of Akt and regulates glucose transporter 4 translocation by insulin in muscle cells. Cell Signal. 2008;20(8):1555–63.

Hare DJ, Faux NG, Roberts BR, Volitakis I, Martins RN, Bush AI. Lead and manganese levels in serum and erythrocytes in Alzheimer's disease and mild cognitive impairment: results from the Australian imaging, biomarkers and lifestyle flagship study of ageing. Metallomics. 2016;8(6):628–32.

Harischandra DS, Jin H, Anantharam V, Kanthasamy A, Kanthasamy AG. α-Synuclein protects against manganese neurotoxic insult during the early stages of exposure in a dopaminergic cell model of Parkinson's disease. Toxicol Sci. 2015;143(2):454–68.

Harris H, Rubinsztein DC. Control of autophagy as a therapy for neurodegenerative disease. Nat Rev Neurol. 2011;8(2):108–17.

Hiney JK, Srivastava VK, Dees WL. Manganese induces IGF-1 and cyclooxygenase-2 gene expressions in the basal hypothalamus during prepubertal female development. Toxicol Sci Off J Soc Toxicol. 2011;121(2):389–96.

Ho C-mJ, Zheng S, Comhair SA, Farver C, Erzurum SC. Differential expression of manganese superoxide dismutase and catalase in lung cancer. Cancer Res. 2001;61(23):8578–85.

Hodge RD, D'Ercole JA, O'Kusky JR. Insulin-like growth factor-I accelerates the cell cycle by decreasing G1 phase length and increases cell cycle reentry in the embryonic cerebral cortex. J Neurosci Off J Soc Neurosci. 2004;24(45):10201–10.

Hölscher C. First clinical data of the neuroprotective effects of nasal insulin application in patients with Alzheimer's disease. Alzheimer's Dementia J Alzheimer's Assoc. 2014;10(1 Suppl):7.

Homolak J, Janeš I, Filipović M. The role of IGF-1 in neurodegenerative diseases. Gyrus. 2015;3(3):162–7.

Horning KJ, Caito SW, Tipps KG, Bowman AB, Aschner M. Manganese is essential for neuronal health. Annu Rev Nutr. 2015;35:71–108.

Hu Y, Rosen DG, Zhou Y, Feng L, Yang G, Liu J, Huang P. Mitochondrial manganese-superoxide dismutase expression in ovarian cancer: role in cell proliferation and response to oxidative stress. J Biol Chem. 2005;280(47):39485–92.

Humbert S, Saudou F. Huntingtin phosphorylation and signaling pathways that regulate toxicity in Huntington's disease. Clin Neurosci Res. 2003;3(3):149–55.

Humbert S, Bryson EA, Cordelières FP, Connors NC, Datta SR, Finkbeiner S, Greenberg ME, Saudou F. The IGF-1/Akt pathway is neuroprotective in Huntington's disease and involves huntingtin phosphorylation by Akt. Dev Cell. 2002;2(6):831–7.

Hurlbert MS, Zhou W, Wasmeier C, Kaddis FG, Hutton JC, Freed CR. Mice transgenic for an expanded CAG repeat in the Huntington's disease gene develop diabetes. Diabetes. 1999;48(3):649–51.

Hurtado-Chong A, Yusta-Boyo MJ, Vergaño-Vera E, Bulfone A, Pablo F, Vicario-Abejón C. IGF-I promotes neuronal migration and positioning in the olfactory bulb and the exit of neuroblasts from the subventricular zone. Eur J Neurosci. 2009;30(5):742–55.

Ismailoglu I, Chen Q, Popowski M, Yang L, Gross SS, Brivanlou AH. Huntingtin protein is essential for mitochondrial metabolism, bioenergetics and structure in murine embryonic stem cells. Dev Biol. 2014;391(2):230–40.

Jang B-CC. Induction of COX-2 in human airway cells by manganese: role of PI3K/PKB, p38 MAPK, PKCs, Src, and glutathione depletion. Toxicology in vitro : an international journal published in association with BIBRA. 2009;23(1):120–6.

Jiu Y-M, Yue Y, Yang S, Liu L, Yu J-W, Wu Z-X, Xu T. Insulin-like signaling pathway functions in integrative response to an olfactory and a gustatory stimuli in *Caenorhabditis elegans*. Protein Cell. 2010;1(1):75–81.

Johansson P, Åberg D, Johansson J-O, Mattsson N, Hansson O, Ahrén B, Isgaard J, Åberg DN, Blennow K, Zetterberg H, Wallin A, Svensson J. Serum but not cerebrospinal fluid levels of insulin-like growth factor-I (IGF-I) and IGF-binding protein-3 (IGFBP-3) are increased in Alzheimer's disease. Psychoneuroendocrinology. 2013;38(9):1729–37.

Johri A, Beal FM. Mitochondrial dysfunction in neurodegenerative diseases. J Pharmacol Exp Ther. 2012;342(3):619–30.

Josefsen K, Nielsen MD, Jørgensen KH, Bock T, Nørremølle A, Sørensen SA, Naver B, Hasholt L. Impaired glucose tolerance in the R6/1 transgenic mouse model of Huntington's disease. J Neuroendocrinol. 2007;20(2):165–72.

Kalia K, Jiang W, Zheng W. Manganese accumulates primarily in nuclei of cultured brain cells. Neurotoxicology. 2008;29(3):466–70.

Kanyo ZF, Scolnick LR, Ash DE, Christianson DW. Structure of a unique binuclear manganese cluster in arginase. Nature. 1996;383(6600):554–7.

Keen CL, Baly DL, Lönnerdal B. Metabolic effects of high doses of manganese in rats. Biol Trace Elem Res. 1984;6(4):309–15.

Kieslich M, Hoche F, Reichenbach J, Weidauer S, Porto L, Vlaho S, Schubert R, Zielen S. Extracerebellar MRI—lesions in ataxia telangiectasia go along with deficiency of the GH/IGF-1 Axis, markedly reduced body weight, high ataxia scores and advanced age. Cerebellum. 2010;9(2):190–7.

Kim J, Moody JP, Edgerly CK, Bordiuk OL, Cormier K, Smith K, Beal FM, Ferrante RJ. Mitochondrial loss, dysfunction and altered dynamics in Huntington's disease. Hum Mol Genet. 2010;19(20):3919–35.

Koh ES, Kim SJ, Yoon HE, Chung JH, Chung S, Park CW, Chang YS, Shin SJ. Association of blood manganese level with diabetes and renal dysfunction: a cross-sectional study of the Korean general population. BMC Endocrine Disorders. 2014;14:24.

Koroshetz WJ, Jenkins BG, Rosen BR, Beal FM. Energy metabolism defects in Huntington's disease and effects of coenzyme Q10. Ann Neurol. 1997;41(2):160–5.

Krishnamurthi R, Stott S, Maingay M, Faull RLM, McCarthy D, Gluckman P, Guan J. N-terminal tripeptide of IGF-1 improves functional deficits after 6-OHDA lesion in rats. Neuroreport. 2004;15(10):1601–4.

Kumar A, Singh S, Kumar V, Kumar D, Agarwal S, Rana M. Huntington's disease: an update of therapeutic strategies. Gene. 2015;556(2):91–7.

Kwakye GF, Li D, Bowman AB. Novel high-throughput assay to assess cellular manganese levels in a striatal cell line model of Huntington's disease confirms a deficit in manganese accumulation. Neurotoxicology. 2011;32(5):630–9.

Kwakye GF, Paoliello MM, Mukhopadhyay S, Bowman AB, Aschner M. Manganese-induced parkinsonism and Parkinson's disease: shared and distinguishable features. Int J Environ Res Public Health. 2015;12(7):7519–40.

Lalić NM, Marić J, Svetel M, Jotić A, Stefanova E, Lalić K, Dragašević N, Miličić T, Lukić L, Kostić VS. Glucose homeostasis in Huntington disease: abnormalities in insulin sensitivity and early-phase insulin secretion. Arch Neurol. 2008;65(4):476–80.

Landles C, Bates GP. Huntingtin and the molecular pathogenesis of Huntington's disease. EMBO Rep. 2004;5(10):958–63.

Larsen NA, Pakkenberg H, Damsgaard E, Heydorn K. Topographical distribution of arsenic, manganese, and selenium in the normal human brain. J Neurol Sci. 1979;42(3):407–16.

Lee B, Pine M, Johnson L, Rettori V, Hiney JK, Dees LW. Manganese acts centrally to activate reproductive hormone secretion and pubertal development in male rats. Reproductive toxicology (Elmsford, NY). 2006;22(4):580–5.

Lee B, Hiney JK, Pine MD, Srivastava VK, Dees LW. Manganese stimulates luteinizing hormone releasing hormone secretion in prepubertal female rats: hypothalamic site and mechanism of action. J Physiol. 2007;578(3):765–72.

Lee S-H, Jouihan HA, Cooksey RC, Jones D, Kim HJ, Winge DR, McClain DA. Manganese supplementation protects against diet-induced diabetes in wild type mice by enhancing insulin secretion. Endocrinology. 2013;154(3):1029–38.

Lee JH, Tecedor L, Chen Y, Monteys A, Sowada MJ, Thompson LM, Davidson BL. Reinstating aberrant mTORC1 activity in Huntington's disease mice improves disease phenotypes. Neuron. 2014;85(2):303–15.

Leyva-Illades D, Chen P, Zogzas CE, Hutchens S, Mercado JM, Swaim CD, Morrisett RA, Bowman AB, Aschner M, Mukhopadhyay S. SLC30A10 is a cell surface-localized manganese efflux transporter, and parkinsonism-causing mutations block its intracellular trafficking and efflux activity. J Neurosci. 2014;34(42):14079–95.

Lim J, Yue Z. Neuronal aggregates: formation, clearance, and spreading. Dev Cell. 2015;32(4):491–501.

Liou J-C, Tsai F-Z, Ho S-Y. Potentiation of quantal secretion by insulin-like growth factor-1 at developing motoneurons in Xenopus cell culture. J Physiol. 2003;553(Pt 3):719–28.

Liu W, Ye P, O'Kusky JR, D'Ercole JA. Type 1 insulin-like growth factor receptor signaling is essential for the development of the hippocampal formation and dentate gyrus. J Neurosci Res. 2009;87(13):2821–32.

Liu Y, Liu F, Grundke-Iqbal I, Iqbal K, Gong CX. Deficient brain insulin signalling pathway in Alzheimer's disease and diabetes. J Pathol. 2011;225(1):54–62.

Lopes C, Ribeiro M, Duarte AI, Humbert S, Saudou F, Pereira de Almeida L, Hayden M, Rego AC. IGF-1 intranasal administration rescues Huntington's disease phenotypes in YAC128 mice. Mol Neurobiol. 2014;49(3):1126–42.

Lou S, Lepak T, Eberly LE, Roth B, Cui W, Zhu X-H, Öz G, Dubinsky JM. Oxygen consumption deficit in Huntington disease mouse brain under metabolic stress. Human Mol Genet. 2016;

Luo X, Suzuki M, Ghandhi SA, Amundson SA, Boothman DA. ATM regulates insulin-like growth factor 1-secretory Clusterin (IGF-1-sCLU) expression that protects cells against senescence. PLoS One. 2014;9(6)

Ma J, Jiang Q, Xu J, Sun Q, Qiao Y, Chen W, Wu Y, Wang Y, Xiao Q, Liu J, Tang H, Chen S. Plasma insulin-like growth factor 1 is associated with cognitive impairment in Parkinson's disease. Dement Geriatr Cogn Disord. 2015;39(5–6):251–6.

Marks DR, Tucker K, Cavallin MA, Mast TG, Fadool DA. Awake intranasal insulin delivery modifies protein complexes and alters memory, anxiety, and olfactory behaviors. J Neurosci. 2009;29(20):6734–51.

Martin LJ. Chapter 11 biology of mitochondria in neurodegenerative diseases. Prog Mol Biol Transl Sci. 2012;107:355–415.

Martin WWR, Wieler M, Hanstock CC. Is brain lactate increased in Huntington's disease? J Neurol Sci. 2007;263(1–2):70–4.

Martin DDO, Ladha S, Ehrnhoefer DE, Hayden MR. Autophagy in Huntington disease and huntingtin in autophagy. Trends Neurosci. 2014;38(1):26–35.

Martinez-Vicente M, Talloczy Z, Wong E, Tang G, Koga H, Kaushik S, de Vries R, Arias E, Harris S, Sulzer D, Cuervo A. Cargo recognition failure is responsible for inefficient autophagy in Huntington's disease. Nat Neurosci. 2010;13(5):567–76.

Maydan M, McDonald PC, Sanghera J, Yan J, Rallis C, Pinchin S, Hannigan GE, Foster LJ, Ish-Horowicz D, Walsh MP, Dedhar S. Integrin-linked kinase is a functional Mn2+−dependent protein kinase that regulates glycogen synthase kinase-3β (GSK-3β) phosphorylation. PLoS One. 2010;5(8)

Metzler M, Gan L, Mazarei G, Graham RK, Liu L, Bissada N, Lu G, Leavitt BR, Hayden MR. Phosphorylation of huntingtin at Ser421 in YAC128 neurons is associated with protection of YAC128 neurons from NMDA-mediated excitotoxicity and is modulated by PP1 and PP2A. J Neurosci. 2010;30(43):14318–29.

Michiorri S, Gelmetti V, Giarda E, Lombardi F, Romano F, Marongiu R, Nerini-Molteni S, Sale P, Vago R, Arena G, Torosantucci L, Cassina L, Russo MA, Dallapiccola B, Valente EM, Casari G. The Parkinson-associated protein PINK1 interacts with Beclin1 and promotes autophagy. Cell Death Differ. 2010;17(6):962–74.

Milakovic T, Johnson GVW. Mitochondrial respiration and ATP production are significantly impaired in striatal cells expressing mutant huntingtin. J Biol Chem. 2005;280(35):30773–82.

Miles PD, Treuner K, Latronica M, Olefsky JM, Barlow C. Impaired insulin secretion in a mouse model of ataxia telangicctasia. Am J Physiol Endocrinol Metab. 2007;293(1):4.

Miyata S, Nakamura S, Nagata H, Kameyama M. Increased manganese level in spinal cords of amyotrophic lateral sclerosis determined by radiochemical neutron activation analysis. J Neurol Sci. 1983;61(2):283–93.

Mochel F, Charles P, Seguin F, Barritault J, Coussieu C, Perin L, Bouc Y, Gervais C, Carcelain G, Vassault A, Feingold J, Rabier D, Durr A. Early energy deficit in Huntington disease: identification of a plasma biomarker traceable during disease progression. PLoS One. 2007;2(7)

Moloney AM, Griffin RJ, Timmons S, O'Connor R, Ravid R, O'Neill C. Defects in IGF-1 receptor, insulin receptor and IRS-1/2 in Alzheimer's disease indicate possible resistance to IGF-1 and insulin signalling. Neurobiol Aging. 2010;31

de la Monte SM, Wands JR. Review of insulin and insulin-like growth factor expression, signaling, and malfunction in the central nervous system: relevance to Alzheimer's disease. J Alzheimers Dis. 2005;7

Morello M, Canini A, Mattioli P, Sorge RP, Alimonti A, Bocca B, Forte G, Martorana A, Bernardi G, Sancesario G. Sub-cellular localization of manganese in the basal ganglia of normal and manganese-treated rats an electron spectroscopy imaging and electron energy-loss spectroscopy study. Neurotoxicology. 2008;29(1):60–72.

Morrison BD, Feltz SM, Pessin JE. Polylysine specifically activates the insulin-dependent insulin receptor protein kinase. J Biol Chem. 1989;264(17):9994–10001.

Nagano I, Ilieva H, Shiote M, Murakami T, Yokoyama M, Shoji M, Abe K. Therapeutic benefit of intrathecal injection of insulin-like growth factor-1 in a mouse model of amyotrophic lateral sclerosis. J Neurol Sci. 2005;235

Nagano I, Shiote M, Murakami T, Kamada H, Hamakawa Y, Matsubara E, Yokoyama M, Morita K, Shoji M, Abe K. Beneficial effects of intrathecal IGF-1 administration in patients with amyotrophic lateral sclerosis. Neurol Res. 2013;27(7):768–72.

Nagata H, Miyata S, Nakamura S, Kameyama M, Katsui Y. Heavy metal concentrations in blood cells in patients with amyotrophic lateral sclerosis. J Neurol Sci. 1985;67(2):173–8.

Naia L, Ferreira IL, Cunha-Oliveira T, Duarte AI, Ribeiro M, Rosenstock TR, Laço MNN, Ribeiro MJ, Oliveira CR, Saudou F, Humbert S, Rego AC. Activation of IGF-1 and insulin signaling pathways ameliorate mitochondrial function and energy metabolism in Huntington's disease human lymphoblasts. Mol Neurobiol. 2015;51(1):331–48.

Naia L, Ribeiro M, Rodrigues J, Duarte AI, Lopes C, Rosenstock TR, Hayden MR and Rego CA (2016). Insulin and IGF-1 regularize energy metabolites in neural cells expressing full-length mutant huntingtin. *Neuropeptides*.

Nakaso K, Ito S, Nakashima K. Caffeine activates the PI3K/Akt pathway and prevents apoptotic cell death in a Parkinson's disease model of SH-SY5Y cells. Neurosci Lett. 2008;432

Narendra D, Tanaka A, Suen D-F, Youle RJ. Parkin-induced mitophagy in the pathogenesis of Parkinson disease. Autophagy. 2009;5(5):706–8.

Neill C. PI3-kinase/Akt/mTOR signaling: impaired on/off switches in aging, cognitive decline and Alzheimer's disease. Exp Gerontol. 2013;48(7):647–53.

Neulen A, Blaudeck N, Zittrich S, Metzler D, Pfitzer G, Stehle R. Mn2+−dependent protein phosphatase 1 enhances protein kinase A-induced Ca2+ desensitisation in skinned murine myocardium. Cardiovasc Res. 2007;74(1):124–32.

Nissenkorn A, Levy-Shraga Y, Banet-Levi Y, Lahad A, Sarouk I, Modan-Moses D. Endocrine abnormalities in ataxia telangiectasia: findings from a national cohort. Pediatr Res. 2016;79(6):889–94.

O'Kusky JR, Ye P, D'Ercole AJ. Insulin-like growth factor-I promotes neurogenesis and synaptogenesis in the hippocampal dentate gyrus during postnatal development. J Neurosci Off J Soc Neurosci. 2000;20(22):8435–42.

Ochaba J, Lukacsovich T, Csikos G, Zheng S, Margulis J, Salazar L, Mao K, Lau AL, Yeung SY, Humbert S, Saudou F, Klionsky DJ, Finkbeiner S, Zeitlin SO, Marsh LJ, Housman DE, Thompson LM, Steffan JS. Potential function for the huntingtin protein as a scaffold for selective autophagy. Proc Natl Acad Sci. 2014;111(47):16889–94.

Offen D, Shtaif B, Hadad D, Weizman A, Melamed E, Gil-Ad I. Protective effect of insulin-like-growth-factor-1 against dopamine-induced neurotoxicity in human and rodent neuronal cultures: possible implications for Parkinson's disease. Neurosci Lett. 2001;316(3):129–32.

Oishi K, Watatani K, Itoh Y, Okano H, Guillemot F, Nakajima K, Gotoh Y. Selective induction of neocortical GABAergic neurons by the PDK1-Akt pathway through activation of Mash1. Proc Natl Acad Sci. 2009;106(31):13064–9.

Oláh J, Klivényi P, Gardián G, Vécsei L, Orosz F, Kovacs GG, Westerhoff HV, Ovádi J. Increased glucose metabolism and ATP level in brain tissue of Huntington's disease transgenic mice. FEBS J. 2008;275(19):4740–55.

Ozdinler HP, Macklis JD. IGF-I specifically enhances axon outgrowth of corticospinal motor neurons. Nat Neurosci. 2006;9(11):1371–81.

Paull TT, Gellert M. The 3′ to 5′ exonuclease activity of Mre 11 facilitates repair of DNA double-strand breaks. Mol Cell. 1998;1(7):969–79.

Peres T, Parmalee NL, Martinez-Finley EJ, Aschner M. Untangling the manganese-α-Synuclein web. Front Neurosci. 2016;10:364.

Peretz S, Jensen R, Baserga R, Glazer PM. ATM-dependent expression of the insulin-like growth factor-I receptor in a pathway regulating radiation response. Proc Natl Acad Sci. 2001;98(4):1676–81.

Peters TL, Beard JD, Umbach DM, Allen K, Keller J, Mariosa D, Sandler DP, Schmidt S, Fang F, Ye W, Kamel F. Blood levels of trace metals and amyotrophic lateral sclerosis. Neurotoxicology. 2016;54:119–26.

Picillo M, Erro R, Santangelo G, Pivonello R, Longo K, Pivonello C, Vitale C, Amboni M, Moccia M, Colao A, Barone P, Pellecchia M. Insulin-like growth factor-1 and progression of motor symptoms in early, drug-naïve Parkinson's disease. J Neurol. 2013;260(7):1724–30.

Podolsky S, Leopold N, Sax D. Increased frequency of diabetes mellitus in patients with Huntington's chorea. Lancet. 1972;299(7765):1356–9.

Pouladi MA, Xie Y, Skotte NH, Ehrnhoefer DE, Graham RK, Kim JE, Bissada N, Yang XW, Paganetti P, Friedlander RM, Leavitt BR, Hayden MR. Full-length huntingtin levels modulate body weight by influencing insulin-like growth factor 1 expression. Hum Mol Genet. 2010;19(8):1528–38.

Prohaska JR. Functions of trace elements in brain metabolism. Physiol Rev. 1987;67(3):858–901.

Pryor WM, Biagioli M, Shahani N, Swarnkar S, Huang W-C, Page DT, MacDonald ME, Subramaniam S. Huntingtin promotes mTORC1 signaling in the pathogenesis of Huntington's disease. Sci Signal. 2014;7(349)

Quadri M, Federico A, Zhao T, Breedveld GJ, Battisti C, Delnooz C, Severijnen L-A, Di Toro ML, Mignarri A, Monti L, Sanna A, Lu P, Punzo F, Cossu G, Willemsen R, Rasi F, Oostra BA, van de Warrenburg BP, Bonifati V. Mutations in SLC30A10 cause parkinsonism and dystonia with Hypermanganesemia, polycythemia, and chronic liver disease. Am J Hum Genet. 2012;90(3):467–77.

Quesada A, Lee BY, Micevych PE. PI3 kinase/Akt activation mediates estrogen and IGF-1 nigral DA neuronal neuroprotection against a unilateral rat model of Parkinson's disease. Dev Neurobiol. 2008;68(5):632–44.

Rauskolb S, Dombert B, Sendtner M. Insulin-like growth factor 1 in diabetic neuropathy and amyotrophic lateral sclerosis. Neurobiol Dis. 2016;

Ravikumar B, Rubinsztein DC. Role of autophagy in the clearance of mutant huntingtin: a step towards therapy? Mol Asp Med. 2006;27(5–6):520–7.

Reddy HP, Mao P, Manczak M. Mitochondrial structural and functional dynamics in Huntington's disease. Brain Res Rev. 2009;61(1):33–48.

Reger MA, Watson GS, Green PS, Wilkinson CW, Baker LD, Cholerton B, Fishel MA, Plymate SR, Breitner JCS, DeGroodt W, Mehta P, Craft S. Intranasal insulin improves cognition and modulates beta-amyloid in early AD. Neurology. 2007;70(6):440–8.

Reyes ET, Perurena OII, Fcstoff BW, Jorgensen R, Moore WV. Insulin resistance in amyotrophic lateral sclerosis. J Neurol Sci. 1984;63(3):317–24.

Ribeiro M, Rosenstock TR, Oliveira AM, Oliveira CR, Rego AC. Insulin and IGF-1 improve mitochondrial function in a PI-3K/Akt-dependent manner and reduce mitochondrial generation of reactive oxygen species in Huntington's disease knock-in striatal cells. Free Radic Biol Med. 2014;74:129–44.

Rickle A, Bogdanovic N, Volkman I, Winblad B, Ravid R, Cowburn RF. Akt activity in Alzheimer's disease and other neurodegenerative disorders. Neuroreport. 2004;15(6):955–9.

Rivera EJ, Goldin A, Fulmer N, Tavares R, Wands JR, de la Monte SM. Insulin and insulin-like growth factor expression and function deteriorate with progression of Alzheimer's disease: link to brain reductions in acetylcholine. J Alzheimers Dis. 2005;8

Roos PM, Lierhagen S, Flaten T, Syversen T, Vesterberg O, Nordberg M. Manganese in cerebrospinal fluid and blood plasma of patients with amyotrophic lateral sclerosis. Exp Biol Med. 2012;237(7):803–10.

Root CM, Ko KI, Jafari A, Wang JW. Presynaptic facilitation by neuropeptide signaling mediates odor-driven food search. Cell. 2011;145(1):133–44.

Rubenstein AH, Levin NW, Elliott GA. Manganese-induced hypoglycaemia. Lancet (London, England). 1962;2(7270):1348–51.

Rui Y-NN XZ, Patel B, Chen Z, Chen D, Tito A, David G, Sun Y, Stimming EF, Bellen HJ, Cuervo AM, Zhang S. Huntingtin functions as a scaffold for selective macroautophagy. Nat Cell Biol. 2015;17(3)

Saavedra A, García-Martínez JM, Xifró X, Giralt A, Torres-Peraza JF, Canals JM, Díaz-Hernández M, Lucas JJ, Alberch J, Pérez-Navarro E. PH domain leucine-rich repeat protein phosphatase 1 contributes to maintain the activation of the PI3K/Akt pro-survival pathway in Huntington's disease striatum. Cell Death Differ. 2009;17(2):324–35.

Saccà F, Quarantelli M, Rinaldi C, Tucci T, Piro R, Perrotta G, Carotenuto B, Marsili A, Palma V, Michele G, Brunetti A, Morra V, Filla A, Salvatore M. A randomized controlled clinical trial of growth hormone in amyotrophic lateral sclerosis: clinical, neuroimaging, and hormonal results. J Neurol. 2012;259(1):132–8.

Saleh N, Moutereau S, Durr A, Krystkowiak P, Azulay J-P, Tranchant C, Broussolle E, Morin F, Bachoud-Lévi A-C, Maison P. Neuroendocrine disturbances in Huntington's disease. PLoS One. 2009;4(3)

Saleh N, Moutereau S, Azulay JP, Verny C, Simonin C, Tranchant C, Hawajri EN, Bachoud-Lévi AC, Maison P, Group H. High insulinlike growth factor I is associated with cognitive decline in Huntington disease. Neurology. 2010;75(1):57–63.

Sarkar S, Rubinsztein DC. Huntington's disease: degradation of mutant huntingtin by autophagy. FEBS J. 2008;275(17):4263–70.

Sarkar S, Ravikumar B, Floto RA, Rubinsztein DC. Rapamycin and mTOR-independent autophagy inducers ameliorate toxicity of polyglutamine-expanded huntingtin and related proteinopathies. Cell Death Differ. 2008;16(1):46–56.

Sasazawa Y, Sato N, Umezawa K, Simizu S. Conophylline protects cells in cellular models of neurodegenerative diseases by inducing mammalian target of rapamycin (mTOR)-independent autophagy. J Biol Chem. 2015;290(10):6168–78.

Sato T, Nakashima A, Guo L, Tamanoi F. Specific activation of mTORC1 by Rheb G-protein in vitro involves enhanced recruitment of its substrate protein. J Biol Chem. 2009;284(19):12783–91.

Saudou F, Humbert S. The biology of huntingtin. Neuron. 2016;89(5):910–26.

Schilling G, Coonfield ML, Ross CA, Borchelt DR. Coenzyme Q10 and remacemide hydrochloride ameliorate motor deficits in a Huntington's disease transgenic mouse model. Neurosci Lett. 2001;315(3):149–53.

Schubert M, Gautam D, Surjo D, Ueki K, Baudler S, Schubert D, Kondo T, Alber J, Galldiks N, Küstermann E. Role for neuronal insulin resistance in neurodegenerative diseases. Proc Nat Acad Sci USA. 2004:101.

Schubert R, Reichenbach J, Zielen S. Growth factor deficiency in patients with ataxia telangiectasia. Clinical & Experimental Immunology. 2005;140(3):517–9.

Shahrabani-Gargir L, Pandita TK, Werner H. Ataxia-telangiectasia mutated gene controls insulin-like growth factor I receptor gene expression in a deoxyribonucleic acid damage response pathway via mechanisms involving zinc-finger transcription factors Sp1 and WT1. Endocrinology. 2004;145(12):5679–87.

Shen F, Cai W-S, Li J-L, Feng Z, Cao J, Xu B. The association between deficient manganese levels and breast cancer: a meta-analysis. Int J Clin Exp Med. 2015;8(3):3671–80.

Skeberdis VA, Lan J, Zheng X, Zukin RS, Bennett MV. Insulin promotes rapid delivery of N-methyl-D- aspartate receptors to the cell surface by exocytosis. Proc Natl Acad Sci U S A. 2001;98(6):3561–6.

Sørensen AS, Fenger K, Olsen JH. Significantly lower incidence of cancer among patients with Huntington disease. Cancer. 1999;86(7):1342–6.

Sorenson EJ, Windbank AJ, Mandrekar JN, Bamlet WR, Appel SH, Armon C, Barkhaus PE, Bosch P, Boylan K, David WS, Feldman E, Glass J, Gutmann L, Katz J, King W, Luciano CA, McCluskey LF, Nash S, Newman DS, Pascuzzi RM, Pioro E, Sams LJ, Scelsa S, Simpson EP, Subramony SH, Tiryaki E, Thornton CA. Subcutaneous IGF-1 is not beneficial in 2-year ALS trial. Neurology. 2008;71(22):1770–5.

Sosa L, Dupraz S, Laurino L, Bollati F, Bisbal M, Cáceres A, Pfenninger KH, Quiroga S. IGF-1 receptor is essential for the establishment of hippocampal neuronal polarity. Nat Neurosci. 2006;9(8):993–5.

Spilman P, Podlutskaya N, Hart MJ, Debnath J, Gorostiza O, Bredesen D, Richardson A, Strong R, Galvan V. Inhibition of mTOR by rapamycin abolishes cognitive deficits and reduces amyloid-β levels in a mouse model of Alzheimer's disease. PLoS One. 2010;5(4)

Srivastava VK, Hiney JK, Dees LW. Prepubertal ethanol exposure alters hypothalamic transforming growth factor-α and erbB1 receptor signaling in the female rat. Alcohol. 2011;45(2):173–81.

Srivastava VK, Hiney JK, Dees WL. Early life manganese exposure upregulates tumor-associated genes in the hypothalamus of female rats: relationship to manganese-induced precocious puberty. Toxicol Sci. 2013;136(2):373–81.

Srivastava VK, Hiney JK, Dees WL. Manganese stimulated Kisspeptin is mediated by the insulin-like growth factor-1/Akt/ mammalian target of rapamycin pathway in the prepubertal female rat. Endocrinology. 2016;

Stansfield KH, Bichell T, Bowman AB, Guilarte TR. BDNF and huntingtin protein modifications by manganese: implications for striatal medium spiny neuron pathology in manganese neurotoxicity. J Neurochem. 2014;131(5):655–66.

Steen E, Terry BM, Rivera EJ, Cannon JL, Neely TR, Tavares R, Xu JX, Wands JR, de la Monte SM. Impaired insulin and insulin-like growth factor expression and signaling mechanisms in Alzheimer's disease–is this type 3 diabetes? J Alzheimer's Dis. 2005;7

Subasinghe S, Greenbaum AL, McLean P. The insulin-mimetic action of Mn2+: involvement of cyclic nucleotides and insulin in the regulation of hepatic hexokinase and glucokinase. Biochem Med. 1985;34(1):83–92.

Sulzer D. Multiple hit hypotheses for dopamine neuron loss in Parkinson's disease. Trends Neurosci. 2007;30(5):244–50.

Suzanne M, Wands JR. Alzheimer's disease is type 3 diabetes—evidence reviewed. J Diabetes Sci Technol. 2008;2

Tabrizi SJ, Blamire AM, Manners DN, Rajagopalan B, Styles P, Schapira AHV, Warner TT. Creatine therapy for Huntington's disease: clinical and MRS findings in a 1-year pilot study. Neurology. 2003;61(1):141–2.

Takeda A. Manganese action in brain function. Brain Res Rev. 2003;41(1):79–87.

Tidball AM, Bryan MR, Uhouse MA, Kumar KK, Aboud AA, Feist JE, Ess KC, Neely DM, Aschner M, Bowman AB. A novel manganese-dependent ATM-p53 signaling pathway is selectively impaired in patient-based neuroprogenitor and murine striatal models of Huntington's disease. Hum Mol Genet. 2015a;24(7):1929–44.

Tidball AM, Bichell T, Bowman AB. Manganese in health and disease. rsc. 2015b:540–73.

Timmons S, Coakley MF, Moloney AM, Neill C. Akt signal transduction dysfunction in Parkinson's disease. Neurosci Lett. 2009;467(1):30–5.

Tong M, Dong M, de la Monte SM. Brain insulin-like growth factor and neurotrophin resistance in Parkinson's disease and dementia with Lewy bodies: potential role of manganese neurotoxicity. J Alzheimer's Dis JAD. 2009;16(3):585–99.

Torres-Aleman I. Targeting insulin-like growth factor-1 to treat Alzheimer's disease. Expert Opin Ther Targets. 2007;11(12):1535–42.

Trejo JL, Llorens-Martín MV, Torres-Alemán I. The effects of exercise on spatial learning and anxiety-like behavior are mediated by an IGF-I-dependent mechanism related to hippocampal neurogenesis. Mol Cell Neurosci. 2007;37(2):402–11.

Truant R, Atwal R, Desmond C, Munsie L, Tran T. Huntington's disease: revisiting the aggregation hypothesis in polyglutamine neurodegenerative diseases. FEBS J. 2008;275(17):4252–62.

Trujillo KM, Yuan SS, Lee EY, Sung P. Nuclease activities in a complex of human recombination and DNA repair factors Rad50, Mre11, and p95. J Biol Chem. 1998;273(34):21447–50.

Valenciano A, Henríquez-Hernández L, Moreno M, Lloret M, Lara P. Role of IGF-1 receptor in radiation response. Transl Oncol. 2014;5(1):1–9.

Vara J, Casado E, de Castro J, Cejas P, Belda-Iniesta C, González-Barón M. PI3K/Akt signalling pathway and cancer. Cancer Treat Rev. 2004;30(2):193–204.

Verbessem P, Lemiere J, Eijnde BO, Swinnen S, Vanhees L, Leemputte VM, Hespel P, Dom R. Creatine supplementation in Huntington's disease: a placebo-controlled pilot trial. Neurology. 2003;61(7):925–30.

Vidal J-SS, Hanon O, Funalot B, Brunel N, Viollet C, Rigaud A-SS, Seux M-LL, le-Bouc Y, Epelbaum J and Duron E (2016). Low serum insulin-like growth factor-I predicts cognitive decline in Alzheimer's disease. Journal of Alzheimer's disease : JAD 52 (2): 641–649.

Vives-Bauza C, Przedborski S. Mitophagy: the latest problem for Parkinson's disease. Trends Mol Med. 2010;17(3):158–65.

Wang X, Fan H, Ying Z, Li B, Wang H, Wang G. Degradation of TDP-43 and its pathogenic form by autophagy and the ubiquitin-proteasome system. Neurosci Lett. 2010;469(1):112–6.

Warby SC, Doty CN, Graham RK, Shively J, Singaraja RR, Hayden MR. Phosphorylation of huntingtin reduces the accumulation of its nuclear fragments. Mol Cell Neurosci. 2009;40(2):121–7.

Wedler FC, Ley BW. Kinetic, ESR, and trapping evidence for in vivo binding of Mn(II) to glutamine synthetase in brain cells. Neurochem Res. 1994;19(2):139–44.

Weydert CJ, Waugh TA, Ritchie JM, Iyer KS, Smith JL, Li L, Spitz DR, Oberley LW. Overexpression of manganese or copper-zinc superoxide dismutase inhibits breast cancer growth. Free Radic Biol Med. 2006;41(2):226–37.

Weydt P, Pineda VV, Torrence AE, Libby RT, Satterfield TF, Lazarowski ER, Gilbert ML, Morton GJ, Bammler TK, Strand AD, Cui L, Beyer RP, Easley CN, Smith AC, Krainc D, Luquet S, Sweet IR, Schwartz MW, Spada AR. Thermoregulatory and metabolic defects in Huntington's disease transgenic mice implicate PGC-1α in Huntington's disease neurodegeneration. Cell Metab. 2006;4(5):349–62.

Williams A, Sarkar S, Cuddon P, Ttofi EK, Saiki S, Siddiqi FH, Jahreiss L, Fleming A, Pask D, Goldsmith P, O'Kane CJ, Floto R, Rubinsztein DC. Novel targets for Huntington's disease in an mTOR-independent autophagy pathway. Nat Chem Biol. 2008;4(5):295–305.

Williams BB, Kwakye GF, Wegrzynowicz M, Li D, Aschner M, Erikson KM, Bowman AB. Altered manganese homeostasis and manganese toxicity in a Huntington's disease striatal cell model are not explained by defects in the iron transport system. Toxicol Sci. 2010a;117(1):169–79.

Williams BB, Li D, Wegrzynowicz M, Vadodaria BK, Anderson JG, Kwakye GF, Aschner M, Erikson KM, Bowman AB. Disease-toxicant screen reveals a neuroprotective interaction between Huntington's disease and manganese exposure. J Neurochem. 2010b;112(1):227–37.

Wolfe DM, Lee J, Kumar A, Lee S, Orenstein SJ, Nixon RA. Autophagy failure in Alzheimer's disease and the role of defective lysosomal acidification. Eur J Neurosci. 2013;37(12):1949–61.

Woźniak-Celmer E, Ołdziej S, Ciarkowski J. Theoretical models of catalytic domains of protein phosphatases 1 and 2A with Zn2+ and Mn2+ metal dications and putative bioligands in their catalytic centers. Acta Biochim Pol. 2001;48(1):35–52.

Xiang Y, Ding N, Xing Z, Zhang W, Liu H, Li Z. Insulin-like growth factor-1 regulates neurite outgrowth and neuronal migration from Organotypic cultured dorsal root ganglion. Int J Neurosci. 2010;121(2):101–6.

Xing C, Yin Y, He X, Xie Z. Effects of insulin-like growth factor 1 on voltage-gated ion channels in cultured rat hippocampal neurons. Brain Res. 2006;1072(1):30–5.

Xing C, Yin Y, Chang R, Gong X, He X, Xie Z. Effects of insulin-like growth factor 1 on synaptic excitability in cultured rat hippocampal neurons. Exp Neurol. 2007;205(1):222–9.

Xiromerisiou G, Hadjigeorgiou GM, Papadimitriou A, Katsarogiannis E, Gourbali V, Singleton AB. Association between AKT1 gene and Parkinson's disease: a protective haplotype. Neurosci Lett. 2008;436

Xu B, Bird VG, Miller WT. Substrate specificities of the insulin and insulin-like growth factor 1 receptor tyrosine kinase catalytic domains. J Biol Chem. 1995;270(50):29825–30.

Xu Y, Liu C, Chen S, Ye Y, Guo M, Ren Q, Liu L, Zhang H, Xu C, Zhou Q. Activation of AMPK and inactivation of Akt result in suppression of mTOR-mediated S6K1 and 4E-BP1 pathways leading to neuronal cell death in in vitro models of Parkinson's disease. Cell Signal. 2014;26

Yamamoto A, Cremona ML, Rothman JE. Autophagy-mediated clearance of huntingtin aggregates triggered by the insulin-signaling pathway. J Cell Biol. 2006;172(5):719–31.

Yang D-S, Stavrides P, Mohan PS, Kaushik S, Kumar A, Ohno M, Schmidt SD, Wesson D, Bandyopadhyay U, Jiang Y, Pawlik M, Peterhoff CM, Yang AJ, Wilson DA, George-Hyslop P, Westaway D, Mathews PM, Levy E, Cuervo AM, Nixon RA. Reversal of autophagy dysfunction in the TgCRND8 mouse model of Alzheimer's disease ameliorates amyloid pathologies and memory deficits. Brain J Neurol. 2011;134(Pt 1):258–77.

Yu HW, Cuervo A, Kumar A, Peterhoff CM, Schmidt SD, Lee J-H, Mohan PS, Mercken M, Farmery MR, Tjernberg LO. Macroautophagy—a novel β-amyloid peptide-generating pathway activated in Alzheimer's disease. J Cell Biol. 2005;171

Zala D, Colin E, Rangone H, Liot G, Humbert S, Saudou F. Phosphorylation of mutant huntingtin at S421 restores anterograde and retrograde transport in neurons. Hum Mol Genet. 2008;17(24):3837–46.

Zhang D, Kanthasamy A, Anantharam V, Kanthasamy A. Effects of manganese on tyrosine hydroxylase (TH) activity and TH-phosphorylation in a dopaminergic neural cell line. Toxicol Appl Pharmacol. 2011;254(2):65–71.

Zhang J, Cao R, Cai T, Aschner M, Zhao F, Yao T, Chen Y, Cao Z, Luo W, Chen J. The role of autophagy dysregulation in manganese-induced dopaminergic neurodegeneration. Neurotox Res. 2013;24(4):478–90.

Zhang Z, Miah M, Culbreth M, Aschner M. Autophagy in neurodegenerative diseases and metal neurotoxicity. Neurochem Res. 2016;41(1–2):409–22.

Zhao W, Chen H, Xu H, Moore E, Meiri N, Quon MJ, Alkon DL. Brain insulin receptors and spatial memory. Correlated changes in gene expression, tyrosine phosphorylation, and signaling molecules in the hippocampus of water maze trained rats. J Biol Chem. 1999;274(49):34893–902.

Zhou T, Chou J, Zhou Y, Simpson DA, Cao F, Bushel PR, Paules RS, Kaufmann WK. Ataxia telangiectasia-mutated dependent DNA damage checkpoint functions regulate gene expression in human fibroblasts. Mol Cancer Res MCR. 2007;5(8):813–22.

Occupational Metal Exposure and Parkinsonism

W. Michael Caudle

Abstract   Parkinsonism is comprised of a host of neurological disorders with an underlying clinical feature of movement disorder, which includes many shared features of bradykinesia, tremor, and rigidity. These clinical outcomes occur subsequent to pathological deficits focused on degeneration or dysfunction of the nigrostriatal dopamine system and accompanying pathological inclusions of alpha-synuclein and tau. The heterogeneity of parkinsonism is equally matched with the complex etiology of this syndrome. While a small percentage can be attributed to genetic alterations, the majority arise from an environmental exposure, generally composed of pesticides, industrial compounds, as well as metals. Of these, metals have received significant attention given their propensity to accumulate in the basal ganglia and participate in neurotoxic cascades, through the generation of reactive oxygen species as well as their pathogenic interaction with intracellular targets in the dopamine neuron. The association between metals and parkinsonism is of critical concern to subsets of the population that are occupationally exposed to metals, both through current practices, such as mining, and emerging settings, like E-waste and the manufacture of metal nanoparticles. This review will explore our current understanding of the molecular and pathological targets that mediate metal neurotoxicity and lead to parkinsonism and will highlight areas of critical research interests that need to be addressed.

Keywords Copper • E-Waste • Iron • Manganese • Manganism • Nanoparticles • Parkinsonism

W.M. Caudle (✉)
Department of Environmental Health, Rollins School of Public Health,
Center for Neurodegenerative Disease, Emory University,
1518 Clifton Rd., NE, Claudia Nance Rollins Building, Room 2033,
Atlanta, GA 30322, USA
e-mail: william.m.caudle@emory.edu

© Springer International Publishing AG 2017
M. Aschner and L.G. Costa (eds.), *Neurotoxicity of Metals*, Advances in Neurobiology 18, DOI 10.1007/978-3-319-60189-2_7

Introduction

The neurotoxicological outcomes related to metal exposure are varied, as metals are capable of migrating to many different brain regions and interacting with a catalogue of different neuronal populations, affecting a variety of intracellular targets and pathways within these cells. This promiscuity introduces a complexity to ascribing neurotoxicological effects to select metals. In light of these complexities, extensive work has highlighted the unique sensitivity of select brain regions to metal exposure and the neurological deficits that arise from these interactions. To this end, the basal ganglia, a region severely damaged in Parkinson disease (PD), appear to be a selective target for metal-induced neurotoxicity. Perhaps not surprising, as with other environmental factors, metal exposure has been suggested to be a risk factor for the development of PD and other parkinsonian-related movement disorders. Recent work has been focused on bringing to light a more specific understanding of how certain metals interact with neuronal targets and intracellular pathways in the basal ganglia to elicit neurological deficits. Thus, this section will present recent findings concerning the potential role and mechanisms of action of metal toxicity in Parkinson disease and parkinsonism and will introduce emerging exposure concerns that may have neurotoxicological implications for the future.

Clinical and Pathological Signs of Parkinsonism

Parkinsonism is a heterogeneous group of neurological disorders that share common pathological features of alteration to the nigrostriatal dopamine circuit, in addition to pathological accumulations of alpha-synuclein or tau, as determined at autopsy (Dickson 2012). These pathological signs give rise to a suite of clinical symptoms focused on the presence of disordered movement, predominantly described in terms of bradykinesia, tremor, and postural instability, in addition to other extrapyramidal deficits and neuropsychiatric signs. With this broad category of movement disorders, three specific disorders, Parkinson disease (PD), progressive supranuclear palsy (PSP), and multiple system atrophy (MSA), have been extensively investigated and used to demonstrate the clinical and pathological heterogeneity that defines parkinsonism. Of these, PD has been the most extensively characterized, with cardinal clinical signs including asymmetrical resting tremor, slowed movement, cogwheel rigidity, and postural instability (Fahn 2003). These motor signs are often accompanied by non-motor deficits in autonomic and neuropsychiatric function, ranging from gastrointestinal and cognitive deficits to olfactory and sleep disturbances (Langston 2006). These clinical manifestations appear to be extensively associated with pathological findings, including damage to the dopaminergic circuit that originates in the substantia nigra pars compacta (SNpc) and sends dopaminergic projections to the striatum (Fahn 2003). While these deficits appear to underlie many of the motor deficits observed in PD, pathological alterations to other brain regions and neurotransmitter systems, including the noradrenergic and serotonergic circuits, have also been demonstrated to be involved in the plethora of non-motor symptoms that define the disease. In addition to alterations in neurotransmitter pathways, PD is pathologically defined by the presence of alpha-synuclein inclusions (Dickson 2012). Although the precise neuronal function of alpha-synuclein is still under investigation, it is well-established that conformational changes and misfolding of this protein are pathogenic mechanism involved in damage to the dopamine neurons in the SNpc.

Multiple system atrophy has also been extensively evaluated and is classified as a major parkinsonian syndrome. Like PD, MSA is clinically described as exhibiting alterations in movement involving the basal ganglia, as well as autonomic dysfunction that precedes the motoric deficits (Gilman et al. 1999). While MSA shares many of the clinical features as PD, a major divergence of these disorders is the responsiveness to the dopamine replacement drug, L-DOPA. Given the extensive loss of dopamine in the nigrostriatal pathway in PD, L-DOPA treatment has served as the "gold standard" of therapeutic intervention for well over 50 years. In contrast, MSA does not respond as robustly to L-DOPA treatment. The reasons for this discrepancy are not known. However, it is interesting to note that while PD and MSA share many clinical and pathological features, one striking difference is the localization of pathogenic alpha-synuclein inclusions in the brain. Such inclusions in MSA are predominantly localized to glia, specifically oligodendrocytes, rather than intraneuronally, as seen with PD (Lantos 1998).

Progressive supranuclear palsy is clinically defined by motor disturbances in addition to other neurological deficits, including dementia and a lack of autonomic participation (Williams et al. 2005). These clinical findings suggest the involvement of other brain regions and neural circuits that are independent of the nigrostriatal dopamine pathology. Indeed, the pathological findings in PSP are significantly more promiscuous than PD or MSA, encompassing a variety of other neuronal nuclei and regions. The involvement of these other brain regions and circuits may also underlie the attenuated response of PSP patients to L-DOPA therapy. Additional support for the pathological complexity of PSP is found in the presence of inclusions of the microtubule protein, tau, to create neurofibrillary tangles and glial inclusions, which are hallmark pathological signs in disorders like Alzheimer disease and dementia (Dickson et al. 2007). Taken in sum, the clinicopathological landscape of parkinsonian disorders comes with an extensive complexity that must be appreciated when discussing the etiology and pathogenic processes that may underlie observed clinical signs.

While parkinsonism clearly has underlying genetic etiology, the majority of cases are idiopathic, suggesting an exogenous contribution to disease etiopathogenesis. To this end, a variety of environmental factors have been associated with parkinsonism, including exposure to pesticides, solvents and other industrial chemicals, as well as metals (Caudle et al. 2012; Hatcher et al. 2008; Caudle 2015). Metal exposure represents an interesting etiological feature of the disease. In one hand, we rely extensively on specific metals to perform various biological functions in the body and the brain. However, if levels of metals are not tightly regulated, they can have severe repercussions to the function of the nervous system. While metals comprise a naturally occurring element of our dietary intake, these exposures provide a relatively low level of exposure to these compounds. In contrast, the inclusion of metals in a variety of industrial and commercial applications has introduced an occupational exposure scenario that exposes subsets of the population to elevated levels of potentially neurotoxic metals. When in excess, metals can damage the brain through a variety of intracellular cascades, most notably through their ability to generate highly reactive molecular species that can target a variety of intracellular components, leading to dysfunction. Within the brain, the basal ganglia appear to be especially vulnerable to the neurotoxic effects of excess metal exposure and the development of parkinsonian disorders.

To specifically address these points, this review will focus on occupational exposures to select metals that have been demonstrated to have significant associations with damage to the basal ganglia and the dopamine circuit seen in parkinsonism. This discussion will appraise the current epidemiological evidence related to these exposures and disease and will further explore these findings, in the context of pathological, clinical, and mechanistic perspectives.

Iron Exposure and Parkinsonism

Iron is one of the most prevalent metals and is utilized in a variety of biological functions, including its incorporation into hemoglobin for oxygen transport, and as a cofactor for enzymatic activity in cytochrome C and catalase. However, alterations in the regulation of iron levels in the basal ganglia leading to excess accumulation have been shown to result in parkinsonian deficits. For example, Friedreich's ataxia, which manifests as a host of neurological deficits including motor dysfunction, is defined by alterations in iron handling in the mitochondria by the protein, frataxin, leading to accumulation (Gomes and Santos 2013). While these findings point to deficits in iron homeostasis following biological levels of iron exposure, other occupational-based studies have evaluated the effect of excess iron exposure, via iron fumes generated by welding activities or iron dust from iron and steel production, as a risk factor for PD and parkinsonism (Gorell et al. 1997, 1999a, b; Rybicki et al. 1999). However, many of these studies were unable to demonstrate iron exposure, on its own, as a factor that underlies PD etiology. Rather, iron exposure that occurred in combination with other metals, including lead or copper, seemed to implicate iron as a risk factor. Interestingly, in many of these studies, copper alone was significantly associated with risk of PD, while iron + copper seemed to elevate this risk. While the occupational data may still be controversial, a more concise discussion of the role of iron in PD can be gleaned from pathological and imaging data. Indeed, a variety of studies have found significantly elevated levels of iron in the putamen and SNpc of PD patients, placing iron at the point source of PD pathology (Gerlach et al. 2006; Hare and Double 2016; Oakley et al. 2007; Sofic et al. 1988). Whether these excess levels arise from an explicit increase in occupational exposure to iron is unclear, but evidence has shown that alteration in intracellular iron homeostasis in the brain and dopamine neurons may underlie iron accumulation and pathology in the basal ganglia. Several proteins are involved in mediating the intracellular dynamics of iron. Iron is predominantly transported into the brain and neurons by the divalent metal transporter 1 (DMT1) as well as through binding to transferrin, which is then trafficked across the neuronal membrane by the transferrin receptor. Once inside the neuron, iron can be stored by ferritin, which regulates the levels of free iron in the cytoplasm and reduces the ability of iron to generate reactive species (Honarmand Ebrahimi et al. 2015; Moos et al. 2007). Thus, alterations at various points in iron homeostasis could underlie iron-mediated neurotoxicity in dopamine neurons. Evidence to support these ideas has shown elevations in iron in the substantia nigra of PD patients that have increased expression of DMT1 and transferrin receptors on dopamine neurons. Additionally, reductions in ferritin also result in an increase of free cytosolic iron. In contrast to these findings, a polymorphism in the transferrin receptor that causes a reduction in its activity has been shown to limit the amount of iron transported into dopamine neurons, serving as a protective mechanism in PD (Dexter et al. 1991; Rhodes et al. 2014).

Dopamine neurons, as well as noradrenergic neurons in the locus coeruleus, have additional means to regulate cytosolic iron levels through the sequestration by neuromelanin (NM). Neuromelanin is a dark-colored pigment that is synthesized from the breakdown products of dopamine and other catecholamines in the midbrain (Sulzer et al. 2000). While the physiological role of NM is still being debated, evidence suggests that it acts as a "sink" for a variety of potentially neurotoxic exogenous and endogenous compounds in dopamine and norepinephrine neurons (Zecca et al. 2002). However, under cellular distress or damage, NM may become detrimental to the neuron by releasing its neurotoxic cargo back into the cell and into the extracellular environment. Additionally, it has been suggested that NM can become overwhelmed or saturated with such species, including iron, causing it to release excess amounts into the cell (Zucca et al. 2004).

The neurotoxicity of elevated iron in dopamine neurons is focused on the ability of cytosolic iron to catalyze the formation of reactive oxygen species, including hydroxyl radicals, through the Fenton reaction. Furthermore, iron in the cytosol can also interact with dopamine to metabolize dopamine into neurotoxic dopamine quinones and other neurotoxic species. These neurotoxic species are highly reactive and can interact with various intracellular components in the dopamine neuron, including DNA, membrane lipids, and proteins, leading to their dysfunction and decrement of the dopamine neuron (Hare and Double 2016). Moreover, iron may also participate in the formation of neurotoxic accumulations of the PD-relevant protein, alpha-synuclein (Uversky et al. 2001). Although the specific function of alpha-synuclein is still under investigation, it is clear that it plays a critical role in synaptic function in dopamine neurons. A key pathological feature of PD is accumulation of neurotoxic alpha-synuclein aggregates in dopamine neurons (Dickson 2012). Although the precise pathway that mediates the formation of these inclusions is vague, extensive work has described the interaction between intracellular metals and alpha-synuclein in accelerating its pathological misfolding into neurotoxic species (Carboni and

Lingor 2015; Lu et al. 2011). Such an interaction may be critical in delineating the role iron exposure plays in parkinsonism. While elevated iron levels and reduced ferritin are seen in the substantia nigra of PD patients, similar alterations in iron handling are not observed in the substantia nigra of patients with MSA. Although both PD and MSA display significant damage to the dopamine system and are pathologically defined by alpha-synuclein accumulations, the localization of these inclusions in these two disorders diverges. While they are exclusively seen in dopamine neurons in the SNpc of PD patients, neuronal inclusions are rarely seen in MSA, instead collecting in oligodendrocytes (Lantos 1998). Thus, these findings could give critical insight into the pathological mechanisms related to PD and MSA and the environmental contribution of iron exposure to each disorder.

Copper Exposure and Parkinsonism

Similar to iron, copper is a critical metal element for several biological functions in the human body. Copper serves as a cofactor for the antioxidant copper/zinc superoxide dismutase (Cu/ZnSOD), which functions to metabolize the reactive oxygen species, superoxide to limit its potential interaction with intracellular targets. Copper is also involved in neurotransmitter synthesis, specifically through its interaction with dopamine beta-hydroxylase, a key enzyme in the synthesis of norepinephrine from dopamine (Harris 2000). Like iron, exposure to elevated levels of copper can also occur in occupational settings, including mining. Indeed, epidemiological evidence supports such exposures as risk factors for the development of PD. Studies performed by Gorell et al. have evaluated workers occupationally exposed to elevated levels of copper over multiple decades. From these studies it was found that occupational exposure for greater than 20 years resulted in a 2.5-fold increased risk for PD. Interestingly, when copper was assessed in the context of combined exposure with other metals, including lead or iron, the risk increased to 5.3- and 3.7-fold, respectively (Gorell et al. 1997, 1999a).

Under physiological conditions copper is bound to ceruloplasmin in the blood. When copper is unbound, it can be transported across biological membranes, including the blood-brain barrier and neuronal membranes via the copper transporter 1 (CTR1) (Hellman and Gitlin 2002). Once inside the neuron, intracellular levels of copper are tightly regulated by additional transporters ATP7A and ATP7B, which function to efflux excess copper from the cell (Hellman and Gitlin 2002). Each of these transport mechanisms is highly expressed in the substantia nigra and targeted for dysfunction, leading to alterations in copper homeostasis. Indeed, Wilson disease is defined by an excess accumulation of copper in the brain and damage to the basal ganglia following a reduction in the expression and function of ATP7A (Bandmann et al. 2015).

In light of these findings, the contribution of copper to dopaminergic pathology is complex, with both elevations and reductions in copper concentrations suggested to contribute to pathogenesis of dopamine neurons. From the context of excess accumulation of copper in the substantia nigra, it has been suggested that copper participates in the formation of reactive species, such as hydroxyl radicals, through the Fenton reaction, which can subsequently damage the dopamine neuron (Oder et al. 1993; Barbeau and Friesen 1970; Hitoshi et al. 1991; Barthel et al. 2003). While this certainly provides a possible mechanism, extensive work has also focused on the interaction of copper with alpha-synuclein and its contribution to conformational changes and protein misfolding leading to the acceleration of neurotoxic alpha-synuclein fibrils (Carboni and Lingor 2015; Dell'Acqua et al. 2016; Valensin et al. 2016). Interestingly, phosphorylation of alpha-synuclein, specifically at serine 129, has been shown to increase the binding affinity of copper with alpha-synuclein and further increase the neurotoxic accumulations of the protein (Lu et al. 2011). While alpha-synuclein has received the greatest attention, DJ-1, another PD-relevant protein, has also been shown to interact with copper (Bjorkblom et al. 2013; Girotto et al. 2014). Unlike alpha-synuclein, DJ-1 appears to serve as a copper-binding protein that participates in additional copper homeostasis. Indeed, mutation of specific residues abolishes this metal-binding function and increases copper-induced neurotoxicity. Interestingly, such an interaction with DJ-1 is independent of its endogenous antioxidant functions.

While each of these studies provides strong evidence for a mechanistic pathway leading to copper-induced dopaminergic neurotoxicity, it is predicated on the idea of excess copper, either due to elevated exposure or dysfunction in proteins that regulate intracellular levels of copper. However, to date, elevated tissue levels of copper have not been observed in PD. In contrast, copper has been found to be reduced or unchanged in both the substantia nigra and serum of patients with PD (Davies et al. 2014, 2016; Montes et al. 2014; Torsdottir et al. 1999, 2006). Additionally, alterations in the expression or function of copper-handling proteins have not been previously associated with PD incidence. This evidence seems to suggest another potential mechanism, by which copper could participate in neurotoxicity. As discussed above, copper is necessary for the enzymatic activity of the antioxidant Cu/ZnSOD, which functions to degrade superoxide that is generated in the neuron. Thus, a reduction in copper in the substantia nigra may increase the vulnerability of these dopamine neurons to oxidative damage that is constantly taking place through the normal biosynthesis and metabolic processes in the dopamine neuron. Moreover, as the transport of iron is tightly mediated by copper transport, a reduction in copper could elicit an accelerated transport of iron into the dopamine neurons (Ayton et al. 2013). In contrast to copper, iron is found to be elevated in the substantia nigra of PD patients (Jin et al. 2011).

Manganese Exposure and Parkinsonism

By far, one of the more interesting discussions related to metal toxicity and parkinsonism relates to the contribution excess exposure to manganese makes to basal ganglia pathology and clinical manifestations related to this pathology. As with other metals already discussed, manganese is an essential cofactor for several enzymes, including superoxide dismutase (SOD), and plays a role in the synthesis and metabolism of neurotransmitters (Schroeder et al. 1966; Hurley et al. 1984; Golub et al. 2005). Similarly, manganese can have detrimental effects on this system through its accumulation and generation of reactive species among other mechanisms (Graham et al. 1978; Cohen 1984). At the root of this argument is whether or not excess manganese exposure results in damage to the nigrostriatal dopamine system similarly to that seen in idiopathic PD, suggesting that it is a possible causative environmental risk factor for PD, or whether such exposures generate a pathologically distinct parkinsonian syndrome, usually referred to as manganism. Work in recent years has addressed these concerns using a spectrum of epidemiological and lab-based studies to delineate the key pathological and clinical features of excess manganese exposure and contrast them with those seen in PD.

Manganism was originally described in 1837 by Dr. James Couper following his examination of patients who had been exposed to excess amount of manganese through the mining of manganese ore. In his clinical assessment, Dr. Couper noted extensive neurological deficits that initially manifested as deficits in neuropsychiatric and cognitive endpoints. Only after these symptoms were expressed did the more familiar motoric dysfunction so often associated with PD, including bradykinesia, tremor, and cogwheel rigidity, present (Gibbs et al. 1999; Huang et al. 1993). Interestingly, excess manganese deposits prominently in the basal ganglia, specifically within the globus pallidus, which is enriched in GABAergic neurons (Erikson and Aschner 2006; Bouabid et al. 2015; Kwakye et al. 2015).

Exposure to manganese is still a critical concern in occupational settings, as elevated exposures can most often occur via mining activities, steel manufacturing, and the inhalation of welding fumes (Hudnell 1999; Huang et al. 1989). While numerous studies have been conducted to address the neurological impacts of manganese exposure in these settings, its role in PD etiology is still controversial, with some suggesting a definitive association and others unable to equate exposure and disease (Santamaria et al. 2007). Similar to iron and copper, manganese is easily transported across biological membranes by the DMT1 as well as transferrin and transferrin receptor. In addition to these transporters, intracellular regulation of manganese is mediated by the SLC30A10 transporter which is critical to maintaining manganese homeostasis (DeWitt et al. 2013). Indeed, studies have found mutation in the SLC30A10 transporter that results in a significant reduction in expression causes an excess buildup of manganese in the basal ganglia, leading to parkinsonism (Quadri et al. 2012). Of note, while patients with this mutation demonstrate clinical manifestations of parkinsonism, this does not appear to be due to loss of the dopamine terminal in the striatum or damage to other aspects of dopaminergic function (Olanow et al. 1996; Shinotoh et al. 1995; Pal et al. 1999; Olanow 2004). This lack of dopamine terminal pathology may explain the lack of response to dopamine replacement with L-DOPA in these patients, suggesting the motor alterations associated with manganese accumulation are independent of dopaminergic losses.

In an effort to better define and delineate manganism, extensive characterization of the pathological and clinical signs has been undertaken. Using a variety of experimental models, including human subjects that have been exposed to manganese and nonhuman primate imaging data, a clearer picture has begun to emerge. As discussed in previous sections, the pathological manifestations of idiopathic PD are well-established, showing extensive damage within the nigrostriatal dopamine system. While this pathology is often highlighted by severe losses in dopamine neurons in the SNpc and accompanied by reduction in dopamine terminals and dopamine content in the striatum, additional pathological features, including loss of VMAT2 and increased D2 receptor expression, provide further evidence for alterations in the integrity of the pre- and postsynaptic dopamine landscape. Subsequent to such dramatic dysfunction of the dopamine circuit, treatment with L-DOPA, which provides dopamine replacement, is key to restoring motor function to PD patients (Dickson 2012; Fahn 2003).

In stark contrast to these alterations, the nigrostriatal dopamine system is relatively spared in patients with elevated manganese exposure (Bouabid et al. 2015; Kwakye et al. 2015; Pal et al. 1999; Guilarte 2010, 2013; Perl and Olanow 2007). Indeed, for the most part, dopamine terminals appear to be intact, showing normal expression and function of DAT and VMAT2, unchanged striatal dopamine content, and a slight reduction in D2 receptors. Perhaps a more telling indictment of the effect manganese has on the dopamine system is the lack of response to L-DOPA, which, again, tends to be used to highlight clinical symptoms that emerge from alterations to the nigrostriatal dopamine system. In light of a paucity of overt dopaminergic pathology, dopamine neurotransmission within this circuit may still be dysfunctional (Guilarte and Gonzales 2015). Work from nonhuman primates exposed to manganese has demonstrated a significant reduction in dopamine release from the presynaptic terminals in the striatum (Guilarte et al. 2006). Although tissue content of dopamine may be unchanged, the ability of the dopamine neuron to release it and utilize it may be compromised through a yet-to-be discovered pathway. Indeed, while critical to the function of the dopamine circuit, dopamine transporters (DAT and VMAT2) comprise a very small sample of proteins involved in mediating normal dopamine neurotransmission. Thus, the alterations in this function could be occurring through a variety of intracellular cascades in the dopamine terminal that remain to be identified.

Further evidence for the delineation between PD and manganism resides in the presence and localization of alpha-synuclein or tau inclusions within the CNS. To date, few studies have evaluated alpha-synuclein expression following manganese exposure in human patients. However, studies using rodent and nonhuman primate models have identified an increase in the expression of alpha-synuclein oligomers in organotypic brain slices acutely treated with manganese. Moreover, alpha-synuclein inclusions were observed in neurons as well as glia in the frontal cortex of nonhuman primates treated with manganese (Cai et al. 2010; Verina et al. 2013; Xu et al. 2014). Interestingly, these findings seem to follow pathological signs routinely observed in MSA and that are used as to delineate MSA from PD. While the impact of these inclusions is still being formulated, current work has found under circumstances of elevated manganese exposure, manganese can facilitate the formation of alpha-synuclein aggregates, suggesting a possible mechanism of action for manganese in the neurotoxicity.

Emerging Metal Exposures and Neurotoxicological Concerns

Metal Nanoparticles and Parkinsonism

The manufacture and use of metal-containing nanoparticles has significantly increased over the last several decades and seemingly integrated into various aspects of our daily lives. Consumer products, such as clothing and cosmetics, structure materials used in building, as well as biomedical imaging and drug delivery have all found extensive use for nanoparticles. Nanoparticles vary in size from 1 to 100 nm and can be covered with metallic coatings, ranging from titanium (Ti), aluminum (Al), iron (Fe), manganese (Mn), copper, (Cu), and gold (Au), among others (Win-Shwe and Fujimaki 2011). Because of their small size, they are able to easily move across biological membranes and deposit in tissue. Indeed, this property makes them ideally suited for therapeutic approaches that target the CNS, which would not otherwise be able to access the brain via conventional delivery systems. While these capabilities have provided new and exciting opportunities, evaluation of the health effects and, more specifically, the neurotoxicological impact of these compounds have not kept pace (Feng et al. 2015; Heusinkveld et al. 2016; Oberdorster et al. 2009).

Indeed, a critical area of research related to nanoparticles is health effects arising from occupational exposures during the manufacturing process. Unfortunately, the use of nanoparticles is still relatively new, and the identification of a highly exposed cohort does not currently exist. However, our understanding of the potential neurotoxic effects of metal nanoparticles has been significantly enhanced through in vitro and in vivo laboratory models. Indeed, given their size, nanoparticles are quickly taken up by the olfactory bulb and transported to the CNS by way of the olfactory nerve, following inhalational exposure. Via this route, nanoparticles have been shown to deposit throughout the brain, including the frontal cortex, striatum, hippocampus, and cerebellum (Elder et al. 2006; Imam et al. 2015). Although inhalational exposure represents the major route of access to the brain, nanoparticles can also be ingested or absorbed across the skin, making their way into the general circulation, and then transported across the blood-brain barrier. Similar to inhalational exposure, ingested nanoparticles have been found to accumulate in specific brain regions. And aligning with our discussion of general metal transport, nanoparticles appear to use redundant mechanisms, including transferrin and the transferrin receptor to gain access to the brain tissue. Once in the brain, nanoparticles are able to access a variety of neural cells, including neurons, astrocytes, and microglia. These targets provide a platform for interesting discussions related to the possible neurotoxic mechanisms, including the generation of reactive oxygen species and neuroinflammation that may underlie nanoparticle interactions with the brain.

Our current understanding of the impact metal nanoparticles may have on the human nervous system has extensively focused on a select group of molecules that utilize metals in the form of iron oxide, manganese oxide, or titanium dioxide for their function. Of these, the neurotoxicity of iron oxide nanoparticles has been established. These nanoparticles are most commonly found in biomedical applications related to brain imaging as well as drug delivery of therapeutic compounds. The iron oxide coating allows them to bind to transferrin and be easily trafficked across the blood-brain barrier via the transferrin receptor. Once inside a biological tissue, these polymers can lose their iron coating, leading to accumulation of iron in the brain. Following an inhalational exposure, these nanoparticles were found to extensively accumulate in the olfactory bulb, hippocampus, striatum, and cortex (Imam et al. 2015). Interestingly, while the striatum is a critical part of the basal ganglia and is enriched in dopaminergic projections from the SNpc, alteration to olfactory function is appreciated as one of the earliest clinical indicators of PD. Thus, these findings suggest that inhalation of metal nanoparticles can deposit in brain regions associated with PD pathology. Additional studies highlighted iron oxide-induced reduction in dopamine using both in vitro and in vivo models of exposure (Imam et al. 2015; Wu et al. 2013). While the mechanisms related to these deficits are not clear, the same groups also identified an increase in reactive oxygen species, in addition to elevations in alpha-synuclein. Another potential neurotoxic mechanism, by which metal nanoparticles could induce damage, is via the activation of neuroinflammation. Findings from a study using inhalational exposure to manganese oxide nanoparticles found the greatest deposition of manganese oxide in the olfactory bulbs and the striatum. In these same brain regions, investigators recorded elevations in markers or neuroinflammation, including glial fibrillary-associated protein and tumor necrosis factor-alpha (Elder et al. 2006). As in many neurodegenerative disorders, inflammation plays a critical role in the pathogenesis of PD. While it has proven difficult to define inflammation as a cause or consequence of dopaminergic neurodegeneration, it is clear that neuroinflammation can participate in both sides of this neurotoxicological equation, resulting in a cyclical, self-propagating cascade that leaves a persistent inflammatory mark on PD (McGeer and McGeer 2004).

E-Waste, Metal Exposure, and Parkinsonism

An emerging health risk that has extensive relevance to our discussion of metal-induced parkinsonism is the contribution of occupational exposure to metals through the recycling or reclamation of electronic waste (E-waste) (Breivik et al. 2014; Ogunseitan et al. 2009; Heacock et al. 2016). In general, E-waste can be simply defined as discarded electronic equipment, including computers, televisions, copiers, cell phones, circuit boards, and semiconductor chips, among other unwanted electronic products. A critical health concern arises when it is appreciated that these products contain a variety of heavy metals, including iron, manganese, copper, and cadmium that, as discussed, can enact severe neurological deficits (Luo et al. 2011; Tsydenova and Bengtsson 2011; Xue et al. 2012). Moreover, the neurological concerns are amplified when the conditions under which these metals are extracted may not endorse health of the workers that have direct inhalational and dermal interactions with these materials. While E-waste disposal and recycling does occur domestically, a vast majority of these products are transported, globally, most often to Africa, Asia, and South America. In many of these settings, there is a lack of environmental health infrastructure in place to instill the appropriate policies and regulations necessary to ensure worker safety and reduce exposure to neurotoxic metals (LaDou and Lovegrove 2008; Leung et al. 2008; Zhang et al. 2012).

With these issues in mind, a critical gap exists in our understanding of the neurotoxicological issues that may arise in workers that are involved in E-waste. In this context, we are missing important exposure assessments of both the work environment and the workers themselves to gain a better understanding of the metals that are being exposed to, the concentrations they are being exposed at, and the potential body burdens of these compounds. While some data does exist that provides evidence that E-waste workers are being exposed to excess levels of metals, these data are minimal and do not provide a comprehensive picture of the exposure landscape (Asante et al. 2012; Julander et al. 2014). Further assessment that needs to evaluate the possible neurotoxicological effects of these exposures has also not been performed. While it is easy to present these shortcomings and resolutions in a simplistic manner, such approaches are far from straightforward, as several considerations need to be appreciated. For example, similar to other occupational settings, workers are not exposed to just one metal. Rather, their exposures most likely represent a mixture of metals. Thus, it becomes necessary to evaluate how these metals may interact to elicit neurological impacts and delineate the biological pathways that may underlie neurotoxicity. Related to this, a variety of other neurotoxic compounds are also part of E-waste, including several persistent organic pollutants, such as brominated flame retardants and dioxins, among others. Teasing out the relative contributions of these other compounds to neurotoxic endpoints will also be critical to elaborating our understanding of metal-mediated neurotoxicity in E-waste workers.

Conclusion

Although the contribution of metal exposure to parkinsonism has been appreciated for decades, our understanding of the various occupational settings of exposure as well as more specific pathological and clinical outcomes has allowed for an enriched discussion of these topics. Significant progress has been made in delineating the molecular targets and cascades of metal exposure that facilitate neurotoxicity in the basal ganglia. This data can now be integrated with epidemiological data being generated from emerging exposure scenarios, such as metal nanoparticles and metals in E-waste to elaborate the landscape of metal neurotoxicity and parkinsonism.

References

Asante KA, Agusa T, Biney CA, Agyekum WA, Bello M, Otsuka M, et al. Multi-trace element levels and arsenic speciation in urine of e-waste recycling workers from Agbogbloshie, Accra in Ghana. Sci Total Environ. 2012;424:63–73.

Ayton S, Lei P, Duce JA, Wong BX, Sedjahtera A, Adlard PA, et al. Ceruloplasmin dysfunction and therapeutic potential for Parkinson disease. Ann Neurol. 2013;73(4):554–9.

Bandmann O, Weiss KH, Kaler SG. Wilson's disease and other neurological copper disorders. Lancet Neurol. 2015;14(1):103–13.

Barbeau A, Friesen H. Treatment of Wilson's disease with L-dopa after failure with penicillamine. Lancet. 1970;1(7657):1180–1.

Barthel H, Hermann W, Kluge R, Hesse S, Collingridge DR, Wagner A, et al. Concordant pre- and postsynaptic deficits of dopaminergic neurotransmission in neurologic Wilson disease. AJNR Am J Neuroradiol. 2003;24(2):234–8.

Bjorkblom B, Adilbayeva A, Maple-Grodem J, Piston D, Okvist M, Xu XM, et al. Parkinson disease protein DJ-1 binds metals and protects against metal-induced cytotoxicity. J Biol Chem. 2013;288(31):22809–20.

Bouabid S, Tinakoua A, Lakhdar-Ghazal N, Benazzouz A. Manganese neurotoxicity: behavioral disorders associated with dysfunctions in the basal ganglia and neurochemical transmission. J Neurochem. 2015;

Breivik K, Armitage JM, Wania F, Jones KC. Tracking the global generation and exports of e-waste. do existing estimates add up? Environ Sci Technol. 2014;48(15):8735–43.

Cai T, Yao T, Zheng G, Chen Y, Du K, Cao Y, et al. Manganese induces the overexpression of alpha-synuclein in PC12 cells via ERK activation. Brain Res. 2010;1359:201–7.

Carboni E, Lingor P. Insights on the interaction of alpha-synuclein and metals in the pathophysiology of Parkinson's disease. Metallomics Integ Biom Sci. 2015;7(3):395–404.

Caudle WM. Occupational exposures and parkinsonism. Handb Clin Neurol. 2015;131:225–39.

Caudle WM, Guillot TS, Lazo CR, Miller GW. Industrial toxicants and Parkinson's disease. Neurotoxicology. 2012;33(2):178–88.

Cohen G. Oxy-radical toxicity in catecholamine neurons. Neurotoxicology. 1984;5(1):77–82.

Davies KM, Bohic S, Carmona A, Ortega R, Cottam V, Hare DJ, et al. Copper pathology in vulnerable brain regions in Parkinson's disease. Neurobiol Aging. 2014;35(4):858–66.

Davies KM, Mercer JF, Chen N, Double KL. Copper dyshomoeostasis in Parkinson's disease: implications for pathogenesis and indications for novel therapeutics. Clin Sci. 2016;130(8):565–74.

Dell'Acqua S, Pirota V, Monzani E, Camponeschi F, De Ricco R, Valensin D, et al. Copper(I) forms a redox-stable 1:2 complex with alpha-Synuclein N-terminal peptide in a membrane-like environment. Inorg Chem. 2016;55(12):6100–6.

DeWitt MR, Chen P, Aschner M. Manganese efflux in parkinsonism: insights from newly characterized SLC30A10 mutations. Biochem Biophys Res Commun. 2013;432(1):1–4.

Dexter DT, Carayon A, Javoy-Agid F, Agid Y, Wells FR, Daniel SE, et al. Alterations in the levels of iron, ferritin and other trace metals in Parkinson's disease and other neurodegenerative diseases affecting the basal ganglia. Brain J Neurol. 1991;114(Pt 4):1953–75.

Dickson DW. Parkinson's disease and parkinsonism: neuropathology. Cold Spring Harb Perspect Med. 2012;2(8)

Dickson DW, Rademakers R, Hutton ML. Progressive supranuclear palsy: pathology and genetics. Brain Pathol. 2007;17(1):74–82.

Elder A, Gelein R, Silva V, Feikert T, Opanashuk L, Carter J, et al. Translocation of inhaled ultrafine manganese oxide particles to the central nervous system. Environ Health Perspect. 2006;114(8):1172–8.

Eriksson KM, Aschner M. Increased manganese uptake by primary astrocyte cultures with altered iron status is mediated primarily by divalent metal transporter. Neurotoxicology. 2006;27(1):125–30.

Fahn S. Description of Parkinson's disease as a clinical syndrome. Ann N Y Acad Sci. 2003;991:1–14.

Feng X, Chen A, Zhang Y, Wang J, Shao L, Wei L. Central nervous system toxicity of metallic nanoparticles. Int J Nanomedicine. 2015;10:4321–40.

Gerlach M, Double KL, Youdim MB, Riederer P. Potential sources of increased iron in the substantia nigra of parkinsonian patients. J Neural Transm Suppl. 2006;70:133–42.

Gibbs JP, Crump KS, Houck DP, Warren PA, Mosley WS. Focused medical surveillance: a search for subclinical movement disorders in a cohort of U.S. workers exposed to low levels of manganese dust. Neurotoxicology. 1999;20(2–3):299–313.

Gilman S, Low PA, Quinn N, Albanese A, Ben-Shlomo Y, Fowler CJ, et al. Consensus statement on the diagnosis of multiple system atrophy. J Neurol Sci. 1999;163(1):94–8.

Girotto S, Cendron L, Bisaglia M, Tessari I, Mammi S, Zanotti G, et al. DJ-1 is a copper chaperone acting on SOD1 activation. J Biol Chem. 2014;289(15):10887–99.

Golub MS, Hogrefe CE, Germann SL, Tran TT, Beard JL, Crinella FM, et al. Neurobehavioral evaluation of rhesus monkey infants fed cow's milk formula, soy formula, or soy formula with added manganese. Neurotoxicol Teratol. 2005;27(4):615–27.

Gomes CM, Santos R. Neurodegeneration in Friedreich's ataxia: from defective frataxin to oxidative stress. Oxidative Med Cell Longev. 2013;2013:487534.

Gorell JM, Johnson CC, Rybicki BA, Peterson EL, Kortsha GX, Brown GG, et al. Occupational exposures to metals as risk factors for Parkinson's disease. Neurology. 1997;48(3):650–8.

Gorell JM, Johnson CC, Rybicki BA, Peterson EL, Kortsha GX, Brown GG, et al. Occupational exposure to manganese, copper, lead, iron, mercury and zinc and the risk of Parkinson's disease. Neurotoxicology. 1999a;20(2–3):239–47.

Gorell JM, Rybicki BA, Cole Johnson C, Peterson EL. Occupational metal exposures and the risk of Parkinson's disease. Neuroepidemiology. 1999b;18(6):303–8.

Graham DG, Tiffany SM, Bell WR Jr, Gutknecht WF. Autoxidation versus covalent binding of quinones as the mechanism of toxicity of dopamine, 6-hydroxydopamine, and related compounds toward C1300 neuroblastoma cells in vitro. Mol Pharmacol. 1978;14(4):644–53.

Guilarte TR. Manganese and Parkinson's disease: a critical review and new findings. Environ Health Perspect. 2010;118(8):1071–80.

Guilarte TR. Manganese neurotoxicity: new perspectives from behavioral, neuroimaging, and neuropathological studies in humans and non-human primates. Front Aging Neurosci. 2013;5:23.

Guilarte TR, Gonzales KK. Manganese-induced parkinsonism is not idiopathic Parkinson's disease: environmental and genetic evidence. Toxicolog Sci Off J Soc Toxicol. 2015;146(2):204–12.

Guilarte TR, Chen MK, McGlothan JL, Verina T, Wong DF, Zhou Y, et al. Nigrostriatal dopamine system dysfunction and subtle motor deficits in manganese-exposed non-human primates. Exp Neurol. 2006;202(2):381–90.

Hare DJ, Double KL. Iron and dopamine: a toxic couple. Brain J Neurol. 2016;139(Pt 4):1026–35.

Harris ED. Cellular copper transport and metabolism. Annu Rev Nutr. 2000;20:291–310.

Hatcher JM, Pennell KD, Miller GW. Parkinson's disease and pesticides: a toxicological perspective. Trends Pharmacol Sci. 2008;29(6):322–9.

Heacock M, Kelly CB, Asante KA, Birnbaum LS, Bergman AL, Brune MN, et al. E-waste and harm to vulnerable populations: a growing global problem. Environ Health Perspect 2016;124(5):550–5.

Hellman NE, Gitlin JD. Ceruloplasmin metabolism and function. Annu Rev Nutr. 2002;22:439–58.

Heusinkveld HJ, Wahle T, Campbell A, Westerink RH, Tran L, Johnston H, et al. Neurodegenerative and neurological disorders by small inhaled particles. Neurotoxicology. 2016;56:94–106.

Hitoshi S, Iwata M, Yoshikawa K. Mid-brain pathology of Wilson's disease: MRI analysis of three cases. J Neurol Neurosurg Psychiatry. 1991;54(7):624–6.

Honarmand Ebrahimi K, Hagedoorn PL, Hagen WR. Unity in the biochemistry of the iron-storage proteins ferritin and bacterioferritin. Chem Rev. 2015;115(1):295–326.

Huang CC, Chu NS, Lu CS, Wang JD, Tsai JL, Tzeng JL, et al. Chronic manganese intoxication. Arch Neurol. 1989;46(10):1104–6.

Huang CC, Lu CS, Chu NS, Hochberg F, Lilienfeld D, Olanow W, et al. Progression after chronic manganese exposure. Neurology. 1993;43(8):1479–83.

Hudnell HK. Effects from environmental Mn exposures: a review of the evidence from non-occupational exposure studies. Neurotoxicology. 1999;20(2–3):379–97.

Hurley LS, Keen CL, Baly DL. Manganese deficiency and toxicity: effects on carbohydrate metabolism in the rat. Neurotoxicology. 1984;5(1):97–104.

Imam SZ, Lantz-McPeak SM, Cuevas E, Rosas-Hernandez H, Liachenko S, Zhang Y, et al. Iron oxide nanoparticles induce dopaminergic damage: in vitro pathways and in vivo imaging reveals mechanism of neuronal damage. Mol Neurobiol. 2015;52(2):913–26.

Jin L, Wang J, Zhao L, Jin H, Fei G, Zhang Y, et al. Decreased serum ceruloplasmin levels characteristically aggravate nigral iron deposition in Parkinson's disease. Brain J Neurol. 2011;134(Pt 1):50–8.

Julander A, Lundgren L, Skare L, Grander M, Palm B, Vahter M, et al. Formal recycling of e-waste leads to increased exposure to toxic metals: an occupational exposure study from Sweden. Environ Int. 2014;73:243–51.

Kwakye GF, Paoliello MM, Mukhopadhyay S, Bowman AB, Aschner M. Manganese-induced parkinsonism and Parkinson's disease: shared and distinguishable features. Int J Environ Res Public Health. 2015;12(7):7519–40.

LaDou J, Lovegrove S. Export of electronics equipment waste. Int J Occup Environ Health. 2008;14(1):1–10.

Langston JW. The Parkinson's complex: parkinsonism is just the tip of the iceberg. Ann Neurol. 2006;59(4):591–6.

Lantos PL. The definition of multiple system atrophy: a review of recent developments. J Neuropathol Exp Neurol. 1998;57(12):1099–111.

Leung AO, Duzgoren-Aydin NS, Cheung KC, Wong MH. Heavy metals concentrations of surface dust from e-waste recycling and its human health implications in southeast China. Environ Sci Technol. 2008;42(7):2674–80.

Lu Y, Prudent M, Fauvet B, Lashuel HA, Girault HH. Phosphorylation of alpha-Synuclein at Y125 and S129 alters its metal binding properties: implications for understanding the role of alpha-Synuclein in the pathogenesis of Parkinson's disease and related disorders. ACS Chem Neurosci. 2011;2(11):667–75.

Luo C, Liu C, Wang Y, Liu X, Li F, Zhang G, et al. Heavy metal contamination in soils and vegetables near an e-waste processing site, South China. J Hazard Mater. 2011;186(1):481–90.

McGeer PL, McGeer EG. Inflammation and neurodegeneration in Parkinson's disease. Parkinsonism Relat Disord. 2004;10(Suppl 1):S3–7.

Montes S, Rivera-Mancia S, Diaz-Ruiz A, Tristan-Lopez L, Rios C. Copper and copper proteins in Parkinson's disease. Oxidative Med Cell Longev. 2014;2014:147251.

Moos T, Rosengren Nielsen T, Skjorringe T, Morgan EH. Iron trafficking inside the brain. J Neurochem. 2007;103(5):1730–40.

Oakley AE, Collingwood JF, Dobson J, Love G, Perrott HR, Edwardson JA, et al. Individual dopaminergic neurons show raised iron levels in Parkinson disease. Neurology. 2007;68(21):1820–5.

Oberdorster G, Elder A, Rinderknecht A. Nanoparticles and the brain: cause for concern? J Nanosci Nanotechnol. 2009;9(8):4996–5007.

Oder W, Prayer L, Grimm G, Spatt J, Ferenci P, Kollegger H, et al. Wilson's disease: evidence of subgroups derived from clinical findings and brain lesions. Neurology. 1993;43(1):120–4.

Ogunseitan OA, Schoenung JM, Saphores JD, Shapiro AA. Science and regulation. The electronics revolution: from e-wonderland to e-wasteland. Science. 2009;326(5953):670–1.

Olanow CW. Manganese-induced parkinsonism and Parkinson's disease. Ann N Y Acad Sci. 2004;1012:209–23.

Olanow CW, Good PF, Shinotoh H, Hewitt KA, Vingerhoets F, Snow BJ, et al. Manganese intoxication in the rhesus monkey: a clinical, imaging, pathologic, and biochemical study. Neurology. 1996;46(2):492–8.

Pal PK, Samii A, Calne DB. Manganese neurotoxicity: a review of clinical features, imaging and pathology. Neurotoxicology. 1999;20(2–3):227–38.

Perl DP, Olanow CW. The neuropathology of manganese-induced parkinsonism. J Neuropathol Exp Neurol. 2007;66(8):675–82.

Quadri M, Federico A, Zhao T, Breedveld GJ, Battisti C, Delnooz C, et al. Mutations in SLC30A10 cause parkinsonism and dystonia with hypermanganesemia, polycythemia, and chronic liver disease. Am J Hum Genet. 2012;90(3):467–77.

Rhodes SL, Buchanan DD, Ahmed I, Taylor KD, Loriot MA, Sinsheimer JS, et al. Pooled analysis of iron-related genes in Parkinson's disease: association with transferrin. Neurobiol Dis. 2014;62:172–8.

Rybicki BA, Johnson CC, Peterson EL, Kortsha GX, Gorell JM. A family history of Parkinson's disease and its effect on other PD risk factors. Neuroepidemiology. 1999;18(5):270–8.

Santamaria AB, Cushing CA, Antonini JM, Finley BL, Mowat FS. State-of-the-science review: does manganese exposure during welding pose a neurological risk? J Toxicol Environ Health. 2007;10(6):417–65.

Schroeder HA, Balassa JJ, Tipton IH. Essential trace metals in man: manganese. A study in homeostasis. J Chronic Dis. 1966;19(5):545–71.

Shinotoh H, Snow BJ, Hewitt KA, Pate BD, Doudet D, Nugent R, et al. MRI and PET studies of manganese-intoxicated monkeys. Neurology. 1995;45(6):1199–204.

Sofic E, Riederer P, Heinsen H, Beckmann H, Reynolds GP, Hebenstreit G, et al. Increased iron (III) and total iron content in post mortem substantia nigra of parkinsonian brain. J Neural Transm. 1988;74(3):199–205.

Sulzer D, Bogulavsky J, Larsen KE, Behr G, Karatekin E, Kleinman MH, et al. Neuromelanin biosynthesis is driven by excess cytosolic catecholamines not accumulated by synaptic vesicles. Proc Natl Acad Sci U S A. 2000;97(22):11869–74.

Torsdottir G, Kristinsson J, Sveinbjornsdottir S, Snaedal J, Johannesson T. Copper, ceruloplasmin, superoxide dismutase and iron parameters in Parkinson's disease. Pharmacol Toxicol. 1999;85(5):239–43.

Torsdottir G, Sveinbjornsdottir S, Kristinsson J, Snaedal J, Johannesson T. Ceruloplasmin and superoxide dismutase (SOD1) in Parkinson's disease: a follow-up study. J Neurol Sci. 2006;241(1–2):53–8.

Tsydenova O, Bengtsson M. Chemical hazards associated with treatment of waste electrical and electronic equipment. Waste Manag. 2011;31(1):45–58.

Uversky VN, Li J, Fink AL. Metal-triggered structural transformations, aggregation, and fibrillation of human alpha-synuclein. A possible molecular NK between Parkinson's disease and heavy metal exposure. J Biol Chem. 2001;276(47):44284–96.

Valensin D, Dell'Acqua S, Kozlowski H, Casella L. Coordination and redox properties of copper interaction with alpha-synuclein. J Inorg Biochem. 2016.

Verina T, Schneider JS, Guilarte TR. Manganese exposure induces alpha-synuclein aggregation in the frontal cortex of non-human primates. Toxicol Lett. 2013;217(3):177–83.

Williams DR, de Silva R, Paviour DC, Pittman A, Watt HC, Kilford L, et al. Characteristics of two distinct clinical phenotypes in pathologically proven progressive supranuclear palsy: Richardson's syndrome and PSP-parkinsonism. Brain J Neurol. 2005;128(Pt 6):1247–58.

Win-Shwe TT, Fujimaki H. Nanoparticles and neurotoxicity. Int J Mol Sci. 2011;12(9):6267–80.

Wu J, Ding T, Sun J. Neurotoxic potential of iron oxide nanoparticles in the rat brain striatum and hippocampus. Neurotoxicology. 2013;34:243–53.

Xu B, Wang F, Wu SW, Deng Y, Liu W, Feng S, et al. Alpha-Synuclein is involved in manganese-induced ER stress via PERK signal pathway in organotypic brain slice cultures. Mol Neurobiol. 2014;49(1):399–412.

Xue M, Yang Y, Ruan J, Xu Z. Assessment of noise and heavy metals (Cr, Cu, Cd, Pb) in the ambience of the production line for recycling waste printed circuit boards. Environ Sci Technol. 2012;46(1):494–9.

Zecca L, Tampellini D, Gatti A, Crippa R, Eisner M, Sulzer D, et al. The neuromelanin of human substantia nigra and its interaction with metals. J Neural Transm (Vienna). 2002;109(5–6):663–72.

Zhang K, Schnoor JL, Zeng EY. E-waste recycling: where does it go from here? Environ Sci Technol. 2012;46(20):10861–7.

Zucca FA, Giaveri G, Gallorini M, Albertini A, Toscani M, Pezzoli G, et al. The neuromelanin of human substantia nigra: physiological and pathogenic aspects. Pigm Cell Res/Sponsored by the European Society for Pigment Cell Research and the International Pigment Cell Society. 2004;17(6):610–7.

Inflammatory Activation of Microglia and Astrocytes in Manganese Neurotoxicity

Ronald B. Tjalkens, Katriana A. Popichak, and Kelly A. Kirkley

Abstract Neurotoxicity due to excessive exposure to manganese (Mn) has been described as early as 1837 (Couper, Br Ann Med Pharm Vital Stat Gen Sci 1:41–42, 1837). Extensive research over the past two decades has revealed that Mn-induced neurological injury involves complex pathophysiological signaling mechanisms between neurons and glial cells. Glial cells are an important target of Mn in the brain, both for sequestration of the metal, as well as for activating inflammatory signaling pathways that damage neurons through overproduction of numerous reactive oxygen and nitrogen species and inflammatory cytokines. Understanding how these pathways are regulated in glial cells during Mn exposure is critical to determining the mechanisms underlying permanent neurological dysfunction stemming from excess exposure. The subject of this review will be to delineate mechanisms by which Mn interacts with glial cells to perturb neuronal function, with a particular emphasis on neuroinflammation and neuroinflammatory signaling between distinct populations of glial cells.

Keywords Manganism • Pattern recognition receptors (PRRs) • Astrogliosis • Glial fibrillary acidic protein (GFAP) • Parkinson's disease (PD)

R.B. Tjalkens (✉)
Program in Cell and Molecular Biology, Colorado State University,
Fort Collins, CO 80523-1680, USA

Professor of Toxicology and Neuroscience, Center for Environmental Medicine, Colorado State University, Fort Collins, CO 80523-1680, USA

Department of Environmental and Radiological Health Sciences, Colorado State University, Fort Collins, CO 80523-1680, USA
e-mail: Ron.Tjalkens@colostate.edu

K.A. Popichak
Program in Cell and Molecular Biology, Colorado State University,
Fort Collins, CO 80523-1680, USA

K.A. Kirkley
Professor of Toxicology and Neuroscience, Center for Environmental Medicine, Colorado State University, Fort Collins, CO 80523-1680, USA

Department of Environmental and Radiological Health Sciences, Colorado State University, Fort Collins, CO 80523-1680, USA

© Springer International Publishing AG 2017
M. Aschner and L.G. Costa (eds.), *Neurotoxicity of Metals*, Advances in Neurobiology 18, DOI 10.1007/978-3-319-60189-2_8

Introduction

Glia represent a diverse class of cells grouped together due their status as non-excitable neural cells that lack the ability to form an action potential and thus transmit electrical signals. Within the central nervous system (CNS), glia represent 90% of all cells and are classified on the basis of morphology, function, and location consisting of astrocytes, microglia, oligodendrocytes, and ependymal cells. Early descriptions of these cells labeled them as "glue" with a primarily passive structural/supportive role. However, with the advent of patch clamping and fluorescent calcium dye techniques from the late 1980s through the early 2000s, researchers have found that the role of these cells is much more extensive and complex (Araque et al. 2001). Glia are essential for neuronal development and survival, as well as for regulating synaptic function, brain metabolism, and cerebral blood flow. These roles are evolutionarily conserved across different phyla, demonstrating the importance of glial cells in regulating neuronal function and pathology in the CNS.

Role of Glia in Manganese Neurotoxicity

Manganese (Mn) neurotoxicity, or manganism, is a neurodegenerative disease of the cerebral cortex and basal ganglia caused by excessive exposure to Mn and is characterized by motor deficits that resemble those seen in idiopathic Parkinson's disease (PD), such as gait disturbances, facial masking, hypoxia and dysphonia, dystonia and action, and postural tremor (Guilarte 2010; Perl and Olanow 2007). However, there are clear neurological distinctions from PD, including a typical lack of resting tremor, distinct gait abnormalities, and differential involvement of neurons in the substantia nigra pars compacta. These PD-like manifestations are due to the neuropathological changes including neuronal loss, atrophy and gliosis within the globus pallidus (GP), substantia nigra pars reticulata (SNpr), and striatum (ST) of exposed individuals (Aschner and Aschner 1991; Sigel 2007). Typically, exposures to high levels of Mn occur occupationally in welders, miners, and steel workers (Hua and Huang 1991); however, the neurological consequences of environmental exposure to low levels of Mn through ingestion of crops with residues of the Mn-containing pesticide Maneb (Santamaria 2008) and well water with high concentrations of Mn (Woolf et al. 2002) are under scrutiny as an important route for nonoccupational exposure to the general population. In particular, there is increased concern with chronic Mn exposure in children due to their lower ability to clear Mn (Collipp et al. 1983); higher levels of iron deficiency, which have shown to elevate brain Mn levels (Aschner and Aschner 2005); and greater absorption of Mn from the GI tract (Neal and Guilarte 2012). Recent epidemiological studies have reported cognitive deficits in children exposed to high levels of Mn in drinking water (Menezes-Filho et al. 2011; Riojas-Rodriguez et al. 2010; Kim et al. 2009), highlighting the need for future studies addressing the long-term consequences of these exposures.

The mechanisms of how Mn exposure leads to specific neurodegenerative changes in the basal ganglia of exposed humans and animals are poorly understood. Elevated levels of Mn are routinely documented in the basal ganglia of exposed humans and animals (Olanow 2004), and experimental evidence has shown that Mn can be directly neurotoxic through inhibition of mitochondrial respiration leading to energy failure and oxidative stress (Zhang et al. 2003) and through excitotoxicity (Centonze et al. 2001). Other established mechanisms such as oxidative stress, glial toxicity, and neuroinflammation are also implicated in the progression of Mn neurotoxicity. Notably, the disorder will continue to progress both clinically and in rodent models of the disease even after cessation of exposure, suggesting ongoing mechanisms linked to progression that may include both unfolded protein stress and neuroinflammation (Sigel 2007; Aschner and Aschner 2005; Filipov and Dodd 2012).

The involvement of glia in Mn-induced neurotoxicity has only received increased attention over the past 20 years as a fundamental mechanism in the progression of Manganism (Filipov and Dodd 2012). Although activated astrocytes and microglia were often noted in post mortem evaluation of Mn-exposed patients (Perl and Olanow 2007), few studies examined the functional consequences or mechanisms of glial activation following exposure to Mn. This was most likely due to the ability of Mn to be directly toxic to neurons through inhibition of mitochondrial respiration and induction of oxidative stress (Zhang et al. 2003) and the historical focus on acute, high-level exposures. A study in 1998 by Spranger et al. (1998) changed the perceptions of glia involvement in Manganism by reporting that exposure to low concentrations of Mn could amplify inflammatory activation of glial cells and enhance neurotoxicity. Other studies have now built upon these initial findings revealing that Mn can exacerbate the effects of LPS and cytokines on activation of both microglia and astrocytes that causes dramatic potentiation in production of TNFα, IL-1β, ROS, and NOS2 expression (Barhoumi et al. 2004; Chen et al. 2006; Filipov et al. 2005; Moreno et al. 2008, 2011). Increased levels of these and other inflammatory genes have also been measured in both rodent (Moreno et al. 2009; Zhao et al. 2009) and nonhuman primate (Verina et al. 2011) studies with deletion or inhibition of these pathways showing neuroprotection (Zhao et al. 2009; Streifel et al. 2012; Zhang et al. 2009).

Neuroinflammation in Manganese Toxicity

Overview of Neuroinflammation in the CNS

It is now clear that Mn exposure, even early in life, can have lasting effects on the neuroinflammatory status of glial cells (Moreno et al. 2009). Thus, neuroinflammatory activation of glia may be a fundamental mechanism in determining long-term neurological outcomes from Mn exposure. Astrocytes and microglia serve a multitude of essential functions within the CNS including integral roles in the innate immune system of the brain (Wyss-Coray and Mucke 2002). In response to foreign or endogenous signals, both astrocytes and microglia adopt an activated phenotype resulting in the release of pro-inflammatory mediators (Craft et al. 2005). This inflammatory system, known as neuroinflammation, is essential in normal tissue repair and in defense against foreign invasion; however, when sustained, this process can become deleterious through the release of neurotoxic factors that amplify underlying disease (Mosley et al. 2006; Glass et al. 2010; Tansey et al. 2007).

In normal circumstances, the neuroinflammatory reaction has auto regulatory mechanisms in place to limit the extent of activation as the process is neither discriminatory or specific (Wyss-Coray and Mucke 2002; Glass et al. 2010). For sustained inflammation to occur, there must be failure of self-resolution mechanisms or the presence of endogenous or environmental factors that are perceived as a threat. There are a variety of factors known to elicit activation of both microglia and astrocytes including products released by injured neurons such as glutamate (Kaushal and Schlichter 2008), ATP (Di Virgilio et al. 2009), and matrix metalloproteinase-3 (Kim et al. 2005); cytokines including interferon gamma (IFNγ), interleukin-1β (IL-1β), and interleukin-6 (IL-6); adhesion molecules; growth factors; blood-derived factors; ionic imbalances; activation of complement products from viruses and bacteria; and presence of reactive oxygen species (Wyss-Coray and Mucke 2002; Sofroniew and Vinters 2010; Gehrmann et al. 1995). Furthermore, new evidence suggests that both microglia and astrocytes express endogenous pattern recognition receptors (PRRs) that respond to a variety of damage-associated molecular patterns (DAMPs) that results in molecular signaling events that promote inflammation and disease progression (Glass et al. 2010). These PRRs become activated in response to signals released by necrotic neurons or other pathologic products produced during disease including oxidized proteins and lipids (Husemann et al. 2002), messenger ribonucleic acid (mRNA), fibronectin, hyaluronic acid, heat shock proteins, amyloid-beta, neuromelanin, and alpha-synuclein (Block and Hong 2005; Gensel et al. 2012; Zhang et al. 2005). The production of inflammatory mediators is further increased by activated glia, leading to a feed-forward cycle of inflammation and further release of neurotoxic mediators of tissue injury.

Activated glia release diverse inflammatory factors including cytokines, chemokines, reactive oxygen species (ROS), and nitric oxide (NO) that are toxic to neurons (Kim et al. 2005; Gonzalez-Scarano and Baltuch 1999). Cytokines such as tumor necrosis factor-alpha (TNFα) and interleukin-6 (IL-6) are often upregulated very quickly in activated glial cells and can directly amplify inflammation through recruitment of both innate and adaptive immune cells, leading to neuronal apoptosis (Gensel et al. 2012; Gonzalez-Scarano and Baltuch 1999). Reactive oxygen species from Mn exposure can also damage neurons directly by increasing lipid peroxidation and mitochondrial dysfunction, causing subsequent energy failure, protein modifications, and DNA damage (Mosley et al. 2006). The formation of peroxynitrite, a by-product of superoxide and NO, is thought to be a major contributor to neuronal-induced cell death through nitration and nitrosylation of tyrosine and serine residues of proteins leading to impairment of normal cellular functions (McCarty 2006). Mn exposure results in significant increases in protein nitrosylation, indicative of nitrosative stress from NO production by glial cells (Moreno et al. 2009). Inhibition or deletion of many of these pathways has shown to be neuroprotective, but often the neuroprotection achieved is dependent on the timing of inhibition as often early downregulation of inflammation has actually worsened neuronal injury (Frank-Cannon et al. 2009). However, mice lacking the inducible form of NO synthase (iNOS/NOS2) are protected from Mn neurotoxicity, demonstrating the importance of this glial inflammatory pathway in the mechanism of neuronal injury (Streifel et al. 2012). Due to the complicated nature of neuroinflammation and the vast majority of implicated factors, systematic and thorough understanding is vital to understanding the implications that may come from targeting this pathway.

Glial Cell Activation in Neuroinflammation

The activation of microglia and astrocytes is one of the universal components of neuroinflammation and is implicated in the progression of neurodegeneration in ischemia, seizure, Alzheimer's disease, multiple sclerosis, amyotrophic lateral sclerosis, Parkinson's disease (PD), and manganism (Mosley et al. 2006; Glass et al. 2010; Block and Hong 2005; Hirsch and Hunot 2009; Vezzani et al. 2013). Since the first early descriptions of activated glia in neurodegenerative diseases, there have been an increasing number of CNS pathologies described as having an association with activated glia. Although the regional pattern of neuroinflammation can vary among different disorders, there are common mechanisms by which activated glial cells sense stress and injury within the CNS and consequently transduce signals that amplify inflammatory activity of surrounding microglia and astrocytes (Glass et al. 2010). Research aimed at elucidating the pathogenesis of neuroinflammation is quickly expanding to understand the importance of this mechanism in the progression of many neuropathologies, including manganism. In this regard, it is useful to compare the role of neuroinflammation in Mn neurotoxicity to that of other better studied disorders, such as Parkinson's disease, to develop an appreciation for the mechanisms that are common to degenerative conditions of the CNS. The molecular regulation of neuroinflammatory gene expression in glial cells shares important commonalities between astrocytes, microglia, and peripheral immune cells such as monocytes.

Glial inflammatory activation is regulated by several different pathways including mitogen-activated protein kinases (MAPKs), activator protein-1 (AP-1), Janus kinase (JAK)/signal transducer and activator of transcription (STAT), and interferon regulator factor families (Glass et al. 2010); nevertheless, the nuclear factor kappa B (NF-κB) appears to be the primary pathway involved in the activation of pro-inflammatory genes (Karin 2005). Deletion of NF-κB is detrimental to the ability of the immune system to initiate immunoprotective responses. Mice deficient in this pathway often succumb to opportunistic infections (Alcamo et al. 2001). Genetic deletion of this pathway in specific glial cells within the CNS has shown to be very neuroprotective with better recovery after spinal cord injury (Brambilla et al. 2005), decreased pathology in mouse models of multiple sclerosis (van Loo et al. 2006), and decreased seizure-induced neuronal death in kainic acid model of seizure (Cho et al. 2008).

NF-κB represents a family of transcription factors that are regulated by inhibitory κBs (IκBs). Upon signal activation, IκBs are phosphorylated by IκB kinase complex (IKK) marking them for polyubiquitination and, ultimately, degradation by the 26s proteasome, thus freeing the transcription factors, located as dimers within the cytosol, to translocate into the nucleus (DiDonato et al. 1997). The IKK complex consists of three different proteins including the two catalytic units IKKα/IKK1 and IKKβ/IKK2 and the regulatory subunit IKKγ. These two catalytic subunits mark the division of the two NF-κB activation pathways known as the classical pathway and the alternative pathway. The classical NF-κB pathway involves the heterodimers of p50 and p65/RelA and is activated by the action of IKKβ/IKK2 of the IKK complex. This pathway is primarily involved in immunoregulation controlling innate immune responses and survival of immune cells. The alternative pathway is primarily involved in the development of secondary lymphoid organs and requires only IKKα/IKK1 and results in the processing of p100 (Karin 1999; Li et al. 2003; Bonizzi and Karin 2004). Deletion of IKKβ/IKK2 and not IKKα/IKK1 recapitulates similar mouse phenotypes as RelA knockout mice with almost complete inhibition of inflammatory responses and thus represents a major target in modulating glia neuroinflammatory activation (Alcamo et al. 2001). As detailed below, the NF-κB pathway is an important target of Mn in glial cells that integrates multiple extra- and intracellular stress signals to activate inflammatory gene expression.

Manganese and Astrocytes

Description and Distribution of Astrocytes

Astrocytes accumulate higher levels of Mn than neurons and are therefore considered an important target cell for transport of Mn into the brain as well as for initiating inflammatory signaling during neuronal stress and injury. Astrocytes encompasses a heterogeneous population of cells that can have vastly different morphological and physiological characteristics depending on their location with the brain (Matyash and Kettenmann 2010). Their morphological forms range from the protoplasmic astrocyte with extensive arborization found in the gray matter to the more rodlike fibrous astrocyte located within the white matter (Sofroniew and Vinters 2010; Perea and Araque 2010). With their extensive processes, they make contacts with neuronal bodies, synapses, axons, blood vessels, and other astrocytes, thereby creating a vast network that allows them to serve a multitude of both structural and important physiological roles within the CNS.

Astrocytes are the most numerous type of cell of the CNS, making up 60–70% of all cells in the brain and also comprise 90% of all glial cells. Astrocytes are found throughout the CNS in a contingent but nonoverlapping manner that comprises distinct microdomains which enable them to make contact with a large number of neurons and with the microvasculature (Sofroniew and Vinters 2010). Astrocytes are morphologically characterized by expression of the intermediate filament proteins glial fibrillary acidic protein (GFAP) and vimentin. Other known markers of astrocytes in the adult brain include glutamine synthetase (GS), S100 calcium-binding protein-β, and glutamate transporters GLT-1/EAAT2 and GLAST/EAAT1 (Kimelberg 2004); however, GFAP has been shown to be the most consistent marker in both physiological and pathological states (O'Callaghan and Sriram 2005).

Functional Roles of Astrocytes Relevant to Manganese Neurotoxicity

The first noted function of astrocytes within the adult CNS was purely structural; astrocytes were described as a scaffold to arrange and contain the neuronal circuitry due to their relative abundance and formation of glial scars in disease. Although it is now known that astrocytes have more complex roles, their formation of a continuous syncytium is still important for the structural integrity of the brain. These vast networks help to create specific micro and macro domains and help to create physical barriers between neuronal synapses (Sofroniew and Vinters 2010). Furthermore, astrocytic end feet are an important component of the glia limitans, a barrier that helps to isolate the brain parenchyma from the vasculature and subarachnoid compartments (Nimmerjahn 2009), as well as the blood-brain barrier (BBB) through the ensheathing of blood vessels throughout the CNS (Carmignoto and Gomez-Gonzalo 2010).

Past their structural roles, astrocytes serve as important facilitators of neuronal homeostasis through nutritive and trophic support. As a primary component of the BBB, astrocytes that surround endothelial cells are enriched in glucose receptors and channels and act as the main vehicle for the movement of glucose and oxygen from the blood to neurons. Astrocytes, but not neurons, are capable of storing glucose in the form of glycogen and of de novo synthesis of glutamate, which forms the basis for the functional metabolic coupling between these two cell types that maintains neuronal homeostasis (Parpura et al. 2012). Glutamate is the primary excitatory neurotransmitter in the brain, and its synaptic concentration is tightly regulated by astrocytes, which rapidly removed glutamate from synapses, where it can be safely transaminated to glutamine for recycling to neurons in the glutamate-glutamine cycle. Eighty percent of glutamate released into the synapse is removed by astrocytes and then converted to glutamine by GS, thereby preventing excitotoxic injury to neurons. This glutamine is released and then taken up by neurons that convert glutamine into glutamate and γ-amino butyric acid (GABA). Additionally, production of lactate by astrocytes is used by neurons to produce pyruvate and generate adenosine triphosphate (ATP) via the tricarboxylic acid cycle (TCA). These metabolically coupled support pathways in astrocytes are critical for neuronal survival and are important targets of Mn during neurotoxic exposures. Notably, Mn exposure results in marked increases in excitatory neurontransmission that likely damages neurons, supported by studies demonstrating the efficacy of the ionotropic glutamate receptor antagonist, MK-801, in preventing neuronal injury from chronic exposure to Mn (Xu et al. 2010).

In addition to being critical for neuronal metabolism, astrocytes are required for normal synaptic transmission through regulation of neurotransmitters, ions, water, and extracellular pH (Sofroniew and Vinters 2010). Astrocytes surround both pre- and postsynaptic terminals to form what is known as the tripartite synapse, allowing astrocytes to not only regulate neurotransmitters but also actively respond to and modulate synaptic plasticity through the release of gliotransmitters (Araque et al. 2001; Perea and Araque 2010; Nedergaard and Verkhratsky 2012; Perea et al. 2009). Astrocytes express a wide assortment of functional neurotransmitters including glutamate, GABA, dopamine, adrenalin/epinephrine, histamine, and glycine, the expression of which varies depending on the local microenvironment to match the physiology of their neuronal neighbors (Parpura et al. 2012). The majority of the neurotransmitter receptors expressed are metabotropic receptors coupled to G-proteins whose activation results in the generation of inositol triphosphate (IP3) and the release of calcium (Ca^{2+}). Astrocytes express at least three types of ionotropic receptors: α-amino-3-hydroxy-5-methyl-isoxazole propionate (AMPA), N-methyl-D-aspartate (NMDA) types of tetrameric glutamate receptors, and P2X trimeric purinoreceptors (Lalo et al. 2008). Activation of glia metabotropic and inotropic receptors results in the generation of Ca^{2+} waves within astrocytes that are propagated between astroglial networks through connexin gap junctions and glia release of ATP and glutamate (Araque et al. 2001; Kim and de Vellis 2005). This intercommunication between astrocytes is dynamic and is influenced by the extent of and frequency of neurotransmitter release which is important in the modulation of synapses in both learning and memory (Perea et al. 2009). It was recently reported that Mn disrupts ATP-dependent Ca^{2+} signaling in astrocytes by inhibiting entry of Ca^{2+} through the plasma membrane subsequent to activation of P2Y purinergic receptors (Streifel et al. 2013), suggesting that Ca^{2+}-dependent homeostatic processes in astrocytes could be an important target of Mn that likely impacts neuronal physiology.

Calcium-based communication between astrocytes not only plays a large role in synaptic plasticity but is vital to the regulation of blood flow in response to neuronal activity known as neurovascular coupling (Sofroniew and Vinters 2010; Carmignoto and Gomez-Gonzalo 2010). In areas of high neuronal activity, elevations in calcium in astrocytes result in release of vasoactive compounds such as nitric oxide (NO), prostaglandin E2 (PGE_2), potassium (K^+), and epoxygenase derivatives (EETs) at astrocytic end feet that results in a dilation or constriction of local vasculature (Nimmerjahn 2009; Mulligan and MacVicar 2004). This control of cerebral blood flow is complex, and the elucidations of how astrocytes cause specific vasodilation versus vasoconstriction in response to neuronal activity are still being fully elucidated. Mn can disrupt ATP-induced Ca^{2+} signaling and intercellular Ca^{2+} waves in astrocytes (Streifel et al. 2012), which could be detrimental to neuronal trophic support, rendering affected brain regions both focally hypoxic and with insufficient metabolic support. In this regard, even low levels of Mn^{2+} can disrupt ATP-dependent calcium signaling in astrocytes, in part through inhibition of TRPC3 cation channels, which could alter the regulation of cerebral blood flow and therefore negatively impact neuronal homeostasis (Streifel et al. 2013). Inhibition of ATP-dependent calcium signaling in primary astrocytes has also been described for the cationic neurotoxicants, 1-methyl-4-phenylpyridinium (MPP^+), and 6-hydroxydopamine, suggesting that disruption of homeostatic calcium signaling in astrocytes may be a common mechanism of injury for structurally diverse compounds affecting dopaminergic brain regions (Streifel et al. 2014).

Astrocytes are thus diverse and important regulators of neuronal metabolism and activity in the developed CNS; likewise, they also play an important role in the developing CNS, through neuronal guidance and synaptogenesis, and in adult neurogenesis (Doetsch 2003). In development, boundaries created by astrocytes help the migration of axons and neuroblasts, whereas the release of thrombospondin from astrocytes directs synapse formation. Furthermore, tagging of formed synapses with complement protein, C1q, helps tags synapses for pruning and removal (Christopherson et al. 2005; Powell and Geller 1999). In the adult CNS, neurogenesis within the subventricular zone of the olfactory bulb and the hippocampus is regulated by secretion of astrocytic factors such as Wnt3, interleukin-1β (IL-1β), interleukin-6, and insulin-like growth factor-binding protein 6 (Parpura et al. 2012). Additionally, astrocytes themselves are believed to be the source of newly generated neurons determined by labeling based lineage tracking experiments (Doetsch 2003). Thus, neuronal generation, function, and continued survival are intimately linked and dependent on the vast and extensive physiology of their astrocytic counterparts.

Role of Astrocytes in Neuroinflammation

Activation of astrocytes is a biological reaction that is documented in most CNS diseases, as measured by increased expression of GFAP and alterations in astrocyte morphology that are considered early indicators of neuropathology (O'Callaghan and Sriram 2005; Parpura et al. 2012). Activation of astrocytes can be neuroprotective through isolation of damage, glutathione production, BBB repair, and release of neurotrophic factors such as neural growth factor and glial-derived growth factor (Sofroniew and Vinters 2010; Block and Hong 2005; Kuno et al. 2006); however, astrogliosis can also be neurotoxic and promote disease progression. Detrimental consequences of astrogliosis include inhibition of axonal regeneration (Block and Hong 2005; Silver and Miller 2004), exacerbation of inflammation via cytokine production (Brambilla et al. 2005, 2009), production of reactive oxygen and nitrogen species (Hamby et al. 2006; Liu et al. 2006; Carbone et al. 2009), and excessive release of glutamate (Takano et al. 2009). Additionally, chronic inflammatory stimulation of astrocytes reduces glial capacity to generate and release neurotrophic mediators and execute normal physiological functions (Parpura et al. 2012). We and others have reported extensively on reactive astrocytosis in rodent models of manganism (Moreno et al. 2009, 2011; Streifel et al. 2012; Liu et al. 2006).

The regulation of astrocyte activation is under the control of many factors including cytokines IL-6, IFNγ, tumor necrosis factor-alpha (TNFα), toll-like receptor activators, neurotransmitters, ATP, reactive oxygen species, hypoxia, glucose deprivation, ammonia, and protein aggregates (Sofroniew and Vinters 2010; Parpura et al. 2012). Frequently, these activators are by-products of already injured neurons or factors released by activated microglia which indicate that astrocyte activation is often later in disease progression (Hirsch and Hunot 2009). Recent studies suggest that α-synuclein may be protective against Mn neurotoxicity, implicating protein misfolding in neurons as an important pathogenic mechanism following exposure to Mn (Harischandra et al. 2015). However, astrogliosis is often more persistent than microgliosis and is believed to be important in amplifying inflammatory processes and thereby inducing greater damage (Saijo et al. 2009). Moreover, in vitro studies have shown that isolated human astrocytes and not microglia are the major source of NO-induced neurotoxicity indicating they may be more significant in neuroinflammatory-induced neuronal death in humans than have been indicated in rodent models (Lee et al. 1993).

Neuroinflammatory Effects of Manganese in Astrocytes

Neuropathology in manganism is associated with robust astrogliosis in the basal ganglia, particularly the globus pallidus, subthalamic nucleus, and substantia nigra pars reticulata (Olanow 2004). Mn preferentially accumulates in astrocytes due to their expression of high-capacity transporters (Sidoryk-Wegrzynowicz and Aschner 2013), and therefore concentrations of Mn in astrocytes can be 50–60 times higher than in neurons, with the highest concentration of Mn found in the mitochondria (Sidoryk-Wegrzynowicz and Aschner 2013; Aschner et al. 1992; Morello et al. 2008). Similar to microglia, astrocytes release inflammatory cytokines and nitric oxide (NO) that influence the progression of neuronal injury during exposure to Mn (Moreno et al. 2008, 2009; Liu et al. 2006). In vitro studies demonstrate that human astrocytes are the primary source of NO-induced neurotoxicity, more so than microglia, suggesting that astrocytes could play a greater role in neuroinflammation-induced neuronal death from Mn than was initially demonstrated in rodent studies (Lee et al. 1993). Another in vitro study demonstrated that Mn inhibits the capacity of astrocytes to promote neuronal differentiation through a mechanism that involves oxidative stress and a reduction in levels of the extracellular matrix protein, fibronectin (Giordano et al. 2009). Oxidative stress in astrocytes also leads to dysfunction in mitochondria and, not surprisingly, a decreased production of ATP that could negatively impact neuronal function and survival (Barhoumi et al. 2004; Chen et al. 2006; Streifel et al. 2012).

Mn causes metabolic changes in astrocytic glucose metabolism by inhibition of the astrocyte-specific enzyme, glutamine synthetase (GS), thereby contributing to downregulation of glutamate transporters and compromising glutamate uptake (Suarez-Fernandez et al. 1999; Verkhratski and Butt 2013; Yin et al. 2007; Verkhratsky et al. 2016). These oxidative changes are consistent with the combined effects of Mn and inflammatory stimuli on mitochondrial function in astrocytes, which promotes greater production of ROS and deprecations in metabolic support of neurons, in addition to the damaging effects of inflammatory gene expression (Barhoumi et al. 2004). Such an effect would predispose neurons to excitotoxic injury due to the presence of excess synaptic glutamate resulting from dysfunction or downregulation of high-affinity astrocytic transporters. Additionally, Mn perturbs ATP-induced Ca^{2+} signaling and intercellular Ca^{2+} waves in astrocytes, which could negatively impact the ability of astrocytes to stimulate Ca^{2+}-dependent increases in local cerebral blood flow in response to synaptic activity (Streifel et al. 2012; Tjalkens et al. 2006). Any failure of the capacity of astrocytes to supply adequate neurotrophic support during chronic overexposure to Mn could therefore lead to loss of trophic support, as well as excitotoxicity and neuronal death.

Mn exposure can also directly stimulate inflammatory gene expression in astrocytes through activation of NF-κB. This results in expression of COX2, NOS2, and multiple inflammatory cytokines and chemokines, leading to enhanced neuronal apoptosis (Araque et al. 2001; Moreno et al. 2008; Streifel et al. 2012; Carbone et al. 2009). Previous research from our laboratory demonstrated that gene deletion of *Nos2* in astrocytes protected against Mn-induced neurological dysfunction in vivo and prevented NF-κB-dependent neuronal injury from activated astrocytes exposed to Mn in vitro (Streifel et al. 2012). Neuroprotection in these studies was associated with decreased production of NO and inflammatory cytokines in Mn-treated astrocytes, highlighting the importance of inflammatory activation of these cells in the progression of neuronal injury from Mn exposure. These data also demonstrate that reactive inflammatory mediators such as NO are important contributors to Mn-induced neuronal dysfunction during exposure to Mn.

Manganese and Microglia

Description and Distribution of Microglia

Microglia are the primary immunological effector cells of the brain, entering the CNS during embryonic development from a monocyte-derived cell type (Kim et al. 2005). As discussed below, microglia represent an important effector cell type during Mn neurotoxicity that respond rapidly with increased production of neuroinflammatory mediators. In the adult brain, microglia have very low rates of division, but their numbers can be replenished by perivascular mononuclear phagocytes (Gehrmann et al. 1995). They are heterogeneous through the adult brain and constitute 10 to 15% of all glial cells with greater numbers located within the gray matter. In particular, the highest concentrations of microglia are found within the olfactory bulb, hippocampus, and basal ganglia, including the substantia nigra, which holds the greatest density of microglia: 12% of all cells. Microglia exist in three different morphological states: a ramified phenotype found proximal to the neuropil, a rod-like state in fiber tracts, and a macrophage-like amoeboid shape in areas with an incomplete BBB (Lawson et al. 1990). Microglia are never at rest and are constantly migrating, but these migration patterns are distinct between different cells and do not overlap (Gehrmann et al. 1995).

Functional Roles of Microglia Relevant to Manganese Neurotoxicity

As with other macrophage-like cells resident in tissues, one of the primary functions of microglia in the CNS is immunosurveillance, and they possess dendritic and phagocytic functions similar to other monocyte-derived cells (Kim et al. 2005; Block and Hong 2005; Gonzalez-Scarano and Baltuch 1999). Microglia constantly move and sample the extracellular environment within their individual domains, clearing up debris via their phagocytic function as they migrate. They express a variety of neurotransmitter receptors, pattern recognition receptors (PRRs) and ionotropic receptors such as P2X7 to sense alterations in brain homeostasis, the presence of foreign materials and neuronal damage (Ransohoff and Perry 2009). Microglia represent the main class of cell involved in antigen presentation and are important in recruitment of immune cells such at T and B lymphocytes to sites of injury (Gehrmann et al. 1995; Gonzalez-Scarano and Baltuch 1999). Notably, microglia express the NOD-like receptor, NLRP3, which is critical to increasing expression of IL-1β via inflammasome activation and which can be potently activated by high levels of extracellular ATP through P2X7 receptors (Surprenant and North 2009). Astrocytes can release high concentrations of ATP in response to neuronal stress and injury, which suppresses neuronal excitability but can also stimulate activation of microglia. Mn may therefore stimulate neuroinflammatory activation of microglia both directly, as well as indirectly through its effects on astrocytes, suggesting several pathways by which glial-glial interactions may stimulate neuroinflammatory injury during Mn exposure.

More recently, research has determined that microglia may also play integral roles in neuronal development and migration. Amoeboid microglia are implicated in synaptic remodeling and regulation of neuronal apoptosis through the release of soluble factors and phagocytic pruning of synapses in late embryonic development (Block and Hong 2005). Furthermore, studies have shown microglia to release growth and neurotrophic factors during synaptogenesis (Nakajima and Kohsaka 1993). However, the role of microglia in mediating both trophic and neurotoxic cell-cell interactions during Mn neurotoxicity is still not well known.

Role of Microglia in Neuroinflammation

Microglia are the primary effectors of the innate immune response within the CNS, with activation occurring early in states of disease and often preceding overt neuropathology (Gehrmann et al. 1995; Hirsch and Hunot 2009). Under physiological conditions, microglia exist in a resting, ramified state releasing both anti-inflammatory and neurotrophic factors while surveying their domains (Streit 2002). However, in the presence of viral or bacterial products (Glass et al. 2010), ATP, changes in ion or neurotransmitter homeostasis (Mastroeni et al. 2009), cytokines such as IFNγ and interleukin-4 (Gehrmann et al. 1995), colony-stimulating factors (CSFs) (Kim and de Vellis 2005), and a list of other pathological products, microglia transform into an activated phenotype, proliferate, and migrate to the site of injury (Block and Hong 2005). Activation occurs in two stages. In the first stage, microglia adopt a rodlike shape and increase expression of major histocompatibility complex II (MHCII) and other inflammatory molecules. In the second stage, microglia morph into an amoeboid cell capable of phagocytosis (Gehrmann et al. 1995; Kim and de Vellis 2005).

Once activated, microglia can be both beneficial and deleterious in disease as they release both pro- and anti-inflammatory factors (Block and Hong 2005). Determining whether microglia neuroinflammatory responses will be helpful or toxic is often predicted by adoption of either the M1 known as "classical activation" or M2, the "alternative activation," phenotype (David and Kroner 2011).The M1 phenotype is primarily inflammatory with microglia upregulating MHCII, CD86, CD32, and CD16 with the production of TNFα, IL-1β, and IL-6. In contrast, the M2 phenotype is more closely associated with tissue repair with increased expression of arginase 1 and CD206, as well as increased release of brain-derived neurotrophic factor (BDNF), insulin-like growth factor-1 (IGF-1), and interleukin-10 (IL-10) (Kigerl et al. 2009).

The classical activated or M1 microglia are known to elicit neuronal death and perpetuate inflammation through release of a variety of cytotoxic substances such as NO, superoxide anion, cytokines, glutamate, prostaglandins, and aspartate (Takano et al. 2009; Mastroeni et al. 2009). They appear to be the major initial sensors of foreign or endogenous signals, secreting inflammatory mediators such as TNFα and IL-1β that can act on astrocytes to induce secondary inflammatory responses (Lee et al. 2012). Furthermore, prevention of microglial activation pharmacologically or genetically often protects against neuroinflammatory pathology, thus placing them as important regulators of inflammatory mechanism in neurodegenerative diseases (Block and Hong 2005; Cho et al. 2008).

Neuroinflammatory Effects of Manganese in Microglia

Resting or ramified microglia can release anti-inflammatory and neurotrophic factors. However, when activated microglia release neuroinflammatory mediators and pro-inflammatory cytokines such TNFα, IL-1β, IL-6, and as well as reactive oxygen and nitrogen species (ROS and RNS), all of which can act on astrocytes to amplify inflammatory responses in the CNS (Liu et al. 2003, 2006; Lee et al. 2012; Chao et al. 1992). Rapid expression of these cytokines in microglia early in CNS injury suggests that such glial inflammatory responses are integral to neuronal injury from a variety of exogenous insults (Gensel et al. 2012; Gonzalez-Scarano and Baltuch 1999). Exposure to Mn causes activation of both microglia (Chang and Liu 1999) and astrocytes (Spranger et al. 1998), resulting in increased production of numerous inflammatory factors that cause neuronal injury (Filipov et al. 2005). Manganese can also enhance the inflammatory effects of other microglial activators, such as lipopolysaccharide (LPS), which amplifies neuronal injury in models of Mn neurotoxicity (Filipov and Dodd 2012; Filipov et al. 2005; Park and Chun 2016). However, the molecular mechanisms underlying these responses to Mn in microglia are still not completely understood. For example, Mn and LPS can synergistically enhance activation and inflammatory gene expression in microglia (Park and Chun 2016), consistent with transformation into a phagocytic phenotype that can lead to neuronal injury (Diaz-Aparicio et al. 2016). This may be directly relevant to Mn-induced neuronal injury in vivo, because blockade of phagocytosis may prevent some forms of inflammatory neurodegeneration (Neher et al. 2011).

Manganese also activates NF-κB in microglia and enhances the effects of LPS, resulting in inflammatory gene expression and production of inflammatory mediators leading to neuronal death (Filipov and Dodd 2012; Filipov et al. 2005; Park and Chun 2016). In BV-2 microglial cells treated with LPS, NF-κB-dependent inflammatory gene expression by pharmacological modulators of the nuclear receptor, Nurr1, which stabilizes nuclear corepressor proteins and reduces binding of NF-κB/p65 to inflammatory gene promoters (Saijo et al. 2009; De Miranda et al. 2015a). Thus, the inflammatory effects of Mn in microglia are tightly regulated both at the level of IKK activation as well as by nuclear proteins that modulate transcriptional activity of inflammatory genes. Evidence for the importance of NR4A1 (Nurr1) in regulating microglial activation was recently reported in studies using a novel pharmacological Nurr1 agonist that prevented glial activation and neuronal injury from the dopaminergic neurotoxicant, MPTP (1-methyl-4-phenyl-1,2,3,6-tetrahydropyridine) (De Miranda et al. 2015b). This data suggest that targeting NR4A receptors in microglia could be a promising avenue for prevention of NF-κB activation in glia and, thus, inflammatory neuronal injury from a variety of neurotoxicants, including Mn.

Another research group recently reported that P2Y12 participates in ischemia-related inflammation by mediating microglial migration and potentiation of neurotoxicity (Webster et al. 2013). Upon inhibition or deficiency of the P2Y12 receptor in BV-2 microglia, there is less NF-κB activation, suggesting an additional anti-inflammatory, neuroprotective benefit of the antiplatelet drug, clopidogrel. Mn potently affects glial signaling through P2Y and P2X receptors (Streifel et al. 2013, 2014), suggesting that mitigating these effects could potentially reduce NF-κB-mediated neuroinflammation in glia following Mn exposure.

Glial-Glial Cross Talk in Mn-Dependent Inflammatory Signaling

Sustained inflammatory activation of microglia is implicated as an important mechanism in the progression of many neurodegenerative diseases including AD, PD, and manganism. Experimental models of PD (Hirsch and Hunot 2009) and manganism (Zhao et al. 2009) have often identified the transition of microglia from a resting to activated phenotype prior to overt neuropathology. Cell culture models show Mn potentiates microglia inflammatory gene expression in response to LPS/cytokine treatment through activation of pathways such as NF-κB and mitogen-activated protein kinase (MAPK) (Filipov and Dodd 2012; Crittenden and Filipov 2011). Removal of microglia or use of antioxidants has shown to reduce neuronal loss indicating microglial activation may serve as a critical step in mediating neuronal injury during Mn exposure and that microglia also likely directly promote activation of astrocytes that then amplify neuronal damage (Zhao et al. 2009; Zhang et al. 2010). Evidence for this is provided by studies indicating that activated microglia can enhance the activation of adjacent astrocytes by releasing factors such TNF and IL-1β that can further magnify neuronal injury (Hirsch and Hunot 2009; Saijo et al. 2009). Microglial responses are often rapid, in contrast to the more delayed activation often seen in astrocytes, suggesting that temporally distinct signaling events are required for a reactive glial phenotype. Underscoring this point, decreased microgliosis in vivo is associated with reduced astrocytosis (Zhang et al. 2010). Despite the known role of microglial-astrocyte cross talk in AD and PD, these important glial-glial interactions are less well understood in Mn neurotoxicity.

Astrocytes serve as the major homeostatic regulator and storage site for Mn in the brain (Araque et al. 2001; Aschner and Aschner 1991) and are a prominent contributor to Mn-stimulated nitric oxide (NO) production through NOS2 (Zhang et al. 2009). We previously reported that Mn enhances the inductive effects of inflammatory cytokines on astrocyte expression of *Nos2* through stimulation of NF-κB (Moreno et al. 2008) and astrocytes activated by exposure to Mn and inflammatory cytokine-induced apoptosis in co-cultured striatal neurons (Streifel et al. 2012). However, without co-treatment with cytokines, astrocytes are unable to cause neuronal apoptosis in response to Mn treatment (Spranger et al. 1998) indicating that microglia are likely required for initiation of neuroinflammatory mechanisms in astrocytes during Mn neurotoxicity.

Using immunopurified cultures of primary microglia and astrocytes, data from our laboratory demonstrate that microglia directly accumulate Mn, which stimulates a mixed inflammatory phenotype characterized by release of IL-6, TNF, CCL2, and CCL5. These experiments also revealed that products from Mn-activated microglia are essential for neuroinflammatory activation of Mn-exposed astrocytes and that NF-κB-dependent release of TNF from microglia is a key signaling event in microglia regulating these glial-glial interactions. To decipher the complex signaling mechanisms likely to influence development of a neuroinflammatory phenotype in Mn neurotoxicity, additional experimentation in primary astrocytes and microglia is required to determine the relative contributions of each cell type to a reactive phenotype following Mn exposure.

Despite the heightened focus on glial involvement in manganism, there are still many unanswered questions regarding mechanisms due to the limited number of in vivo studies and the inability of Mn to be a very potent glial activator in the absence of other inflammatory factors (Park and Chun 2016).As with other disorders of the CNS with a neuroinflammatory component, most studies into glial involvement in manganism have used single or mixed cultures of microglia or astrocytes, with few studies examining cell-cell interactions. The studies described in this review suggest that complicated signaling mechanisms between microglia and astrocytes likely underlie development of a neuroinflammatory phenotype Mn neurotoxicity that ultimately results in the progression of neuronal injury leading to psychological and motor manifestations of manganism. Microglia produce a large number of pro-inflammatory factors that could amplify the activation state of astrocytes, including TNFα and IL-1β, as well as numerous cytokines and chemokines.

It will be important in future studies to determine which of these factors are most relevant to the cell-cell interactions underlying the damaging effects of neuroinflammation following exposure to Mn. With limited or no treatment options for interdicting neuroinflammation in the CNS, it will be imperative to better identify underlying mechanisms in order to develop better therapies. This is particularly of concern, given the recent appreciation for how Mn exposures in children can lead to persistent adverse neurological affects. Thus, there is a need for a more systematic and comprehensive look at glial involvement and the potential importance of this response in chronic exposures.

Conclusion

Neuroinflammatory activation of glial cells is an important mechanism in Mn neurotoxicity and in other degenerative conditions of the CNS. Studies in the last several decades have redefined the importance of astrocytes and microglia to neuronal development, homeostasis, and survival, transforming our understanding of the role of these cells from inert structural components to important components of brain physiology and pathology. More specifically, the importance of astrocytes and microglia to neuronal survival has received increased attention, as these two glial types are the most often altered during disease states and are now known to be fundamental components of the innate immune system of the brain. Inflammatory activation of glia, or neuroinflammation, is a classic and conserved marker of neuropathology and is implicated in the progression, and possibly initiation, of several CNS disorders including seizure, Parkinson's disease, and manganism. Yet, much of the above information on neuroinflammation has been gleaned from rodent modeling with few studies utilizing translational or environmental relevant models to examine these important mechanisms. Furthermore, because glial activation can also serve either neuroprotective or neurotoxic functions, there exists a need to better understand the timeline and pathways of glial activation with a more extensive focus on the relative contributions of different glial types and the dynamics of glial-to-glial signaling. By examining specific glial-derived mechanisms in several neurodegenerative diseases, we may better understand the implications of neuroinflammation for CNS pathology and discover new potential targets for therapeutic intervention.

References

Alcamo E, et al. Targeted mutation of TNF receptor I rescues the RelA-deficient mouse and reveals a critical role for NF-kappa B in leukocyte recruitment. J Immunol. 2001;167:1592–600.
Araque A, Carmignoto G, Haydon PG. Dynamic signaling between astrocytes and neurons. Annu Rev Physiol. 2001;63:795–813. doi:10.1146/annurev.physiol.63.1.795.

Aschner M, Aschner JL. Manganese neurotoxicity: cellular effects and blood-brain barrier transport. Neurosci Biobehav Rev. 1991;15:333–40.

Aschner JL, Aschner M. Nutritional aspects of manganese homeostasis. Mol Asp Med. 2005;26:353–62. doi:10.1016/j.mam.2005.07.003.

Aschner M, Gannon M, Kimelberg HK. Manganese uptake and efflux in cultured rat astrocytes. J Neurochem. 1992;58:730–5.

Barhoumi R, Faske J, Liu X, Tjalkens RB. Manganese potentiates lipopolysaccharide-induced expression of NOS2 in C6 glioma cells through mitochondrial-dependent activation of nuclear factor kappaB. Brain Res Mol Brain Res. 2004;122:167–79. doi:10.1016/j.molbrainres.2003.12.009.

Block ML, Hong JS. Microglia and inflammation-mediated neurodegeneration: multiple triggers with a common mechanism. Prog Neurobiol. 2005;76:77–98. doi:10.1016/j.pneurobio.2005.06.004.

Bonizzi G, Karin M. The two NF-kappaB activation pathways and their role in innate and adaptive immunity. Trends Immunol. 2004;25:280–8. doi:10.1016/j.it.2004.03.008.

Brambilla R, et al. Inhibition of astroglial nuclear factor kappaB reduces inflammation and improves functional recovery after spinal cord injury. J Exp Med. 2005;202:145–56. doi:10.1084/jem.20041918.

Brambilla R, et al. Transgenic inhibition of astroglial NF-kappa B improves functional outcome in experimental autoimmune encephalomyelitis by suppressing chronic central nervous system inflammation. J Immunol. 2009;182:2628–40. doi:10.4049/jimmunol.0802954.

Carbone DL, Popichak KA, Moreno JA, Safe S, Tjalkens RB. Suppression of 1-methyl-4-phenyl-1,2,3,6-tetrahydropyridine-induced nitric-oxide synthase 2 expression in astrocytes by a novel diindolylmethane analog protects striatal neurons against apoptosis. Mol Pharmacol. 2009;75:35–43. doi:10.1124/mol.108.050781.

Carmignoto G, Gomez-Gonzalo M. The contribution of astrocyte signalling to neurovascular coupling. Brain Res Rev. 2010;63:138–48. doi:10.1016/j.brainresrev.2009.11.007.

Centonze D, Gubellini P, Bernardi G, Calabresi P. Impaired excitatory transmission in the striatum of rats chronically intoxicated with manganese. Exp Neurol. 2001;172:469–76. doi:10.1006/exnr.2001.7812.

Chang JY, Liu LZ. Manganese potentiates nitric oxide production by microglia. Brain Res Mol Brain Res. 1999;68:22–8.

Chao CC, Hu S, Molitor TW, Shaskan EG, Peterson PK. Activated microglia mediate neuronal cell injury via a nitric oxide mechanism. J Immunol. 1992;149:2736–41.

Chen CJ, et al. Manganese modulates pro-inflammatory gene expression in activated glia. Neurochem Int. 2006;49:62–71. doi:10.1016/j.neuint.2005.12.020.

Cho IH, et al. Role of microglial IKKbeta in kainic acid-induced hippocampal neuronal cell death. Brain. 2008;131:3019–33. doi:10.1093/brain/awn230.

Christopherson KS, et al. Thrombospondins are astrocyte-secreted proteins that promote CNS synaptogenesis. Cell. 2005;120:421–33. doi:10.1016/j.cell.2004.12.020.

Collipp PJ, Chen SY, Maitinsky S. Manganese in infant formulas and learning disability. Ann Nutr Metab. 1983;27:488–94.

Couper J. On the effects of black oxide of manganese when inhaled into the lungs. *British Annals of Medicine, Pharmacy, Vital Statistics, and General Science*. 1837;1:41–2.

Craft JM, Watterson DM, Van Eldik LJ. Neuroinflammation: a potential therapeutic target. Expert Opin Ther Targets. 2005;9:887–900. doi:10.1517/14728222.9.5.887.

Crittenden PL, Filipov NM. Manganese modulation of MAPK pathways: effects on upstream mitogen activated protein kinase kinases and mitogen activated kinase phosphatase-1 in microglial cells. J Appl Toxicol. 2011;31:1–10. doi:10.1002/jat.1552.

David S, Kroner A. Repertoire of microglial and macrophage responses after spinal cord injury. Nat Rev Neurosci. 2011;12:388–99. doi:10.1038/nrn3053.

De Miranda BR, et al. The Nurr1 activator 1,1-Bis(3′-Indolyl)-1-(p-Chlorophenyl)methane blocks inflammatory Gene expression in BV-2 microglial cells by inhibiting nuclear factor kappaB. Mol Pharmacol. 2015a;87:1021–34. doi:10.1124/mol.114.095398.

De Miranda BR, et al. Novel para-phenyl substituted diindolylmethanes protect against MPTP neurotoxicity and suppress glial activation in a mouse model of Parkinson's disease. Toxicol Sci. 2015b;143:360–73. doi:10.1093/toxsci/kfu236.

Di Virgilio F, Ceruti S, Bramanti P, Abbracchio MP. Purinergic signalling in inflammation of the central nervous system. Trends Neurosci. 2009;32:79–87. doi:10.1016/j.tins.2008.11.003.

Diaz-Aparicio I, Beccari S, Abiega O, Sierra A. Clearing the corpses: regulatory mechanisms, novel tools, and therapeutic potential of harnessing microglial phagocytosis in the diseased brain. Neural Regen Res. 2016;11:1533–9. doi:10.4103/1673-5374.193220.

DiDonato JA, Hayakawa M, Rothwarf DM, Zandi E, Karin M. A cytokine-responsive IkappaB kinase that activates the transcription factor NF-kappaB. Nature. 1997;388:548–54. doi:10.1038/41493.

Doetsch F. The glial identity of neural stem cells. Nat Neurosci. 2003;6:1127–34. doi:10.1038/nn1144.

Filipov NM, Dodd CA. Role of glial cells in manganese neurotoxicity. J Appl Toxicol. 2012;32:310–7. doi:10.1002/jat.1762.

Filipov NM, Seegal RF, Lawrence DA. Manganese potentiates in vitro production of proinflammatory cytokines and nitric oxide by microglia through a nuclear factor kappa B-dependent mechanism. Toxicol Sci. 2005;84:139–48. doi:10.1093/toxsci/kfi055.

Frank-Cannon TC, Alto LT, McAlpine FE, Tansey MG. Does neuroinflammation fan the flame in neurodegenerative diseases? Mol Neurodegener. 2009;4:47. doi:10.1186/1750-1326-4-47.

Gehrmann J, Matsumoto Y, Kreutzberg GW. Microglia: intrinsic immuneffector cell of the brain. Brain Res Brain Res Rev. 1995;20:269–87.

Gensel JC, Kigerl KA, Mandrekar-Colucci SS, Gaudet AD, Popovich PG. Achieving CNS axon regeneration by manipulating convergent neuro-immune signaling. Cell Tissue Res. 2012;349:201–13. doi:10.1007/s00441-012-1425-5.

Giordano G, Pizzurro D, VanDeMark K, Guizzetti M, Costa LG. Manganese inhibits the ability of astrocytes to promote neuronal differentiation. Toxicol Appl Pharmacol. 2009;240:226–35. doi:10.1016/j.taap.2009.06.004.

Glass CK, Saijo K, Winner B, Marchetto MC, Gage FH. Mechanisms underlying inflammation in neurodegeneration. Cell. 2010;140:918–34. doi:10.1016/j.cell.2010.02.016.

Gonzalez-Scarano F, Baltuch G. Microglia as mediators of inflammatory and degenerative diseases. Annu Rev Neurosci. 1999;22:219–40. doi:10.1146/annurev.neuro.22.1.219.

Guilarte TR. Manganese and Parkinson's disease: a critical review and new findings. Environ Health Perspect. 2010;118:1071–80. doi:10.1289/ehp.0901748.

Hamby ME, Hewett JA, Hewett SJ. TGF-beta1 potentiates astrocytic nitric oxide production by expanding the population of astrocytes that express NOS-2. Glia. 2006;54:566–77. doi:10.1002/glia.20411.

Harischandra DS, Jin H, Anantharam V, Kanthasamy A, Kanthasamy AG. Alpha-Synuclein protects against manganese neurotoxic insult during the early stages of exposure in a dopaminergic cell model of Parkinson's disease. Toxicol Sci. 2015;143:454–68. doi:10.1093/toxsci/kfu247.

Hirsch EC, Hunot S. Neuroinflammation in Parkinson's disease: a target for neuroprotection? Lancet Neurol. 2009;8:382–97. doi:10.1016/S1474-4422(09)70062-6.

Hua MS, Huang CC. Chronic occupational exposure to manganese and neurobehavioral function. J Clin Exp Neuropsychol. 1991;13:495–507. doi:10.1080/01688639108401066.

Husemann J, Loike JD, Anankov R, Febbraio M, Silverstein SC. Scavenger receptors in neurobiology and neuropathology: their role on microglia and other cells of the nervous system. Glia. 2002;40:195–205. doi:10.1002/glia.10148.

Karin M. How NF-kappaB is activated: the role of the IkappaB kinase (IKK) complex. Oncogene. 1999;18:6867–74. doi:10.1038/sj.onc.1203219.

Karin M. Inflammation-activated protein kinases as targets for drug development. Proc Am Thorac Soc. 2005;2:386–390.; discussion 394–385. doi:10.1513/pats.200504-034SR.

Kaushal V, Schlichter LC. Mechanisms of microglia-mediated neurotoxicity in a new model of the stroke penumbra. J Neurosci. 2008;28:2221–30. doi:10.1523/JNEUROSCI.5643-07.2008.

Kigerl KA, et al. Identification of two distinct macrophage subsets with divergent effects causing either neurotoxicity or regeneration in the injured mouse spinal cord. J Neurosci. 2009;29:13435–44. doi:10.1523/JNEUROSCI.3257-09.2009.

Kim SU, de Vellis J. Microglia in health and disease. J Neurosci Res. 2005;81:302–13. doi:10.1002/jnr.20562.

Kim YS, et al. Matrix metalloproteinase-3: a novel signaling proteinase from apoptotic neuronal cells that activates microglia. J Neurosci. 2005;25:3701–11. doi:10.1523/JNEUROSCI.4346-04.2005.

Kim Y, et al. Co-exposure to environmental lead and manganese affects the intelligence of school-aged children. Neurotoxicology. 2009;30:564–71. doi:10.1016/j.neuro.2009.03.012.

Kimelberg HK. The problem of astrocyte identity. Neurochem Int. 2004;45:191–202. doi:10.1016/j.neuint.2003.08.015.

Kuno R, et al. The role of TNF-alpha and its receptors in the production of NGF and GDNF by astrocytes. Brain Res. 2006;1116:12–8. doi:10.1016/j.brainres.2006.07.120.

Lalo U, et al. P2X1 and P2X5 subunits form the functional P2X receptor in mouse cortical astrocytes. J Neurosci. 2008;28:5473–80. doi:10.1523/JNEUROSCI.1149-08.2008.

Lawson LJ, Perry VH, Dri P, Gordon S. Heterogeneity in the distribution and morphology of microglia in the normal adult mouse brain. Neuroscience. 1990;39:151–70.

Lee SC, Liu W, Dickson DW, Brosnan CF, Berman JW. Cytokine production by human fetal microglia and astrocytes. Differential induction by lipopolysaccharide and IL-1 beta. J Immunol. 1993;150:2659–67.

Lee DJ, Hsu MS, Seldin MM, Arellano JL, Binder DK. Decreased expression of the glial water channel aquaporin-4 in the intrahippocampal kainic acid model of epileptogenesis. Exp Neurol. 2012;235:246–55. doi:10.1016/j.expneurol.2012.02.002.

Li ZW, Omori SA, Labuda T, Karin M, Rickert RC. IKK beta is required for peripheral B cell survival and proliferation. J Immunol. 2003;170:4630–7.

Liu B, Gao HM, Hong JS. Parkinson's disease and exposure to infectious agents and pesticides and the occurrence of brain injuries: role of neuroinflammation. Environ Health Perspect. 2003;111:1065–73.

Liu X, Sullivan KA, Madl JE, Legare M, Tjalkens RB. Manganese-induced neurotoxicity: the role of astroglial-derived nitric oxide in striatal interneuron degeneration. Toxicol Sci. 2006;91:521–31. doi:10.1093/toxsci/kfj150.

Mastroeni D, et al. Microglial responses to dopamine in a cell culture model of Parkinson's disease. Neurobiol Aging. 2009;30:1805–17. doi:10.1016/j.neurobiolaging.2008.01.001.

Matyash V, Kettenmann H. Heterogeneity in astrocyte morphology and physiology. Brain Res Rev. 2010;63:2–10. doi:10.1016/j.brainresrev.2009.12.001.

McCarty MF. Down-regulation of microglial activation may represent a practical strategy for combating neurodegenerative disorders. Med Hypotheses. 2006;67:251–69. doi:10.1016/j.mehy.2006.01.013.

Menezes-Filho JA, Novaes Cde O, Moreira JC, Sarcinelli PN, Mergler D. Elevated manganese and cognitive performance in school-aged children and their mothers. Environ Res. 2011;111:156–63. doi:10.1016/j.envres.2010.09.006.

Morello M, et al. Sub-cellular localization of manganese in the basal ganglia of normal and manganese-treated rats an electron spectroscopy imaging and electron energy-loss spectroscopy study. Neurotoxicology. 2008;29:60–72. doi:10.1016/j.neuro.2007.09.001.

Moreno JA, Sullivan KA, Carbone DL, Hanneman WH, Tjalkens RB. Manganese potentiates nuclear factor-kappaB-dependent expression of nitric oxide synthase 2 in astrocytes by activating soluble guanylate cyclase and extracellular responsive kinase signaling pathways. J Neurosci Res. 2008;86:2028–38. doi:10.1002/jnr.21640.

Moreno JA, Streifel KM, Sullivan KA, Legare ME, Tjalkens RB. Developmental exposure to manganese increases adult susceptibility to inflammatory activation of glia and neuronal protein nitration. Toxicol Sci. 2009;112:405–15. doi:10.1093/toxsci/kfp221.

Moreno JA, Streifel KM, Sullivan KA, Hanneman WH, Tjalkens RB. Manganese-induced NF-kappaB activation and nitrosative stress is decreased by estrogen in juvenile mice. Toxicol Sci. 2011;122:121–33. doi:10.1093/toxsci/kfr091.

Mosley RL, et al. Neuroinflammation, Oxidative Stress and the Pathogenesis of Parkinson's Disease. Clin Neurosci Res. 2006;6:261–81. doi:10.1016/j.cnr.2006.09.006.

Mulligan SJ, MacVicar BA. Calcium transients in astrocyte endfeet cause cerebrovascular constrictions. Nature. 2004;431:195–9. doi:10.1038/nature02827.

Nakajima K, Kohsaka S. Functional roles of microglia in the brain. Neurosci Res. 1993;17:187–203.

Neal AP, Guilarte TR. Mechanisms of heavy metal neurotoxicity: lead and manganese. Drug Metab toxicol. 2012;5

Nedergaard M, Verkhratsky A. Artifact versus reality--how astrocytes contribute to synaptic events. Glia. 2012;60:1013–23. doi:10.1002/glia.22288.

Neher JJ, et al. Inhibition of microglial phagocytosis is sufficient to prevent inflammatory neuronal death. J Immunol. 2011;186:4973–83. doi:10.4049/jimmunol.1003600.

Nimmerjahn A. Astrocytes going live: advances and challenges. J Physiol. 2009;587:1639–47. doi:10.1113/jphysiol.2008.167171.

O'Callaghan JP, Sriram K. Glial fibrillary acidic protein and related glial proteins as biomarkers of neurotoxicity. Expert Opin Drug Saf. 2005;4:433–42. doi:10.1517/14740338.4.3.433.

Olanow CW. Manganese-induced parkinsonism and Parkinson's disease. Ann N Y Acad Sci. 2004;1012:209–23.

Park E, Chun HS. Melatonin attenuates manganese and lipopolysaccharide-induced inflammatory activation of BV2 microglia. Neurochem Res. 2016; doi:10.1007/s11064-016-2122-7.

Parpura V, et al. Glial cells in (patho)physiology. J Neurochem. 2012;121:4–27. doi:10.1111/j.1471-4159.2012.07664.x.

Perea G, Araque A. GLIA modulates synaptic transmission. Brain Res Rev. 2010;63:93–102. doi:10.1016/j.brainresrev.2009.10.005.

Perea G, Navarrete M, Araque A. Tripartite synapses: astrocytes process and control synaptic information. Trends Neurosci. 2009;32:421–31. doi:10.1016/j.tins.2009.05.001.

Perl DP, Olanow CW. The neuropathology of manganese-induced parkinsonism. J Neuropathol Exp Neurol. 2007;66:675–82. doi:10.1097/nen.0b013e31812503cf.

Powell EM, Geller HM. Dissection of astrocyte-mediated cues in neuronal guidance and process extension. Glia. 1999;26:73–83.

Ransohoff RM, Perry VH. Microglial physiology: unique stimuli, specialized responses. Annu Rev Immunol. 2009;27:119–45. doi:10.1146/annurev.immunol.021908.132528.

Riojas-Rodriguez H, et al. Intellectual function in Mexican children living in a mining area and environmentally exposed to manganese. Environ Health Perspect. 2010;118:1465–70.

Saijo K, et al. A Nurr1/CoREST pathway in microglia and astrocytes protects dopaminergic neurons from inflammation-induced death. Cell. 2009;137:47–59. doi:10.1016/j.cell.2009.01.038.

Santamaria AB. Manganese exposure, essentiality & toxicity. Indian J Med Res. 2008;128:484–500.

Sidoryk-Wegrzynowicz M, Aschner M. Role of astrocytes in manganese mediated neurotoxicity. BMC Pharmacol Toxicol. 2013;14:23. doi:10.1186/2050-6511-14-23.

Sigel, A. S., H.; Sigel, R.K.O. *Metal Ions in Life Sciences*. (Wiley, 2007).

Silver J, Miller JH. Regeneration beyond the glial scar. Nat Rev Neurosci. 2004;5:146–56. doi:10.1038/nrn1326.

Sofroniew MV, Vinters HV. Astrocytes: biology and pathology. Acta Neuropathol. 2010;119:7–35. doi:10.1007/s00401-009-0619-8.

Spranger M, et al. Manganese augments nitric oxide synthesis in murine astrocytes: a new pathogenetic mechanism in manganism? Exp Neurol. 1998;149:277–83. doi:10.1006/exnr.1997.6666.

Streifel KM, Moreno JA, Hanneman WH, Legare ME, Tjalkens RB. Gene deletion of nos2 protects against manganese-induced neurological dysfunction in juvenile mice. Toxicol Sci. 2012;126:183–92. doi:10.1093/toxsci/kfr335.

Streifel KM, Miller J, Mouneimne R, Tjalkens RB. Manganese inhibits ATP-induced calcium entry through the transient receptor potential channel TRPC3 in astrocytes. Neurotoxicology. 2013;34:160–6. doi:10.1016/j.neuro.2012.10.014.

Streifel KM, et al. Dopaminergic neurotoxicants cause biphasic inhibition of purinergic calcium signaling in astrocytes. PLoS One. 2014;9:e110996. doi:10.1371/journal.pone.0110996.

Streit WJ. Microglia as neuroprotective, immunocompetent cells of the CNS. Glia. 2002;40:133–9. doi:10.1002/glia.10154.

Suarez-Fernandez MB, et al. Aluminum-induced degeneration of astrocytes occurs via apoptosis and results in neuronal death. Brain Res. 1999;835:125–36.

Surprenant A, North RA. Signaling at purinergic P2X receptors. Annu Rev Physiol. 2009;71:333–59. doi:10.1146/annurev.physiol.70.113006.100630.

Takano T, Oberheim N, Cotrina ML, Nedergaard M. Astrocytes and ischemic injury. Stroke. 2009;40:S8–12. doi:10.1161/STROKEAHA.108.533166.

Tansey MG, McCoy MK, Frank-Cannon TC. Neuroinflammatory mechanisms in Parkinson's disease: potential environmental triggers, pathways, and targets for early therapeutic intervention. Exp Neurol. 2007;208:1–25. doi:10.1016/j.expneurol.2007.07.004.

Tjalkens RB, Zoran MJ, Mohl B, Barhoumi R. Manganese suppresses ATP-dependent intercellular calcium waves in astrocyte networks through alteration of mitochondrial and endoplasmic reticulum calcium dynamics. Brain Res. 2006;1113:210–9. doi:10.1016/j.brainres.2006.07.053.

van Loo G, et al. Inhibition of transcription factor NF-kappaB in the central nervous system ameliorates autoimmune encephalomyelitis in mice. Nat Immunol. 2006;7:954–61. doi:10.1038/ni1372.

Verina T, Kiihl SF, Schneider JS, Guilarte TR. Manganese exposure induces microglia activation and dystrophy in the substantia nigra of non-human primates. Neurotoxicology. 2011;32:215–26. doi:10.1016/j.neuro.2010.11.003.

Verkhratski AN, Butt A. Glial physiology and pathophysiology. Chichester: Wiley-Blackwell; 2013.

Verkhratsky A, Steardo L, Parpura V, Montana V. Translational potential of astrocytes in brain disorders. Prog Neurobiol. 2016;144:188–205. doi:10.1016/j.pneurobio.2015.09.003.

Vezzani A, Aronica E, Mazarati A, Pittman QJ. Epilepsy and brain inflammation. Exp Neurol. 2013;244:11–21. doi:10.1016/j.expneurol.2011.09.033.

Webster CM, et al. Microglial P2Y12 deficiency/inhibition protects against brain ischemia. PLoS One. 2013;8:e70927. doi:10.1371/journal.pone.0070927.

Woolf A, Wright R, Amarasiriwardena C, Bellinger D. A child with chronic manganese exposure from drinking water. Environ Health Perspect. 2002;110:613–6.

Wyss-Coray T, Mucke L. Inflammation in neurodegenerative disease--a double-edged sword. Neuron. 2002;35:419–32.

Xu B, Xu ZF, Deng Y. Protective effects of MK-801 on manganese-induced glutamate metabolism disorder in rat striatum. Exp Toxicol Pathol. 2010;62:381–90. doi:10.1016/j.etp.2009.05.007.

Yin Z, et al. Methylmercury induces oxidative injury, alterations in permeability and glutamine transport in cultured astrocytes. Brain Res. 2007;1131:1–10. doi:10.1016/j.brainres.2006.10.070.

Zhang S, Zhou Z, Fu J. Effect of manganese chloride exposure on liver and brain mitochondria function in rats. Environ Res. 2003;93:149–57.

Zhang W, et al. Aggregated alpha-synuclein activates microglia: a process leading to disease progression in Parkinson's disease. FASEB J. 2005;19:533–42. doi:10.1096/fj.04-2751com.

Zhang P, et al. Microglia enhance manganese chloride-induced dopaminergic neurodegeneration: role of free radical generation. Exp Neurol. 2009;217:219–30. doi:10.1016/j.expneurol.2009.02.013.

Zhang P, Lokuta KM, Turner DE, Liu B. Synergistic dopaminergic neurotoxicity of manganese and lipopolysaccharide: differential involvement of microglia and astroglia. J Neurochem. 2010;112:434–43. doi:10.1111/j.1471-4159.2009.06477.x.
Zhao F, et al. Manganese induces dopaminergic neurodegeneration via microglial activation in a rat model of manganism. Toxicol Sci. 2009;107:156–64. doi:10.1093/toxsci/kfn213.

Aluminum and Alzheimer's Disease

Maria Teresa Colomina and Fiona Peris-Sampedro

Abstract Aluminum (Al) is one of the most extended metals in the Earth's crust. Its abundance, together with the widespread use by humans, makes Al-related toxicity particularly relevant for human health.

Despite some factors influence individual bioavailability to this metal after oral, dermal, or inhalation exposures, humans are considered to be protected against Al toxicity because of its low absorption and efficient renal excretion. However, several factors can modify Al absorption and distribution through the body, which may in turn progressively contribute to the development of silent chronic exposures that may lately trigger undesirable consequences to health. For instance, Al has been recurrently shown to cause encephalopathy, anemia, and bone disease in dialyzed patients. On the other hand, it remains controversial whether low doses of this metal may contribute to developing Alzheimer's disease (AD), probably because of the multifactorial and highly variable presentation of the disease.

This chapter primarily focuses on two key aspects related to Al neurotoxicity and AD, which are metabolic impairment and iron (Fe) alterations. We discuss sex and genetic differences as a plausible source of bias to assess risk assessment in human populations.

Keywords Neurodegeneration • Aluminum • Transferrin (Tf) • Iron (Fe) • Sex differences • Glucose homeostasis

M.T. Colomina (✉) • F. Peris-Sampedro
Research in Neurobehavior and Health (NEUROLAB), Universitat Rovira i Virgili, Tarragona, Spain

Department of Psychology and Research Center for Behavior Assessment (CRAMC), Universitat Rovira i Virgili, Tarragona, Spain
e-mail: mariateresa.colomina@urv.cat

© Springer International Publishing AG 2017
M. Aschner and L.G. Costa (eds.), *Neurotoxicity of Metals*, Advances in Neurobiology 18, DOI 10.1007/978-3-319-60189-2_9

Aluminum in the Environment and Human Exposure

Al stands as the most abundant metallic element and ranks third in abundance among the Earth's crust constituents. Although natural processes and acidic rain redistributes it in the nature, thus contributing to the natural occurrence of the metal in food and water, growing industrialization has been responsible for increasing the presence of Al in the environment.

To date, no physiological functions for Al have been described in mammals and, therefore, it has sometimes been regarded as not presenting a significant health hazard. In addition to the insoluble nature of the metal and its low absorption, an efficient renal elimination prevents Al accumulation in the body, thereby reducing the risk of acute human toxicity under non-pathological conditions. Despite human natural barriers (i.e., skin, gastrointestinal barrier, lungs, etc.) protect general population from environmental exposures, patients suffering from chronic renal failure may be at risk of Al toxicity (Fenwick et al. 2005).

Al has been extensively used in the industry, and it is currently added to a vast number of products available to everyone, including drinking water, many processed foods, infant formulae, cosmetics, toothpaste, antiperspirants, and various medical preparations and medicines (for a review, see Bondy 2016). These widespread uses make human exposure to Al practically unavoidable. Therefore, once presumed Al is ubiquitous in the environment, it is not so unreasonable to expect a wide range of sources of exposure.

Considering the general population, food and water represent the most common form of human exposure to the metal (Bondy 2016; Crisponi et al. 2013). The concentration of Al in food is extremely variable, due both to the original content and to food interaction with the material it contacts when stored or cooked. For example, when food or beverages are stored in Al-derived packaging formats, Al content is five to seven times higher compared to the same type of food from other containers (Duggan et al. 1992). Even though the Al content in most plant food is low (i.e., less than 25 μg/g of dry food weight), relatively high levels of Al have been reported in some spices, such as marjoram and thyme, soy-based milk formulas, tea leaves, and coffee beans (Burrell and Exley 2010; Crisponi et al. 2013; Malik et al. 2008). As for animal-derived food, Al levels in some dairy products (i.e., milk, cheese, etc.) have been found to be beyond the permissible limits (AI-Ashmawy 2011). The increasing contamination of rivers and seas has also prompted the accumulation and storage of Al in such crustaceans as crayfishes (Woodburn et al. 2011).

On the other hand, some data have endorsed that both the inhalation of Al particles and dermal absorption upon contact with the skin may also account, although to a lesser extent, for the body burden of Al (Darbre 2005; Pauluhn 2009).

Although Al total intake considerably varies upon country, place of residence, and diet composition, several authors have estimated Al typical adult intake to be ranged from 3 to 12 mg/day (Domingo et al. 2011; Krewski et al. 2007). Al absorption, though, is generally low, being almost 95% of the total Al ingested directly excreted through feces. In point of fact, the total Al absorption may vary from 0.01 to 1% of the total metal intake (Moore et al. 2000; Wilhelm et al. 1990). The presence of certain compounds in the diet such as citrate, lactate, ascorbate, gluconate, succinate, tartrate, malate, and oxalate can increase the rate of absorption of Al (Krewski et al. 2007). Likewise, low plasma levels of magnesium (Mg) and Fe (Cannata et al. 1991), as well as enhanced vitamin D status, may increase Al absorption (Krewski et al. 2007; Schwalfenberg and Genuis 2015). Therefore, Al bioavailability is highly dependent on individual differences, fact that merits to be taken into account to control confounding variables in both experimental and epidemiological studies.

Once absorbed, Al has a half-life of several hours in the blood. Indeed, Al is primarily bound to plasma transferrin (Tf) (i.e., 90%) and, to a lesser extent, to low molecular weight molecules, such as citrate (i.e., 10%) (Ohman and Martin 1994). Even though the mechanisms through which Al enter the brain are not yet fully understood, this process seems to be governed by two different mechanisms. Firstly, Al can enter the brain from blood. As a matter of fact, it is well known for more than 25 years that transferrin can mediate Al transport across the blood-brain barrier (BBB) by transferrin receptor (TfR)-mediated endocytosis of Al transferrin (Bondy 2016; Yokel and McNamara 2001). On the other hand, though, there is evidence to suggest that a second mechanism transporting Al citrate across the BBB into the brain independently of transferrin may exist. Indeed, transferrin concentration is very low in cerebrospinal fluid (CSF), and presumably brain extracellular fluid, whereas the citrate concentration is higher in CSF than in plasma, favoring Al citrate as the predominant Al species in brain extracellular fluid (Martin and Bruce 1997; Yokel and McNamara 2001). Although most brain Al is quickly removed, some experimental research evidenced that its half-life may be about 150 days in rats (Yokel et al. 2001).

At physiological pH, Al exhibits the trivalent oxidation state (i.e., Al^{3+}), which is crucial in determining the physicochemical properties and biological interactions inherent to the metal.

The main mechanism of Al toxicity involves the disruption of the homeostasis of metals, such as Mg, calcium (Ca), and Fe (Harris et al. 1996; Yokel and McNamara 2001). Indeed, the physical and chemical properties of Al (i.e., the small radius of Al^{3+}, its affinity to oxygen atoms, carboxylate, deprotonated hydroxyl and phosphate groups, etc.) allow it to effectively mimic these metals in their respective biological functions and trigger biochemical abnormalities, thereby defining Al individual's toxicokinetics (Yokel and McNamara 2001).

Aluminum in the Brain: Molecular and Functional Interactions

Al is unequally distributed in brain areas and neural cells. Indeed, Al accumulates in glia largely than in neurons (Oshiro et al. 2000). Experimental studies in rats and mice showed that Al accumulates in the brain cortex, hippocampus, and cerebellum (Bellés et al. 1998; Esparza et al. 2003; García et al. 2009) after either parenteral or oral exposures. Accordingly, several authors have measured brain levels of Al in AD patients and non-demented controls, and both the hippocampus and the amygdala stand as the most relevant areas containing Al. Despite data are not always consistent (Akatsu et al. 2012), statistical treatments and the control of confounding variables have been found to influence the statistical significance of the result (Rusina et al. 2011).

It is well known that such metal ions as Al are able to interact with different proteins to induce conformational changes that eventually result in misfolding, aggregation, or oligomerization, thus leading to an altered turnover and removal of the protein. Protein misfolding and aggregation is a key pathophysiological mechanism on AD. Hence, an increased interest on the possible contribution of Al on the amyloid (Aβ) cascade hypothesis for AD has generated a deal of research. Unfortunately, results have not always been consistent. Indeed, many studies reported that Al promotes the expression of the precursor (APP) of the Aβ protein, increases the levels of β-40 and β-42 fragments in the brain, and boosts the aggregation of Aβ protein (Zatta et al. 2009; Bolognin et al. 2011; Praticò et al. 2002; Banks et al. 2006). Other in vivo approaches, though, did not replicate the results previously reported for the Aβ pathway (Ribes et al. 2010; Akiyama et al. 2012). Further, it has been shown that Al promotes both the phosphorylation and the aggregation of highly phosphorylated proteins such as tau protein (Leterrier et al. 1992; Nübling et al. 2012). According to this, Al has been detected in neuritic plaques and tangle-bearing neurons, pointing at the involvement of this metal in the pathogenesis of AD (Miu et al. 2003; Yumoto et al. 2009). Moreover, Al has also been related to altered synaptic plasticity in the hippocampus of rats when chronically and orally administered at high doses (Colomina et al. 2002).

Aluminum and Glucose Metabolism

Despite the high prevalence of Al in the environment, there is a gap of knowledge on its interaction with physiological systems. An escalating body of experimental research has demonstrated so far that Al inhibits a vast number of ATP-dependent reactions, thereby interfering with energy-dependent processes (Caspers et al. 1994; Joshi et al. 1994; Kaizer et al. 2007; Silva et al. 2005; Singla and Dhawan 2012). Nonetheless, the exact mechanisms remain to be fully unraveled. Thus, Al^{3+} binds to ATP 107 times more tightly than does Mg^{2+}, but upon in vivo testing not every ATP-dependent reactions are inhibited (Joshi 1990).

Many experimental approaches have also endorsed that Al exposure may impair glucose utilization, upon altering activities of glucose-metabolizing enzymes, such as glucose-6-phosphate dehydrogenase (G6PD), hexokinase, or glutamate dehydrogenase (Dua et al. 2010; Joshi et al. 1994). Strikingly, G6PD enzyme has been shown to reduce its activity in the presence of Al in such brain regions that are also affected in AD (Joshi et al. 1994; Jovanović et al. 2014). In point of fact, accumulated evidence indicates that AD is a metabolic neurodegenerative disease. Thus, impaired cerebral glucose metabolism represents an invariant pathophysiological feature in AD, and its occurrence mostly precedes cognitive dysfunction and pathological alterations. Therefore, delving into the consequences associated with abnormal cerebral glucose metabolism will provide valuable clues for treatment strategies as well as ideal diagnostic approaches in AD.

Aluminum and Iron Interactions

Over the last years, a considerable amount of literature has grown up shedding light on the disruption of Fe homeostasis by Al. Fe, an essential trace metal, displays a great deal of biological functions, including oxygen transport and exchange, metabolic protein synthesis, and enzyme cofactor (Aisen et al. 2001). Because of its biological importance and high redox potential, Fe is strictly regulated by Tf, transferrin receptor (TfR), and ferritin. Thus, increases in TfR allow more Fe to enter the cell, while a decrease in ferritin levels enables more free iron to reach the respiratory chain and other Fe-requiring systems. Under Fe-replete conditions, TfR decreases and levels of ferritin increase, thereby allowing the metal to be stored in a complex with ferritin, which prevents iron-mediated oxidative stress (Aisen et al. 2001). Several in vitro studies have reported that Al exhibits the same effect on TfR and ferritin as Fe does in a deprived status. Thus, Al increases the number of TfRs, which leads to an increase in Fe absorption, and also decreases ferritin, which might result in higher levels of free Fe. These effects of Al on Fe homeostasis might explain the increases in oxidative stress and inflammation in both in vitro and in vivo upon exposure to Al (Kim et al. 2007).

The total body burden of Fe has been found to increase with age in a sex-dependent manner (Joshi et al. 1994). While males progressively increase Fe levels from 300 to 1800 mg between 20–25 and 80–90 years, Fe stores in women remain at 300 mg from 20 to 25 years until the premenopausal state. Then, Fe stores begin to increase until reaching 1300 mg at the age of 80–90 years (Joshi et al. 1994). Furthermore, women increase Fe storage parameters from premenopause to postmenopause. Strikingly, such Fe increases correlate with the increase in HOMA-R index, which indicated insulin resistance (Van den Bosch et al. 2001).

In view of the influence of Fe status in Al absorption, we speculated that this sex-related pattern of storage can influence the onset and course of neurodegeneration. The sharp increase in Fe stores from middle age to elderly shows some parallelisms in temporal patterns observed for AD prevalence in women. According to Fe status, young women, which display low levels of Fe storage, would be more able to store Al but protected from Al-Fe interactions, and therefore from oxidative stress. By contrast, postmenopausal women would have higher Fe stores, but Al deposits would remain high because of a long-life exposure and efficient storage favored by moderated Fe levels in serum.

Aluminum and Oxidative Stress

As previously stated, no biological role for Al is known yet. However, it is well accepted that it can induce severe toxic manifestations in mammals. Given the non-dividing nature of neurons, the brain has sometimes been considered to be the most vulnerable tissue to the toxic effects of Al (Kumar and Gill 2014). Indeed, a constellation of experimental research has highlighted neuropathological, neurobehavioral, neurophysical, and neurochemical changes upon Al administration (Akiyama et al. 2012; Colomina et al. 2002; Verstraeten et al. 1997; Ribes et al. 2008, 2010).

Further, the brain is particularly sensitive to oxidative stress due to an increased level of free radicals and decreased level of antioxidants following toxic insult (Kumar and Gill 2014). Several authors have suggested that Al exert a strong prooxidant activity despite its non-redox status (Exley 2004; Kumar and Gill 2014; Yuan et al. 2012).

To date, there are described many potential mechanisms underlying Al-related prooxidant toxicity, of which the effect of Al on Fe homeostasis is of special interest (Ward et al. 2001; Wu et al. 2012). As a matter of fact, the interaction between both agents generates labile Fe from Fe-containing enzymes and proteins, thereby increasing the intracellular pool of free Fe, which in turn leads to the formation of reactive oxygen species (ROS).

On the other hand, Al oxidative toxicity has also been related to increased lipid peroxidation, decreased membrane fluidity, and oxidized high-density lipoprotein (Johnson et al. 2005; Kaneko et al. 2007; Silva et al. 2002; Zatta et al. 2003). For example, Al has been shown to potentiate the free radical damage initiated by Fe^{3+} in lipid peroxidation, probably by facilitating the action of OH^- radicals in the membrane of phospholipids (Zatta et al. 2003). Other mechanisms, such as the formation of superoxide Al^{3+} semi-reduced radical, have been suggested to explain Al prooxidant effects (Ruipérez et al. 2012; Exley 2004). In general terms, the interaction with lipid substrates as well as with other prooxidant metals or elements such as O^{2-} are subjects of study in this regard (for review, see Exley 2004). Additionally, Al^{3+} decreases the activity of some antioxidant enzymes such as catalase, superoxide dismutase, and glutathione peroxidase (Fattoretti et al. 2003; Sánchez-Iglesias et al. 2009), thus aggravating neuronal damage induced by oxidative stress.

An Al-dependent oxidative environment is also characterized by a sharp decrease in mitochondrial activity (Han et al. 2013). Specifically, Al^{3+} disrupts mitochondrial bioenergetics and decreases the respiratory efficiency and the capacity of the mitochondria to modulate and control the energy production through the phosphorylation system (Iglesias-González et al. 2016).

Aluminum and Neurotransmission

Several studies have indicated Al is able to disrupt the cholinergic system, which is in turn implicated in AD pathogenesis. Both in vivo and in vitro studies have consistently shown changes in acetylcholinesterase (AChE) activity, as well as in ACh-evoked neurotransmission (Yokel et al. 1994; Bielarczyk et al. 1998; Szutowicz et al. 2000; Yellamma et al. 2010). Accordingly, the group of Petronijević found activity changes in AChE and lipid peroxides in a series of different studies with high Al doses administered to Gerbils (Mićić et al. 2003; Vučetić-Arsić et al. 2013). Despite the possible relevance of this pathway, few research has evaluated cholinergic function together with other parameters of interest. Strikingly, estradiol administration has shown to ameliorate alterations in cholinergic parameters and oxidative stress induced by Al intoxication (Mohamd et al. 2011).

Moreover, the neurotransmitters serotonin (5-HT) and dopamine (DA) (Abu-Taweel et al. 2012), as well as glutamate and aspartate (Liu et al. 2010), have been reported to decrease upon Al exposure. It is well known that neurotransmitter systems are modulated by sex hormones. In this sense, differences between sexes have been reported for the septo-hippocampal cholinergic system (Mitsushima 2011), monoamines 5-HT, and DA, as well as for glutamate and GABA (Barth et al. 2015). Therefore, the effects of Al in neurotransmitter systems could be different depending on sex, but no data exist so far in this regard.

Worldwide Advices and Al Regulations

Needless to say, to date, the detrimental effects of Al to human health are well established, and increasing eagerness to regulate its uses has become noticeable. Thus, many relevant regulatory agencies, including the Agency for Toxic Substances and Disease Registry (ATSDR), the US Environmental Protection Agency (EPA), the World Health Organization (WHO), the Food and Agriculture Organization (FAO), and the European Food Safety Authority (EFSA), have published a great body of reports on Al toxicity. However, there is still widespread mistrust about its potential deleterious effects upon silent chronic exposures. By way of example, there is still no convincing evidence to associate the Al found in food and drinking water with increased risk of AD. Neither is there clear evidence to suggest increased risk of AD nor some types of cancer upon using Al-containing antiperspirant or cosmetics. At most, the US FDA has warned that increasing Al concentration in an antiperspirant product may result in skin irritation. Further, even if adverse effects to vaccines with Al adjuvants have occurred, recent controlled trials found that the immunologic response to certain vaccines containing Al was no greater, and in some cases less than, that after identical vaccination without Al adjuvants (for a review, see Willhite et al. 2014).

The FAO expert committee on food additives and food contaminants had originally recommended a tolerable weekly intake (TWI) of Al of 7 mg/kg, which was lately reduced to 1 mg/kg upon reconsidering the reproductive and neurological detrimental effects of the metal (FAO 2006). In Europe, the EFSA stated in 2011 a TWI equivalent to 280 µg/kg per day (Anon 2011). Nonetheless, some authors have questioned these values since the EFSA assumed back then that gastrointestinal uptake of all ingested Al materials was equivalent to that measured for Al citrate (Willhite et al. 2014). On the other hand, the WHO recommended a maximum drinking water concentration of 0.2 mg Al/L (WHO 2004; WHO 2010).

Given that it has not yet been established which levels of Al are safe upon chronic exposures in human populations, an effort is needed to demand more regulations for the use of this metal in drinking water, dairy products, pharmaceuticals, and occupational exposures.

Alzheimer's Disease and Environmental Al Exposures

Alzheimer's disease (AD) is one of the most devastating neurodegenerative diseases, accounting for more than 80% of dementia cases in the elderly. It is a complex neurodegenerative disorder characterized by a neurological progressive impairment affecting several cognitive domains, behavior, and personality. Clinical phenotype is accompanied by three main neuropathological hallmarks: diffuse loss of neurons, neuronal cytoskeletal alterations or neurofibrillary tangles (NFT) produced by hyperphosphorylated tau protein aggregations, and extracellular Aβ protein deposits or senile plaques (Torreilles and Touchon 2002).

Two forms have been described for AD: the familial form, which is less frequent (1–5%) and mainly genetic, and late-onset AD (LOAD), which is most prevalent and heterogeneous in both onset and progression (Ridge et al. 2013).

The familial forms of the disease are mostly associated with mutations exhibiting an autosomal dominant pattern of inheritance. Thus, three mutated human genes encoding for (1) APP and the enzymes related to APP processing, (2) presenilin 1 (PSEN1), (3) and presenilin 2 (PSEN2) are crucial to the establishment of the disease (Levy-Lahad et al. 1998; Schellenberg et al. 1992). All of these genetic mutations lead to abnormal processing of APP and give rise to the Aβ cascade hypothesis. Although crucial, this hypothesis fails to explain by itself all the molecular, cellular, and clinical events observed in the different forms of AD. Before the identification of familiar forms, anatomical-pathological and biochemical studies of postmortem human brains from AD patients described deficits in the cholinergic system. In addition, considerable pieces of experimental and human studies have supported that a dysfunctional cholinergic system is sufficient to produce learning and memory deficits (Muir 1997). Thenceforth, the earliest cholinergic hypothesis of AD emerged. Degeneration of cholinergic neurons in the basal telencephalon (i.e., Meynert nucleus and medial septum nucleus) innervating the hippocampus, amygdala, and frontal cortex has been associated with severe cognitive deficits implicated in AD (Muir 1997). Moreover, pharmacological interventions with cholinergic agonists have endorsed the contribution of this system to cognitive decline (Giacobini 2003).

On the other hand, environmental risk factors accumulating over years are constantly challenging the integrity of the brain and thereby contributing together with risk genetic factors to the onset and progression of LOAD. Accordingly, *APOE4* genotype is the largest genetic risk for AD accounting for approximately 60% cases (Higgins et al. 1997). Indeed, being a carrier of one *ε4* allele increases the risk for AD in 2–3-folds, whereas the risk rises about 12-fold when carrying two *ε4* alleles (Roses 1996). Interestingly, emerging lines of evidence supported an *APOE4*-sex interaction in humans. Women carrying *ε4* have been shown to display more pronounced AD-like changes in neuroimaging, neuropathological, and neuropsychological measures than men (Beydoun et al. 2013; Ungar et al. 2014). As for environmental factors, it is worth asking for these agents and to which extent they are contributing to the onset and progression of the disease. In this sense, the hypothesis of Al and AD has become the subject of intense debate over the last decades.

The putative link between dietary Al and neurodegenerative disorders has been addressed in a large volume of clinical (Wills and Savory 1989; Yumoto et al. 2009), occupational (Riihimäki et al. 2000), and epidemiological surveys (Flaten 1990; Rondeau et al. 2008). Moreover, some anatomopathological findings in the brain of AD patients (Walton and Wang 2009; Yumoto et al. 2009) and some experimental studies (Praticò et al. 2002; Ribes et al. 2008; Ribes et al. 2010; Walton and Wang 2009) have also provided links between Al and AD. A collection of different studies performed until 2014 are reviewed in Willhite et al. (2014).

While some experimental results have not been widely replicable, epidemiological studies showed considerable consistent associations. In a recent meta-analysis of epidemiological studies, Wang et al. (2016) assessed the relation between Al exposure and AD. They included 8 studies and a total population of 10,567 individuals, the source of Al exposure they evaluated was drinking water and occupational exposure, and the follow-up duration from the cohort studies ranged between 8 and 48 years. The primary result of this meta-analysis was a significant association between Al exposure and the risk for AD (OR = 1.71, 95% CI, 1.35–2.18). Further, the authors also reported differences between groups exposed at 100 µg/L or higher Al concentrations in drinking water and those exposed to levels below 100 µg/L (OR = 1.95, 95% CI, 1.47–2.59). They concluded a possible link between Al exposure and AD (Wang et al. 2016). Authors also highlighted the importance of obtaining data from long-term Al exposure from food consumption to establish a possible dose-dependent link between Al and AD. The results from this study are in line with existing literature, thus indicating the importance of time of exposure and the exposure level in chronic Al exposure.

Notwithstanding the numerous scientific efforts and our actual knowledge of mechanisms involved in Al neurotoxicity, there is still no consensus on the real implication of Al and AD. Probably controlling for confounding factors in both epidemiological and experimental approaches would help to disentangle this complex relation. Remarkably, no information exists on sex possible differences in Al neurotoxicity.

Have Sex Differences Been Overlooked in AD and Al Toxicity?

Needless to say, experimental investigations involving male individuals are to date much more abundant than those using females. The female's estrous cycle is often singled out as the driving reason researchers prefer to use male subjects, but this selective discrimination is to blame for the lack of empirical data regarding the differences between both sexes. Nowadays, it is well recognized that they differ in such several behavioral processes as emotion (Girbovan and Plamondon 2013), impulsivity (Bayless et al. 2012; Weafer and de Wit 2014), basal activity (Simpson and Kelly 2012), learning and memory (Jonasson 2005; Li and Singh 2014), or attention (Bayless et al. 2012).

AD prevalence varies by age, sex, ethnicity, and geographic region, suggesting environmental and genetic factors may play an important modulating role (Mazure and Swendsen 2016). Indeed, as it has been suggested on many occasions, sex differences are evident when analyzing the prevalence and severity of AD. In fact, clinical and preclinical studies have shown that women not only are more prone to develop AD than men but also show significantly age-related faster decline and greater deterioration of cognition than they actually do (Cornutiu 2015; Li and Singh 2014). Some investigations have also described sex-genetic interactions. As previously stated, *APOE4* confers greater AD risk associated with tau pathology in women (Altmann et al. 2014). Similarly, the development of Aβ pathology in several transgenic mouse models of AD is greater in females (Maynard et al. 2006). These sex differences are also evident for metal brain levels in Cu and Mn, suggesting natural sex differences may contribute to the increased propensity of females to develop AD (Maynard et al. 2006).

Despite early epidemiological studies have clearly related differences among sexes as for AD onset (Gao et al. 1998), current clinical AD research sometimes considers males and females having equal risk toward developing the disease (Altmann et al. 2014).

Conclusions

Upon taken together Al implication in oxidative stress, mitochondrial dysfunction, Fe and Ca dyshomeostasis, neuroinflammation, microtubule alterations, cholinergic disruption, as well as compromised axonal transport and Aβ aggregation, Al implication in cognitive deficits and neurodegeneration is undeniable. Therefore, the contribution of Al to AD is plausible. However, it has not yet been established which levels of Al are needed, which factors are essential, or how long the exposure to it must be to induce functional brain deficits. It is not unrealistic to hypothesize that some populations may be protected against Al exposure or show some kind of resistance to it, a point that it is important to take into account as a possible source of bias in epidemiological studies. The major challenge for future researchers is identifying which variables are needed to be controlled in epidemiological studies and further designing more focused and translational experimental studies. The exposure pattern including time of exposure, dose-response effects, and the time elapsed between exposure and cognitive evaluations are of special importance. The identification of factors contributing to either resilience or exacerbated vulnerability to Al neurotoxicity must be taken into account in epidemiological and experimental studies. Clearly, research is needed to establish sex and age Al-related interactions, as no data exist so far in this regard.

References

Abu-Taweel GM, Ajarem JS, Ahmad M. Neurobehavioral toxic effects of perinatal oral exposure to aluminum on the developmental motor reflexes, learning, memory and brain neurotransmitters of mice offspring. Pharmacol Biochem Behav. 2012;101(1):49–56.

AI-Ashmawy MAM. Prevalence and public health significance of aluminum residues in milk and some dairy products. J Food Sci. 2011;76(3):T73–6.

Aisen P, Enns C, Wessling-Resnick M. Chemistry and biology of eukaryotic iron metabolism. Int J Biochem Cell Biol. 2001;33(33):940–59.

Akatsu H, et al. Transition metal abnormalities in progressive dementias. Biometals. 2012;25(2):337–50.

Akiyama H, et al. Long-term oral intake of aluminium or zinc does not accelerate Alzheimer pathology in AβPP and AβPP/tau transgenic mice. Neuropathology. 2012;32(4):390–7.

Altmann A, et al. Sex modifies the APOE-related risk of developing Alzheimer disease. Ann Neurol. 2014;75(4):563–73.

Anon. Statement of EFSA on the Evaluation of a new study related to the bioavailability of aluminium in food. EFSA Journal. 2011;9(5):2157.

Banks WA, et al. Aluminum complexing enhances amyloid β protein penetration of blood–brain barrier. Brain Res. 2006;1116(1):215–21.

Barth C, Villringer A, Sacher J. Sex hormones affect neurotransmitters and shape the adult female brain during hormonal transition periods. Front Neurosci. 2015;9

Bayless DW, et al. Sex differences in attentional processes in adult rats as measured by performance on the 5-choice serial reaction time task. Behav Brain Res. 2012;235(1):48–54.

Bellés M, et al. Silicon reduces aluminum accumulation in rats: relevance to the aluminum hypothesis of Alzheimer disease. Alzheimer Dis Assoc Disord. 1998;12(2):83–7.

Beydoun MA, et al. Apolipoprotein E ε4 allele interacts with sex and cognitive status to influence all-cause and cause-specific mortality in U.S. older adults. J Am Geriatr Soc. 2013;61(4):525–34.

Bielarczyk H, Tomaszewicz M, Szutowicz A. Effect of aluminum on acetyl-CoA and acetylcholine metabolism in nerve terminals. J Neurochem. 1998;70(3):1175–81.

Bolognin S, et al. Aluminum, copper, iron and zinc differentially alter amyloid-Aβ1–42 aggregation and toxicity. Int J Biochem Cell Biol. 2011;43(6):877–85.

Bondy SC. Low levels of aluminum can lead to behavioral and morphological changes associated with Alzheimer's disease and age-related neurodegeneration. Neurotoxicology. 2016;52:222–9.

Burrell, S.-A.M. & Exley, C., 2010. There is (still) too much aluminium in infant formulas. *BMC Pediatrics*, 10(1), p.63.

Cannata JB, et al. Role of iron metabolism in absorption and cellular uptake of aluminum. Kidney Int. 1991;39(4):799–803.

Caspers ML, Dow MJ, Mei-Jun F, Jacques PS, Kwaiser TM. Aluminum-induced alterations in [3H]Ouabain binding and ATP hydrolysis catalyzed by the rat brain synaptosomal (Na+ + K+)-ATPase? Mol Chem Neuropathol. 1994;22(1):43–55.

Colomina MT, et al. Influence of age on aluminum-induced neurobehavioral effects and morphological changes in rat brain. Neurotoxicology. 2002;23(6):775–81.

Cornutiu G. The epidemiological scale of Alzheimer's disease. J Clin Med Res. 2015;7(9):657–66.

Crisponi G, et al. The meaning of aluminium exposure on human health and aluminium-related diseases. Biomol Concepts. 2013;4(1)

Darbre PD. Aluminium, antiperspirants and breast cancer. J Inorg Biochem. 2005;99(9):1912–9.

Domingo JL, Gómez M, Colomina MT. Oral silicon supplementation: an effective therapy for preventing oral aluminum absorption and retention in mammals. Nutr Rev. 2011;69(1)

Dua R, Kumar V, Sunkaria A, Gill KD. Altered glucose homeostasis in response to aluminium phosphide induced cellular oxygen deficit in rat. Indian J Exp Biol. 2010;48(7):722–30.

Duggan JM, et al. Aluminium beverage cans as a dietary source of aluminium. Med J Aust. 1992;156(9):604–5.

Esparza JL, et al. Aluminum-induced pro-oxidant effects in rats: protective role of exogenous melatonin. J Pineal Res. 2003;35(1):32–9.

Exley C. The pro-oxidant activity of aluminum. Free Radic Biol Med. 2004;36(3):380–7.

FAO, Evaluations of the Joint FAO/WHO Expert Committee on Food Additives (JECFA), Geneva, 2006.

Fattoretti P, et al. The effect of chronic aluminum(III) administration on the nervous system of aged rats: clues to understand its suggested role in Alzheimer's disease. J Alzheimer's Dis JAD. 2003;5(6):437–44.

Fenwick S, et al. In end-stage renal failure, does infection lead to elevated plasma aluminium and neurotoxicity? Implications for monitoring. Ann Clin Biochem. 2005;42(2):149–52.

Flaten TP. Geographical associations between aluminium in drinking water and death rates with dementia (including Alzheimer's disease), Parkinson's disease and amyotrophic lateral sclerosis in Norway. Environ Geochem Health. 1990;12(1–2, 152):–167.

Gao S, et al. The relationships between age, sex, and the incidence of dementia and Alzheimer disease: a meta-analysis. Arch Gen Psychiatry. 1998;55(9):809–15.

García T, et al. Evaluation of the protective role of melatonin on the behavioral effects of aluminum in a mouse model of Alzheimer's disease. Toxicology. 2009;265(1–2):49–55.

Giacobini E. Cholinergic function and Alzheimer's disease. Int J Geriatr Psychiatry. 2003;18(Suppl 1):S1–5.

Girbovan C, Plamondon H. Environmental enrichment in female rodents: considerations in the effects on behavior and biochemical markers. Behav Brain Res. 2013;253:178–90.

Han S, et al. How aluminum, an intracellular ROS generator promotes hepatic and neurological diseases: the metabolic tale. Cell Biol Toxicol. 2013;29(2):75–84.

Harris WR, et al. Speciation of aluminum in biological systems. J Toxicol Environ Health. 1996;48(6):543–68.

Higgins GA, et al. Apolipoprotein E and Alzheimer's disease: a review of recent studies. Pharmacol Biochem Behav. 1997;56(4):675–85.

Iglesias-González J, et al. Effects of Aluminium on rat brain mitochondria bioenergetics: an in vitro and in vivo study. Mol Neurobiol. 2016:1–8.

Johnson VJ, et al. Decreased membrane fluidity and hyperpolarization in aluminum-treated PC-12 cells correlates with increased production of cellular oxidants. Environ Toxicol Pharmacol. 2005;19(2):221–30.

Jonasson Z. Meta-analysis of sex differences in rodent models of learning and memory: a review of behavioral and biological data. Neurosci Biobehav Rev. 2005;28(8):811–25.

Joshi JG. Aluminum, a neurotoxin which affects diverse metabolic reactions. Biofactors. 1990;2(3):163–9.

Joshi JG, et al. Iron and aluminum homeostasis in neural disorders. Environ Health Perspects. 1994;102:207–13.

Jovanović MD, Jelenković A, Stevanović ID, Bokonjić D, Colić M, Petronijević N, Stanimirović DB. Protective effects of glucose-6-phosphate dehydrogenase on neurotoxicity of aluminium applied into the CA1 sector of rat hippocampus. Indian J Med Res. 2014;139(6):864–72.

Kaizer RR, Maldonado PA, Spanevello RM, Corrêa MC, Gonçalves JF, Becker LV, Morsch VM, Schetinger MRC. The effect of aluminium on NTPDase and 5′-nucleotidase activities from rat synaptosomes and platelets. Int J Dev Neurosci. 2007;25(6):381–6.

Kaneko N, Sugioka T, Sakurai H. Aluminum compounds enhance lipid peroxidation in liposomes: insight into cellular damage caused by oxidative stress. J Inorg Biochem. 2007;101(6):967–75.

Kim Y, et al. Aluminum stimulates uptake of non-transferrin bound iron and transferrin bound iron in human glial cells. Toxicol Appl Pharmacol. 2007;220(3):349–56.

Krewski, D. et al., Human health risk assessment for aluminium, aluminium oxide, and aluminium hydroxide, 2007.

Kumar V, Gill KD. Oxidative stress and mitochondrial dysfunction in aluminium neurotoxicity and its amelioration: a review. Neurotoxicology. 2014;41:154–66.

Leterrier JF, et al. A molecular mechanism for the induction of neurofilament bundling by aluminum ions. J Neurochem. 1992;58(6):2060–70.

Levy-Lahad E, Tsuang D, Bird TD. Recent advances in the genetics of Alzheimer's disease. J Geriatr Psychiatry Neurol. 1998;11(2):42–54.

Li R, Singh M. Sex differences in cognitive impairment and Alzheimer's disease. Front Neuroendocrinol. 2014;35(3):385–403.

Liu Y, et al. Memory performance, brain excitatory amino acid and acetylcholinesterase activity of chronically aluminum exposed mice in response to soy isoflavones treatment. Phytother Res. 2010;24(10):1451–6.

Malik J, et al. Determination of certain micro and macroelements in plant stimulants and their infusions. Food Chem. 2008;111(2):520–5.

Martin RB, Bruce R. In: Yasui M, et al., editors. Mineral and metal neurotoxicology; 1997.

Maynard CJ, et al. Gender and genetic background effects on brain metal levels in APP transgenic and normal mice: implications for Alzheimer β-amyloid pathology. J Inorg Biochem. 2006;100(5–6):952–62.

Mazure CM, Swendsen J. Sex differences in Alzheimer's disease and other dementias. Lancet Neurology. 2016;15(5):451–2.

Mićić DV, Petronijević ND, Vucetić SS. Superoxide dismutase activity in the mongolian gerbil brain after acute poisoning with aluminum. Journal of Alzheimer's disease : JAD. 2003;5(1):49–56.

Mitsushima D. Sex differences in the septo-hippocampal cholinergic system in rats: behavioral consequences. Curr Top Behav Neurosci. 2011;8:57–71.

Miu AC, et al. A behavioral and histological study of the effects of long-term exposure of adult rats to aluminum. Int J Neurosci. 2003;113(9):1197–211.

Mohamd EM, et al. Windows into estradiol effects in Alzheimer's disease therapy. Eur Rev Med Pharmacol Sci. 2011;15(10):1131–40.

Moore PB, et al. Absorption of aluminium-26 in Alzheimer's disease, measured using accelerator mass spectrometry. Dement Geriatr Cogn Disord. 2000;11(2):66–9.

Muir JL. Acetylcholine, aging, and Alzheimer's disease. Pharmacol Biochem Behav. 1997;56(4):687–96.

Neha S, Dhawan DK. Regulatory role of zinc during aluminium-induced altered carbohydrate metabolism in rat brain. J Neurosci Res. 2012;90(3):698–705.

Nübling G, et al. Synergistic influence of phosphorylation and metal ions on tau oligomer formation and coaggregation with α-synuclein at the single molecule level. Mol Neurodegener. 2012;7:35.

Ohman LO, Martin RB. Citrate as the main small molecule binding Al3+ in serum. Clin Chem. 1994;40(4):598–601.

Oshiro S, et al. Glial cells contribute more to iron and aluminum accumulation but are more resistant to oxidative stress than neuronal cells. Biochim Biophys Acta (BBA) - Mol Basis Dis. 2000;1502(3):405–14.

Pauluhn J. Pulmonary toxicity and fate of agglomerated 10 and 40 nm aluminum Oxyhydroxides following 4-week inhalation exposure of rats: toxic effects are determined by agglomerated, not primary particle size. Toxicol Sci. 2009;109(1):152–67.

Praticò D, et al. Aluminum modulates brain amyloidosis through oxidative stress in APP transgenic mice. FASEB J Off Publ Feder Am Soc Exp Biol. 2002;16(9):1138–40.

Ribes D, et al. Effects of oral aluminum exposure on behavior and neurogenesis in a transgenic mouse model of Alzheimer's disease. Exp Neurol. 2008;214(2)

Ribes D, et al. Impaired spatial learning and unaltered neurogenesis in a transgenic model of alzheimer's disease after oral aluminum exposure. Curr Alzheimer Res. 2010;7(5)

Ridge PG, Ebbert MTW, Kauwe JSK. Genetics of Alzheimer's disease. Biomed Res Int. 2013;2013:254954.

Riihimäki V, et al. Body burden of aluminum in relation to central nervous system function among metal inert-gas welders. Scand J Work Environ Health. 2000;26(2):118–30.

Rondeau V, et al. Aluminum and silica in drinking water and the risk of Alzheimer's disease or cognitive decline: findings from 15-year follow-up of the PAQUID cohort. Am J Epidemiol. 2008;169(4):489–96.

Roses, A.D., 1996. Apolipoprotein E and Alzheimer's disease a rapidly expanding field with medical and epidemiological consequences. *Annals of the New York Academy of Sciences*, 802(1 Apolipoprotein), pp.50–57.

Ruipérez F, et al. Pro-oxidant activity of aluminum: promoting the Fenton reaction by reducing Fe(III) to Fe(II). J Inorg Biochem. 2012;117:118–23.

Rusina R, et al. Higher aluminum concentration in Alzheimer's disease after box–cox data transformation. Neurotox Res. 2011;20(4):329–33.

Sánchez-Iglesias S, et al. Brain oxidative stress and selective behaviour of aluminium in specific areas of rat brain: potential effects in a 6-OHDA-induced model of Parkinson's disease. J Neurochem. 2009;109(3):879–88.

Schellenberg GD, et al. Genetic linkage evidence for a familial Alzheimer's disease locus on chromosome 14. Science (New York, NY). 1992;258(5082):668–71.

Schwalfenberg GK, Genuis SJ. Vitamin D, essential minerals, and toxic elements: exploring interactions between nutrients and toxicants in clinical medicine. Sci World J. 2015;2015:1–8.

Silva VS, et al. Aluminum accumulation and membrane fluidity alteration in synaptosomes isolated from rat brain cortex following aluminum ingestion: effect of cholesterol. Neurosci Res. 2002;44(2):181–93.

Silva VS. Effect of chronic exposure to aluminium on isoform expression and activity of Rat (Na+/K+)ATPase. Toxicol Sci. 2005;88(2):485–94.

Simpson J, Kelly JP. An investigation of whether there are sex differences in certain behavioural and neurochemical parameters in the rat. Behav Brain Res. 2012;229(1):289–300.

Szutowicz A, et al. Acetyl-CoA metabolism in cholinergic neurons and their susceptibility to neurotoxic inputs. Metab Brain Dis. 2000;15(1):29–44.

Torreilles F, Touchon J. Pathogenic theories and intrathecal analysis of the sporadic form of Alzheimer's disease. Prog Neurobiol. 2002;66(3):191–203.

Ungar L, Altmann A, Greicius MD. Apolipoprotein E, gender, and Alzheimer's disease: an overlooked, but potent and promising interaction. Brain Imaging Behav. 2014;8(2):262–73.

Van den Bosch G, et al. Determination of iron metabolism-related reference values in a healthy adult population. Clin Chem. 2001;47(8)

Verstraeten SV, et al. Myelin is a preferential target of aluminum-mediated oxidative damage. Arch Biochem Biophys. 1997;344(2):289–94.

Vučetić-Arsić S, et al. Oxidative stress precedes mitochondrial dysfunction in gerbil brain after aluminum ingestion. Environ Toxicol Pharmacol. 2013;36(3):1242–52.

Walton JR, Wang M-X. APP expression, distribution and accumulation are altered by aluminum in a rodent model for Alzheimer's disease. J Inorg Biochem. 2009;103(11):1548–54.

Wang Z, et al. Chronic exposure to aluminum and risk of Alzheimer's disease: a meta-analysis. Neurosci Lett. 2016;610:200–6.

Ward RJ, Zhang Y, Crichton RR. Aluminium toxicity and iron homeostasis. J Inorg Biochem. 2001;87(1):9–14.

Weafer J, de Wit H. Sex differences in impulsive action and impulsive choice. Addict Behav. 2014;39(11):1573–9.

WHO, Guidelines for drinking-water quality, Geneve, 2004.

WHO, Aluminium in drinking-water. Background document for development of WHO Guidelines for Drinking-water Quality, Geneva, 2010.

Wilhelm M, Jäger DE, Ohnesorge FK. Aluminium toxicokinetics. Pharmacol Toxicol. 1990;66(1):4–9.

Willhite CC, et al. Systematic review of potential health risks posed by pharmaceutical, occupational and consumer exposures to metallic and nanoscale aluminum, aluminum oxides, aluminum hydroxide and its soluble salts. Critic Rev Toxicol. 2014;44(Suppl 4):1–80.

Wills MR, Savory J. Aluminum and chronic renal failure: sources, absorption, transport, and toxicity. Crit Rev Clin Lab Sci. 1989;27(1):59–107.

Woodburn K, et al. Accumulation and toxicity of aluminium-contaminated food in the freshwater crayfish, *Pacifastacus leniusculus*. Aquat Toxicol. 2011;105(3–4):535–42.

Wu Z, et al. Aluminum induces neurodegeneration and its toxicity arises from increased iron accumulation and reactive oxygen species (ROS) production. Neurobiology of Aging. 2012;33(1):199. e1–199.e12.

Yellamma K, Saraswathamma S, Kumari BN. Cholinergic system under aluminium toxicity in rat brain. Toxicol Int. 2010;17(2):106.

Yokel RA, McNamara PJ. Aluminium toxicokinetics: an updated minireview. Pharmacol Toxicol. 2001;88(4):159–67.

Yokel RA, Allen DD, Meyer JJ. Studies of aluminum neurobehavioral toxicity in the intact mammal. Cell Mol Neurobiol. 1994;14(6):791–808.

Yokel RA, et al. Entry, half-life, and desferrioxamine-accelerated clearance of brain aluminum after a single (26) Al exposure. Toxicolog Sci Off J Soc Toxicol. 2001;64(1):77–82.

Yuan C-Y, Lee Y-J, Hsu G-S. Aluminum overload increases oxidative stress in four functional brain areas of neonatal rats. J Biomed Sci. 2012;19(1):51.

Yumoto S, et al. Demonstration of aluminum in amyloid fibers in the cores of senile plaques in the brains of patients with Alzheimer's disease. J Inorg Biochem. 2009;103(11):1579–84.

Zatta P, et al. The role of metals in neurodegenerative processes: aluminum, manganese, and zinc. Brain Res Bull. 2003;62(1):15–28.

Zatta P, et al. Alzheimer's disease, metal ions and metal homeostatic therapy. Trends Pharmacol Sci. 2009;30(7):346–55.

Copper and Alzheimer's Disease

Zoe K. Mathys and Anthony R. White

Abstract Alzheimer's disease (AD) is the most common form of adult neurodegeneration and is characterised by progressive loss of cognitive function leading to death. The neuropathological hallmarks include extracellular amyloid plaque accumulation in affected regions of the brain, formation of intraneuronal neurofibrillary tangles, chronic neuroinflammation, oxidative stress, and abnormal biometal homeostasis. Of the latter, major changes in copper (Cu) levels and localisation have been identified in AD brain, with accumulation of Cu in amyloid deposits, together with deficiency of Cu in some brain regions. The amyloid precursor protein (APP) and the amyloid beta (Aβ) peptide both have Cu binding sites, and interaction with Cu can lead to potentially neurotoxic outcomes through generation of reactive oxygen species. In addition, AD patients have systemic changes to Cu metabolism, and altered Cu may also affect neuroinflammatory outcomes in AD. Although we still have much to learn about Cu homeostasis in AD patients and its role in disease aetiopathology, therapeutic approaches for regulating Cu levels and interactions with Cu-binding proteins in the brain are currently being developed. This review will examine how Cu is associated with pathological changes in the AD brain and how these may be targeted for therapeutic intervention.

Keywords Copper • Alzheimer's disease • Ceruloplasmin • Reactive oxygen species • Amyloid precursor protein • Tau • Neuroinflammation • Clioquinol • PBT-2

Z.K. Mathys
Department of Pathology, The University of Melbourne, Parkville, VIC 3010, Australia

A.R. White, Ph.D. (✉)
Department of Pathology, The University of Melbourne, Parkville, VIC 3010, Australia

Cell and Molecular Biology, QIMR Berghofer Medical Research Institute, 300 Herston Road, Herston, QLD 4006, Australia

Royal Brisbane Hospital, Locked Bag 2000, Herston, QLD 4029, Australia
e-mail: Tony.White@qimrberghofer.edu.au

© Springer International Publishing AG 2017
M. Aschner and L.G. Costa (eds.), *Neurotoxicity of Metals*, Advances in Neurobiology 18, DOI 10.1007/978-3-319-60189-2_10

Background

Alzheimer's disease (AD) is a neurodegenerative disorder that is the most common form of dementia, affecting approximately 47 million people worldwide (Prince et al. 2015). Advanced ageing is a notable risk factor for AD, whereby almost 50% of cases are found in individuals older than 85 years. Fewer than 5% of cases are due to genetic mutations with the remaining 95% resulting from occurrence without known familial genetic mutation (Ceccom et al. 2012). With advancements in the medical industry and prolonged life expectancy, the prevalence of AD is predicted to increase immensely (Ballard et al. 2011). AD patients typically display symptoms of early memory loss, personality and behavioural changes and deficits in sensory and motor functions. Macroscopically, the brain of AD patients presents synaptic and neuronal loss resulting in brain atrophy, affecting regions including the entorhinal cortex, hippocampus, basal forebrain and amygdala.

The complex aetiology of AD has been widely studied but is yet to be fully understood. It is associated with the key hallmarks of extracellular amyloid peptide accumulation, intracellular tau hyperphosphorylation, neuroinflammation and oxidative stress. In recent years there has been a substantial focus on the role of transition metals, particularly iron (Fe), zinc (Zn) and copper (Cu) in AD aetiology. These metals bind to amyloid-β peptide (Aβ), accelerate Aβ aggregation and consequently promote neurotoxic plaque formation. Fe and Cu are also likely to be involved in promotion of oxidative stress and neuroinflammatory changes in the AD brain (Choo et al. 2013; Pratico 2008; Sayre et al. 2008). In addition, cognitive decline in AD has been associated with the interference of the processing and function of the amyloid-β precursor protein (APP) and the phosphorylation and aggregation of the microtubule-associated protein (MAP), tau, both of which are associated with altered metal homeostasis (Crouch et al. 2009).

Copper Homeostasis

Although there are key roles for a number of metals in AD, this review will focus on Cu. This metal is a key trace element, necessary for all oxygen-requiring processes. Cu is concentrated at high levels in the brain for metabolic needs and additional functions. Specifically, Cu is an essential cofactor that readily binds to enzymes, shifting between the Cu(II) and Cu(I) oxidative states (Hung et al. 2010). The redox capacity of Cu is biochemically important for biological energy metabolism (cytochrome c oxidase), iron metabolism (ceruloplasmin), antioxidative activity (copper zinc superoxide dismutase, SOD1), neurotransmitter synthesis (dopamine-β-monooxygenase), neuropeptide synthesis (peptidylglycine-α-amdinating enzyme) and neuronal myelination (Davies et al. 2013; Scheiber et al. 2014). As a consequence of its redox activity, Cu can also induce oxidative stress through the production of reactive oxygen species (ROS) and its ability to bind with molecular oxygen (McCord and Fridovich 1969). Tight regulation of Cu therefore exists through the duodenal absorption, uptake and excretion from cells and sequestration within cells in order to prevent both excess Cu accumulation and Cu deficiency. This process further prevents abnormal Cu-oxygen interactions (Kaden et al. 2011). The trafficking and transportation of Cu is highly important for the maintenance of Cu homeostasis. In plasma, ceruloplasmin (Cp), albumin and transcuprein are major Cu-binding proteins (Choo et al. 2013). Subcellular Cu transportation involves the Ctr1 protein for the transit into cells across the cell membrane and the Cu ATP7a/b transporters for Cu exportation. In addition, cytosolic Cu chaperones aid delivery and include the Cox17 system in the mitochondria (Amaravadi et al. 1997), copper chaperone for SOD (CCS) (Culotta et al. 1997) and the ATOX1 system in trans-Golgi network (Klomp et al. 1997). These critical cellular mechanisms are essential for maintaining normal neuronal health and function.

Copper in the Brain

The brain contains approximately 7.3% of total body Cu content (Hung et al. 2010). The brain possesses disproportionately low levels of antioxidants, thereby making it highly susceptible to oxidative stress induced by the redox nature of Cu (Hung et al. 2013). Cu homeostasis and transport must therefore be tightly regulated in the brain in order to maintain neuronal health. In the cortex and hippocampus, Cu is released into the synaptic cleft of glutamatergic synapses as an essential component of neuronal transmission (Bush and Tanzi 2008; Zheng et al. 2010). Upon synaptic depolarisation, Cu and Zn are released into the synaptic cleft at micromolar concentrations estimated at approximately 15 µM (Hung et al. 2010; Kardos et al. 1989), whereby excitatory and inhibitory neurotransmission can be modulated. Free ionic Cu is released at NMDA-responsive synapses, and Cu efflux is thought to be associated with the activation of NMDA receptors. This release of Cu may act as a post-translational mechanism to modulate the extracellular s-nitrosylation of NMDA receptors (Bush and Tanzi 2008). Impaired Cu regulation could promote the glutamatergic dysfunction that is present in the AD brain (Ayton et al. 2015). Cu also has the ability to block GABAergic and AMPAergic neurotransmission on rat olfactory bulb neurons (Trombley and Shepherd 1996) and AMPAergic neurotransmission on rat cortical neurons (Weiser and Wienrich 1996). However, more recently it has been established that Cu acts as more than just a negative modulator of neurotransmission. Following 3 hours of Cu exposure, AMPAergic neurotransmission has been seen to increase. Cu therefore demonstrates a unique biphasic mechanism in neurotransmission. Furthermore, it has been demonstrated that Cu has acute inhibitory effects on long-term potentiation (Opazo et al. 2014).

Copper Levels in the Brain in AD

Dyshomeostasis of Cu levels in the brain is a common characteristic in neurodegenerative diseases, including AD. Studies have indicated that mis-localisation of Cu can be observed in the brain of AD patients, and some regions may be in excess while others Cu deficient. This Cu imbalance has extensive effects on neuronal function and is significantly linked to cognitive deficits and AD pathology (Mao et al. 2012; Rembach et al. 2013; Squitti 2014). It has been widely demonstrated that Cu accumulates within Aβ plaques, where Cu levels have been observed at ~390 μM. This is a substantial increase when compared to the brain of normal age-matched control patients, where Cu is found at a concentration of ~79 μM (Mot et al. 2011). Moreover, tissue surrounding the Aβ plaque (high in Cu) has been demonstrated to present with substantially lower levels of Cu, engendering local Cu deficiency (Zheng et al. 2010; Mot et al. 2011). Rembach et al. (2013) established that Cu levels in the frontal cortex are significantly lower than age-matched healthy controls, specifically confined to the soluble fraction. Post-mortem examination is required to measure Cu concentration in the brain and detect Aβ plaques (Hung et al. 2013; Rembach et al. 2013). In order to determine Cu levels in living patients, an ancillary method must be used (e.g. serum Cu) that may in the future be a diagnostic tool for AD (Wang et al. 2015).

A large proportion of the literature has focussed on measuring Cu levels in the serum and cerebrospinal fluid (CSF) of living AD patients. Cu transport in the serum can be found in the form of non-ceruloplasmin-bound Cu (non-Cp-Cu) or bound to Cp or albumin. It is the uptake of the free Cu ion that passes the brain barriers (the blood-brain barrier (BBB) and the blood-cerebrospinal fluid barrier) and is distributed to the CSF and brain parenchyma. The choroid plexus tightly regulates the transportation of Cu into the CSF, whereas movement into the parenchyma may appear more readily in the cerebral capillaries (Choi and Zheng 2009). Meta-analysis of Cu serum levels has indicated that AD patients display higher Cu serum levels (particularly non-Cp-Cu) compared to healthy non-diseased controls (Wang et al. 2015; Squitti 2012; Ventriglia et al. 2012). Interestingly, analysis revealed that Cu levels in the CSF showed no discrepancy between AD patients and controls (Bucossi et al. 2011). The reason for differential plasma Cu but not CSF Cu in AD patients compared to healthy controls is not yet understood.

Copper and Amyloid Aggregation

The aggregation of Aβ is thought to be a major contributor in AD pathology, which may be explained by the amyloid cascade hypothesis (Henry et al. 2010). The amyloid hypothesis suggests that aggregated and oligomerised Aβ is the major contributing factor to synaptic and neuronal degeneration in AD. Amyloid deposition and subsequent senile plaque formation can occur through age-related changes in amyloid generation and clearance and are also induced by mutations in genes such as amyloid precursor protein (APP), presenilin 1 (PSEN1) and presenilin 2 (PSEN2). Originally it was believed that plaque formation led to neuronal cell death and AD (Reitz 2012). However, over recent years this hypothesis has been modified to suggest that it is the soluble aggregated forms of Aβ and not the endpoint plaques per se that drive neuronal death. With many clinical trials currently focussed on inhibition or removal of amyloid, the veracity of the amyloid hypothesis will soon be established.

Aβ peptides are metabolic products, generated by the proteolysis of APP (Henry et al. 2010). APP is a ubiquitously expressed, transmembrane glycoprotein that accumulates at nerve terminals (Buxbaum et al. 1998) and is thought to be involved in axonal transport, vesicular trafficking, cell adhesion, neuronal survival, apoptosis and perhaps protein folding and degradation (Hung et al. 2010; Cottrell et al. 2005). APP is approximately 110–140 kDa in size, and heterogeneity emerges due to alternative exon splicing and post-translational modifications (Selkoe 2001). Two Cu-binding domains exist within APP, one localised in the Aβ region (Fig. 1) and the other in the N-terminus (Fig. 2) (Hung et al. 2010). The ligands His147, His151, Tyr168 and Met170 are required for high-affinity Cu binding to the N-terminal domain (Barnham et al. 2003; Kong et al. 2007) (Fig. 2). Cu(II) reductase activity is present within the Cu-binding domain of APP, which may further contribute to free radical formation and is sufficient to promote copper-mediated neurotoxicity (Hung et al. 2010).

The enzymatic cleavage of APP occurs via either the amyloidogenic or non-amyloidogenic pathways. The non-amyloidogenic pathway requires α-secretase and γ-secretase, whereas amyloidogenic processing of APP occurs by β-secretase and γ-secretase (Hung et al. 2010; Ayton et al. 2015). The activity of β-secretase-BACE1 in the amyloidogenic pathway is regulated by Cu. Cu binds to BACE1 in the Cu(I) binding site present in the C-terminal domain of BACE1, whereby BACE1 mRNA expression is upregulated and hence decreases β-cleavage of APP. This upregulation and increased activity of BACE1 has been observed in copper-deficient fibroblasts (Cater et al. 2008). Therefore, altered levels of intracellular Cu may influence APP metabolism. Increased intracellular Cu elevated the secretion of the

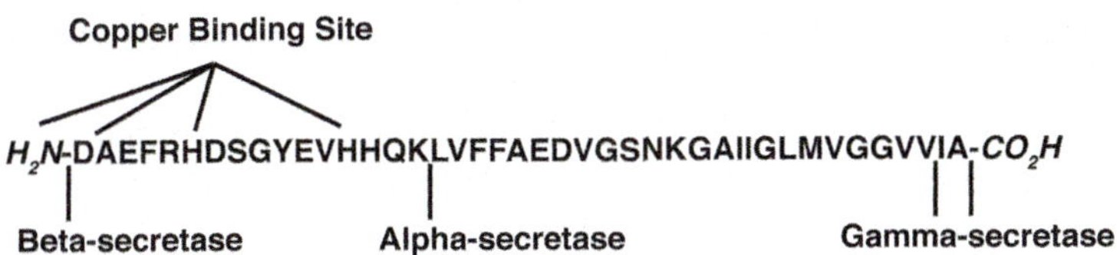

Fig. 1 Schematic of amyloid beta peptide showing amino acid residues involved in binding of copper. Secretase cleavage sites are also shown

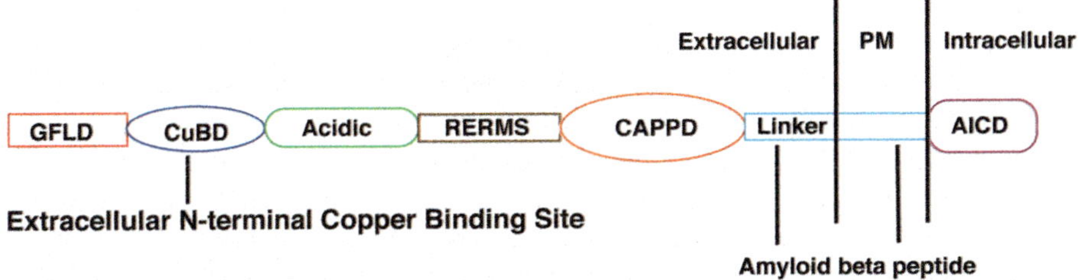

Fig. 2 Schematic of amyloid precursor protein (APP) showing different domains including the N-terminal copper-binding domain. *GFLD* growth factor-like domain, *CuBD* copper-binding domain, *RERMS* amino acid sequence associated with growth-promoting activity, *CAPPD* central APP domain, *AICD* APP intracellular domain, *PM* plasma membrane

α-cleaved APP, whereas the β-cleaved APP production and subsequent secretion were increased in Cu-deficient cells (Cater et al. 2008). Copper has therefore been shown to modulate APP not only via the Cu-binding domain but also through its processing and hence control of Aβ production. Regulation of Aβ subsequently effects Aβ-induced neurotoxicity (Barnham et al. 2003).

In both healthy and AD patients, heterogeneous forms of Aβ have been identified. Extensive research has focussed on the 40-amino acid form of Aβ (Aβ_{1-40}) and the Aβ species ending with a C-terminal residue of 42 (Aβ_{1-42}). Aβ_{1-40} accounts for ~90% of secreted Aβ, and the remaining 10% of total Aβ secreted from cells is mostly Aβ_{1-42}, which aggregates more readily to form plaque deposits (Citron et al. 1996; Small and McLean 1999; White et al. 2006a). Extracellular plaque formation is primarily composed of aggregated Aβ peptides, both insoluble fibrillar Aβ and soluble Aβ oligomers. Aβ_{1-40} is the predominant soluble isoform of Aβ (Ahuja et al. 2015) with increasing evidence suggesting a close relationship with cognitive impairment (Lue et al. 1999). Alternatively, Aβ_{1-42} has a higher propensity to aggregate, which could enhance toxic outcomes. Lue et al. (1999) interestingly espouse that soluble Aβ (or 1–40, 1–42) has the ability to impact a wider area of neurons and synapses in comparison to insoluble Aβ and, thus, may have a greater role in neurotoxicity.

Metals such as Cu, Fe and Zn play a significant role in the formation of soluble Aβ oligomers, and in particular Cu(II)-Aβ interactions are a major driver of peptide aggregation. The oligomer species of Aβ generated is dependent on the molar ratio of Cu(II) to Aβ. At sub-equimolar ratios, amyloid-like aggregates form that are highly stable and resistant to sodium dodecyl sulphate (SDS). Conversely, at supra-equimolar ratios of Cu(II) to Aβ, soluble, less stable, neurotoxic oligomers are formed. Cu(II)-induced aggregates may be spontaneously formed at Cu(II)/peptide ratios of 0.25:1; however, amyloid formation is not subsequently accelerated as compared to Aβ_{1-42} in the absence of Cu(II) (Smith et al. 2007). The formation of a His bridge between Cu(II) ions leads to the generation of His-bridged Aβ oligomers that are highly toxic (Smith et al. 2006; Huang et al. 2004). His modification has been demonstrated to reduce the amount of Cu-mediated Aβ_{1-40} aggregation, but aggregation is not entirely abolished. These data suggest that the interaction with non-His residues on the peptide is partially responsible for the aggregation of this particular form of Aβ (Atwood et al. 1998). The tendency of Aβ peptides to self-aggregate is dependent on the specific oligomer of Aβ, as analysis has proposed that peptide precipitation may be mediated by the high-affinity Cu(II) binding site present in Aβ_{1-42} oligomers. In comparison, Aβ_{1-40} is less self-aggregating and may be explained by the presence of a lower-affinity Cu(II) binding site (Atwood et al. 2000). Conclusively, it is well established that nucleated aggregation, where aggregation follows from a pre-aggregated seed, is considerably accelerated in the presence of Cu(II) (Huang et al. 2004) and heightened by mild acidic conditions similar to that in AD brains (Atwood et al. 1998).

Cu and Tau

Tau is a microtubule-associated protein (MAP) expressed abundantly in the CNS, predominantly in neurons, and at lower levels in astrocytes and oligodendrocytes (LoPresti et al. 1995). Normal tau plays a significant role in axonal microtubule (MT) organisation, specifically axonal growth and development of neuronal polarity. This is achieved through axonal stabilising of MTs by promoting MT assembly. In addition, recent findings established a multifunctional role of tau, recognising its interaction with synaptic vesicles as a key factor in neurotransmission (Hung et al. 2013; de Calignon et al. 2012; Liu et al. 2012; Pooler et al. 2013). The regulation of tau is controlled by specific protein kinases and phosphatases, which mediate post-translational phosphorylation (Higuchi et al. 2002). Highly phosphorylated tau aggregates, sequestering normal tau and disrupting MTs. This characteristic of tau links directly to the aetiology and pathogenesis of neurodegenerative diseases such as AD (Khlistunova et al. 2006).

A relationship between Cu and tau has been established to suggest that the promoter of the tau gene (MAPT) may be regulated by the Cu-responsive transcription factor, Sp1 (Heicklen-Klein and Ginzburg 2000; Song et al. 2008). Sp1 is also a key regulator of BACE1 processing of APP to produce Aβ (Christensen et al. 2004). Overexpression of tau inhibits kinesin-dependent transport of peroxisomes, increasing the vulnerability of cells to oxidative stress and hence degeneration. Additionally, inhibiting APP transport into axons and dendrites induces cell body accumulation of APP (Stamer et al. 2002). Therefore, a clear link between Cu-responsive overregulation of tau and APP may be important. However, critical evidence in the literature demonstrates that despite the role of tau in APP and Aβ-induced cognitive decline, Aβ accumulation most likely precedes and drives the accumulation of tau neurofibrillary tangles (NFT) (Götz et al. 2001; Hu et al. 2008; Lewis et al. 2001).

Structural analysis of tau protein postulates that the microtubule-binding domain (MBD) comprises of four highly conserved 18-amino-acid repeats, R1, R2, R3 and R4. The MBD is well recognised for its role in the formation of paired helical filaments (PHFs), promoting tau aggregation and formation of NFTs in vitro. Neurotoxicity to neurons may be induced by abnormal tau aggregation; hence, several studies have been conducted to determine the aggregation process. More specifically, the relationship between these repeat regions and Cu(II) has been investigated (Ma et al. 2005, 2006; Zhou et al. 2007). Cu(II) binding to repeat regions is a pH-dependent and stoichiometrically determined process. The R2 and R3 peptides adopt a monomeric α-helical structure in the presence of Cu(II). The helical structure induces PHF formation that aggregates to form NFTs (Ma et al. 2005, 2006). Furthermore, R3 peptide may additionally form a β-sheet structure in 1 mol eq. of Cu(II). Interestingly, Cu(II) binding to R1 peptide has been shown to delay the onset and level of R1 aggregation. It is suggested that Cu(II) coordination affects the electrostatic surface of R1 peptide, hence regulating in vivo aggregation of tau protein (Zhou et al. 2007). The present literature therefore suggests that Cu(II) binding to repeat of the MBD influences peptide aggregation, hence encapsulating an important feature of AD, the presence of NFTs. Moreover, tau demonstrates redox activity when bound to Cu inappropriately, which further contributes to oxidative stress (Hung et al. 2010; Su et al. 2007). R2 peptides have the capacity to reduce Cu(II) to Cu(I), favouring the generation of hydrogen peroxide (H_2O_2) and hence subsequent ROS (Su et al. 2007).

Cu and Oxidative Stress in AD

Oxidative stress is a predominant feature in AD and ageing brains, whereby conditions of dyshomeostasis lead to the generation of ROS. ROS production can be facilitated by the high redox nature of Cu. Oxidative stress may therefore be induced through participation in Fenton and Haber-Weiss reactions, Aβ-Cu(II) binding, diminished glutathione (GSH) levels and reduced expression of Cu-dependent enzymes (Hung et al. 2010; Mot et al. 2011; Halliwell and Gutteridge 1984). ROS may interact with biomolecules and engender irreversible oxidative modification such as lipid peroxidation, protein oxidation and nucleic acid cleavage leading to cellular impairment (Hung et al. 2010; Halliwell and Gutteridge 1984).

Cu is thought to play a key causative role in oxidative stress-induced neurodegeneration by Fenton and Haber-Weiss reactions, where it directly catalyses the formation of ROS (Hung et al. 2010; Ahuja et al. 2015). This is a two-step process involving the reaction between cupric ion (Cu(II)) and superoxide anion radical ($O_2^{\bullet-}$) or biological reductants such as ascorbic acid or glutathione (GSH) to produce the reduced cuprous ion (Cu(I)). Cu(I) in the presence of H_2O_2 can catalyse the formation of highly unstable hydroxyl ($^{\bullet}$OH) radicals (Ahuja et al. 2015; Halliwell and Gutteridge 1984; Barbusiński 2009). Reactive $^{\bullet}$OH radicals may react with biomolecules close to the site of formation, exacerbating oxidative stress in this region (Jomova and Valko 2011). In addition, the capacity of Cu to induce DNA damage and oxidation of bases is induced by ROS production via the Fenton reaction (Moriwaki et al. 2008):

$$Cu\left(II\right)+O_2^{\bullet-} \rightarrow Cu\left(I\right)+O_2 \tag{1}$$

$$Cu\left(I\right)+H_2O_2 \rightarrow Cu\left(II\right)+^{\bullet}OH+OH^- \left(\textit{Fenton reaction}\right) \tag{2}$$

Amyloid deposits present in AD patients exhibit high levels of copper and oxidative stress markers. Cu(II) noticeably potentiates Aβ neurotoxicity, promoting the greatest toxic effect for $Aβ_{1-42}$ compared to $Aβ_{1-40}$ (Jomova and Valko 2011; Cuajungco et al. 2000). This is based on the peptide's ability to reduce Cu(II) to Cu(I) and therefore mediate O_2-dependent H_2O_2 production. The redox potential of Cu hence plays a significant role in exacerbating and facilitating Aβ-induced oxidative stress and subsequent neuronal death in AD (Huang et al. 1999). Interestingly, further oxidative damage may be mediated by dityrosine cross-linking between Aβ peptides. Cross-linking can be induced through direct attack of ˙OH radicals on Aβ peptides (Barnham et al. 2004; Galeazzi et al. 1999; White et al. 2006b), and the oxidative environment causes the accumulation of multi-protein aggregates that have a greater resistance to clearance (White et al. 2006b; Perry et al. 2002).

Elevated levels of Cu diminish GSH, an antioxidant present in cells. GSH is a substrate for enzymes that remove ROS. Dyshomeostasis of Cu and hence reduced levels of GSH may induce an oxidative environment that enhances the production and cytotoxic effects of ROS (Ahuja et al. 2015).

In AD patients, brain tissue can present with elevated Cu levels as well as Cu-deficient regions. The expression of Cu-dependent enzymes (SOD1 and ATOX1) was considerably reduced in multiple microarray studies, reinforcing that neurons may be Cu deficient (Myhre et al. 2013). In addition to its antioxidant role, SOD1 also has anti-inflammatory functions through ROS detoxification. Therefore, reduced expression and hence activity of SOD1 would heighten toxic ROS accumulation and exacerbate both oxidative stress and chronic inflammation (Choo et al. 2013).

Cu and Inflammation

Inflammation is a significant pathological factor present in neurodegenerative diseases such as AD. Inflammation is normally a protective response in the brain involving interactions between cells and mediators to prevent cell injury. The resident macrophages of the brain parenchyma, microglia, exhibit protective effects such as monitoring the local microenvironment and responding to disturbances. Slight disruptions in the microenvironment cause morphological and functional changes in microglia, which may further contribute to neuroinflammation present in the brain (Bamberger et al. 2003; Minghetti 2005). During such periods of dyshomeostasis in AD, inflammation may be triggered by the accumulation of abnormal protein aggregates or by pro- and anti-inflammatory cytokine imbalances (Minghetti 2005; Wyss-Coray and Mucke 2002). Copper can play a central role in toxic and protective inflammatory reactions (Choo et al. 2013).

The binding of Cu with Aβ induces Aβ deposition and neurotoxicity through ROS generation (Barnham et al. 2004). The production of ROS alone contributes to the neurotoxic inflammatory environment. Additionally, the presence of activated inflammatory cells, such as microglia, surrounding Aβ plaques contributes significantly to the chronic inflammatory response (Choo et al. 2013; Dickson et al. 1988; McGeer et al. 1989; Rozemuller et al. 1989). Microglia participate in the clearance of senile plaques through phagocytosis or Aβ plaque degradation (DeWitt et al. 1998; Shaffer et al. 1995). The ultimate outcome of this 'beneficial' microglial response is unknown. Paradoxically, phagocytosis of Aβ may stimulate immune activation and release of pro-inflammatory mediators (Choo et al. 2013). Interestingly, astrocytes are closely associated with microglial phagocytosis and prevent clearance and slow degradation of amyloid plaque materials. These materials can therefore persist and further contribute to AD (DeWitt et al. 1998; Shaffer et al. 1995). Additionally, within close vicinity of the amyloid plaques, increased levels of complement, cytokines, chemokines and free radicals have been observed, promoting a self-propagating toxic cycle leading to neurodegeneration in AD (Minghetti 2005).

Cu displays both pro- and anti-inflammatory characteristics. It is well established that Cu induces peripheral secretion of IL-6 (Schmalz et al. 1998) and IL-8, the latter specifically through NF-κB activation (Kennedy et al. 1998). However, little evidence exists depicting the direct relationship of Cu to neuroinflammation. Current literature has reported that cholesterol and Cu may synergistically interact to generate Cu-induced neurotoxicity through oxidative stress-mediated apoptosis in AD (Choo et al. 2013; Lu et al. 2006). Conversely, it has been suggested that Cu can contribute to the development of an anti-inflammatory microglial phenotype (M2). A specific study indicated that the BV2 microglial cell line exposed to LPS induced an inflammatory phenotype (M1); however, the combination of both Cu(I) and LPS may lead to a shift towards an M2-like phenotype. Cu alone, in the absence of LPS, has not shown any effect on either M1 or M2 phenotypes. The studies propose that the shift from M1 to M2 is due to the redox state of NO, which may be due to Cu(I). Furthermore, the absence of NO is proposed to be a factor in the adoption of the M2 microglial state (Choo et al. 2013; Bamberger et al. 2003).

Non-ceruloplasmin Cu in Plasma

Non-ceruloplasmin Cu (non-Cp-Cu) may also be referred to as "free Cu" and is simply defined as serum Cu not bound to ceruloplasmin (Cp). It is well established that AD patients display elevated levels of non-Cp-Cu (Squitti et al. 2005). Furthermore, this increase may be representative of total serum copper (Squitti 2014). Free Cu levels correlate with cognitive function (measured by Mini-Mental State Examination (MMSE)), and higher level of free Cu is a predictor of severe cognitive decline, worsening MMSE outcomes, in patients with AD (Salustri et al. 2010; Squitti et al. 2009). Findings suggest that non-Cp-Cu may be a predictor for the progression of mild cognitive impairment to AD (Squitti et al. 2014). Additionally, alterations in electroencephalographic (EEG) rhythms have been observed in patients with elevated non-Cp-Cu, more specifically the slowing of cortical EEG rhythms (Zappasodi et al. 2008).

The liver plays a central role in Cu storage, Cu movement and Cu coordination into Cp. It has been hypothesised that free Cu may arise from impaired transfer of Cu into the secretory pathway of hepatocytes. Moreover, research has been undertaken to determine whether there is a correlation between free Cu and liver function in AD. A negative correlation was found between free copper and markers of liver function. This study had multiple limitations and hence further investigation is required (Squitti et al. 2007). In addition, genetic defects associated with the Cu efflux pump, ATP7b, may cause altered loading of Cu into Cp. The transmembrane domain of the ATP7b ionic pump is associated with AD and increases the amount of non-Cp-Cu released into the circulation (Squitti 2012; Squitti et al. 2008).

A substantial number of meta-analyses have been conducted on Cu levels and specifically non-Cp-Cu levels. These meta-analyses in fact have demonstrated that the fraction of non-Cp-Cu in circulation is increased and as a whole Cu dyshomeostasis consists of decreased Cu in the brain (Schrag et al. 2011) and an increase in the blood (Wang et al. 2015). These studies undoubtedly support the correlation between altered Cu levels and AD pathogenesis. However, limited research has been conducted on the direct mechanisms in which elevated non-Cp-Cu levels effects AD patients.

Therapeutic Considerations

As eluded to in this review, metal ion dyshomeostasis is central to AD pathogenesis and hence has been a target for therapeutic interventions. Currently there is no clinical cure for AD. However, the development and investigation of therapeutics such as dietary Cu supplementation, Cu chelation and Cu complexes have been considered in AD.

Dietary supplementation of Cu has been studied as a therapeutic approach for AD. The brains of APP transgenic mice have lower levels of Cu and display reduced activity of Cu/Zn-SOD1 in comparison to wild-type mice. Following oral treatment of Cu, APP transgenic mice exhibited restored SOD1 activity to normal levels and an increase in bioavailable brain Cu levels and decreased $A\beta_{1-40}$ and $A\beta_{1-42}$. These mice did not present a premature death phenotype (Bayer et al. 2003). It was suspected that in AD patients Cu intake may stabilise cognition. A randomised, double-blinded, placebo-controlled phase II clinical trial in patients with mild AD was conducted to evaluate the efficacy of oral Cu supplementation for 12 months. AD patients were treated with either Cu-(II)-orotate-dihydrate (8 mg Cu daily) or the placebo, and no significant differences were observed in primary outcome measures. The results demonstrated that although long-term oral intake of 8 mg Cu is well tolerated by AD patients, it is not therapeutic and has no effect on AD progression (Kessler et al. 2008). However, it should be noted that if Cu regulation is abnormal in AD (as supported by the evidence discussed here), then supplementation with dietary Cu is unlikely to have any impact. To overcome this, therapeutics are needed that bypass faulty Cu-handling processes.

Clioquinol (CQ; 5-chloro-7-iodo-8-hydroxyquinoline) (Fig. 3) is a small lipophilic, metal-protein attenuating compound (MPAC) that has demonstrated therapeutic potential in neurodegenerative diseases such as AD (Di Vaira et al. 2004). Conflicting hypotheses exist with regard to the approach CQ undertakes to interfere with metal homeostasis. Initially CQ was regarded as a metal chelator, suggesting that it may lead to alteration of Cu or Zn levels in specific diseased brains regions (Hegde et al. 2009; Treiber et al. 2004). CQ was later considered a metal-protein attenuating compound (MPAC), a substance that may influence and restore metal ion homeostasis without greatly effecting overall Cu regulation (Grossi et al. 2009). However, it is now termed an ionophore, acting as a Cu carrier facilitating Cu transport across membranes (White et al. 2006b; Grossi et al. 2009; Filiz et al. 2008;

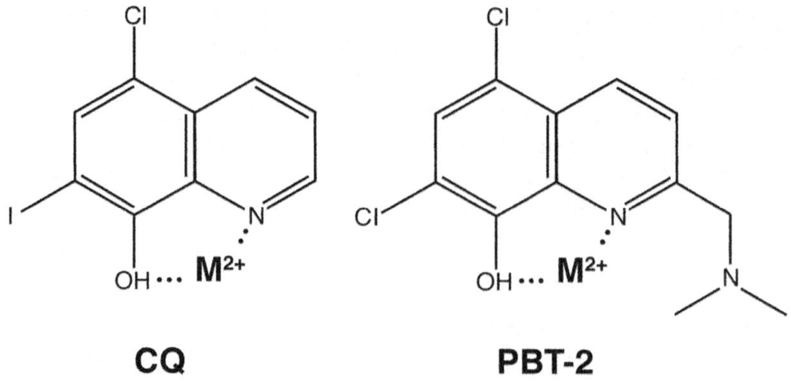

CQ   **PBT-2**

Fig. 3 Schematic structure of the metal ionophores clioquinol (CQ) and PBT-2

Caragounis et al. 2007). The beneficial effects of CQ were reported in a nine-week study of oral CQ administration in AD mice. Results depicted reduced Aβ deposition and improved cognitive performance (Cherny et al. 2001). In a phase II clinical trial of CQ, 32 patients were recruited in this double-blind, placebo-controlled, parallel-group randomised study. The CQ group demonstrated improved cognitive performance and a decline in plasma Aβ_{1-42} concentration. However, in this study the cognitive benefit of CQ was only illustrated in the more severely affected subjects (Ritchie et al. 2003). Following some controversial published data on CQ, it has recently been withdrawn from human clinical studies (Hegde et al. 2009). Furthermore, a derivative of CQ, another hydroxyquinoline ligand (PBT-2) (Fig. 3) has been clinically tested in a human phase IIa double-blinded trial that demonstrated lowered CSF Aβ_{1-42} levels and improvement in two areas of a neuropsychological test battery (Lannfelt et al. 2008). PBT-2 and CQ have the ability to block H_2O_2 generation through the Aβ-Cu complex. PBT-2 can also decrease interstitial brain Aβ and improve cognitive performance to a greater degree than CQ. In addition, PBT-2 has outperformed CQ as an ionophore and shows increased BBB permeability (Adlard et al. 2008). Additional large and longer studies are required to further determine the beneficial effects of PBT-2 on AD patients.

Cu complexes of bis(thiosemicarbazone) (BTSC) are able to bind Cu(II) and Zn(II) to form stable, lipophilic complexes that cross membranes and specifically the BBB (Duncan and White 2012; Green et al. 1988). In addition, Cu complexes of BTSCs (Cu(II)(btsc)s) have been investigated as a potential therapeutic of AD. Treatment of APP-CHO cells with Cu(II)(btsc)s specifically, glyoxalbis(N (4)-methyl-3-thiosemicarbazonato)copper(II) (Cu(II)(gtsm)) demonstrated increased intracellular Cu levels and a reduction in secreted Aβ levels (Donnelly et al. 2008). Furthermore, neurotoxic pathways are regulated by glycogen synthase kinase 3β (GSK3β) and have been targeted as a therapeutic in AD. Treatment with Cu(II)(gtsm) in APP/PS1 transgenic AD mice resulted in reduced active GSK3β, lower abundance of Aβ trimers and phosphorylated tau and conclusively reversed cognitive deficits in the APP/PS1 transgenic mice (Crouch et al. 2009). Additionally, despite pyrrolidine dithiocarbamate (PDTC) being classified as a Cu-chelating compound, it has also demonstrated in APP/PS1 mice to down regulate the GSK3β pathway and thus improve spatial learning. This study also presented that Cu levels in the brain increased with PDTC treatment and moreover reduced tau phosphorylation. It remains unknown whether PDTC binds and prevents metal binding of Aβ (Malm et al. 2007).

Future Directions and Conclusions

The present review gives an insight into the role of Cu in the complex neurodegenerative disorder, AD. Considerable evidence exists to demonstrate that altered Cu homeostasis in the brain is a key factor in AD. This is likely to involve mislocalisation of Cu rather than excess or deficiency per se. Increased Cu may exacerbate some subtypes, but mostly Cu changes are part of the disease, leading to loss of Cu function in key cell types and Cu-mediated toxicity in other cells or locations including amyloid aggregation. Copper-based therapeutics may be developed for Cu chelation and need to progress to address Aβ-Cu interactions, control tau-Cu and decrease oxidative stress and neuroinflammation in the brain. Ultimately, further research should contribute to ameliorating the deleterious effects of Cu dyshomeostasis in AD patients.

References

Adlard PA, Cherny RA, Finkelstein DI, Gautier E, Robb E, Cortes M, et al. Rapid restoration of cognition in Alzheimer's transgenic mice with 8-hydroxy quinoline analogs is associated with decreased interstitial Aβ. Neuron. 2008;59(1):43–55.

Ahuja A, Dev K, Tanwar RS, Selwal KK, Tyagi PK. Copper mediated neurological disorder: visions into amyotrophic lateral sclerosis, Alzheimer and Menkes disease. J Trace Elem Med Biol. 2015;29:11–23.

Amaravadi R, Glerum DM, Tzagoloff A. Isolation of a cDNA encoding the human homolog of COX17, a yeast gene essential for mitochondrial copper recruitment. Hum Genet. 1997;99(3):329–33.

Atwood CS, Moir RD, Huang X, Scarpa RC, Bacarra NME, Romano DM, et al. Dramatic aggregation of Alzheimer Aβ by Cu (II) is induced by conditions representing physiological acidosis. J Biol Chem. 1998;273(21):12817–26.

Atwood CS, Scarpa RC, Huang X, Moir RD, Jones WD, Fairlie DP, et al. Characterization of copper interactions with alzheimer amyloid beta peptides: identification of an attomolar-affinity copper binding site on amyloid beta1-42. J Neurochem. 2000;75(3):1219–33.

Ayton S, Lei P, Bush AI. Biometals and their therapeutic implications in Alzheimer's disease. Neurotherapeutics. 2015;12(1):109–20.

Ballard C, Gauthier S, Corbett A, Brayne C, Aarsland D, Jones E. Alzheimer's disease. Lancet (London, England). 2011;377(9770):1019–31.

Bamberger ME, Harris ME, McDonald DR, Husemann J, Landreth GE. A cell surface receptor complex for fibrillar β-amyloid mediates microglial activation. J Neurosci. 2003;23(7):2665–74.

Barbusiński K. Fenton reaction-controversy concerning the chemistry. Ecolog Chem Eng Sci. 2009;16(3):347–58.

Barnham KJ, McKinstry WJ, Multhaup G, Galatis D, Morton CJ, Curtain CC, et al. Structure of the Alzheimer's disease amyloid precursor protein copper binding domain a regulator of neuronal copper homeostasis. J Biol Chem. 2003;278(19):17401–7.

Barnham KJ, Haeffner F, Ciccotosto GD, Curtain CC, Tew D, Mavros C, et al. Tyrosine gated electron transfer is key to the toxic mechanism of Alzheimer's disease beta-amyloid. FASEB J. 2004;18(12):1427–9.

Bayer TA, Schäfer S, Simons A, Kemmling A, Kamer T, Tepests R, et al. Dietary Cu stabilizes brain superoxide dismutase 1 activity and reduces amyloid Aβ production in APP23 transgenic mice. Proc Natl Acad Sci. 2003;100(24):14187–92.

Bucossi S, Ventriglia M, Panetta V, Salustri C, Pasqualetti P, Mariani S, et al. Copper in Alzheimer's disease: a meta-analysis of serum,plasma, and cerebrospinal fluid studies. J Alzheimer's Dis. 2011;24(1):175–85.

Bush AI, Tanzi RE. Therapeutics for Alzheimer's disease based on the metal hypothesis. Neurotherapeutics. 2008;5(3):421–32.

Buxbaum JD, Thinakaran G, Koliatsos V, O'Callahan J, Slunt HH, Price DL, et al. Alzheimer amyloid protein precursor in the rat hippocampus: transport and processing through the perforant path. J Neurosci. 1998;18(23):9629–37.

Caragounis A, Du T, Filiz G, Laughton KM, Volitakis I, Sharples RA, et al. Differential modulation of Alzheimer's disease amyloid beta-peptide accumulation by diverse classes of metal ligands. Biochem J. 2007;407(3):435–50.

Cater MA, McInnes KT, Li Q-X, Volitakis I, La Fontaine S, Mercer JF, et al. Intracellular copper deficiency increases amyloid-β secretion by diverse mechanisms. Biochem J. 2008;412(1):141–52.

Ceccom J, Coslédan F, Halley H, Francès B, Lassalle JM, Meunier B. Copper chelator induced efficient episodic memory recovery in a non-transgenic Alzheimer's mouse model. PLoS One. 2012;7(8):e43105.

Cherny RA, Atwood CS, Xilinas ME, Gray DN, Jones WD, McLean CA, et al. Treatment with a copper-zinc chelator markedly and rapidly inhibits β-amyloid accumulation in Alzheimer's disease transgenic mice. Neuron. 2001;30(3):665–76.

Choi B-S, Zheng W. Copper transport to the brain by the blood-brain barrier and blood-CSF barrier. Brain Res. 2009;1248:14–21.

Choo XY, Alukaidey L, White AR, Grubman A. Neuroinflammation and copper in Alzheimer's disease. Int J Alzheimers Dis. 2013;2013:145345.

Christensen MA, Zhou W, Qing H, Lehman A, Philipsen S, Song W. Transcriptional regulation of BACE1, the β-amyloid precursor protein β-secretase, by Sp1. Mol Cell Biol. 2004;24(2):865–74.

Citron M, Diehl TS, Gordon G, Biere AL, Seubert P, Selkoe DJ. Evidence that the 42-and 40-amino acid forms of amyloid β protein are generated from the β-amyloid precursor protein by different protease activities. Proc Natl Acad Sci. 1996;93(23):13170–5.

Cottrell BA, Galvan V, Banwait S, Gorostiza O, Lombardo CR, Williams T, et al. A pilot proteomic study of amyloid precursor interactors in Alzheimer's disease. Ann Neurol. 2005;58(2):277–89.

Crouch PJ, Hung LW, Adlard PA, Cortes M, Lal V, Filiz G, et al. Increasing Cu bioavailability inhibits Abeta oligomers and tau phosphorylation. Proc Natl Acad Sci U S A. 2009;106(2):381–6.

Cuajungco MP, Goldstein LE, Nunomura A, Smith MA, Lim JT, Atwood CS, et al. Evidence that the β-amyloid plaques of Alzheimer's disease represent the redox-silencing and entombment of Aβ by zinc. J Biol Chem. 2000;275(26):19439–42.

Culotta VC, Klomp LW, Strain J, Casareno RLB, Krems B, Gitlin JD. The copper chaperone for superoxide dismutase. J Biol Chem. 1997;272(38):23469–72.

Davies KM, Hare DJ, Cottam V, Chen N, Hilgers L, Halliday G, et al. Localization of copper and copper transporters in the human brain. Metallomics Integ Biometal Sci. 2013;5(1):43–51.

de Calignon A, Polydoro M, Suárez-Calvet M, William C, Adamowicz DH, Kopeikina KJ, et al. Propagation of tau pathology in a model of early Alzheimer's disease. Neuron. 2012;73(4):685–97.

DeWitt DA, Perry G, Cohen M, Doller C, Silver J. Astrocytes regulate microglial phagocytosis of senile plaque cores of Alzheimer's disease. Exp Neurol. 1998;149(2):329–40.

Di Vaira M, Bazzicalupi C, Orioli P, Messori L, Bruni B, Zatta P. Clioquinol, a drug for Alzheimer's disease specifically interfering with brain metal metabolism: structural characterization of its zinc (II) and copper (II) complexes. Inorg Chem. 2004;43(13):3795–7.

Dickson DW, Farlo J, Davies P, Crystal H, Fuld P, Yen S-H. Alzheimer's disease. A double-labeling immunohistochemical study of senile plaques. Am J Pathol. 1988;132(1):86.

Donnelly PS, Caragounis A, Du T, Laughton KM, Volitakis I, Cherny RA, et al. Selective intracellular release of copper and zinc ions from bis(thiosemicarbazonato) complexes reduces levels of Alzheimer disease amyloid-beta peptide. J Biol Chem. 2008;283(8):4568–77.

Duncan C, White AR. Copper complexes as therapeutic agents. Metallomics. 2012;4(2):127–38.

Filiz G, Price KA, Caragounis A, Du T, Crouch PJ, White AR. The role of metals in modulating metalloprotease activity in the AD brain. Eur Biophys J. 2008;37(3):315–21.

Galeazzi L, Ronchi P, Franceschi C, Giunta S. In vitro peroxidase oxidation induces stable dimers of beta-amyloid (1-42) through dityrosine bridge formation. Amyloid. 1999;6(1):7–13.

Götz J, Chen F, Van Dorpe J, Nitsch R. Formation of neurofibrillary tangles in P301L tau transgenic mice induced by Aβ42 fibrils. Science. 2001;293(5534):1491–5.

Green MA, Klippenstein DL, Tennison JR. Copper(II) bis(thiosemicarbazone) complexes as potential tracers for evaluation of cerebral and myocardial blood flow with PET. J Nucl Med. 1988;29(9):1549–57.

Grossi C, Francese S, Casini A, Rosi MC, Luccarini I, Fiorentini A, et al. Clioquinol decreases amyloid-β burden and reduces working memory impairment in a transgenic mouse model of Alzheimer's disease. J Alzheimers Dis. 2009;17(2):423–40.

Halliwell B, Gutteridge J. Oxygen toxicity, oxygen radicals, transition metals and disease. Biochem J. 1984;219(1):1.

Hegde ML, Bharathi P, Suram A, Venugopal C, Jagannathan R, Poddar P, et al. Challenges associated with metal chelation therapy in Alzheimer's disease. J Alzheimers Dis. 2009;17(3):457–68.

Heicklen-Klein A, Ginzburg I. Tau promoter confers neuronal specificity and binds Sp1 and AP-2. J Neurochem. 2000;75(4):1408–18.

Henry W, Querfurth H, LaFerla F. Mechanisms of disease Alzheimer's disease. New Engl J Med. 2010;362:329–44.

Higuchi M, Lee VM-Y, Trojanowski JQ. Tau and axonopathy in neurodegenerative disorders. NeuroMolecular Med. 2002;2(2):131–50.

Hu M, Waring JF, Gopalakrishnan M, Li J. Role of GSK-3beta activation and alpha7 nAChRs in Abeta(1-42)-induced tau phosphorylation in PC12 cells. J Neurochem. 2008;106(3):1371–7.

Huang X, Cuajungco MP, Atwood CS, Hartshorn MA, Tyndall JD, Hanson GR, et al. Cu (II) potentiation of Alzheimer Aβ neurotoxicity correlation with cell-free hydrogen peroxide production and metal reduction. J Biol Chem. 1999;274(52):37111–6.

Huang X, Atwood CS, Moir RD, Hartshorn MA, Tanzi RE, Bush AI. Trace metal contamination initiates the apparent auto-aggregation, amyloidosis, and oligomerization of Alzheimer's Aβ peptides. J Biol Inorg Chem. 2004;9(8):954–60.

Hung YH, Bush AI, Cherny RA. Copper in the brain and Alzheimer's disease. J Biol Inorg Chem. 2010;15(1):61–76.

Hung YH, Bush AI, La Fontaine S. Links between copper and cholesterol in Alzheimer's disease. Front Physiol. 2013;4:111.

Jomova K, Valko M. Advances in metal-induced oxidative stress and human disease. Toxicology. 2011;283(2):65–87.

Kaden D, Bush AI, Danzeisen R, Bayer TA, Multhaup G. Disturbed copper bioavailability in Alzheimer's disease. Int J Alzheimers Dis. 2011;2011:345614.

Kardos J, Kovacs I, Hajos F, Kalman M, Simonyi M. Nerve endings from rat brain tissue release copper upon depolarization. A possible role in regulating neuronal excitability. Neurosci Lett. 1989;103(2):139–44.

Kennedy T, Ghio AJ, Reed W, Samet J, Zagorski J, Quay J, et al. Copper-dependent inflammation and nuclear factor-kappaB activation by particulate air pollution. Am J Respir Cell Mol Biol. 1998;19(3):366–78.

Kessler H, Bayer TA, Bach D, Schneider-Axmann T, Supprian T, Herrmann W, et al. Intake of copper has no effect on cognition in patients with mild Alzheimer's disease: a pilot phase 2 clinical trial. J Neural Transm. 2008;115(8):1181–7.

Khlistunova I, Biernat J, Wang Y, Pickhardt M, von Bergen M, Gazova Z, et al. Inducible expression of Tau repeat domain in cell models of tauopathy aggregation is toxic to cells but can be reversed by inhibitor drugs. J Biol Chem. 2006;281(2):1205–14.

Klomp LW, Lin S-J, Yuan DS, Klausner RD, Culotta VC, Gitlin JD. Identification and functional expression of HAH1, a novel human gene involved in copper homeostasis. J Biol Chem. 1997;272(14):9221–6.

Kong G-W, Adams JJ, Cappai R, Parker MW. Structure of Alzheimer's disease amyloid precursor protein copper-binding domain at atomic resolution. Acta Crystallogr Sect F: Struct Biol Cryst Commun. 2007;63(10):819–24.

Lannfelt L, Blennow K, Zetterberg H, Batsman S, Ames D, Harrison J, et al. Safety, efficacy, and biomarker findings of PBT2 in targeting Abeta as a modifying therapy for Alzheimer's disease: a phase IIa, double-blind, randomised, placebo-controlled trial. Lancet Neurol. 2008;7(9):779–86.

Lewis J, Dickson DW, Lin WL, Chisholm L, Corral A, Jones G, et al. Enhanced neurofibrillary degeneration in transgenic mice expressing mutant tau and APP. Science. 2001;293(5534):1487–91.

Liu L, Drouet V, Wu JW, Witter MP, Small SA, Clelland C, et al. Trans-synaptic spread of tau pathology in vivo. PLoS One. 2012;7(2):e31302.

LoPresti P, Szuchet S, Papasozomenos SC, Zinkowski RP, Binder LI. Functional implications for the microtubule-associated protein tau: localization in oligodendrocytes. Proc Natl Acad Sci. 1995;92(22):10369–73.

Lu J, Zheng Y-L, Wu D-M, Sun D-X, Shan Q, Fan S-H. Trace amounts of copper induce neurotoxicity in the cholesterol-fed mice through apoptosis. FEBS Lett. 2006;580(28–29):6730–40.

Lue L-F, Kuo Y-M, Roher AE, Brachova L, Shen Y, Sue L, et al. Soluble amyloid β peptide concentration as a predictor of synaptic change in Alzheimer's disease. Am J Pathol. 1999;155(3):853–62.

Ma QF, Li YM, Du JT, Kanazawa K, Nemoto T, Nakanishi H, et al. Binding of copper (II) ion to an Alzheimer's tau peptide as revealed by MALDI-TOF MS, CD, and NMR. Biopolymers. 2005;79(2):74–85.

Ma Q, Li Y, Du J, Liu H, Kanazawa K, Nemoto T, et al. Copper binding properties of a tau peptide associated with Alzheimer's disease studied by CD, NMR, and MALDI-TOF MS. Peptides. 2006;27(4):841–9.

Malm TM, Iivonen H, Goldsteins G, Keksa-Goldsteine V, Ahtoniemi T, Kanninen K, et al. Pyrrolidine dithiocarbamate activates Akt and improves spatial learning in APP/PS1 mice without affecting beta-amyloid burden. J Neurosci. 2007;27(14):3712–21.

Mao X, Ye J, Zhou S, Pi R, Dou J, Zang L, et al. The effects of chronic copper exposure on the amyloid protein metabolisim associated genes' expression in chronic cerebral hypoperfused rats. Neurosci Lett. 2012;518(1):14–8.

McCord JM, Fridovich I. Superoxide dismutase. An enzymic function for erythrocuprein (hemocuprein). J Biol Chem. 1969;244(22):6049–55.

McGeer PL, Akiyama H, Itagaki S, McGeer EG. Immune system response in Alzheimer's disease. Can J Neurol Sci. 1989;16(4 Suppl):516–27.

Minghetti L. Role of inflammation in neurodegenerative diseases. Curr Opin Neurol. 2005;18(3):315–21.

Moriwaki H, Osborne MR, Phillips DH. Effects of mixing metal ions on oxidative DNA damage mediated by a Fenton-type reduction. Toxicol In Vitro. 2008;22(1):36–44.

Mot AI, Wedd AG, Sinclair L, Brown DR, Collins SJ, Brazier MW. Metal attenuating therapies in neurodegenerative disease. Expert Rev Neurother. 2011;11(12):1717–45.

Myhre O, Utkilen H, Duale N, Brunborg G, Hofer T. Metal dyshomeostasis and inflammation in Alzheimer's and Parkinson's diseases: possible impact of environmental exposures. Oxidative Med Cell Longev. 2013;2013:726954.

Opazo CM, Greenough MA, Bush AI. Copper: from neurotransmission to neuroproteostasis. Front Aging Neurosci. 2014;6:143.

Perry G, Cash AD, Smith MA. Alzheimer disease and oxidative stress. Biomed Res Int. 2002;2(3):120–3.

Pooler AM, Phillips EC, Lau DH, Noble W, Hanger DP. Physiological release of endogenous tau is stimulated by neuronal activity. EMBO Rep. 2013;14(4):389–94.

Pratico D. Evidence of oxidative stress in Alzheimer's disease brain and antioxidant therapy: lights and shadows. Ann N Y Acad Sci. 2008;1147:70–8.

Prince M, Wimo A, Guerchet M, Ali G, Wu Y, Prina M. World Alzheimer report 2015 [Internet]. London. 2015. Available from: http://www.alz.co.uk/research/WorldAlzheimerReport2015.pdf

Reitz C. Alzheimer's disease and the amyloid cascade hypothesis: a critical review. Int J Alzheimers Dis. 2012;2012:369808.

Rembach A, Hare DJ, Lind M, Fowler CJ, Cherny RA, McLean C, et al. Decreased copper in Alzheimer's disease brain is predominantly in the soluble extractable fraction. Int J Alzheimers Dis. 2013;2013:623241.

Ritchie CW, Bush AI, Mackinnon A, Macfarlane S, Mastwyk M, MacGregor L, et al. Metal-protein attenuation with iodochlorhydroxyquin (clioquinol) targeting Aβ amyloid deposition and toxicity in Alzheimer disease: a pilot phase 2 clinical trial. Arch Neurol. 2003;60(12):1685–91.

Rozemuller JM, Eikelenboom P, Pals ST, Stam FC. Microglial cells around amyloid plaques in Alzheimer's disease express leucocyte adhesion molecules of the LFA-1 family. Neurosci Lett. 1989;101(3):288–92.

Salustri C, Barbati G, Ghidoni R, Quintiliani L, Ciappina S, Binetti G, et al. Is cognitive function linked to serum free copper levels? A cohort study in a normal population. Clin Neurophysiol. 2010;121(4):502–7.

Sayre LM, Perry G, Smith MA. Oxidative stress and neurotoxicity. Chem Res Toxicol. 2008; 21(1):172–88.

Scheiber IF, Mercer JF, Dringen R. Metabolism and functions of copper in brain. Prog Neurobiol. 2014; 116:33–57.

Schmalz G, Schuster U, Schweikl H. Influence of metals on IL-6 release in vitro. Biomaterials. 1998;19(18):1689–94.

Schrag M, Mueller C, Oyoyo U, Smith MA, Kirsch WM. Iron, zinc and copper in the Alzheimer's disease brain: a quantitative meta-analysis. Some insight on the influence of citation bias on scientific opinion. Prog Neurobiol. 2011;94(3):296–306.

Selkoe DJ. Alzheimer's disease: genes, proteins, and therapy. Physiol Rev. 2001;81(2):741–66.

Shaffer LM, Dority MD, Gupta-Bansal R, Frederickson RC, Younkin SG, Brunden KR. Amyloid β protein (Aβ) removal by neuroglial cells in culture. Neurobiol Aging. 1995;16(5):737–45.

Small DH, McLean CA. Alzheimer's disease and the amyloid β protein. J Neurochem. 1999; 73(2):443–9.

Smith DP, Smith DG, Curtain CC, Boas JF, Pilbrow JR, Ciccotosto GD, et al. Copper-mediated amyloid-β toxicity is associated with an intermolecular histidine bridge. J Biol Chem. 2006;281(22):15145–54.

Smith DP, Ciccotosto GD, Tew DJ, Fodero-Tavoletti MT, Johanssen T, Masters CL, et al. Concentration dependent Cu^{2+} induced aggregation and Dityrosine formation of the Alzheimer's disease amyloid-β peptide. Biochemistry. 2007;46(10):2881–91.

Song I-S, Chen HH, Aiba I, Hossain A, Liang ZD, Klomp LW, et al. Transcription factor Sp1 plays an important role in the regulation of copper homeostasis in mammalian cells. Mol Pharmacol. 2008;74(3):705–13.

Squitti R. Copper dysfunction in Alzheimer's disease: from meta-analysis of biochemical studies to new insight into genetics. J Trace Elem Med Biol. 2012;26(2):93–6.

Squitti R. Copper subtype of Alzheimer's disease (AD): meta-analyses, genetic studies and predictive value of non-ceruloplasmim copper in mild cognitive impairment conversion to full AD. J Trace Elem Med Biol. 2014;28(4):482–5.

Squitti R, Pasqualetti P, Dal Forno G, Moffa F, Cassetta E, Lupoi D, et al. Excess of serum copper not related to ceruloplasmin in Alzheimer disease. Neurology. 2005;64(6):1040–6.

Squitti R, Ventriglia M, Barbati G, Cassetta E, Ferreri F, Dal Forno G, et al. 'Free' copper in serum of Alzheimer's disease patients correlates with markers of liver function. J Neural Transm. 2007;114(12):1589–94.

Squitti R, Quattrocchi CC, Salustri C, Rossini PM. Ceruloplasmin fragmentation is implicated in 'free' copper deregulation of Alzheimer disease. Prion. 2008;2(1):23–7.

Squitti R, Bressi F, Pasqualetti P, Bonomini C, Ghidoni R, Binetti G, et al. Longitudinal prognostic value of serum "free" copper in patients with Alzheimer disease. Neurology. 2009;72(1):50–5.

Squitti R, Ghidoni R, Siotto M, Ventriglia M, Benussi L, Paterlini A, et al. Value of serum nonceruloplasmin copper for prediction of mild cognitive impairment conversion to Alzheimer disease. Ann Neurol. 2014;75(4):574–80.

Stamer K, Vogel R, Thies E, Mandelkow E, Mandelkow E-M. Tau blocks traffic of organelles, neurofilaments, and APP vesicles in neurons and enhances oxidative stress. J Cell Biol. 2002;156(6):1051–63.

Su X-Y, Wu W-H, Huang Z-P, Hu J, Lei P, Yu C-H, et al. Hydrogen peroxide can be generated by tau in the presence of Cu (II). Biochem Biophys Res Commun. 2007;358(2):661–5.

Treiber C, Simons A, Strauss M, Hafner M, Cappai R, Bayer TA, et al. Clioquinol mediates copper uptake and counteracts copper efflux activities of the amyloid precursor protein of Alzheimer's disease. J Biol Chem. 2004;279(50):51958–64.

Trombley PQ, Shepherd GM. Differential modulation by zinc and copper of amino acid receptors from rat olfactory bulb neurons. J Neurophysiol. 1996;76(4):2536–46.

Ventriglia M, Bucossi S, Panetta V, Squitti R. Copper in Alzheimer's disease: a meta-analysis of serum, plasma, and cerebrospinal fluid studies. J Alzheimers Dis. 2012;30(4):981–4.

Wang Z-X, Tan L, Wang H-F, Ma J, Liu J, Tan M-S, et al. Serum iron, zinc, and copper levels in patients with Alzheimer's disease: a replication study and meta-analyses. J Alzheimers Dis. 2015;47(3):565–81.

Weiser T, Wienrich M. The effects of copper ions on glutamate receptors in cultured rat cortical neurons. Brain Res. 1996;742(1–2):211–8.

White AR, Barnham KJ, Bush AI. Metal homeostasis in Alzheimer's disease. Expert Rev Neurother. 2006a;6(5):711–22.

White AR, Du T, Laughton KM, Volitakis I, Sharples RA, Xilinas ME, et al. Degradation of the Alzheimer disease amyloid β-peptide by metal-dependent up-regulation of metalloprotease activity. J Biol Chem. 2006b;281(26):17670–80.

Wyss-Coray T, Mucke L. Inflammation in neurodegenerative disease—a double-edged sword. Neuron. 2002;35(3):419–32.

Zappasodi F, Salustri C, Babiloni C, Cassetta E, Del Percio C, Ercolani M, et al. An observational study on the influence of the APOE-ε4 allele on the correlation between 'free'copper toxicosis and EEG activity in Alzheimer disease. Brain Res. 2008;1215:183–9.

Zheng Z, White C, Lee J, Peterson TS, Bush AI, Sun GY, et al. Altered microglial copper homeostasis in a mouse model of Alzheimer's disease. J Neurochem. 2010;114(6):1630–8.

Zhou L-X, Du J-T, Zeng Z-Y, Wu W-H, Zhao Y-F, Kanazawa K, et al. Copper (II) modulates in vitro aggregation of a tau peptide. Peptides. 2007;28(11):2229–34.

Uranium and the Central Nervous System: What Should We Learn from Recent New Tools and Findings?

Céline Dinocourt

Abstract Increasing industrial and military use of uranium has led to environmental pollution, which may result in uranium reaching the brain and causing cerebral dysfunction. A recent literature review has discussed data published over the last 10 years on uranium and its effects on brain function (Dinocourt C, Legrand M, Dublineau I, et al., Toxicology 337:58–71, 2015). New models of uranium exposure during neonatal brain development and the emergence of new technologies (omics and epigenetics) are of value in identifying new specific targets of uranium. Here we review the latest studies of neurogenesis, epigenetics, and metabolic dysfunctions and the identification of new biomarkers used to establish potential pathophysiological states of neurodevelopmental and neurodegenerative diseases.

Keywords Omics • Epigenetics • Neurogenesis • Brain • Uranium

Introduction

Uranium, a naturally occurring heavy metal, is found as a mixture of three isotopes: uranium 238, uranium 235, and uranium 234. Both natural and anthropogenic processes contribute to its distribution throughout the environment (Bleise et al. 2003) and raise concerns about health risks associated with human chronic exposure to low dose of uranium (ATSDR 2013). Uranium's effects on the central nervous system are well documented in the literature (see review (Dinocourt et al. 2015; Jiang and Aschner 2006)). Although the exact mechanisms by which uranium enters brain structures are unknown, it induces changes in neurobehavioral functions as locomotion, sleep-wake cycle, learning, and memory and induces depressive-like behavior. Several mechanistic pathways have been investigated, and numerous "classic" analyses of gene or protein expression

C. Dinocourt (✉)
Institut de Radioprotection et de Sûreté Nucléaire (IRSN), Direction de la Stratégie du Développement et des Partenariats, Service Programmes et Stratégies scientifiques, BP 17, F-92262 Fontenay-aux-Roses, France
e-mail: celine.dinocourt@irsn.fr

© Springer International Publishing AG 2017
M. Aschner and L.G. Costa (eds.), *Neurotoxicity of Metals*, Advances in Neurobiology 18, DOI 10.1007/978-3-319-60189-2_11

have focused on imbalance between pro-/antioxidant systems, neurotransmitter pathways, and neurophysiological properties. Most of these studies have focused principally on mechanisms in uranium-exposed adult animals, though new data on developing animals are beginning to be published. The need to improve knowledge of the effects of uranium on brain development has been confirmed by means of new models of uranium exposure during gestation. Over recent years, new technologies, like omics and epigenetics, have been used to study new specific targets of uranium. This short review covers the mechanistic pathways of uranium, in terms of its chemical properties as a heavy metal, with a major focus on the latest published findings.

How Does Uranium Act on Brain Function?

Although uranium accumulation in the brain is low (Dublineau et al. 2014; Lerebours et al. 2009; Paquet et al. 2006), the landscape of uranium-induced effects shows that the central nervous system can be largely affected by uranium. The molecular targets of uranium's effects are multiple and include DNA binding and transcriptional and posttranscriptional effects that may influence enzyme activity as well as gene and protein expression.

To date, the neurotoxicity of uranium has been examined by classic studies of oxidative stress responses, the metabolism of neurotransmitter pathways, and electrophysiological properties in animal models as rodents (see review (Dinocourt et al. 2015)) and zebrafish as a neurotoxicological model (Linney et al. 2004). Briefly:

Oxidative Stress Responses Uranium induces oxidation of brain lipids (Briner and Davis 2002; Briner and Murray 2005; Ghosh et al. 2007; Lestaevel et al. 2009, 2015; Linares et al. 2007), modulates antioxidant defense in rats (Lestaevel et al. 2009, 2015; Linares et al. 2007) and zebrafish (Barillet et al. 2007, 2011; Lerebours et al. 2009), and modulates nitric oxide production (Abou-Donia et al. 2002; Lestaevel et al. 2009).

Metabolism of Neurotransmitter Pathways Concentrations of breakdown enzymes (acetylcholinesterase [AChE]), M1 muscarinic acetylcholine receptor, and acetylcholine in the cholinergic system, the most studied, are more affected by uranium in the cerebral cortex than in the hippocampus in rats exposed at adult stages (Bensoussan et al. 2009; Bussy et al. 2006) and in juvenile rats contaminated from birth (Lestaevel et al. 2013). In the zebrafish model, increased AChE activity associated with gene induction involves a neuronal response of the cholinergic system (Barillet et al. 2007; Lerebours et al. 2009). The dopaminergic system is also modulated after uranium exposure more specifically in the hypothalamus, cortex, and striatum, but apparently not in the hippocampus (Bussy et al. 2006). Finally, the expression of genes involved in glutamate synthesis is induced transiently in the brain of zebrafish exposed to uranium (Lerebours et al. 2009).

Neurophysiological Properties Uranium alters neuronal excitability in the hippocampus. In vitro studies have shown a decrease of depolarization-evoked glutamate after uranium exposure (Tomsig and Suszkiw 1996). Pellmar et al. demonstrated that the efficacy of synaptic transmission and glutamatergic release is impaired in the hippocampus of uranium-exposed rat brains (Pellmar et al. 1999). These electrophysiological changes point to a decrease in neuronal excitability.

Uranium and Brain Development: Insights from a New Model of Uranium Exposure During Gestation

As with toxicants, exposure to uranium in early life may cause later health effects (Grandjean and Landrigan 2014). The effects of uranium during the neonatal period are limited to developmental toxicity and fetal development (Domingo 2001; Paternain et al. 1989) and neurobehavioral disturbances in animals exposed during brain development (Albina et al. 2005; Briner and Abboud 2002). The neurons of the central nervous system are produced during development by neurogenesis, the principal steps of which—cell proliferation/death, differentiation, migration, and synapse formation (Gotz and Huttner 2005)—may be disrupted by uranium, resulting in alteration of the integrity of neural networks, causing neurological disturbances in children as well as adults, as shown, for example, after methylmercury neonatal exposure (Ceccatelli et al. 2013).

Recent experimental studies have focused on specific processes of neurogenesis, i.e., proliferation and differentiation after uranium exposure during prenatal and postnatal development of the rat brain (Legrand et al. 2016a, b). These studies focused on the telencephalon and more specifically on the hippocampus. During prenatal development, hippocampal regions begin to form from gestational day (GD) 15 to GD21, and the dentate granular cells appear mostly in fetuses after GD18 (Bayer 1980). At birth, Ammon's horn is well organized when the dentate gyrus is immature. Granule cells of 85% are generated between postnatal day (PND) 0 and PND 21 in the dentate gyrus (Altman and Bayer 1990). Specific markers linked to the stages of hippocampal development have been used to study cell proliferation and neuronal differentiation processes: two at prenatal stages, GD 13 and 18, and three at postnatal stages, PND 0, 5, and 21. Bromodeoxyuridine (BrdU) incorporation has been used to study cell proliferation (S phase) (Legrand et al. 2016a), and Wnt, Notch signaling, and the pro-/anti-neurogenic bHLH (basic helix-loop-helix) genes, as well as doublecortin, a marker of immature neurons, have been examined because of their involvement in hippocampal development, in maintaining neural progenitor cell proliferation, and in neuronal determination or neuronal differentiation pathways, respectively (Legrand et al. 2016b). The major results of these studies were (1) a decrease in Wnt3a staining in the hippocampal neuroepithelium of GD13 embryos from exposed dams, (2) an increase in BrdU staining cells in the dentate neuroepithelium and a decrease in staining intensity for Notch1 and an increase for Mash1 in the hippocampal neuroepithelium of GD18 fetuses of dams exposed to uranium, and (3) a decrease in the number of BrdU-labeled cells in the granule cell layer of the dentate gyrus and an increase in the number of doublecortin-positive cells in the granular cell layer in PND 21 rats of exposed dams.

Taken together, these data strongly suggest that uranium affects various signaling pathways of neuronal cell proliferation early in neuronal determination or induction at specific times of the prenatal and postnatal stages in the hippocampus during its development. However, these effects on neurogenesis processes do not disrupt the development of the hippocampus, as shown by the organization of its layers (Dinocourt et al. 2017; Legrand et al. 2016a). These results do not allow us to conclude that uranium has a real impact on the development of the hippocampus, and further studies are needed of other markers and of other steps of cell migration and synaptogenesis, i.e., how the neuronal network is built up when it is exposed to uranium. Indeed, even if the layers of the hippocampus are well organized, no data are yet available on synaptogenesis and the organization of neuronal networks after uranium exposure during brain development. The neurophysiological properties of the networks might thus be disrupted. Supporting this hypothesis, preliminary results suggest that adult neurophysiological properties are impaired after exposure to uranium during postnatal development in rats (Dinocourt et al. 2014; Dinocourt et al. 2015).

Uranium and the Central Nervous System: New Technologies Need to Serve New Insights

Neuroepigenetic Mechanisms

Chemical modifications of DNA and of its associated proteins that do not involve a change in DNA sequence, i.e., epigenetic alterations, are possible targets of heavy metals (Cheng et al. 2012). More importantly, several lines of evidence indicate that epigenetic mechanisms in the central nervous system are essential for regulating various neuronal functions and play a critical role in cognitive behavior, as learning and memory (see review (Rudenko and Tsai 2014)). Three principal types of molecular mechanisms mediating epigenetic regulation of gene expression (DNA methylation, histone modification, and expression of noncoding RNAs) can regulate neuronal function and thus are potential targets for the effects of uranium. Two early studies of DNA methylation during uranium-induced lung cancer and leukemia (Miller et al. 2009; Su et al. 2006) were followed by two recent reports showing, after chronic uranium exposure, changes of global DNA methylation in various organs—the gonads in rodents (Elmhiri et al. 2016) and the brain in zebrafish (Gombeau et al. 2016). Gombeau et al. showed for the first time that uranium increases genome-wide cytosine methylation in the brain, while it decreases DNA methylation at specific sites, CpG islands, mainly found in promoter regions of genes, with possible consequences in terms of gene silencing (Hon et al. 2012). These alterations in cytosine methylation patterns depend on uranium concentration and duration of exposure and were greater in males than in females (Gombeau et al. 2016). The authors hypothesize that uranium could modulate genomic DNA methylation by oxidative stress, and it has been reported that oxidative stress induces epigenetic alterations (Franco et al. 2008; Valinluck et al. 2004). The neurotoxicity of uranium via oxidative stress has been clearly demonstrated in rodents (see review (Dinocourt et al. 2015)) as well as in zebrafish (Barillet et al. 2007; Lerebours et al. 2009). They also suggest that DNA methylation could be affected by DNA methyltransferase activity, as already shown for other heavy metals like cadmium (Huang et al. 2008; Takiguchi et al. 2003). Interestingly, these DNA alterations were also found in the nonexposed progeny of uranium-exposed parents (Gombeau et al. personal communication).

Metabolic Fingerprint in the Brain

Extensive literature data show the molecular and physiological effects of various metabolic pathways in the brains of animals chronically exposed to low-dose uranium (Dinocourt et al. 2015, 2017; Legrand et al. 2016a, b; Lerebours et al. 2010). However, these results are generally difficult to interpret. Alternative approaches to assessment of chronic exposure to low doses of uranium include metabolomics (Grison et al. 2013, 2016), which provides a quantitative analysis of metabolic networks, through the simultaneous quantification of free low-molecular-weight metabolites (<1000 Da) and profiling of metabolic phenotypes. A recent metabolomics study of the effect of uranium exposure on brain metabolism used metabolites of the cerebrospinal fluid (CSF) in rats chronically exposed to uranium from birth to discriminate between rats exposed or not exposed and also between the sexes (Lestaevel et al. 2016). Among the 86 most discriminatory CSF metabolites, 7 discriminated control versus exposed female rats, 7 discriminated control versus exposed male rats, and 4 discriminated control versus exposed rats independently of gender. Two of these metabolites belong to arginine and proline metabolism (N2-succinyl-L-arginine, N4-acetylaminobutanoate). N-methylsalsolinol, the dopamine-derived tetrahydroisoquinoline derivative, may be neurotoxic to dopaminergic metabolism (Naoi et al. 1997) and may be involved in the pathogenesis of Parkinson's disease (Maruyama et al. 1996). Thus, considering the metabolome as a the metabolic state of a given physiologic state, metabolomics may be able to reveal biochemical pathways involved in biological mechanisms as well as potential biomarkers after exposure to uranium (Kaddurah-Daouk and Krishnan 2009).

These data point to new potential targets of uranium that should be investigated. First, the epigenetic approach will help to elucidate the underlying genetic mechanisms. Epigenetic regulation of gene expression includes DNA methylation, histone modification, and expression of noncoding RNAs. The latter is important because these noncoding microRNAs play a key role as regulators of synaptic functions and cognitive function and in neurodegenerative diseases (Gapp et al. 2014; Mouradian 2012; Rudenko and Tsai 2014). Furthermore, high-throughput methods such as next-generation sequencing should be used to identify methylation changes at specific gene locations, to clarify the links between histone modifications, DNA methylation, and transcriptomic alterations (Hirst and Marra 2010). And metabolomics can detect genomic, transcriptomic, and proteomic changes. Together they can be used to elucidate complex, heterogeneous mechanisms and to explain how uranium exposure might affect physiological pathways. For example, proteomics has been used to explore sets of serum proteins after chronic exposure by ingestion of uranium (Petitot et al. 2016). Moreover, metabolomics from brain tissue can be used to explore cerebral function and dysfunction (Gonzalez-Riano et al. 2016) and need to be investigated after uranium exposure.

Conclusion

The experimental work reported here and in a previous review (Dinocourt et al. 2015) clearly shows that exposure to uranium during critical periods can affect brain development. Disturbances in developmental processes involve several mechanistic pathways (oxidative stress, biochemical, neurochemical and neurophysiological pathways, neurogenesis) by which uranium may act to disrupt synaptic integrity in neural networks and might disturb brain function in adulthood (Dinocourt et al. 2017; Houpert et al. 2007; Legrand et al. 2016a, b; Lestaevel et al. 2013, 2015, 2016). These data highlight the need to investigate closely the mechanisms by which developmental processes are altered and the long-term consequences of uranium exposure during development. Furthermore, new technologies can be used in radiotoxicology to identify new potential targets of uranium and biomarkers of neurodevelopmental and neurodegenerative disorders. Omics and epigenetic analysis of the brain may help to detect the appearance of signs of neurological dysfunction after uranium exposure if the link between altered biological pathways and pathophysiological states can be established.

Acknowledgments We thank Christelle Adam-Guillermin and Maamar Souidi for their critical reading, specifically on epigenetic and metabolic pathways, of the manuscript.

References

Abou-Donia MB, Dechkovskaia AM, Goldstein LB, et al. Uranyl acetate-induced sensorimotor deficit and increased nitric oxide generation in the central nervous system in rats. Pharmacol Biochem Behav. 2002;72:881–90.

Agency for Toxic Substances and Disease Registry (ATSDR). Toxicological profile for Uranium. Atlanta: U.S. Department of Health and Human Services, Public Health Service; 2013.

Albina ML, Bellés M, Linares V, et al. Restraint stress does not enhance the uranium-induced developmental and behavioral effects in the offspring of uranium-exposed male rats. Toxicology. 2005;215:69–79.

Altman J, Bayer SA. Migration and distribution of two populations of hippocampal granule cell precursors during the perinatal and postnatal periods. J Comp Neurol. 1990;301(3):365–81.

Barillet S, Adam C, Palluel O, et al. Bioaccumulation, oxidative stress, and neurotoxicity in *Danio rerio* exposed to different isotopic compositions of uranium. Environ Toxicol Chem. 2007;26:497–505.

Barillet S, Adam-Guillermin C, Palluel O, et al. Uranium bioaccumulation and biological disorders induced in zebrafish (*Danio rerio*) after a depleted uranium waterborne exposure. Environ Pollut. 2011;159:495–502.

Bayer SA. Development of the hippocampal region in the rat. II. Morphogenesis during embryonic and early postnatal life. J Comp Neurol. 1980;190(1):115–34.

Bensoussan H, Grancolas L, Dhieux-Lestaevel B, et al. Heavy metal uranium affects the brain cholinergic system in rat following sub-chronic and chronic exposure. Toxicology. 2009;261:59–67.

Bleise A, Danesi PR, Burkart W. Properties, use and health effects of depleted uranium (DU): a general overview. J Environ Radioact. 2003;64:93–112.

Briner W, Abboud B. Behavior of juvenile mice chronically exposed to depleted uranium. In: Khassanova L, Collery P, Maymard I, Khassanova Z, Etienne JC, editors. Metal ions in biology and medicine. Paris: John Libby Eurotext; 2002. p. 353–6.

Briner W, Davis D. Lipid oxidation and behavior are correlated in depleted uranium exposed mice. In: Khassanova L, Collery P, Maymard I, Khassanova Z, Etienne JC, editors. Metal ions in biology and medicine. Paris: John Libby Eurotext; 2002. p. 59–63.

Briner W, Murray J. Effects of short-term and long-term depleted uranium exposure on open-field behavior and brain lipid oxidation in rats. Neurotoxicol Teratol. 2005;27:135–44.

Bussy C, Lestaevel P, Dhieux B, et al. Chronic ingestion of uranyl nitrate perturbs acetylcholinesterase activity and monoamine metabolism in male rat brain. Neurotoxicology. 2006;27:245–52.

Ceccatelli S, Bose R, Edoff K, et al. Long-lasting neurotoxic effects of exposure to methylmercury during development. J Intern Med. 2013;273(5):490–7.

Cheng TF, Choudhuri S, Muldoon-Jacobs K. Epigenetic targets of some toxicologically relevant metals: a review of the literature. J Appl Toxicol. 2012;32:643–53.

Dinocourt C, Stefani J et al. Reduced carbachol-induced beta/gamma oscillations in CA3 region of hippocampus after post-natal contamination of uranium in adult rat. Meeting abstract, Neurosciences, Washington, DC. 2014. November 2014.

Dinocourt C, Legrand M, Dublineau I, et al. The neurotoxicology of uranium. Toxicology. 2015; 337:58–71.

Dinocourt C, et al. Chronic exposure to uranium from gestation: Effects on behavior and neurogenesis in adulthood. Int J Environ Res Public Health. 2017;14(5):536.

Domingo JL. Reproductive and developmental toxicity of natural and depleted uranium: a review. Reprod Toxicol. 2001;15:603–9.

Dublineau I, Souidi M, Gueguen Y, et al. Unexpected lack of deleterious effects of uranium on physiological systems following a chronic oral intake in adult rat. Biomed Res Int. 2014; doi:10.1155/2014/181989.

Elmhiri G, Gloaguen C, Kereselidze D, et al. Multigenerational effects of chronic low-dose natural uranium contamination: epigenetic inheritance of methylation signature. Toxicol Lett. 2016;259S:S73–S247. http://dx.doi.org/10.1016/j.toxlet.2016.07.293

Franco R, Schoneveld O, Georgakilas AG, et al. Oxidative stress, DNA methylation and carcinogenesis. Cancer Lett. 2008;266(1):6–11.

Gapp K, Jawaid A, Sarkies P, et al. Implication of sperm RNAs in transgenerational inheritance of the effects of early trauma in mice. Nat Neurosci. 2014;17(5):667–9.

Ghosh S, Kumar A, Pandey BN, et al. Acute exposure of uranyl nitrate causes lipid peroxidation and histopathological damage in brain and bone of Wistar rat. J Environ Pathol Toxicol Oncol. 2007;26:255–61.

Gombeau K, Pereira S, Ravanat JL, et al. Depleted uranium induces sex- and tissue-specific methylation patterns in adult zebrafish. J Environ Radioact. 2016;154:25–33.

Gonzalez-Riano C, Garcia A, Barbas C. Metabolomics studies in brain tissue: a review. J Pharm Biomed Anal. 2016;130:141–68.

Gotz M, Huttner WB. The cell biology of neurogenesis. Nature reviews. Mol Cell Biol. 2005; 6(10):777–88.

Grandjean P, Landrigan PJ. Neurobehavioural effects of developmental toxicity. Lancet Neurol. 2014;13:330–8.

Grison S, Favé G, Maillot M, et al. Metabolomics identifies a biological response to chronic low-dose natural uranium contamination in urine samples. Metabolomics. 2013;9(6):1168–80.

Grison S, Favé G, Maillot M, et al. Metabolomics reveals dose effects of low-dose chronic exposure to uranium in rats: identification of candidate biomarkers in urine samples. Metabolomics. 2016;12(10):154.

Hirst M, Marra MA. Next generation sequencing based approaches to epigenomics. Brief Funct Genomics. 2010;9(5–6):455–65.

Hon GC, Hawkins RD, Caballero OL, et al. Global DNA hypomethylation coupled to repressive chromatin domain formation and gene silencing in breast cancer. Genome Res. 2012;22(2):246–58.

Houpert P, Frelon S, Lestaevel P, et al. Parental exposure to enriched uranium induced delayed hyperactivity in rat offspring. Neurotoxicology. 2007;28:108–13.

Huang D, Zhang Y, Qi Y, et al. Global DNA hypomethylation, rather than reactive oxygen species (ROS), a potential facilitator of cadmium-stimulated K562 cell proliferation. Toxicol Lett. 2008;179(1):43–7.

Jiang G, Aschner M. Neurotoxicity of depleted uranium: reasons for increased concern. Biol Trace Elem Res. 2006;110:1–17.

Kaddurah-Daouk R, Krishnan KR. Metabolomics: a global biochemical approach to the study of central nervous system diseases. Neuropsychopharmacology. 2009;34:173–86.

Legrand M, Elie C, Stefani J, et al. Cell proliferation and cell death are disturbed during prenatal and postnatal brain development after uranium exposure. Neurotoxicology. 2016a;52:34–45.

Legrand M, Lam S, Anselme I et al. Exposure to depleted uranium during development affects neuronal differentiation in the hippocampal dentate gyrus and induces depressive-like behavior in offspring. Neurotoxicology. 2016b. http://dx.doi.org/10.1016/j.neuro.2016.09.006

Lerebours A, Gonzalez P, Adam C, et al. Comparative analysis of gene expression in brain, liver, skeletal muscles, and gills of zebrafish (*Danio rerio*) exposed to environmentally relevant waterborne uranium concentrations. Environ Toxicol Chem. 2009;28:1271–8.

Lerebours A, Adam-Guillermin C, Brèthes D, et al. Mitochondrial energetic metabolism perturbations in skeletal muscles and brain of zebrafish (*Danio rerio*) exposed to low concentrations of waterborne uranium. Aquat Toxicol. 2010;100(1):66–74.

Lestaevel P, Romero E, Dhieux B, et al. Different pattern of brain pro-/anti-oxidant activity between depleted and enriched uranium in chronically exposed rats. Toxicology. 2009;258:1–9.

Lestaevel P, Bensoussan H, Dhieux B, et al. Cerebral cortex and hippocampus respond differently after post-natal exposure to uranium. J Toxicol Sci. 2013;38:803–11.

Lestaevel P, Dhieux B, Delissen O, et al. Uranium modifies or not behavior and antioxidant status in the hippocampus of rats exposed since birth. J Toxicol Sci. 2015;40:99–107.

Lestaevel P, Grison S, Favé G, et al. Assessment of the central effects of natural uranium via behavioural performances and the cerebrospinal fluid metabolome. Neural Plast. 2016; doi:10.1155/2016/9740353.

Linares V, Sanchez DJ, Belles M, et al. Pro-oxidant effects in the brain of rats concurrently exposed to uranium and stress. Toxicology. 2007;236:82–91.

Linney E, Upchurch L, Donerly S. Zebrafish as a neurotoxicological model. Neurotoxicol Teratol. 2004;26(6):709–18.

Maruyama W, Abe T, Tohgi H, et al. A dopaminergic neurotoxin, (R)-N-methylsalsolinol, increases in parkinsonian cerebrospinal fluid. Ann Neurol. 1996;40:119–22.

Miller AC, Stewart M, Rivas R. DNA methylation during depleted uranium-induced leukemia. Biochimie. 2009;91:1328–30.

Mouradian MM. MicroRNAs in Parkinson's disease. Neurobiol Dis. 2012;46:279–84.

Naoi M, Maruyama W, Dostert P, et al. N-methyl-(R)-salsolinol as a dopaminergic neurotoxin: from an animal model to an early marker of Parkinson's disease. J Neural Transm. 1997;50(Suppl):89–105.

Paquet F, Houpert P, Blanchardon E, et al. Accumulation and distribution of uranium in rats after chronic exposure by ingestion. Health Phys. 2006;90:139–47.

Paternain JL, Domingo JL, Ortega A, et al. The effects of uranium on reproduction, gestation, and postnatal survival in mice. Ecotoxicol Environ Saf. 1989;17:291–6.

Pellmar TC, Keyser DO, Emery C, et al. Electrophysiological changes in hippocampal slices isolated from rats embedded with depleted uranium fragments. Neurotoxicology. 1999;20:785–92.

Petitot F, Frelon S, Chambon C, et al. Proteome changes in rat serum after a chronic ingestion of enriched uranium: toward a biological signature of internal contamination and radiological effect. Toxicol Lett. 2016;257:44–59.

Rudenko A, Tsai LH. Epigenetic regulation in memory and cognitive disorders. Neuroscience. 2014;264:51–63.

Su S, Jin Y, Zhang W, et al. Aberrant promoter methylation of p16(INK4a) and O(6)-methylguanine-DNA methyltransferase genes in workers at a Chinese uranium mine. J Occup Health. 2006;48(4):261–6.

Takiguchi M, Achanzar WE, Qu W, et al. Effects of cadmium on DNA-(Cytosine-5) methyltransferase activity and DNA methylation status during cadmium-induced cellular transformation. Exp Cell Res. 2003;286(2):355–65.

Tomsig JL, Suszkiw JB. Metal selectivity of exocytosis in alpha-toxin-permeabilized bovine chromaffin cells. J Neurochem. 1996;66:644–50.

Valinluck V, Tsai HH, Rogstad DK, et al. Oxidative damage to methyl-CpG sequences inhibits the binding of the methyl-CpG binding domain (MBD) of methyl-CpG binding protein 2 (MeCP2). Nucleic Acids Res. 2004;32(14):4100–8.

Neurotoxicity of Metal Mixtures

V.M. Andrade, M. Aschner, and A.P. Marreilha dos Santos

Abstract Metals are the oldest toxins known to humans. Metals differ from other toxic substances in that they are neither created nor destroyed by humans (Casarett and Doull's, Toxicology: the basic science of poisons, 8th edn. McGraw-Hill, London, 2013). Metals are of great importance in our daily life and their frequent use makes their omnipresence and a constant source of human exposure. Metals such as arsenic [As], lead [Pb], mercury [Hg], aluminum [Al] and cadmium [Cd] do not have any specific role in an organism and can be toxic even at low levels. The Substance Priority List of Agency for Toxic Substances and Disease Registry (ATSDR) ranked substances based on a combination of their frequency, toxicity, and potential for human exposure. In this list, As, Pb, Hg, and Cd occupy the first, second, third, and seventh positions, respectively (ATSDR, Priority list of hazardous substances. U.S. Department of Health and Human Services, Public Health Service, Atlanta, 2016). Besides existing individually, these metals are also (or mainly) found as mixtures in various parts of the ecosystem (Cobbina SJ, Chen Y, Zhou Z, Wub X, Feng W, Wang W, Mao G, Xu H, Zhang Z, Wua X, Yang L, Chemosphere 132:79–86, 2015). Interactions among components of a mixture may change toxicokinetics and toxicodynamics (Spurgeon DJ, Jones OAH, Dorne J-L, Svendsen C, Swain S, Stürzenbaum SR, Sci Total Environ 408:3725–3734, 2010) and may result in greater (synergistic) toxicity (Lister LJ, Svendsen C, Wright J, Hooper HL, Spurgeon DJ, Environ Int 37:663–670, 2011). This is particularly worrisome when the components of the mixture individually attack the same organs. On the other hand, metals such as manganese [Mn], iron [Fe], copper [Cu], and zinc [Zn] are essential metals, and their presence in the body below or above homeostatic levels can also lead to disease states (Annangi B, Bonassi S, Marcos R, Hernández A, Mutat Res 770(Pt A):140–161, 2016). Pb, As, Cd, and Hg can induce Fe, Cu, and Zn dyshomeostasis, potentially triggering neurodegenerative disorders, such as Alzheimer's disease (AD) and Parkinson's disease (PD). Additionally,

V.M. Andrade • A.P. Marreilha dos Santos (✉)
Research Institute for Medicines (iMed.ULisboa), Faculty of Pharmacy,
Universidade de Lisboa, Lisboa, Portugal
e-mail: apsantos@ff.ul.pt; apsantos@ff.ulisboa.pt; paula.marreilha@clix.pt

M. Aschner
Department of Molecular Pharmacology, Albert Einstein College of Medicine,
Bronx, NY, USA

© Springer International Publishing AG 2017
M. Aschner and L.G. Costa (eds.), *Neurotoxicity of Metals*, Advances in Neurobiology 18, DOI 10.1007/978-3-319-60189-2_12

changes in heme synthesis have been associated with neurodegeneration, supported by evidence that a decline in heme levels might explain the age-associated loss of Fe homeostasis (Atamna H, Killile DK, Killile NB, Ames BN, Proc Natl Acad Sci U S A 99(23):14807–14812, 2002).

The sources, disposition, transport to the brain, mechanisms of toxicity, and effects in the central nervous system (CNS) and in the hematopoietic system of each one of these metals will be described. More detailed information on Pb, Mn, Al, Hg, Cu, and Zn is available in other chapters. A major focus of the chapter will be on Pb toxicity and its interaction with other metals.

Keywords Metal neurotoxicity • Metal mixtures • Hematopoietic toxicity • Metal interactions

Metals in the Environment

Heavy metal exposure can occur through contaminated air, food, and water or in hazardous occupations. While in the developed world the levels of heavy metal contamination in the environment have decreased in recent decades, developing countries in Asia and Africa continue to experience high levels of metal pollution, in particular in urban environments.

This contamination largely derives from anthropogenic sources, such as the combustion of leaded gasoline or unregulated industrial emissions. There is also a significant problem with metal contamination from mining, which results in elevated metal levels in water and air. Another major and relatively new source of metal contamination in developing countries is the practice of electronic waste recycling. Unfortunately, primitive and unsafe methods are used for the extraction of the precious metals, resulting in contamination of the local environment of highly toxic metals such as Hg and Pb (Neal and Guilarte 2013).

Environmental Exposure to Metal Mixtures

Along with actual apprehensions pertaining to human exposure to metals, it is well recognized that environmental exposures are not observed to single chemicals. The truth is that exposure to mixtures is the environmental reality in the present chemically sophisticated world (Simmons 1995). Mixtures of metals naturally occur, but metals are also often introduced into the environment as mixtures (Fairbrother et al. 2007). These mixtures are ubiquitous in air, water, and soil (Simmons 1995), and thus, people are exposed to them either concurrently or sequentially, by various routes of exposure and from a variety of sources to a large numbers of toxicants at low doses that may result in similar or dissimilar effects over exposure periods that can range from short term to a lifetime (ATSDR 2000; Pohl et al. 1997). Accordingly, the US Environmental Protection Agency (EPA) (ATSDR 2000) recommends greater efforts on understanding the combined toxic effects of metals (Fairbrother et al. 2007). Thus far the studies have largely focused on exposures to single metals (Kortenkamp and Faust 2009; Pohl et al. 1997), with few addressing chronic exposures to low levels of metal mixtures (Feron et al. 1995).

Fortunately, studies on mixtures have accelerated, incorporating more knowledge of specific modes of toxicological action and greater use of statistical methods and mathematical models (ATSDR 2000). Even so, predicting the health consequences of multiple chemical exposures is still a challenge (Pohl et al. 1997), because their study incorporates the understanding of interactions at several levels. These interactions may change toxicokinetics and toxicodynamics (Spurgeon et al. 2010). Therefore mixtures can influence adverse health effects sometimes, resulting in greater (synergistic) toxicity (Lister et al. 2011); this is particularly worrisome when the components of the mixture individually attack the same organs or, combined, overwhelm a particular mechanism that the organism uses to defend itself against toxic substances. Low doses that might not individually cause health effects, in concert, may become a public health issue (Calderon et al. 2003). It is reported that exposure to metal mixtures at concentrations below environmental quality guideline levels for individual components resulted in adverse effects that were attributed to interactions among the constituents (Yen Le et al. 2013). This issue was recently recognized by the EPA as a key gap in metal risk assessment (Abboud and Wilkinson 2013); thus a need exists for the research into the toxicity of metals, very especially with regard to metal mixtures in trace levels (Kim et al. 2009).

Criteria to Select Metal Mixtures

Given the almost infinite number of chemical mixtures, regulators are faced with a problem as to which chemicals should be chosen for assessment and regulation (Kortenkamp and Faust 2009), being important to prioritize them for research efforts (ATSRD 2004). Components produced and emitted together from industrial processes or present together in the same environmental or human body compartment are certainly to be considered. Chemicals thought to exhibit their effects through common mechanisms have been often grouped together based on similarities in chemical structure or derived from mechanistic considerations. Recently it has been argued that grouping criteria should focus on common adverse outcomes, with less emphasis on similarity of mechanisms (Kortenkamp and Faust 2009) and on chemicals having great potential impact on human health (ATSDR 2004). Reproductive, carcinogenic, and neurotoxic effects are considered potentially important health endpoints in epidemiological studies of complex mixtures, particularly when such mixtures contain trace metals (Shy 1993). Exposure to neurotoxic agents represents indeed a concern of high priority in modern society, given the ever increasing reported frequency of neurological diseases (Lucchini and Zimmerman 2009), with special concern to the induced long-term effects (Emerit et al. 2004).

Metals present unique environmental and public health issues, since these elements possess several particularities that should be taken into account when accessing the risks of their exposure; this includes the transformation into species with different valence states and the conversion between inorganic and organic forms. All these forms may possess different behaviors in the organism such as absorption, distribution, transformation, and excretion and/or different toxicities. Some metals are nutritionally essential elements at low levels but are toxic at higher levels (e.g., manganese [Mn], iron [Fe], copper [Cu], and zinc [Zn]), while others have no known biological functions (e.g., Pb, arsenic [As], cadmium [Cd], and Hg). Because metals naturally occur in the environment, many organisms developed specific mechanisms for its uptake and deposition, as well as mechanisms to regulate their accumulation, especially the accumulation of essential metals; additionally, the bioaccumulation of metals is tissue specific. All these characteristics can impact the use and interpretation of bioaccumulation data and the toxicity of metals (Fairbrother et al. 2007).

Neurotoxicity of Metals

Neurotoxicity may be defined as any adverse effect, permanent or reversible, on the structure or function of the central and/or peripheral nervous system originated by a biological, chemical, or physical agent that diminishes the ability of an organism to survive, reproduce, or adapt to its environment (Costa 1998; Costa and Manzo 1995). The nervous system can compensate for the toxic effects caused by low doses of neurotoxicants, but a prolonged and lifetime exposure even to the very low levels may lead to delayed neurodegenerative effects (Lucchini and Zimmerman 2009), with a progressive loss of neural tissues (Rachakonda et al. 2004). Thus, neurotoxic effects can be seen in later stages of life, yet the cause of these effects may be related to events occurring decades earlier. The properties that clinically identify them may bear no more than a superficial resemblance to those manifestations marking their prior stages, and this is why the earliest stages of such diseases may be confused with some other sources, such as aging (Weiss 2006). Neurotoxicity is a sensitive endpoint due to the unique and critical role of the nervous system in the control of body function, including other organs and systems, such as the endocrine and the immune system. The limited ability of neurons to regenerate after injury explains neurodegenerative disease-related loss of function, as neurons die, and the regenerative capacity is limited (Emerit et al. 2004; Mutti 1999). These disorders do not have cures (Rachakonda et al. 2004) rather they are gradually progressive, and the ability of its victims to function effectively and efficiently will be impaired at stages of the disease far earlier than its eventual detection (Weiss 2006). Thus concern exists that in the near future, low-dose long-term metal exposure may give rise to a society with lifelong loss of intelligence and motor capacities and permanent psychological disturbances (Kakkar and Jaffery 2005). These effects can produce reduction of economic productivity, and when this reduction occurs widely across a society, the resulting economic impacts may be even greater than the costs of metal pollution control itself (Landrigan et al. 2006).

Chronic exposure to low levels of metals is a contributor to neurological disease in multiple populations around the world (Christensen 1995; Witholt et al. 2000; Wright and Baccarelli 2007). Several studies demonstrate increased levels of metals in critical brain areas of neurodegenerative disease patients (Migliore and Coppedè 2009). The brain may at times compensate for the effects of an individual chemical itself acting on a particular target system; inversely, when multiple targets or functional sites within one system are impacted by different mechanisms (such as in multi-metal exposures), homeostatic capabilities may be impaired, thereby leading to cumulative damage (Lucchini and Zimmerman 2009). The actual public health concern on the potential for exacerbated cognitive and behavioral deficits resulting from children's exposure to multiple toxic metals provides an example; investigations on the effects on cognition of at least two metals together suggesting that combinations of metals may result in increased toxicity at this level (Kordas et al. 2010). Even so, the effect of mixture interactions on neurotoxicity remains largely unknown (Tiffany-Castiglioni et al. 2006).

Sources, Routes, Disposition, Toxicity, and Mechanisms of Individual Metals

Lead

A recent assessment on the global health impacts of contaminants identified Pb among the six most toxic pollutants threatening human health (Csavina et al. 2012). Pb has many industrial uses including battery manufacture, solders, pigments, and radiation shielding. Its use as an additive in household paint has ceased, but Pb-containing paint is still found in properties built before the 1960s. Occupational exposure usually occurs by inhalation of Pb dust or fumes in Pb industries. Nonoccupational Pb exposure usually involves ingestion, such as "traditional" remedies (adults). Cooking with Pb-glazed earthenware and contaminated soil or water are other potential sources (Bradberry 2016). Concerning Pb disposition once absorbed independently of the route of exposure, 99 percent of circulating Pb is bound to erythrocytes and is dispersed into the soft tissues – brain, liver, renal cortex, aorta, lungs, spleen, teeth, and bones (Patrick 2006). The major route of excretion of absorbed Pb is the kidney.

The mechanisms of Pb neurotoxicity are no doubt complex and numerous. Pb traverses the blood–brain barrier (BBB), accumulates in the brain, and preferentially damages the prefrontal cerebral cortex, hippocampus, and cerebellum (Kwong et al. 2004).

At the biochemical level, one of the most important mechanisms of Pb toxicity is the mimicking of calcium [Ca] action and/or disruption of Ca homeostasis. Pb may also substitute for Zn in some enzymes and in Zn-finger proteins. The fetus and infant may have increased vulnerability to Pb's neurotoxicity due in part to the immaturity of the BBB and to the lack of the high-affinity Pb-binding protein in astroglia, which sequester Pb. In addition, Pb affects virtually every neurotransmitter system in the brain, including the glutamatergic, dopaminergic, and cholinergic systems (Pohl et al. 2011).

At present Pb entrance pathways in the brain remain elusive, but the most studied candidate to date is the divalent metal ion transporter 1 (DMT1), and studies recently demonstrated that DMT1 is present in endothelial cells of the BBB (Huang et al. 2011).

As the structural basis of blood–cerebrospinal fluid barrier (BCB), epithelial cells in the choroid plexus (CP) are targets for Pb. Pb is known to accumulate in the CP; however, the mechanism of Pb uptake in the choroidal epithelial cells remains unknown. The CP, a major component of the BCB, has been shown to be involved in Pb-induced neurotoxicity (Shi and Zheng 2007). Animal studies showed accumulation of Pb in the CP at concentrations 57- and 70-fold greater than the brain cortex and cerebrospinal fluid, respectively (Zheng et al. 1991). Even though Pb serves no nutritional requirements, pathways for Pb transport, such as DMT1, Ca^{2+} channels, endocytosis, and anion changers, have been identified in other tissues or cells (Song et al. 2016). Pb can also exert changes on the hematopoietic system, and actually, among the most important enzymes disrupted by Pb are those involved in heme synthesis. Inhibition of delta-aminolevulinate dehydratase (ALAD) leads to accumulation of delta-aminolevulinic acid (ALA) (Bradberry 2016). Higher blood levels of Pb disturb hemoglobin (Hb) synthesis and, therefore, decrease its concentration.

Neurological effects are one of the most sensitive endpoints of Pb exposure, and children are particularly vulnerable. Exposure to high Pb levels produces encephalopathy with signs such as hyperirritability, ataxia, convulsions, stupor, and coma. In children, exposure to low Pb levels has been associated with neurobehavioral effects including impaired cognitive ability and IQ deficits. In Pb workers, reported neurobehavioral effects include malaise, forgetfulness, irritability, lethargy, headache, fatigue, impotence, decreased libido, dizziness, weakness, and paresthesia. Pb exposure in workers has also been associated with neuropsychological effects, increased prevalence and severity of white matter lesions, changes in nerve conduction velocity and postural balance, and alterations of somatosensory evoked potentials (Pohl et al. 2011).

Manganese

Mn is an essential metal ion for life since it is a cofactor for a wide variety of enzymes (Casarett and Doull's's 2013). There are inorganic and organic Mn compounds, with the inorganic forms being the most common in the environment. Uses of Mn include Fe and steel production, manufacture of dry cell batteries, manufacture of glass, textile bleaching, and oxidizing agent for electrode coating in welding rods. Organic compounds of Mn are present in the fuel additive, in methylcyclopentadienyl manganese tricarbonyl (MMT), and in fungicides (e.g., maneb and mancozeb).

Mn is naturally present in food, with the highest concentrations typically found in soya, nuts, cereals, legumes, fruits, grains, and tea; it is also present at low levels in drinking water. Higher inhalation exposures may be experienced in occupational settings such as Mn mines, foundries, smelters, and battery manufacturing facilities (Santamaria 2008).

The route of exposure can influence the distribution, metabolism, and potential for neurotoxicity of Mn-containing compounds (Anderson 1999; Roels et al. 2012). The oral route is considered to be less important for risk assessment purposes, because Mn is poorly absorbed from the gastrointestinal (GI) tract (5%).

Mn is transported in plasma bound to a gamma 1-globulin, transferrin (TRF), and is widely distributed in the body concentrating in mitochondria, so that tissues rich in these organelles, including the brain, pancreas, liver, kidneys, and intestines, have the highest Mn concentrations. Mn readily crosses the BBB and its half-life in the brain is longer than in the whole body (Casarett and Doull's 2013). The inhalation route is more efficient than ingestion at delivering Mn to the brain (Gianutsos et al. 1985) due to greater Mn absorption from the lungs and slower clearance of absorbed Mn from the circulation. Another efficient inhalation route is olfactory transport, an often overlooked route of direct delivery of chemicals from the nose to the brain (Brenneman et al. 2000). The principal route of excretion of Mn is in the feces (Casarett and Doull's's 2013).

A mechanism for the neurotoxicity of Mn has not been clearly established. A suggested mechanism of Mn neurotoxicity is the increase in auto-oxidation or turnover of intracellular catecholamines including dopamine (DA), norepinephrine, and epinephrine. This leads to the increased production of free radicals, reactive oxygen species (ROS), and other cytotoxic metabolites, along with a depletion of cellular antioxidant defense mechanisms. Other potential mechanisms include the potential substitution for Ca by Mn, the possibility that a transport mechanism for Mn is linked to the DA reuptake carrier, the inhibition of brain mitochondrial oxidative phosphorylation by Mn, and the involvement of Mn in complex interactions with other minerals (Pohl et al. 2011).

The transport of Mn into the brain ultimately depends on its ability to cross the BBB. Similarly to the case of Fe, the transport of Mn across the BBB and its cellular uptake can happen through TRF-dependent and TRF-independent mechanisms (Quintanar 2008). Mn(II) from the bloodstream can also be directly transported by DMT1, or it can cross the cellular membrane using glutamate-activated ionic channels that would normally transport Ca into the cell. Increased plasma Cu and Mn concentrations may lead to brain deposits and CNS damage. It was recently suggested that Mn enters the CNS predominantly through the BCB and that high Mn concentration impairs the integrity of this barrier (Dusek et al. 2015).

Even less information is available pertaining to eventual Mn effects in the heme biosynthetic pathway; one study suggested that Mn can actually interfere at this level (Qato and Maines 1985). More specifically, Mn(II) seems to inhibit aminolevulinic acid synthase (ALAS) activity in the brain and liver. In Maines' study (1980) the inhibition of liver and erythrocyte ALAD by Mn was observed, while in another work a competitive inhibition of ferrochelatase (FECH) by Mn was exhibited (Hift et al. 2011).

Inhalation of high levels of Mn (as seen in occupational studies) can lead to a syndrome of disabling neurological effects in humans called manganism with symptoms like tremors, difficulty in walking, and facial muscle spasms. Initial symptoms of Mn toxicity that can progress into manganism include irritability, aggressiveness, and hallucinations. Effects similar to the preclinical neurological effects and mood effects seen in occupational studies have also been associated with environmental exposures to Mn in air. In addition, there is evidence that oral exposure to Mn may produce similar neurological effects as reported for inhalation exposure. Exposure to excess levels of Mn in drinking water has been associated with subtle learning and behavioral deficits in children (Pohl et al. 2011).

Aluminum

Al is the most abundant metal and the third most abundant element in the Earth's crust. Due to its high reactivity, Al is not found in the free state in nature (Casarett and Doull's's 2013).

As per World Health Organization (WHO) reports, humans get inevitably exposed to Al through food, cooking utensils, deodorants, and antacids (Kaur et al. 2006) and for purifying water and vaccine adjuvants apart from occupational exposure in gun, automobile, aerospace, and defense-related factories (Sinczuk-Walczaki et al. 2003; Singh and Goel 2015), where inhalation absorption dominates (Buchta et al. 2005). Al overload in dialysis patients has also been reported (Abreo et al. 1990).

It has been reported that only 10% of the ingested Al is absorbed in the GI tract (Gorsky et al. 1979). Al has been shown to accumulate in various mammalian tissues such as the brain, bone, liver, and kidney (Wills et al. 1993; Sahin et al. 1994). Al uptake in the brain is much slower as compared to other organs, but once gained access into the brain, Al distributes into the various regions, namely, the medial striatum, corpus callosum, and cingulate bundle (Kumar and Gill 2014).

After inhalation, Al is distributed into the whole organism. It is excreted only by renal elimination (Buchta et al. 2005).

Although the mechanism of Al-induced neurotoxicity remains elusive, recent reports suggest elevated oxidative and inflammatory stress markers (Kumar et al. 2009) to be majorly responsible for disruption of intraneuronal metal homeostasis (Julka and Gill 1995) as well as axonal transport and long-term potentiation (Wenting et al. 2014). Thus, involvement of multiple mechanisms in Al-induced neurotoxicity warrants multi-targeted approach for effective treatment (Singh and Goel 2015).

It causes oxidative damage by binding to prooxidant metals like Fe and Cu and modulates their ability to promote metal-based oxidative events. Also, Al can directly compete with and even substitute several other essential metals in vivo. Strong evidence suggests that Al forms Al-superoxide anion complex, which is a more potent oxidant than superoxide anion (Oteiza et al. 1993; Nehru and Anand 2005).

Al promotes accumulation of insoluble amyloid β-protein and aggregation of hyper-phosphorylated tau protein, which comprises neurofibrillary tangles (NFTs)

(Kawahara and Kato-Negishi 2011) and causes detrimental changes to cholinergic neurotransmission (Sehti et al., Sethi et al. 2009).

Animal studies indicate that Al exposure can affect the permeability of the BBB, cholinergic activity, signal transduction pathways, and lipid peroxidation (LPO), impair neuronal glutamate–nitric oxide–cyclic GMP pathway, and interfere with the metabolism of essential trace elements (Pohl et al. 2011).

Al gains access to the brain through TRF-mediated transport, which subsequently leads to neurotoxicity (Yokel 2006; Singh and Goel 2015). Al is capable of crossing the BBB (Banks and Kastin 1985; Exley 2001), which leads to an increase of Al concentration in the hippocampus (Struys-Ponsar et al. 1997), cortex, singulated bundles, and corpus callosum (Sethi et al. 2009).

Al induces changes in hemato-biochemical parameters in vivo (Ghorbel et al. 2015). It has been demonstrated that Al overload affects two enzymes involved in heme formation, ALAS (the rate-limiting enzyme) and ALAD, as well as the major enzyme of heme degradation, HO (heme oxygenase). Despite, Al increases ALAS activity rather than ALAD, suggesting that Al might promote heme formation. The catabolism of heme prevails over its synthesis (Lin et al. 2013). It has also been proved that Al(III) overdose leads to microcytic anemia, due to its capacity to interfere in heme synthesis, whether by affecting the protoporphyrin biosynthesis or by interfering with Fe metabolism (Bazzoni et al. 2005).

Experimental evidence of Al-induced neurotoxicity subsists since 1965, whereby administration of Al has been reported to induce formation of NFTs in rabbits similar to that found in AD (Klatzo et al. 1965), increasing the risk of neurodegenerative diseases such as AD, Parkinsonism, ALS, etc. (Becaria et al. 2003). Various studies have indicated neuropathological, neurobehavioral, neurophysical, and neurochemical changes following Al exposure (Miu et al., Miu et al. 2003; Colomina et al. 2002; Kaur et al. 2006; Walton 2012). Al concentrations are also elevated in DA-related brain regions of PD patients (Yasui et al. 1992), showing a correlation between PD and Al exposure.

Also, in patients with reduced renal function, prolonged dialysis with Al-containing dialysates has produced a neurotoxicity syndrome (dialysis dementia) characterized by the gradual loss of motor, speech, and cognitive functions (Pohl et al. 2011).

Mercury

Hg is a heavy metal that exists in three chemical forms, metallic or elemental (Hg0), inorganic (Hg^{1+} and Hg^{2+}), and organic, mostly as methylmercury (MeHg) (Park and Zheng 2012; Hsu-Kim et al. 2013). The general population is primarily exposed to MeHg through their diet (particularly seafood) and Hg0 from dental amalgams (Clarkson et al. 2007). A few studies have reported substantial human exposure to inorganic Hg, which may come from the use of personal products, such as skin-lightening cosmetics (Al-Saleh et al. 2016). Because of its antibacterial/antifungal properties, the organic Hg compound thiomersal is used as a preservative in medical preparations (Lohren et al. 2015). In recent years, man-made MeHg contamination has decreased considerably, due to improvements in industrial manufacturing and efforts to minimize the release of Hg in the environment. However, Hg remains a global pollutant, and there are regions in the world, primarily in developing countries, where the levels of environmental contamination remain high (Ceccatelli et al. 2010).

Hg vapor emitted from amalgam dental fillings is the major source of Hg vapor affecting the general public. Elemental Hg vapor is primarily distributed in the kidneys and oxidizes into inorganic Hg that is predominately excreted in the urine (Al-Saleh et al. 2016). The elemental Hg can then be converted into inorganic Hg in the body which can accumulate in the brain. Ingested MeHg is nearly completely absorbed in the GI tract. Organic Hg deposits in various organs, including the blood, brain, and kidney. More than 90% of blood MeHg is in the red blood cells (RBCs) where MeHg appears to be bound to cysteine residues in Hb. Following MeHg exposure, Hg compounds are excreted mainly via the kidney and the GI tract. Demethylation of MeHg, occurring mostly in the liver, is a key step in the excretion process. Both MeHg and the inorganic Hg formed in the liver are excreted in the bile conjugated with glutathione (GSH) and related compounds. However, MeHg undergoes enterohepatic recirculation. The halftime of excretion varies in different species (70 days in humans) (Ceccatelli et al. 2010).

MeHg distributes to all the areas of the brain by crossing the BBB through mechanisms that are not fully characterized. It is possible that neutral amino acid carrier systems are used for the transport of MeHg–cysteine complexes. Demethylation of MeHg seems to take place in the brain. The formed inorganic Hg has a very long half-life in the brain, especially in the thalamus and pituitary. Inorganic Hg produced by demethylation may be sequestered by metallothioneins (MTs), a family of cysteine-rich proteins that binds with high affinity to metals (Cd, Zn, and Hg) (Ceccatelli et al. 2010).

The cytotoxicity of MeHg has been attributed to three major mechanisms: (1) perturbation of intracellular Ca^{2+} levels; (2) induction of oxidative stress (OS) either by overproduction of ROS or by reduced oxidative defense capacity; and (3) interactions with sulfhydryl groups, thus forming complexes with thiol-containing compounds [2] targeting proteins and peptides containing cysteine and methionine. Uncontrolled release of Ca^{2+} from the mitochondria has been reported to occur during OS. The level of ROS increases after exposure to MeHg in brain tissue and in various in vitro neuronal models. MeHg is accumulated in the mitochondria, where it decreases the rate of oxygen consumption, alters the electron transport chain by impairing complex III, and induces loss of the mitochondrial membrane potential. MeHg is also known to interfere with the uptake of cystine, the key precursor of GSH synthesis, via XAG transporters in astrocytes (Ceccatelli et al. 2010).

Previous studies indicate that the BBB is significantly more sensitive to organic Hg species as compared to inorganic compounds (Lohren et al. 2015). However, iHg compounds (e.g., HgCl2) can act as a direct BBB toxicant, increasing thus its permeability in rodents (Zheng et al. 2003). In the human body, Hg ions including MeHg (CH_3Hg^+) are preferably conjugated to reduced SH groups including cysteine and GSH. The disposition of Hg is regulated by the availability of ligands as well as the ability of the resulting complexes to serve as substrates for a variety of transporters (Ballatori 2002). MeHg-L-cysteine has some structural similarity to the amino acid methionine (Hoffmeyer et al. 2006). Thus, the amino acid transporters, which carry methionine into cells, actually transport MeHg-L-cysteine across membranes (Kerper et al. 1992; Simmons-Willis et al. 2002). Once MeHg has entered the cell, it binds to GSH. The conjugate is a substrate for ATP-binding cassette (ABC) transporters that mediate cellular efflux of glutathione S-conjugates (Strak et al. 2016).

An increase in the urinary concentration of specific porphyrins has been described as a biomarker of prolonged exposure to all forms of Hg (Bowers et al. 1992; Woods et al. 1991; Woods 1995) based upon selective interference with the fifth (uroporphyrinogen decarboxylase) (Woods et al. 1984) and sixth (coproporphyrinogen oxidase) (Woods and Southern 1989) enzymes of the heme biosynthetic pathway in kidney cells, a principal target of Hg. Hg induces a specific change in the urinary porphyrin excretion pattern characterized by increased concentrations of pentacarboxyporphyrin and coproporphyrin, along with the appearance of an atypical porphyrin identified empirically as keto-isocoproporphyrin (Heyer et al. 2006).

The neurotoxicity of MeHg was first recognized in adults during the Minamata outbreaks in 1953 (Ekino et al. 2007), but many subsequent studies reported its toxicity in fetal neurodevelopment (Grandjean and Herz 2011). Data for the neurodevelopmental risk of MeHg at low levels are however still limited due to the different interpretations or study designs (Al-Saleh et al. 2016). All signs and symptoms of toxicity in adults are confined mostly to the nervous system where it affects primarily the granule layer of the cerebellum and the visual cortex of the cerebrum (Kaur et al. 2006).

Exposure to Hg produces neurological and behavioral effects in humans. Adverse neurological effects following acute inhalation of high concentrations of Hg vapor include a number of cognitive, personality, sensory, and motor disturbances. In addition, chronic inhalation exposure has produced signs of neurotoxicity including tremors, unsteady walking, irritability, poor concentration, short-term memory deficits, tremulous speech, blurred vision, performance decrements in psychomotor skills, paresthesias, and decreased nerve conduction (Pohl et al. 2011).

Developmental exposure to MeHg can have long-term consequences, supporting the hypothesis of an increased risk for neurodegenerative disorders later in life (Ceccatelli et al. 2010).

Copper

Cu is an essential metal for all living organisms and is a component of many metalloproteins such as the antioxidant enzyme Cu–Zn superoxide dismutase (SOD) and cytochrome oxidase. Cu salts are used in fungicides, algicides, fertilizers, electroplating, dyes, inks, disinfectants, and wood preservatives (Bradberry 2016). It is mainly used in electric and electronic industry. Mining also contributes to environmental contamination, in soil and water (Angelovicová and Fazekasová 2014).

The brain concentrates heavy metals including Cu for metabolic use. As a cofactor of several enzymes and/or as structural component, Cu is involved in many physiological pathways in the brain (Scheiber et al. 2014). In general, Cu contents are higher in the gray matter (1.6–6.5 mg/g wet weight) than in the white matter (0.9–2.5 mg/g wet weight); Cu is enriched in the locus coeruleus and the substantia nigra, which both are pigmented tissues and contain catecholaminergic cells. Both brain Cu content and distribution change during development, with age and in neurodegenerative diseases (Scheiber et al. 2014).

Although Cu is an essential element, it also plays a role in the pathogenesis of neurodegenerative disease such as AD (Lu et al. 2006). Free reduced Cu(I) can bind to SH groups and inactivates enzymes such as glucose-6-phosphate dehydrogenase and glutathione reductase. In addition, Cu can interact with oxygen species (e.g., superoxide anions, hydrogen peroxide) and catalyze the production of reactive toxic hydroxyl radicals (Bradberry 2016). Cumulative evidence has implied that an imbalanced Cu homeostasis in the brain contributes to the pathogenesis of neurodegenerative disorders such as idiopathic Parkinson's disease (IPD), AD, and familial amyotrophic lateral sclerosis (ALS). Increased concentration of redox available Cu has been reported in PD CSF, and its concentration was correlated with motor impairment (Dusek et al. 2015).

Brain Cu is derived from peripheral Cu that is transported across the BBB and/or the BCB, which separate the brain interstitial space from blood and CSF, respectively (Zheng and Monnot 2012). At both barriers Cu is transported primarily as free ion. Although the Cu uptake into cerebral capillaries is much slower than into the CP, the Cu acquired by cerebral capillaries appears to be more readily transported into the brain parenchyma than Cu from the CP to the CSF. In fact, recent evidence indicates that the role of the BCB in brain Cu homeostasis is rather to export Cu from the CSF to the blood than to import Cu. The BBB represents the major route for the transport of Cu from the blood circulation into the brain parenchyma, where Cu is utilized and subsequently released into the CSF via the brain interstitial fluid. The Cu in the CSF can be taken up by choroid epithelial cells, from where it may be stored or exported to the blood. Thus, while the BBB determines the influx of Cu into the brain, the BCB contributes to the maintenance of the Cu homeostasis in the brain extracellular fluids (Gunshin et al. 1997).

The Cu transporter Ctr1 is likely to be the major pathway for Cu entry into brain cells. Experimental evidence was provided that DMT1, which is also expressed in brain cells, is involved in apical Cu uptake by intestinal cells. However, while some authors defend that DMT1 clearly can transport Cu (Garrick et al. 2006), others consider that DMT1 is a Fe-preferring transporter that does not transport Cu (Illing et al. 2012). The alternative Cu transport could be mediated by members of the ZIP (ZRT-/IRT-like protein) family of metal transporters (Scheiber et al. 2014).

The synthesis of hemoproteins may also be affected by Cu deficiency. It is well known that Cu deficiency can lead to anemias which might be explained by an intracellular defect in heme biosynthesis at FECH (Wagner and Tephly 1975).

Alterations of Cu homeostasis have also been associated with neurodegenerative diseases such as prion, AD, PD, and Huntington's disease (Scheiber et al. 2014). In AD Ab peptides have been shown to bind with high affinity (Atwood et al. 2000), and senile plaques are strongly enriched in Cu (Lovell et al. 1998). In addition, Cu ions induce the precipitation of Ab peptides in vitro. These observations suggest that Cu triggers the formation of plaques in the brain (Atwood et al. 2000; Tougu et al. 2011).

In PD, strong and growing evidence suggests abnormalities in Cu homeostasis. Parkinsonism is a frequent symptom in neurological Wilson's disease, which is an inherited disorder of Cu metabolism that is characterized by excessive deposition of Cu in the liver, brain, and other tissues (Lorincz 2010). Cu has been demonstrated to bind to both soluble and membrane-bound α-synuclein with high affinity (Dudzik et al. 2012) and to accelerate aggregation of soluble a-synuclein (Davies et al. 2013).

Zinc

Zn is a nutritionally essential metal, and a deficiency results in severe health consequences. At the other extreme, excess of Zn is relatively uncommon and occurs only at very high levels. Zn is ubiquitous in the environment, so that it is present in most foodstuffs, water, and air. The principal industrial uses of Zn include its applications as a corrosion protector for Fe and steel, application in batteries, and production of metal alloys, brass, and bronze. Zn oxide is the most widely used compound in industry, in the production of paints, plastics, cosmetics, pharmaceuticals, textiles, and electrical and electronic equipment (Peakall and Burger 2003).

The distribution of endogenous Zn is high in the bone, testis, and liver (Yasuno et al. 2011). High concentrations of Zn are also found in the brain and in pancreatic cells. In the brain, Zn highly occurs in the hippocampus, amygdala, and cortex (Kozlowski et al. 2009).

The molecular mechanisms by which Zn^{2+} triggers neuronal injury have not been elucidated clearly. There are several possibilities in which Zn exerts their adverse effects, including impairment of mitochondrial superoxide production (Sensi et al. 2000), disruption of metabolic enzyme activity (Sensi and Jeng 2004), and activation of p38 and voltage-dependent potassium channels (Zhu et al. 2013). Deregulation of neuronal Zn(II) homeostasis is believed to be strictly connected to mitochondrial dysfunction and OS, making the cation a possible contributor to the activation of pathophysiological pathways involved in brain aging and/or neurodegeneration. Zn, in its ionic form, can also exert important modulatory effects on neurotransmission and synaptic function, as well as regulate many signaling pathways (Kozlowski et al. 2009).

Zn permeability for TRPM7 channels is fourfold higher than that of Ca. Recently, TRPM7 channel is reported to play an important role for Zn2þ-mediated neuronal injury and may represent a novel target for neurological disorders where Zn2þ toxicity plays an important role (Kim et al. 2016a, b).

MTs exert a critical role in buffering cytosolic Zn(II). MT-3 seems to be particularly relevant to neuronal Zn(II) homeostasis in critical brain regions such as the hippocampus where it is abundantly present in the same hippocampal glutamatergic terminals that are also strongly enriched in vesicular Zn(II) (Kozlowski et al. 2009). Zn is selectively stored and released from presynaptic vessels of neurons found primarily in the mammalian cerebral cortex (Nriagu 2007).

The most common effects associated with long-term excessive Zn intakes (ranging from 150 mg/day to 1–2 g/day) have included sideroblastic anemia, hypochromic microcytic anemia, leukopenia, lymphadenopathy, neutropenia, hypocupremia, and hypoferremia. Changes in serum lipid profile, serum ferritin, and erythrocyte SOD activity have been reported in a number of patients who have ingested high doses of Zn. Zn is required for the activity of ALAD which plays a protective role in heme biosynthesis (Nriagu 2007).

Concerning neurotoxic effects, an interesting body of scientific literature suggests that Zn is a neurotoxin. There is evidence indicating that the readily available Zn^{2+} could, in certain pathological states, induce neuronal injury. Exposure of mature cortical neuronal cultures to several hundred μM concentrations of Zn^{2+} induced neuronal death. These findings indicate that Zn dyshomeostasis is likely a key modulator of neuronal injury. A previous study has reported that Zn^{2+} induces neurotoxicity in a concentration- and time-dependent manner (Zhu et al. 2012). Because the Zn-releasing neurons also release glutamate, they are sometimes referred to as "gluzinergic" neurons. Zn can modulate the overall excitability of the brain possibly through its effects on glutamate, gamma-aminobutyric acid (GABA) receptors of this network (Nriagu 2007).

Arsenic

Arsenic is particularly difficult to characterize as a single element because its chemistry is so complex and there are many different As compounds. It may be trivalent or pentavalent and is widely distributed in nature. The most common inorganic trivalent As compounds are As trioxide, sodium arsenite, and As trichloride (Casarett and Doull's 2013).

Arsenic is one of the oldest poisons known to men and its applications throughout history are wide and varied. The catastrophe of As toxicity, caused by As-contaminated water, has already been reported in many countries. Yet, an estimated 100 million people worldwide are exposed to excessive amounts of As via drinking water (in the range of ppm) (Watanabe and Hirano 2013; Krüger et al. 2009). The atmospheric deposition of As through the burning of charcoal and activities of metal foundry are examples of human activities that contributes to As environmental contamination (O'Neil 1995), being the excessive use of pesticides and fertilizers and mining the factors that most contribute to As soil contamination (Adriano 2001).

Both arsenate (pentavalent inorganic As) and arsenite are well absorbed by oral and inhalation routes. Absorption by the dermal route has not been well characterized but is low compared to other routes (Casarett and Doull's).

Once absorbed, arsenates are partially reduced to arsenites, yielding a mixture of As(III) and As(V) in the blood. As(III) undergoes methylation primarily in the liver to form monomethylarsonic acid (MMA) and dimethylarsinic acid (DMA). The rate and relative proportion of methylation production varies among species. Most inorganic As is promptly excreted in the urine as a mixture of As(III), As(V), MMA, and

DMA. Smaller amounts are excreted in feces. In most species, including humans, ingested organic arsenical compounds such as MMA and DMA undergo limited metabolism, do not readily enter the cell, and are primarily excreted unchanged in the urine (ATSDR 2007). Some of these metabolites are more potent and toxic than the originally ingested inorganic form of As (iAs), including mono- and dimethylated arsenicals (Watanabe and Hirano 2013; Krüger et al. 2009).

All forms of As, including inorganic and methylated arsenicals, accumulate in many parts of the brain, with the highest accumulation in the pituitary (Sanchez-Pena et al. Sánchez-Peña et al. 2010).

The metabolism of iAs consumes GSH, which is the main antioxidant in the CNS (Dringen 2000). Arsenic may induce OS by cycling between oxidation states of metals such as Fe or by interacting with antioxidants and increasing inflammation, resulting in the accumulation of free radicals in cells. Inadequate GSH availability may modulate iAs biotransformation and determine disease (Ramos-Chávez et al. 2015).

The role of OS as the leading mechanism in As-induced neurological defects has been widely supported by in vitro and in vivo studies. OS may be the initiating mechanism for As-induced neurotoxicity. Arsenic-induced DNA damage and apoptosis in neuronal cells may follow an intrinsic mitochondrial apoptotic pathway, mediating through increased intracellular Ca that triggers mitochondrial stress and generation of ROS (Flora 2011).

In occupationally As-exposed subjects, a positive correlation between compromised subjective neurological symptoms, visual evoke potential, electroneurographic and electroencephalographic results, and As concentration in air and urine was established (Flora 2011).

Studies in murine models have demonstrated that iAs crosses the BBB and is methylated in different brain regions that express the As 3 methyltransferase (AS3MT) enzyme (Ramos-Chávez et al. 2015). It was identified in two uptake pathways: aquaglyceroporin (AQP) channels, in particular the liver isoform AQP9, and the glucose permease (GLUT1) conduct trivalent $As(OH)_3$ and $CH_3As(OH)_2$ which both have oxidative status of +3. It was proposed that GLUT1 is the major pathway for the movement of trivalent inorganic and methylated As into the brain and heart, where AQPs are not abundantly expressed, and this uptake may contribute to cardiovascular disease and neurotoxicity (Jiang et al. 2010)

Considerable evidence supports the observation that As can influence many aspects of the heme system. Previous research has shown that As decreases heme metabolism and can bind to Hb, resulting in lower Hb concentrations. Arsenic has been shown to alter erythrocyte morphology and induce erythrocyte death. Arsenic also depresses bone marrow, which can lead to anemia (Hb < 120 g/L in nonpregnant adults), leukopenia, and thrombocytopenia. Several studies have shown that As alters heme metabolism and contributes to lower Hb concentrations (Kile et al. 2016).

Epidemiological studies show that As can cause serious neurological effects after inhalation or oral exposure. Common effects seen in humans orally exposed to As are peripheral and/or central neuropathy (Pohl et al. 2011). A recently published meta-analysis focused on the negative impact of As exposure on intelligence measured by IQ tests (Rodriguez-Barranco et al. 2013). A 2007 study found a significant association between urinary As concentrations greater than 50 µg/L and poor scores on tests measuring visual-spatial reasoning, language and vocabulary, memory, intelligence, and math skills in 6–8-year-old children from Mexico (Rosado et al. 2007).

Exposure to high levels of As produces mainly CNS effects, and exposure to low levels produces mainly peripheral nervous system effects (Pohl et al. 2011). According to Naujokas et al. (2013), exposure to low As concentrations has been shown to increase susceptibility to cognitive dysfunction. The mechanism of As-induced neurological changes has not been determined. However, some of the neurological effects of high-level oral exposure are thought to be the result of direct cytotoxicity. In addition, animal studies have shown altered neurotransmitter concentrations in some areas of the brain after oral exposure to As (Pohl et al. 2011).

Iron

The major scientific and medical interest in Fe is as an essential metal, but toxicologic considerations are important in terms of Fe deficiency, accidental acute exposures, and chronic Fe overload. Environmental exposure to Fe does occur, for example, from drinking water, Fe pipes, and cookware (Rush et al. 2009). But further evidence indicates that overload of Fe, which is released from Hb, also contributes to brain injury after intracerebral hemorrhage. The toxicity resulting from Fe deposition in neurons is primarily mediated by the increasing of Fe that can generate radical species via the Fenton's reaction since radical species are frequently associated with cytotoxicity, which is the initiating stimuli for cell death (Dai et al. 2013).

The total amount of Fe in the body is mainly present in the form of Hb (60–70%), myoglobin, cytochromes, and other Fe-containing enzymes (10%) as well as in ferritin and hemosiderin (Appel et al. 2001).

Fe absorption is accomplished by enterocytes in the proximal small intestine, near the gastroduodenal junction. Its access to the circulation is modulated by transport via both the apical and basolateral membranes, which is operated by specific transporter proteins and accessory enzymes. Nonheme Fe is first reduced to Fe(II), which is transported by DMT1 (Skjørringe et al. 2015). Some of the absorbed Fe is stored in enterocytes' ferritin, and some is exported to the circulation by ferroportin (FP). Absorbed Fe is rapidly delivered to TRF, which under normal conditions accounts for nearly all serum Fe. In normal human subjects, plasma TRF is only approximately 30% saturated. The absence of a regulated mechanism for Fe excretion determines the necessity of a tight balance between the sites of Fe absorption, uptake, transport, storage, and utilization for maintenance of Fe homeostasis (Kozlowski et al. 2009). After the liver, the brain contains the highest quantity of Fe, ca. 60 mg of nonheme Fe distributed uniquely according to brain structures. Fe is the most abundant trace element in the brain where it is essential for normal brain development and function (Ward et al. 2014). It plays a crucial role for many processes including oxygen transport, the synthesis of DNA and RNA, and the formation of myelin and development of the neuronal dendritic tree (Lieu et al. 2001). The substantia nigra and globus pallidus in normal adult human brain can contain Fe levels, which exceed hepatic levels, in the range of 3.3–3.8 mM Fe. These high brain Fe concentrations can be attributed primarily to the rapid rate of oxidative metabolism necessary to maintain ionic membrane gradients, axonal transport, and synaptic transmission (Kozlowski et al. 2009). Postmortem and in vivo magnetic resonance imaging studies have shown that Fe accumulation follows an exponential saturation function with only little changes after the fourth to fifth decades of life (Pirpamer et al. 2016).

Ferrous Fe showed higher influx into cells than ferric Fe and induced more ROS production which resulted in higher susceptibility of neuron death. The types of neuronal cell death which were induced by Fe overload were testified as necrosis, apoptosis, and autophagic cell death, relying on the level of Fe dosage (Dai et al. 2013).

Two main events linked to increased ROS generation have been identified in the degenerating substantia nigra: (a) increased Fe levels and (b) reduced antioxidant defenses. Fe deposition seems to be a specific hallmark of PD (Rubio-Osornio et al. 2013).

The toxicity of Fe deposition in neurons is primarily mediated by the increasing of Fe that can generate radical species via the Fenton's reaction, since radical species are frequently associated with cytotoxicity, which is the initiating stimulus for cell death (Dai et al. 2013).

Brain Fe unpaired regulation may also result from the disrupted expression of brain Fe metabolism proteins induced by nongenetic factors. These currently undetermined factors may disrupt normal control mechanisms of protein expression and lead to Fe imbalance in the brain, inducing then OS and neuronal death in some neurodegenerative disorders. Fe accumulation in the brain occurs gradually over time with concurrent increases in ferritin. Fe overload results in a large increase in the chelatable free Fe pool, which is too large to be sequestered by ferritin within cells. Fe toxicity, largely based on Fenton chemistry, mainly affects the mitochondrial inner membrane respiratory complexes (Kozlowski et al. 2009).

The transport of Fe across the BBB must be regulated, but the permeation mechanism has not been completely clarified so far. The uptake of TRF-bound Fe by TFR-mediated endocytosis from the blood into cerebral endothelial cells is no different in nature from the uptake into other cell types. After permeation across the BBB or blood–CSF barrier, Fe is likely to bind quickly to the TRF secreted by the oligodendrocytes and CP epithelial cells, which, diversely from what happens in other tissues, becomes fully saturated with Fe. The excess Fe will bind to other transporters, including small molecules like citrate or ascorbate. The widespread distribution of TFR in neurons clearly indicates that neurons can acquire Fe by means of TFR-mediated uptake of TF–Fe (Kozlowski et al. 2009).

Heme is an Fe-containing porphyrin that functions as a cofactor in a wide array of cellular processes. The terminal step of heme biosynthesis, which occurs in the mitochondrial matrix, is the insertion of Fe into protoporphyrin IX (Korolnek and Hamza 2014). Further evidence indicates that overload of Fe, which is released from Hb, also contributes to brain injury after intracerebral hemorrhage (Dai et al. 2013). Fe can be released from the breakdown of Hb following aneurysm or blood disease (Rush et al. 2009).

Fe deposition has also been associated with inflammatory, neurodegenerative, and cerebral small vessel disease. Even in normal elderly persons, elevated levels of Fe relate to worse cognitive performance (Pirpamer et al. 2016). Syndromes with neurodegeneration with brain Fe accumulation (NBIA) are a group of neurodegenerative disorders characterized by abnormalities in brain Fe metabolism with excess Fe accumulation in the globus pallidus and to a lesser degree in the substantia nigra and adjacent areas (Schneider 2016).

The brain of Alzheimer's diseased humans is characterized by the accumulation of Fe within senile plaques (ca. 1 mM) and NFTs and also by lowered expression of TRF receptor. As a consequence, these brains are subject to high levels of OS. Fe may also promote a deposition and may affect the enzymatic processing of the amyloid precursor protein. As for PD, DA cell loss and disease progression are accompanied by the accumulation of high Fe concentrations that are particularly associated with aggregation of alpha-synuclein (especially the mutated form found in familial PD) within Lewy bodies. An increased Fe content can be detected in the substantia nigra of most PD patients, and up to a 255% increase in intracellular Fe concentration has been observed in postmortem PD brains. Together with Fe accumulation, the lowered expression of ferritin within the substantia nigra of PD patients results into OS and decreased GSH levels, thus directly contributing to DA neuronal toxicity (Kozlowski et al. 2009).

Cadmium

Cd is a toxic nonessential transition metal classified as a human carcinogenic (ATSDR 2004). It is characterized by a long half-life in humans (Jin et al. 1998) with a low rate of excretion from the body. The Cd content of the body increases with age in industrialized societies, from less than 1 µg in the newborn to 15–20 mg in adults (Notarachille et al. 2014).

There are several sources of human exposure to Cd, including employment in primary metal industries, production of certain batteries, some electroplating processes, and consumption of tobacco (ATSDR 2012).

Cd is poorly absorbed after oral ingestion. The estimated absorption of Cd is less than 5% from the GI tract in humans and about 1% in animals. Once absorbed, the movement of Cd from blood to tissue is rapid. Over 60% of the body burden of Cd is localized in the liver and kidney. However, the factors that influence absorption and tissue distribution of Cd are not well understood (Liu and Klaassen 1996).

Cd can also be absorbed by the inhalation and dermal routes regardless its chemical form (chloride, carbonate, oxide, sulfide, sulfate, or other forms), although dermal route of exposure is relatively insignificant and of low concern (Wester et al. 1992). Cd can reach the CNS being uptaken from the nasal mucosa or olfactory pathways (Lafuente and Esquifino 1999). Cd is not known to undergo direct metabolic conversions. It has a high affinity for the SH groups of albumin and MT. The interaction between Cd and MT plays a critical role in the toxicokinetics and toxicity; Cd is retained in both organs, liver and kidney, bound mainly to MT. It has a retention halftime of 73 days in the liver and a lifetime in the kidneys. Since a small fraction of the Cd presented to the GI tract is absorbed, most of the oral dose is excreted via the feces. The amount of Cd excreted in urine represents only a small fraction of the total body burden unless renal damage is present. Absorption of Cd is also influenced by metal ions such as Zn, Fe, Ca, and chromium (Cr) (ATSDR 2012).

Cd-induced injury in the cerebral microvessels is thought to be associated with OS. Following in vivo Cd exposure, there was an early increase followed by a later decrease in microvessel enzymes involved in cellular redox reactions, such as SOD, glutathione peroxidase, and catalase. Thus, a depletion of microvessel antioxidant defense systems and a resultant increase in LPO may provoke microvessel damage (Shukla et al. 1996). Cd significantly increases the levels of LPO in parietal cortex, striatum, and cerebellum as compared to a control group in developing rats exposed to Cd (Méndez-Armenta et al. 2003).

Under normal conditions and by oral route, Cd barely reaches the brain in adults due to the presence of the BBB (Yang et al. 2016). Differently, concerning inhalation pathway, Cd is transported along the primary olfactory neurons to their terminal in the olfactory bulbs, thereby bypassing the intact BBB, the olfactory route could therefore be a likely way to reach the brain and should be taken into account for occupational risk assessments for this metal (Tjälve and Henriksson 1999; Bondier et al. 2008).

In the brain, Cd tends to accumulate in the CP at concentrations much greater than those found in the CSF and elsewhere in brain tissues. A postmortem human study revealed that Cd concentration in the CP was about two to three times higher than that found in the brain cortex (Zheng 2001). Due to differences in the BBB integrity (Antonio et al. 2002), Cd is thus more toxic to newborn and young rats than to adult rats. Cd can increase the permeability of the BBB in rats (Shukla et al. 1987) and penetrate and accumulate in the brain of developing and adult rats, leading to brain intracellular accumulation, cellular dysfunction, and cerebral edema (Mendez-Armenta and Ríos Méndez-Armenta and Ríos 2007).

Cd causes anemia through three known mechanisms, i.e., hemolysis due to a deformity of peripheral RBCs, Fe deficiency through the competition with duodenal Fe absorption, and hypoproduction of erythropoietin (Horiguchi et al. 1994, 2010), an erythroid-specific glycoprotein hormone produced mainly from the kidney that promotes RBC formation (Ebert and Bunn 1999). However, there has been rare investigation on the direct toxicity of Cd to erythroid precursors (Wang et al. 2013).

In addition to liver and kidney target organs, the CNS is also subjected to Cd toxicity (Lafuente and Esquifino 1999). Cd can enter the brain parenchyma and neurons, causing neurological alterations in humans and animal models by inhibiting thiol-containing enzymes and decreasing serotonin and acetylcholine levels (Jomova and Valko 2011).

Exposure to Cd also severely affects the function of the nervous system (López Alonso et al. 2003) with symptoms including headache and vertigo, olfactory dysfunction, parkinsonian-like symptoms, slowing of vasomotor functioning, peripheral neuropathy, decreased equilibrium, decreased ability to concentrate, and learning disabilities (Kim et al. 2005; Monroe and Halvorsen 2006).

Concerning biochemical changes of CNS in response to Cd, it can inhibit the release of acetylcholine, probably by interfering with Ca metabolism (Desi 1998). Cd can affect the degree and balance of excitation/inhibition in synaptic neurotransmission as well as the antioxidant levels in animal brain (Mendez-Armenta and Ríos Méndez-Armenta and Ríos 2007).

Moreover, there are studies showing the neurotoxicity of Cd at µM range on cell culture models such as neurons and glial cells (Lopez et al. 2006).

Metal Mixture Neurotoxicity

In all environmental media, mixtures of neurotoxic metals naturally occur, and metals are often introduced into the environment as mixtures (Fairbrother et al. 2007). In this context, Pb, As, Hg, and Cd are metals/metalloids included in a WHO's list of ten chemicals of major public concern (Prüss-Ustün et al. 2016), with Pb and As being among the leading toxic agents detected in the environment (Järup 2003). The four metals are thought to exhibit their neurotoxic effects (Pohl et al. 2011; Kaur et al. 2006) through common mechanisms, such as the generation of ROS (Patrick 2006; Flora 2011; Kaur et al. 2006; Méndez-Armenta et al. 2003) and interaction with essential metals (Pohl et al. 2011; Lin et al. 2013; Freitas Fonseca et al. 2014; Nriagu 2007). Hence, interactions among Pb, As, Hg, and Cd will be addressed, and emphasis will be also given to the interactions of these metals with Fe, Cu, and Zn. Indeed, a mechanistic relationship is established between the dyshomeostasis of these essential metals and OS, as well as associations between both conditions and neurodegeneration (Kozlowski et al. 2009).

Mechanisms Associated with Interactions of Lead with other Metals

Some populations at high risk for Pb toxicity are also overexposed to other metals through environmental pollution (Shukla et al. 1987), most of them nearby areas of industrial activity. Namely, chronic exposure to low levels of Pb and Cd through food, water, and air is described as common in industrial areas (Markiewicz-Górka et al. 2015). Other authors mention that sources of co-exposure to Pb and As are primarily through soil and dust deriving from pyrometallurgical nonferrous metal production or mining activity (Csavina 2012). It is also described that Pb, Al, and Zn are constituents, in major or trace amounts, of raw materials and wastes associated with the production of alumina (Phol et al. 2011). These examples illustrate that some populations are exposed to potential worrisome levels of metal mixtures. Accordingly, children residing near ore smelters have been shown to be exposed to Pb, As, Cd, Zn, and Cu (Shukla et al. 1987), while the levels of Pb, as well as As,

Hg, Cd, Mn, Zn, and Cu, in whole blood of residents from mining communities were found to exceed the permissible WHO guideline values (Obiri et al. 2016).

Other co-exposure scenarios to the general population are not rare, since Pb can occur simultaneously with As, Hg, and Cd in various parts of the ecosystem (Cobbina et al. 2015). Mixtures of metals can be present in the air, being an example the determination of urban metal levels in Pakistan which led to find that Pb and also Cd, Zn, Fe, and Cu are among the dominant contributors in indoor particulates, while Pb, Zn, and Fe constituted the major components in outdoor particulates. The excessive concentrations of Pb, Cd, and Zn were associated with automobile emissions. Even so, the Cd and Cu levels estimated in this study were considerably lower than those for Mexico City, in Mexico (Nazir et al. 2011). The contamination of mixtures of heavy metals in the aquatic environment has also attracted global attention owing to its abundance, persistence, and environmental toxicity. A study performed in a river in Bangladesh showed that the levels of Pb, As, and Cd indicated that water was not safe for drinking and/or cooking (Ali et al. 2016). Metal emissions can also contaminate the soil, with mixtures of Pb, Mn, Zn, and Cu occurring in this environmental compartment (Roneya and Colman 2004). Furthermore, heavy metals prone to bioaccumulate in the food chain might be dangerous to human health. In corroboration, the metal contents in plants and soil show significant correlations for Pb, Cd, Cu, and Zn, while the contents of the metals in vegetables often exceed those allowable for normal human and animal consumption. It has been estimated that if an adult consumed 2 kg potatoes, 2 kg tomatoes, and 1 kg carrots in a week, his/her food would exceed by 12% the maximum allowable level (MAL) for Cd; the daily maximum allowable rate of ingested Pb could be reached by consuming 880 g of vegetables (equal parts of potatoes, tomatoes, carrots, and cucumbers) (Islam et al. 2007). Another study in Korean pregnant women showed that co-exposure to Pb and Hg could come from frequent fish and cereal or vegetable consumption; while fish consumption was positively associated with Hg levels in cord blood, cereal and vegetable consumptions were positively associated with Pb levels (Kim et al. 2016a, b). Pb can also migrate from artisanal Al cookware and enter food at unacceptable levels that can significantly contribute to a child or adult's body burden of this metal (Weidenhamer et al. 2014).

- Disposition in the Brain

There is a general recognition that each mixture component may affect the disposition of other mixture components in the CNS. Being the brain a common target organ for Pb and other metals, potential additive or synergic effects induced by metal mixtures are expected (Mendez-Armenta and Ríos Méndez-Armenta and Ríos 2007).

Concerning Pb interactions with As, the intraperitoneal administration of both metals resulted in increased brain regional Pb levels in rats, accompanied by a significant decrease of As levels in some other regions (Mejía et al. 1997). In the same way as As, Cd can increase the level of Pb in certain brain regions with a magnitude greater than that observed after Cd exposure alone; the accumulation of Cd in several brain regions also increased. It is plausible the possibility that the co-exposure of Pb and Cd can damage the BBB. In this study, the levels of blood Pb decreased suggesting that blood Pb level as a diagnostic tool for Pb toxicity in co-exposed conditions is of insignificant value (Shukla et al. 1987). Again, when Pb is administrated together with Hg or with As/Hg, Pb brain accumulation can increase by 83.6% and 76.1%, respectively (Cobbina et al. 2015).

Pb can also interact with essential metals, such as Mn, resulting in changes in Pb disposition. Even very low doses of Mn when administrated together with Pb can raise striatal Mn, and also Pb (Suchla and Chandra 1987), as well as augments of Pb concentration in the whole brain of adult rodents and in pups after their co-exposure during gestation and/or lactation (Mejía et al. 1997). Overall, these informations indicate that Mn in the presence of Pb increases its distribution and/or retention in the brain (Pohl et al. 2011). The presence of excess Mn in the brain might increase the affinity of brain tissue to bind Pb, as it was suggested by in vitro results (Kalia et al. 1984; Shukla et al. Shukla et al. 1987; Wright and Bacarelli 2007). Analogous outcomes arise when Pb is administrated in vivo through a ternary metal mixture of Pb/As/Mn. Increased levels of Pb in the brain as compared with single exposures to each one of these metals, including the exposure to Pb alone, were already observed. Again, blood Pb levels fail to reflect the increased Pb deposition in the brain, raising concern that blood Pb levels may underestimate risk associated with co-exposure to Pb and other metals (Andrade et al. 2014b).

Fe, Zn, and Cu are also essential metals, however, considered protective of the toxic effects of Pb (Pohl et al. 2011; Klauder and Petering 1975). Interactions of Pb with Fe are broadly referred, and at the brain level, while Pb exposure damages the integrity of the BBB in young animals, Fe supplement may prevent against Pb-induced BBB disruption, with significantly reduced Pb levels in this organ (Wang et al. 2007). Other studies demonstrated that Pb exposure significantly increased Pb concentrations in cerebral cortex and low Fe dose significantly reduced the cortex Pb levels. Remarkably, Fe high dose increased the cortex Pb levels (Zhi et al. 2015). Cu has been shown to impede the GI absorption of Pb, with in vivo studies showing that at higher supplemental Cu doses and higher Cu/Pb dose ratios, supplemental Cu can decrease blood, liver, and kidney concentrations of Pb; however, levels of Pb in the brain were not affected (Pohl et al. 2011). Concomitantly, other studies with ternary and quaternary mixtures led to observe that the exposure to the mixtures Pb/Hg/Cd and Pb/Hg/As/Cd increased brain Cu by 31.9% and 40.8%, respectively (Cobbina et al. 2015). By its turn, Zn can also reduce Pb toxicity due to its antagonistic effect on Pb absorption from the GI tract (Basha et al. 2003). Furthermore, hippocampal Pb levels decreased, as compared to exposure to Pb alone, when rodents were co-exposed to Pb and Zn (Basha et al. 2003; Piao et al. 2007). The co-exposure to Pb, As, Cd, and Hg can result in brain Zn reduction by 65.1%, which was attributable to mimicry of Zn by complexes of toxic metals like As in the mixture (Cobbina et al. 2015).

The mechanisms responsible for the elevation of Pb in the brain, when the metal is in the presence of As, Cd, and/or Hg, are not elucidated. Nevertheless respecting to essential metals, their transporters can be "hijacked" by nonessential metals possessing physicochemical similarities (Martinez-Finley et al. 2012), modifying their uptake and accumulation (Spurgeon et al. 2010). Namely, DMT1 is present in several tissues including the BBB endothelium [Wang et al. 2011) and in both glia and neurons. This transporter is also most likely the major pathway by which Pb is transported into the brain (Wang et al. 2011); it is also involved in the uptake of Hg and Cd at least in intestinal cells (Vazquez et al. 2015; Tallkvist et al. 2001). Gu et al. (2009) also observed an effect greater than additive on DMT1 protein synthesis, enhancing transport of ions in the developing rat brain resulting from exposure to both Pb and Cd orally (von Stackelberg et al. 2013). Plausibly competition for DMT1 may have a relevant role in the increased deposition of Pb in the brain, when in the presence of Cd, Hg, and/or Mn. Differently GLUT1 may be a major pathway uptake of As in the epithelial cells of the BBB (Liu et al. 2006), and hence, the mechanisms leading to increased Pb brain levels after co-exposure to Pb and As remain to be elucidated (Andrade et al. 2014b).

DMT1 is also involved in active transport of Mn, Fe, Cu, and Zn, although Zn shows a different relative capacity (Espinoza et al. 2012; Garrick et al. 2006). These essential metals are harmful in oversupply, and thus, DMT1's role in their homeostasis is very relevant (Garrick et al. 2006).

Deficiencies in Fe can contribute to increased brain Pb levels; during periods of low Fe stores, expression of DMT1 in the duodenum is greatly increased, allowing not only increased Fe absorption but also Pb absorption (Cobbina et al. 2015). DMT1 regulation in the duodenum is sensitive to levels of Fe uptake, and the transporter has a much higher affinity for Fe over Pb (Wang et al. 2011). Nevertheless, the possibility that the expression of other Fe influx transport protein might contribute to increased brain Pb accumulation is not excluded (Zhu et al. 2013). On the other hand, Pb may limit Fe absorption, however through a different mechanism, one similar to Cd which downregulates the expression of DMT1 to 70% of controls when added to human intestinal cells (Kwong et al. 2004). Concomitantly, Gu et al. (2009) have reported that exposures to Pb and Cd synergistically increase DMT1 protein expression in the CNS of developing rats. Largely studies are still needed to clarify the affinities of different metals to DMT1, in order to provide a better understanding regarding interactions among metals during its transport in different tissues. According to a report of an in vitro study, the following order for DMT1 transport affinities is Mn >?Cd >?Fe > Pb > Zn with, as mentioned by the authors, doubts pertaining to Cd and Fe and uncertainty on where to place Cu (Garrick et al. 2006).

Another study proposes a different order: $Cd^{2+} > Fe^{2+} > Mn^{2+} \gg Zn^{2+}$ (Illing et al. 2012). Besides, while some authors defend that DMT1 clearly can transport Cu (Garrick et al. 2006), others consider that DMT1 is a Fe-preferring transporter that does not transport Cu (Illing et al. 2012). It is also very likely that additional mechanisms contribute to changes in the disposition of Pb, when the metal is present with other metals. Actually once within the cells, a particular metal may occupy abundant binding sites (Kalia et al. 1984) on metalloproteins or target molecules, modifying the compartmentalization of other metals, thus leading to aberrant binding and toxicity (Spurgeon et al. 2010). Furthermore a given metal may induce overexpression of transporters and/or binding proteins that alter the uptake of other metals (Kalia et al. 1984; Molina et al. 2011). In this perspective emphasis will be given to MTs, which are Zn-rich metal-binding proteins (Nordberg and Nordberg 2009). Zn induces the expression of a MT that has been shown to sequester Pb in vitro, protecting cells against its cytotoxicity (Roneya and Colman 2004). Many investigators proposed that Pb and Zn compete for similar binding sites on a MT-like transport protein and prevent the absorption of Pb from the GI tract (Basha et al. 2003). Studies on the ability of metals to displace Zn from MTs indicated that Cd had the highest capacity to displace Zn from hepatic MT, followed by Pb, Cu, and Hg. Differently, As had a limited ability to displace Zn from MT, while Al, Fe, and Mn had no effect on Zn binding (Walkess et al. 1984).

Overall, the simultaneous exposure of Pb to As, Mn, and/or Hg induces increased accumulation of Pb in the brain, while in a different way when Pb is in the presence of Zn and/or Cu, its accumulation decreases.

- Mechanisms of Toxicity

In the same way as other metals, Pb may exert its toxic action by multi-mechanisms. Emphasis should be given to OS, since it is an important convergent point on the mechanisms of metal toxicity (Whittaker et al. 2010) representing a pathway that leads to the destruction of cells, including neurons and vascular cells in the CNS (Chong et al. 2005). While OS is a major mechanism of Pb-induced toxicity (Flora et al. 2012), exposures to As (Jomova and Valko 2011), Al (Kumar and Gill 2014), Cd (Mendez-Armenta et al., Méndez-Armenta et al. 2003), and Hg (Kaur et al. 2006) are also associated with excessive ROS production. Additionally, Pb exposure can lead to dyshomeostasis of essential metals in the brain, such as Mn (Pohl et al. 2011), Fe (Dai et al. 2013), Cu (Bradberry 2016), and Zn (McCord and Aizenman 2014). The dyshomeostasis of these metals is also known to induce OS (Pohl et al. 2011; Dai et al. 2013; Bradberry 2016; (McCord and Aizenman 2014) and has been associated with the induction of neurotoxicity.

It is expected that co-exposure to mixtures of Pb with other nonessential metals results in increased generation of ROS and/or decreases in the levels of antioxidants in the brain. Indeed, previous in vivo assays concerning exposure to metals' mixtures representative of groundwater contamination in different parts of India involved the administration of a mixture of metals that included Pb, As, Cd, Mn, and Fe; their concentrations were equal to their WHO maximum permissible limit (MPL). In this work, increased LPO and decreased GSH level and activities of antioxidants in the brain (Jadhav et al. 2007) were observed. Similarly, the in vivo exposure to Pb, Hg, and Cd at the MPLs for each metal stipulated in the National Standard of The Republic of China for Municipal Water Standards resulted in a significant reduction in the activities of the brain antioxidant enzymes SOD and catalase and increases in malondialdehyde which is a marker of LPO (Cobbina et al. 2015). Elevated levels of brain nitric oxide synthase, an indicator of nitric oxide (NO), also suggested the induction of nitrosative stress (NS); NS is a parallel process, similar to OS but with involvement of ROS, known to create major toxicities to the brain and already associated with several neurodegenerative conditions (Klandorf and Van Dyke 2012, Cobbina et al. 2015). Still considering OS, in vitro works showed that astrocytes treated with Pb, As, and Cd triggered ROS generation, resulting in apoptosis of the mixture-treated astrocytes greater than would have been predicted by the individual metal treatment. Other toxicological evidences exist indicating that Mn interacts with Pb and Cd greater than additively, increasing the formation of ROS (von Stackelberg et al. 2013).

Another mechanism through Pb exerts CNS toxicity is the disruption of the normal physiological balance of trace metals in the brain, and essential metals may become compartmentally toxic by accumulation at levels that exceed the normal metal-buffering capacity within the cell (Zhou et al. 2014).

Cumulative evidences have implied that an imbalanced Fe, Cu, and Zn homeostasis in the CNS correlates with the pathogenesis of neurodegenerative disorders (Zheng et al. 2014; Molina et al. 2011; Szewczyk 2013). Actually, Pb has the ability to substitute other bivalent cations like Fe (Flora et al. 2012) and provoke Fe elevation in the brain tissue of Pb-exposed rodents. It is proposed that Pb influences cellular Fe influx or efflux, but changes in brain Fe levels might not be the result of an effect on DMT1 (Zhou et al. 2014) and rather on ferroportin 1 (FP1). This transporter might directly contribute to Fe efflux from neurons when overexpressed, thus preventing cellular Fe accumulation; Pb exposure might downregulate FP1 expression inducing cellular Fe accumulation in cells (Zhou et al. 2014). In agreement, Pb-induced increased Fe content in the old-aged rats' brain might be the result of the decrease of the expression of FP1. Furthermore, the effect of Pb on FP1 expression is regulated at transcriptional and posttranscriptional levels. Overall the perturbation in Fe homeostasis may contribute to the neurotoxic consequences induced by Pb exposure, and FP1 may play a role in Pb-induced Fe accumulation in the brain (Zhu et al. 2013).

Concerning Cu, Pb can induce Cu accumulation in brain tissue as it is reported in in vitro and in vivo studies. Excess Cu concentration is detrimental to cells due to free Cu capability to form toxic free radicals, resulting in OS in neuronal cells. Metals such as Pb (and also Cd and Hg) can affect SH integrity within the MT metal-binding sites. It was hypothesized that exposure of MT to transition metals such as Cu, in combination with certain heavy metals with higher affinity for MT (Pb, Cd, and Hg), could lead to a release of transition metals and, hence, potentiate metal-dependent OS. Interestingly, there might be a bifunctional role of MT in protecting against and enhancing Cu- dependent OS depending upon exposure to Cu/ heavy metal (Fabisiak et al. 1999). Still respecting to Cu, it is proposed that Pb upregulates the expression of the Cu transporter CTR1, which plays a major role in cellular Cu uptake and is abundant in the BBB, BCB, and brain parenchyma. Concurrently, Pb might downregulate the expression of P-type Cu-transporting ATPase (ATP7A), which is a major component of the intracellular Cu homeostasis apparatus. A consequent reduction of Cu efflux from the cells may thereby result in disturbed Cu homeostasis in the brain (Zheng et al. 2014). Less information is available on Pb-induced Zn dyshomeostasis. Even so, a large percentage of behavior-disordered persons exhibit an imbalance in levels of Cu and Zn in urine, blood, and other tissues, being suggested that Pb-induced imbalance in Zn (as well as Cu) may mediate insidious neurotoxic effect of Pb (Ademuyiwa et al. 2010).

- Effects in the Hematopoietic System

Pb directly affects the hematopoietic system through reduction of the life span of circulating erythrocytes, by increasing the fragility of cell membranes. This metal can also restrain the synthesis of Hb, by inhibiting various key enzymes involved in the heme synthesis pathway. The combined aftermath of these two processes leads to anemia (Flora et al. 2012). Eight enzymes catalyze the reactions leading to heme synthesis (Hift et al., 2011; Kauppinen 2005), and most importantly, those enzymes have been shown to be specifically susceptible to impairment by a variety of toxic agents (Bleiberg et al. 1967; Quintanilla-Vega et al. 1996), including metals other than Pb. Blood ALAD and Zn protoporphyrin have been demonstrated to be sensitive for metal interactions at low doses (Wang and Fowler 2008). When interferences occur with the enzymes of the heme biosynthesis, there is characteristically an excessive accumulation and excretion of ALA and/or porphyrins (Adhikari et al. 2006; Guolo et al. 1996). Because individual porphyrins differ by the side-chain substituents, different metals may induce specific and different changes in porphyrin excretion patterns (Woods et al. 2009). Therefore, co-exposure of Pb with other metals with the capability to interfere with this metabolic pathway may certainly result in other characteristic urinary porphyrin profiles.

Indeed, Pb/As co-exposure can lead to synergistic inhibition of blood ALAD as compared to a single exposure either to Pb or As (Wang and Fowler 2008), while an additive effect on coproporphyrin excretion, but without alteration on ALA or uroporphyrin excretion, was already noticed (Fowler and Mahaffey 1978; Mahaffey et al. 1981). By its turn (in sequence), it is reported that while Cd itself do not markedly alter urinary excretion of porphyrins, simultaneous treatment with Pb and Cd increases in the urinary excretion of these heme precursors. In addition, the decrease in urinary ALA excretion observed after concomitant administration of Pb and Cd may result from Cd inhibition of the formation of active metabolites of vitamin D, which appear to play a role in Pb absorption (Fowler and Mahaffey 1978).

When the metals Pb, As, and Cd are administrated as a ternary mixture, interactions also occur, with additive increases in the urinary excretion of porphyrins, along with greater blood Zn protoporphyrin levels, than those observed with single exposures (Fowler and Mahaffey 1978). Another study established that exposure to the same mixture increases ALA, Fe, and Cu levels in vivo. The authors also observed that increases in ALA were followed by statistically significant increases in kidney Cu (Cobbina et al. 2015). Increased RBCs were also noted after administration of Pb or Cd (or As), and more cells were observed when the three metals were concomitantly administered. Despite increased numbers of circulating RBCs, Hb and hematocrit were reduced, especially with the Pb–Cd combination (Mahaffey and Fowler 1977). It was speculated that both Cd and Pb could affect Hb through interference with Fe metabolism, which is an essential element for Hb production. Cd can induce anemia by competing with Fe absorption, and in the same way, Pb is taken up by the Fe absorption machinery secondarily blocking Fe through competitive inhibition (Kwong et al. 2004).

Similarly to Pb, the mechanism by which excess Al induces anemia seems to be a reversible block in heme synthesis, due either to a defect in porphyrin synthesis or to impaired Fe utilization. The exposure to Al can induce an increase in the percentage of uroporphyrin and a decrease in coproporphyrin in urine (Nasiadek et al. 2001). However, Al can have an inhibitory effect in vitro, whereas in in vivo studies it activates the enzyme; Pb and Al together depress the enzyme activity in an additive way that can be reactivated by the addition of Zn (Abdulla et al. 1979). In line, Hg body burden was already shown to correlate with changes in urinary porphyrin profiles and with specific neurobehavioral deficits (Woods 1996). However these outcomes and the lack of studies pertaining to eventual Pb interactions with Hg on the heme synthesis pathway reveal the pertinence of studies on this matter. It is also known that Mn interferes with Fe homeostasis (Molina et al. 2011) and can in vitro inhibit FECH (Hift et al., 2011). This enzyme catalyzes the terminal step in the biosynthesis of heme, converting protoporphyrin IX into heme B through the insertion of Fe (Korolnek and Hamza 2014). More recently, it was demonstrated that Mn induces modifications in the excretion of porphyrins by increasing total porphyrins and modifying their profile. Results obtained after co-exposure to Pb, As, and Mn were suggestive that combined effects of the three metals resulted in higher heme synthesis disorders than those induced by exposure to each one of the metals alone (Andrade et al. 2014a). However, there is a lack of works clarifying the eventual interactions between Pb and Mn at this level.

Both Pb poisoning and Fe deficiency are capable of independently producing anemia. In the erythrocyte, both conditions affect the same cellular processes resulting in similar outcomes, with evidence for a synergistic relationship. During anemia, Pb becomes even more concentrated in RBC, seeming reasonable that the increased toxicity to erythrocytes during Fe deficiency anemia and Pb poisoning is partly due to an increased concentration of Pb in these cells (Kwong et al. 2004). Combined effects of Pb and Fe on heme synthesis are also described; when the pathway is inhibited at the final step, Zn instead of Fe is incorporated into protoporphyrin resulting in elevated levels of Zn protoporphyrin, which is a measure of heme synthesis inhibition. Indeed Fe deficiency and Pb poisoning are each one capable of inhibiting heme synthesis at this final step, and accordingly the concentrations of Zn protoporphyrin are dramatically higher in the presence of both conditions than in either Fe deficiency or Pb poisoning alone (Kwong et al. 2004). Furthermore, FECH is especially sensitive to low Fe levels in the presence of Pb poisoning. Actually Pb accumulates at mitochondria where apparently interferes with transport of Fe, the site of FECH. Concomitantly, while the levels of urinary ALA are usually not affected by Fe deficiency, they are elevated in Pb poisoning due to Pb's inhibition of ALAD. The simultaneous occurrence of Fe deficiency and elevated Pb levels result in an almost twofold increase in urinary ALA, than with the presence of elevated Pb levels alone (Kwong et al. 2004). Pertaining to Pb interactions with Zn, orally exposed children benefitted of a protective effect of Zn on the hematopoietic effects of Pb, in the same way as demonstrated by in vivo works, even with Pb at higher doses. There are mechanistic evidences that Zn excess protects and reactivates Pb-inhibited ALAD, which is a Zn-containing enzyme (Pohl et al. 2011). Actually Zn may induce proteins that sequester Pb and donate Zn to ALAD (Roneya and Colman 2004). Indeed, supplemental Zn can protect against the Pb-induced increases in Zn protoporphyrin and urinary ALA excretion. However, these protective effects were seen at higher but not lower Pb doses and when basal levels of Zn in the diet were adequate (Roneya and Colman 2004). Additionally, effects of the mixed exposure to Pb, Zn, and Cu on workers' erythrocytes include inhibition of ALAD and increased free erythrocyte protoporphyrin in comparison with values in unexposed population; these effects are characteristic of Pb. Thus, characteristic hematological effects of Pb were seen in workers co-exposed to low levels of Zn and Cu. However, whether or not Zn and Cu afforded some protection against these effects cannot be determined from this study (ATSDR 2004).

Interestingly, associations exist among Pb poisoning, hematological disarrays, and neurotoxicity. Pb's inhibition of heme synthesis and disruption of protoporphyrin utilization appear to have a role in the demyelination of neuronal axons and the development of peripheral neuropathy, which is not unusual in adult's Pb intoxication. Pb-intoxicated individuals have a high concentration of blood protoporphyrins and excrete an excess of protoporphyrins that are thought to be precursors of a substance essential for myelin maintenance (Kwong et al. 2004). Moreover, ALA which accumulates due to Pb exposure resembles GABA, and stimulation of GABA receptors is thought to be an important mechanism of Pb-induced neurotoxicity (Bradberry 2016).

- Neurotoxic Effects

Neurotoxic effects are considered among the most potentially important health endpoints in epidemiological studies of complex mixtures, most particularly when such mixtures contain trace metals (Shy 1993). Not many reports are found respecting specifically to the neurotoxic effects induced by mixtures of metals. Even so, the mixture of As and Pb already showed to induce neuropsychological effects in children living in Morales (Mexico) despite that no conclusive results were obtained regarding interactions between these elements in this work (Carrizales et al. 2006). In turn, studies with rodents showed that while Pb alone increased movement and vertical activity and Cd alone decreased movement and increased rest time, Cd antagonized Pb-induced effects in rats co-exposed to both metals (Pohl et al. 2011). A synergistic effect from a mixture of Pb, As, and Cd on the developmental neurotoxicity was also observed with respect to glial and neuronal functions (Zhu et al. 2014). Other studies with the ternary mixture of the same metals revealed reductions in myelin thickness and axon density of the optic nerve and a decrease in thickness of nerve fiber, plexiform layer, and retinal ganglion cell counts in the retina (Cobbina et al. 2015). Concerning Hg, exposure to low Pb and Hg doses for 60 days can affect the learning and memory abilities of mice, evidenced by marginally high escape latency and low swimming speeds in the Morris water maze test. Similar trends were observed after exposure to the mixture Pb/Hg/Cd (Cobbina et al. 2015). In addition, animal studies provide compelling evidence that exposure to both Pb and Mn leads to synergistic neurological effects (Wright and Baccarelli 2007), with references to adversely affect cognitive function in an additive manner (Pohl et al. 2011). Decreased learning of conditioned avoidance responses to either Pb or Mn alone is described, and also gestational exposure to the same mixture reduces brain weight to a greater extent than either each metal alone (Wright and Baccarelli 2007). Despite an unaware of clinical data investigating joint exposures to Pb and Mn, Mexican children co-exposed to Pb and Mn had greater negative effects on cognitive performance than children exposed only to Pb (Neal and Guilarte 2013). Cognitive deficits were shown to be associated not only with Fe deficiency but also with blood Pb concentration in children, and it was proposed that Fe deficiency may affect verbal IQ directly and/or via increased blood Pb concentration due to Fe deficiency (Jeong et al. 2015). With respect to Zn, it is reported a protective effect of this metal against inhibition of smooth muscle contractility by Pb. Nevertheless, some authors consider that these studies do not provide evidence of potentiation, but rather a protective effect, or no effect, of Zn on Pb neurotoxicity (Pohl et al. 2011). Based on animal studies and mechanistic understanding, the predicted direction of joint toxic action for neurological effects is less than additive for the effects of Zn on Pb, and a similar direction is reported for Cu (Roneya and Colman 2004). A series of studies on Japanese workers exposed to Pb, Zn, and Cu at a gun metal foundry revealed that Zn and Cu may antagonize some of the neurological effects of Pb in occupationally exposed adults. Indirect evidence of potential antagonism by Zn and Cu or by Zn in Pb inhibition of peripheral nerve conduction velocity was found in some of these studies, but not in others. Other information from this series of studies include indirect evidence of potential antagonism by Zn and Cu of Pb inhibition of central nerve conduction and potential antagonism by Zn of Pb inhibition of autonomic nervous function (ATSDR 2004).

References

Abboud P, Wilkinson KJ. Role of metal mixtures (Ca, Cu and Pb) on Cd bioaccumulation and phytochelatin production by *Chlamydomonas reinhardtii*. Environ Pollut. 2013;179:33–8.

Abdulla M, Svensson S, Haeger-Aronsen B. Antagonistic effects of zinc and aluminum on lead inhibition of delta-aminolevulinic acid dehydratase.Arch. Environ Health. 1979;34(6):464–9.

Abreo K, Glass J, Sella M. Aluminum inhibits hemoglobin synthesis but enhances iron uptake in friend erythroleukemia cells. Kidney Int. 1990;37:677–81.

Ademuyiwa A, Agarwal R, Chandra R, Behari JR. Effects of sub-chronic low-level lead exposure on the homeostasis of copper and zinc in rat tissues. J Trace Elem Med Biol. 2010;24:207–11.

Adhikari A, Penatti CAA, Resende RR, Ulrich H, Britto LRG, Bechara EJH. 5-Aminolevulinate and 4, 5-dioxovalerate ions decrease GABAA receptor density in neuronal cells, synaptosomes and rat brain. Brain Res. 2006;1093:95–104.

Adriano DC. Trace elements in terrestrial environments. New York: Eds. Springer; 2001. 867p.

Agency for Toxic Substances and Disease Registry. Supplementary guidance for conducting health risk assessment of chemical mixtures, Risk Assessment Forum U.S. Atlanta: U.S. Department of Health and Human Services, Public Health Service; 2000.

Agency for Toxic Substances and Disease Registry. Priority list of hazardous substances. Atlanta: U.S. Department of Health and Human Services, Public Health Service. 2016. https://www.atsdr.cdc.gov/spl/, 5th September 2016, 5 pm.

Agency for Toxic Substances and Disease Registry (ATSDR). Interaction profile for: lead, manganese, zinc and copper. Atlanta: U.S. Department of Health and Human Services, Public Health Service; 2004.

Agency for Toxic Substances and Disease Registry (ATSDR). Arsenic CAS# 7440–38-2. Atlanta, GA: U.S. Department of Health and Human Services, Public Health Service; 2007.

Agency for Toxic Substances and Disease Registry (ATSDR). Toxicological profile for Cadmium. Atlanta: U.S. Department of Health and Human Services, Public Health Service; 2012.

Ali MM, Alia ML, Islam S, Rahman Z. Preliminary assessment of heavy metals in water and sediment of Karnaphuli River. Bangladesh Environ Earth Sci. 2016;73:1837–48.

Al-Saleh I, Nester M, Abduljabbar M, Al-Rouqi R, Eltabache C, Al-Rajudi T, Elkhati R. Mercury (Hg) exposure and its effects on Saudi breastfed infant's neurodevelopment. Int J Hyg Environ Health. 2016;219:129–41.

Anderson D. Factors contributing to biomarker responses in exposed workers. Mutat Res. 1999;428:197–202.

Andrade V, Mateus ML, Batoréu MC, Aschner M, Marreilha dos Santos AP. Changes in rat urinary porphyrin profiles predict the magnitude of the neurotoxic effects induced by a mixture of lead, arsenic and manganese. Neurotoxicology. 2014a;45:168–77.

Andrade V, Mateus ML, Santos D, Aschner M, Batoréu MC, Marreilha dos Santos AP. Arsenic and manganese alter lead deposition in the rat. Biol Trace Elem Res. 2014b;158(3):384–91.

Angelovicová L, Fazekasová D. Contamination of the soil and water environment by heavy metals in the former mining area of Rudňany (Slovakia). Soil Water Res. 2014;9(1):18–24.

Annangi B, Bonassi S, Marcos R, Hernández A. Biomonitoring of humans exposed to arsenic, chromium, nickel, vanadium, and complex mixtures of metals by using the micronucleus test in lymphocytes. Mutat Res. 2016;770(Pt A):140–61.

Antonio MT, López N and Leret ML. Pb and Cd poisoning during development alters cerebellar and striatal function in rats. Toxicology. 2002; 176: 59–66.

Appel MJ, Kuper CF, Woutersen RA. Disposition, accumulation and toxicity of iron fed as iron (II) sulfate or as sodium iron EDTA in rats. Food Chem Toxicol. 2001;39:261–9.

Atamna H, Killile DK, Killile NB, Ames BN. Heme deficiency may be a factor in the mitochondrial and neuronal decay of aging. Proc Natl Acad Sci U S A. 2002;99(23):14807–12.

Atwood CS, Scarpa RC, Huang X, Moir RD, Jones WD, Fairlie DP, Tanzi R, Bush AI. Characterization of copper interactions with Alzheimer amyloid b peptides: identification of an Attomolar-affinity copper binding site on amyloid b1-42. J Neurochem. 2000;43(2):560–8.

Ballatori N. Transport of toxic metals by molecular mimicry. Environ Health Perspect. 2002;110(5):689–94.

Banks WA, Kastin AJ. The aluminum-induced increase in blood-brain barrier permeability to delta-sleep-inducing peptide occurs throughout the brain and is independent of phosphorus and acetylcholinesterase levels. Psychopharmacology. 1985;86(1–2):84–9.

Basha R, Wei W, Brydie M, Razmiafshari M, Zawia NH. Lead-induced developmental perturbations in hippocampal Sp1 DNA-binding are prevented by zinc supplementation: *in vivo* evidence for Pb and Zn competition. Int J Devl Neuroscience. 2003;21:1–12.

Bazzoni GB, Bollini AN, Hernandez GN, Contini MC, Chiarotto MM, Rasia ML. *In vivo* effect of aluminium upon the physical properties of the erythrocyte membrane. J Inorg Biochem. 2005;99:822–7.

Becaria A, Campbell A, Bondy SC. Aluminum and copper interact in the promotion of oxidative but not in? Ammatory events: implications for Alzheimer's disease. J Alzheimers Dis. 2003;5:31–8.

Bleiberg J, Wallen M, Brodkin R, Applebaum IL. Industrially acquired porphyria. Arch Dermatol. 1967;80:793–7.

Bondier JR, Michel G, Propper A. Harmful effects of cadmium on olfactory system in mice. Inhal Toxicol. 2008;20(13):1169–77.

Bowers MA, Aicher LD, Davis HA, Woods JS. Quantitative determination of porphyrins in rat and human urine and evaluation of urinary porphyrin profiles during mercury and lead exposures. J Lab Clin Med. 1992;120:272–81.

Bradberry SM. Metals (cobalt, copper, lead, mercury). Medicine. 2016;44(3):182–4.

Brenneman KA, Wong BA, Buccellato MA, Costa ER, Gross EA, Dorman DC. Direct olfactory transport of inhaled manganese (54MnCl2) to the rat brain: toxicokinetic investigations in a unilateral nasal occlusion model. Toxicol Appl Pharmacol. 2000;169:238–48.

Buchta M, Kiesswetter E, Schaper M, Zschiesche W, Schaller KH, Kuhlmann A, Letzel S. Neurotoxicity of exposures to aluminium welding fumes in the truck trailer construction industry. Environ Toxicol Pharmacol. 2005;19:677–85.

Calderon J, Ortiz-Perez D, Yanez L, Díaz-Barriga F. Human exposure to metals. Pathways of exposure, biomarkers of effect, and host factors. Ecotoxicol Environ Saf. 2003;56:93–103.

Carrizales L, Razoa I, Tellez-Hernandez J, Torres-Nerioa R, Torres A, Batres LE, Cubillas A-C, Díaz-Barriga F. Exposure to arsenic and lead of children living near a copper-smelter in San Luis Potosi, Mexico: Importance of soil contamination for exposure. Environ Res. 2006;101:1–10.

Casarett & Doull's. Toxicology: the basic science of poisons. 8th ed. London: McGraw-Hill; 2013.

Ceccatelli S, Daréb E, Moors M. Methylmercury-induced neurotoxicity and apoptosis. Chem Biol Interact. 2010;188:301–8.

Chong ZZ, Li F, Maiese K. Oxidative stress in the brain: novel cellular targets that govern survival during neurodegenerative disease. Prog Neurobiol. 2005;75:207–46.

Christensen JM. Human exposure to toxic metals: factors influencing interpretation of biomonitoring results. Sci Total Environ. 1995;166:89–135.

Clarkson TW, Vyas JB, Ballatori N. Mechanisms of mercury disposition in the body. Am J Ind Med. 2007;50(10):757–64.

Cobbina SJ, Chen Y, Zhou Z, Wub X, Feng W, Wang W, Mao G, Xu H, Zhang Z, Wua X, Yang L. Low concentration toxic metal mixture interactions: effects on essential and non-essential metals in brain, liver, and kidneys of mice on sub-chronic exposure. Chemosphere. 2015;132:79–86.

Colomina MT, Roig JL, Sánchez DJ, Domingo JL. Influence of age on aluminum-induced neurobehavioral effects and morphological changes in rat brain. Neurotoxicology. 2002;23(6):775–81.

Costa LG. Biochemical and molecular neurotoxicology: relevance to biomarker development, neurotoxicity testing and risk assessment. Toxicol Lett. 1998;102-103:417–21.

Costa LG, Manzo L. Biochemical markers of neurotoxicity: research epidemiological applications. Toxicol Lett. 1995;77(1–3):137–44.

Csavina J, Field J, Taylor MP, Gao S, Landázuri A, Betterton EA, Sáez AE. A review on the importance of metals and metalloids in atmospheric dust and aerosol from mining operations. Sci Total Environ. 2012;433:58–73.

Dai M-C, Zhong Z-H, Sun Y-H, Sun Q-F, Wang Y-T, Yang G-Y, Bian L-G. Curcumin protects against iron induced neurotoxicity in primary cortical neurons by attenuating necroptosis. Neurosci Lett. 2013;536:41–6.

Davies KM, Hare DJ, Cottam V, Chen N, Hilgers L, Halliday G, et al. Localization of copper and copper transporters in the human brain. Metallomics. 2013;5:43–51.

Desi I, Nagymajtenyi L and Schulz H. Behavioural and neurotoxicological changes caused by cadmium treatment of rats during development. J Appl Toxicol. 1998; 18(1), 63–70, 1998.

Dringen R. Metabolism and functions of glutathione in brain. Prog Neurobiol. 2000;62:649–71.

Dudzik CG, Walter ED, Abrams BS, Jurica MS, Millhauser GL. Coordination of copper to the membrane-bound form of alpha-synuclein. Biochemistry. 2012;52:53–60.

Dusek P, Roosc PM, Litwin T, Schneider SA, Flaten TP, Aaseth J. The neurotoxicity of iron, copper and manganese in Parkinson's and Wilson's diseases. J Trace Elem Med Biol. 2015;31:193–203.

Ebert BL, Bunn HF. Regulation of the erythropoietin gene. Blood. 1999;94:1864–77.

Ekino S, Susa M, Ninomiya T, Imamura K, Kitamura T. Minamata disease revisited: an update on the acute and chronic manifestations of methyl mercury poisoning. J Neur Sci. 2007;262:131–44.

Emerit J, Edeas M, Bricaire F. Neurodegenerative diseases and oxidative stress. Biomed Pharmacother. 2004;58:39–46.

Espinoza A, Le Blanc S, Olivares M, Pizarro F, Ruz M, Arredondo M. Iron, copper, and zinc transport: inhibition of divalent metal transporter 1 (DMT1) and human copper transporter 1 (hCTR1) by shRNA. Biol Trace Elem Res. 2012;146(2):281–6.

Exley C. Aluminum and Alzheimer's disease. J Alzheimers Dis. 2001;3(6):551–2.

Fabisiak JP, Pearce LL, Borisenko GG, Tyhurina YY, Tyurin VA, Razzack J, Lazo JS, Pitt BR, Kagan VE. Bifunctional anti/prooxidant potential of metallothionein: redox signaling of copper binding and release. Antioxid Redox Signal. 1999;1:349–64.

Fairbrother A, Wenste R, Sappington K, Wood W. Framework for metals risk assessment. Ecotoxicol Environ Saf. 2007;68:145–227.

Feron VJ, Groten JP, Jonker D, Cassee FR, van Bladeren PJ. Toxicology of chemical mixtures: challenges for today and the future. Toxicology. 1995;105:415–27.

Flora SJS. Arsenic-induced oxidative stress and its reversibility. Free Radic Biol Med. 2011;51:257–81.

Flora G, Gupta D, Tiwari A. Toxicity of lead: a review with recent updates. Iternterdis Toxicol. 2012;5(2):47–58.

Fowler BA, Mahaffey KR. Interactions among lead, cadmium, and arsenic in relation to porphyrin excretion patterns. Environ Health Perspect. 1978;25:87–90.

Freitas Fonseca M, De Souza Hacon S, Grandjean P, Choi AL, Rodrigues Bastos W. Iron status as a covariate in methylmercury-associated neurotoxicity risk. Chemosphere. 2014;100:89–96.

Garrick MD, Singleton S, Vargas F, Kuo HC, Zhao L, Knopfel M, Davidson T, Costa M, Paradkar P, Roth JA, Garrick LM. DMT1: which metals does it transport? Biol Res. 2006;39:79–85.

Ghorbel I, Maktouf S, Kallel C, Chaabouni SE, Boudawara T, Zeghal N. Disruption of erythrocyte antioxidant defense system, hematological parameters, induction of pro-inflammatory cytokines and DNA damage in liver of co-exposed rats to aluminium and acrylamide. Chem Biol Interact. 2015;236:31–40.

Gianutsos G, Seltzer MD, Saymeh R, Wu M-LW, Michel RG. Brain manganese accumulation following systemic administration of different forms. Arch Toxicol. 1985;57:272–5.

Gorsky JE, Dietz AA, Spencer H, Osis D. Metabolic balance of aluminum studied in six men. Clin Chem. 1979;25(10):1739–43.

Grandjean P, Herz KT. Methylmercury and brain development: imprecision and underestimation of developmental neurotoxicity in humans. Mt Sinai J Med. 2011;78(1):107–18.

Gu C, Chen S, Xu X, Zheng L, Li Y, Wu K, Liu J, Qi Z, Han D, Chen G, Huo X. Lead and cadmium synergistically enhance the expression of divalent metal transporter 1 protein in central nervous system of developing rats. Neurochem Res. 2009;34:1150–6.

Gunshin H, Mackenzie B, Berger UV, Gunshin Y, Romero MF, Boron WF, Nussberger S, Gollan JL, Hediger MA. Cloning and characterization of a mammalian proton-coupled metal-ion transporter. Nature. 1997;388:482–8.

Guolo M, Stella AM, Melito V, Parera V, Batle AMC. Altered 5-aminolevulinic acid metabolism leading to pseudophorphyria in hemodialysed patients. lnt J Biochem Cell Bid. 1996;28:311–7.

Heyer NJ, Bittner AC Jr, Echeverria D, Woods JS. A cascade analysis of the interaction of mercury and coproporphyrinogen oxidase (CPOX) polymorphism on the heme biosynthetic pathway and porphyrin production. Toxicol Lett. 2006;161:159–66.

Hift RJ, Thunell S, Brun A. Drugs in porphyria: from observation to a modern algorithm-based system for the prediction of porphyrogenicity. Pharmacol Ther. 2011;132(2):158–69.

Hoffmeyer RE, Singh SP, Doonan CJ, Ross ARS, Hughes RJ, Pickering IJ, George GN. Molecular mimicry in mercury toxicology. Chem Res Toxicol. 2006;19(6):753–9.

Horiguchi H, Teranishi H, Niiya K, Aoshima K, Katoh T, Sakuragawa N, Kasuya M. Hypoproduction of erythropoietin contributes to anemia in chronic cadmium intoxication: clinical study on Itai-itai disease in Japan. Arch Toxicol. 1994;68(10):632–6.

Horiguchi H, Aoshima K, Oguma R, Sasaki S, Miyamoto K, Hosoi Y, Katoh T, Kayama F. Latest status of cadmium accumulation and its effects on kidneys, bone, and erythropoiesis in inhabitants of the formerly cadmium-polluted Jinzu River Basin in Toyama, Japan, after restoration of rice paddies. Int Arch Occup Environ Health. 2010;83:953–70.

Hsu-Kim H, Kucharzyk KH, Zhang T, Deshusses MA. Mechanisms regulating mercury bioavailability for methylating microorganisms in the aquatic environment: a critical review. Environ Sci Technol. 2013;47(6):2441–56.

Huang P, Chen C, Wang H, Li G, Jing H, Han Y, Li N, Xiao Y, Yu Q, Liu Y, Wang P, Shi Z, Sun Z. Manganese effects in the liver following subacute or subchronic manganese chloride exposure in rats. Ecotoxicol Environ Saf. 2011;74:615–22.

Illing AC, Shawki A, Cunningham CL, Mackenzie B. Substrate profile and metal-ion selectivity of human divalent metal-ion transporter-1. J Biol Chem. 2012;287(36):30485–96.

Islam E, Yang X, He Z and Mahmood Q. Assessing potential dietary toxicity of heavy metals in selected vegetables and food crops. J Zhejiang Univ Sci B. 2007; 8(1): 1–13.

Jadhav S, Sarkar S, Patil R, Tripathi H. Effects of subchronic exposure via drinking water to a mixture of eight water-contaminating metals: a biochemical and histopathological study in male rats. Arch Environ Con Tox. 2007;53:667–77.

Järup L. Hazards of heavy metal contamination. Br Med Bull. 2003;68:167–82.

Jeong KS, Park H, Hac E, Hong Y-C, Hae M, Park H, Kimf B-N, Leeg SJ, Lee KY, Kim JH, Kim Y. Evidence that cognitive deficit in children is associated not only with iron deficiency, but also with blood lead concentration: a preliminary study. J Trace Elem Med Biol. 2015;29:336–41.

Jiang X, McDermott JR, Ajees AA, Rosen BP, Liu Z. Trivalent arsenicals and glucose use different translocation pathways in mammalian GLUT1. Metallomics. 2010;2(3):211–9.

Jin T, Lu J and Nordberg M.Toxicokinetics and biochemistry of cadmium with special emphasis on the role of metallothionein. Neurotoxicology. 1998; 19(4-5):529-35.

Jomova K, Valko M. Advances in metal-induced oxidative stress and human disease. Toxicology. 2011;283(2–3):65–87.

Julka D, Gill KD. Development of a possible peripheral marker for aluminum neurotoxicity. Med Sci Res. 1995;23:311–4.

Kakkar P, Jaffery FN. Biological markers for metal toxicity. Environ Toxicol Pharmacol. 2005;19:335–49.

Kalia K, Chandra SV, Viswanathan PN. Effect of 54Mn and lead interaction on their binding with tissue proteins: in vitro studies. Ind Health. 1984;22:207–18.

Kauppinen R. Porphyrias Lancet. 2005;365:241–52.

Kaur A, Joshi K, Minz RW, Gill KD. Neurofilament phosphorylation and disruption: a possible mechanism of chronic aluminium toxicity in Wistar rats. Toxicology. 2006;219(1–3):1–10.

Kawahara M, Kato-Negishi M. Link between aluminum and the pathogenesis of Alzheimer's disease: the integration of the aluminum and amyloid Cascade hypotheses. Int J Alzheimers Dis. 2011;2011:276393.

Kerper LE, Ballatori N, Clarkson TW. Methylmercury transport across the blood-brain barrier by an amino acid carrier. Am J Phys. 1992;262:761–5.

Kile ML, Fara JM, Ronnenberg AG, Quamruzzaman Q, Rahman M, Mostofa G, Afroz S, Christiani DC. A cross sectional study of anemia and iron deficiency as risk factors for arsenic-induced skin lesions in Bangladeshi women. BMC Public Health. 2016;16:158.

Kim S, Moon C, Eun S, Ryu P, Jo S. Identification of ASK1, MKK4, JNK, c-Jun, and caspase-3 as a signaling cascade involved in cadmium-induced neuronal cell apoptosis. Biochem Biophys Res Commun. 2005;328:326–34.

Kim Y, Kim BN, Hong Y-C, Shin M-S, Yoo H-J, Kim J-W, Bhang S-Y, Cho S-C. Co-exposure to environmental lead and manganese affects the intelligence of school-aged children. Neurotoxicology. 2009;30:564–71.

Kim JH, Lee SJ, Kim SY, Choi G, Lee JJ, Kim HJ, Kim S, Park J, Moon HB, Choi K, Kim S, Choi SR. Association of food consumption during pregnancy with mercury and lead levels in cord blood. Sci Total Environ. 2016a;29(563–564):118–24.

Kim Y, Oh HG, Cho YY, Kwon O-H, Park MK, Chung S. Stress hormone potentiates Zn^{2+}-induced neurotoxicity via TRPM7 channel in dopaminergic neuron. Biochem Biophys Res Commun. 2016b;470:362–7.

Klandorf H, Van Dyke K. Oxidative and nitrosative stresses: their role in health and disease in man and birds. Oxidative stress – molecular mechanisms and biological effects. (Chapter 3). Ed. Volodymyr Lushchak and Halyna M. Semchyshyn. 2012. ISBN 978-953-51-0554-1, Published: April 25, 2012 under CC BY 3.0 license.

Klatzo I, Wisniewski H, Streicher E. Experimental production of neurofibrillary degeneration: 1. Light microscopic observations. J Neuropathol Exp Neurol. 1965;24:187–99.

Klauder DS, Petering HG. Protective value of dietary copper and iron against some toxic effects of lead in rats. Environ Health Perspect. 1975;12:77–80.

Kordas K, Queirolo EI, Ettinger AS, Wright RO, Stoltzfus RJ. Prevalence and predictors of exposure to multiple metals in preschool children from Montevideo. Uruguay Sci Total Environ. 2010;408:4488–94.

Korolnek T, Hamza I. Like iron in the blood of the people: the requirement for heme trafficking in iron metabolism. Front Pharmacol. 2014;4;5,126.

Kortenkamp and Faust. State of the art report on mixture toxicity – final report. UE Comission. 2009. http://ec.europa.eu/environment/chemicals/effects/pdf/report_mixture_toxicity.pdf, 13th June 2014, 2 p.m.

Kozlowski H, Janicka-Klos A, Brasun J, Gaggelli E, Valensin D, Valensin G. Copper, iron, and zinc ions homeostasis and their role in neurodegenerative disorders (metal uptake, transport, distribution and regulation). Coord Chem Rev. 2009;253:2665–85.

Krüger K, Straub H, Hirner AV, Hippler J, Binding N, Musshoff U. Effects of monomethylarsonic and monomethylarsonous acid on evoked synaptic potentials in hippocampal slices of adult and young rats. Toxicol Appl Pharmacol. 2009;236(1):115–23.

Kumar V, Gill KD. Oxidative stress and mitochondrial dysfunction in aluminium neurotoxicity and its amelioration: a review. Neurotoxicology. 2014;41:154–66.

Kumar A, Dogra S, Prakash A. Protective effect of curcumin (*Curcuma longa*), against aluminium toxicity: possible behavioral and biochemical alterations in rats. Behav Brain Res. 2009;205(2):384–90.

Kwong WT, Friello P, Semba RD. Interactions between iron deficiency and lead poisoning: epidemiology and pathogenesis. Sci Total Environ. 2004;330:21–37.

Lafuente A, Esquifino A. Cadmium effects on hypothalamic activity and pituitary hormone secretion in the male. Toxicol Lett. 1999;110(3):209–18.

Landrigan P, Nordberg M, Lucchini R, Nordberg G, Grandjean P, Iregren A, Alessio L. The declaration of Brescia on prevention of the neurotoxicity of metals. Am J Ind Med. 2006;50(10):709–11.

Lieu PT, Heiskala M, Peterson PA, Yang Y. The roles of iron in health and disease. Mol Asp Med. 2001;22:1–87.

Lin C-Y, Hsiao W-C, Huang C-J, Kao C-F, Hsua G-S-W. Heme oxygenase-1 induction by the ROS-JNK pathway plays a role in aluminum-induced anemia. J Inorg Biochem. 2013;128:221–8.

Lister LJ, Svendsen C, Wright J, Hooper HL, Spurgeon DJ. Modelling the joint effects of a metal and a pesticide on reproduction and toxicokinetics in Lumbricid earthworms. Environ Int. 2011;37:663–70.

Liu J, Klaassen CD. Absorption and distribution of cadmium in Metallothionein-I transgenic mice. Fund Appl Toxicol. 1996;29:294–300.

Liu Z, Sanchez MA, Jiang X, Boles E, Landfear SM, Rosen BP. Mammalian glucose permease GLUT1 facilitates transport of arsenic trioxide and methylarsonous acid. Biochem Biophys Res Commun. 2006;351(2):424–30.

Lohren H, Pieper I, Blagojevic L, Bornhorst J, Galla H-J, Schwerdtle T. Neurotoxicity of organic and inorganic mercury species – effects on and transfer across the blood-cerebrospinal fluid barrier, cytotoxic effects in target cells. Perspect Sci. 2015;3:21–2.

López Alonso M, Prieto Montaña F, Miranda M, Castillo C, Hernández J, Benedito JL. Cadmium and lead accumulation in cattle in NW Spain. Vet Hum Toxicol. 2003;45(3):128–30.

Lopez E, Arce C, Oset-Gasque MJ, Canadas S, González MP. Cadmium induces reactive oxygen species generation and lipid peroxidation in cortical neurons in culture. Free Rad Biol Med. 2006;40:940–51.

Lorincz MT. Neurologic Wilson's disease. Ann N Y Acad Sci. 2010;1184:173–87.

Lovell MA, Robertson JD, Teesdale WJ, Campbell JL, Markesbery WR. Copper, iron and zinc in Alzheimer's disease senile plaques. J Neurol Sci. 1998;158(1):47–52.

Lu J, Zheng Y-L, Wu D-M, Sun D-X, Shan Q, Fan S-H. Trace amounts of copper induce neurotoxicity in the cholesterol-fed mice through apoptosis. FEBS Lett. 2006;580:6730–40.

Lucchini R, Zimmerman N. Lifetime cumulative exposure as a threat for neurodegeneration: need for prevention strategies on a global scale. Neurotoxicology. 2009;30(6):1144–8.

Mahaffey KR, Fowler BA. Effects of concurrent Administration of Lead, cadmium, and arsenic in the rat. Environ Health Perspect. 1977;19:165–71.

Mahaffey KR, Capar SG, Gladen BC, Fowler BA. Concurrent exposure to lead, cadmium, and arsenic. Effects on toxicity and tissue metal concentrations in the rat. J Lab Clin Med. 1981;98(4):463–81.

Maines MD. Regional distribution of the enzymes of haem biosynthesis and the inhibition of 5-aminolaevulinate synthase by manganese in the rat brain. Biochem J. 1980;190:315–21.

Markiewicz-Górka I, Januszewska L, Michalak A, Prokopowicz A, Januszewska E, Pawlas N, Pawlas K. Effects of chronic exposure to lead, cadmium, and manganese mixtures on oxidative stress in rat liver and heart. Arh Hig Rada Toksikol. 2015;66(1):51–62.

Martinez-Finley EJ, Chakraborty S, Fretham SJB, Aschner M. Cellular transport and homeostasis of essential and nonessential metals. Metallomics. 2012;4(7):593–605.

McCord MC, Aizenman E. The role of intracellular zinc release in aging, oxidative stress, and Alzheimer's disease. Front Aging Neurosci. 2014;6:77.

Mejía JJ, Diáz-Barriga F, Calderón J, Ríos C, Jiménez-Capdeville ME. Effects of lead-arsenic combined exposure on central Monoaminergic systems. Neurotoxicol Teratol. 1997;19(6):489–97.

Méndez-Armenta M, Ríos C. Cadmium neurotoxicity. Environ Toxicol Pharmacol. 2007;23(3):350–8.

Méndez-Armenta M, Villeda-Hernández J, Barroso-Moguel R, Nava-Ruiz C, Jiménez-Capdeville ME, Rios C. Brain regional lipid peroxidation and metallothionein levels of developing rats exposed to cadmium and dexamethasone. Toxicol Lett. 2003;144(2):151–7.

Migliore L, Coppedè F. Environmental-induced oxidative stress in neurodegenerative disorders and aging. Mut Res. 2009;674:73–84.

Miu AC, Andreescu CE, Vasiu R, Olteanu AI. A behavioral and histological study of the effects of long-term exposure of adult rats to aluminum. Int J Neurosci. 2003;113(9):1197–211.

Molina RM, Phattanarudee S, Kim J, Thompson K, Wessling-Resnick M, Maher TJ, Brain JD. Ingestion of Mn and Pb by rats during and after pregnancy alters iron metabolism and behavior in offspring. Neurotoxicology. 2011;32(4):413–22.

Monroe RK, Halvorsen SW. Cadmium blocks receptor-mediated Jak/STAT signaling in neurons by oxidative stress. Free Radic Biol Med. 2006;41(3):493–502.

Mutti A. Biological monitoring in occupational and environmental toxicology. Toxicol Lett. 1999;108:77–89.

Nasiadek M, Chmielnicka J, Subdys J. Analysis of urinary Porphyrins in rats exposed to aluminum and iron. Ecotoxicol Environ Saf. 2001;48(1):11–7.

Naujokas MF, Anderson B, Ahsan H, Aposhian HV, Graziano JH, Thompson C, Suk WA. The broad scope of health effects from chronic arsenic exposure: update on a worldwide public health problem. Environ Health Perspect. 2013;121(3):295–302.

Nazir R, Shaheen N, Shah MH. Indoor/outdoor relationship of trace metals in the atmospheric particulate matter of an industrial area. Atmos Res. 2011;101:765–72.

Neal AP, Guilarte TR. Mechanisms of heavy metal neurotoxicity: lead and manganese. Toxicol Res (Camb). 2013;(2):99–114.

Nehru B, Anand P. Oxidative damage following chronic aluminium exposure in adult and pup rat brains. J Trace Elem Med Biol. 2005;19(2–3):203–8.

Nordberg M, Nordberg GF. Metallothioneins: historical development and overview. Met Ions Life Sci. 2009;5:1–29.

Notarachille G, Arnesano F, Calò V, Meleleo D. Heavy metals toxicity: effect of cadmium ions on amyloid beta protein 1-42. Possible implications for Alzheimer's disease. Biometals. 2014;27(2):371–88.

Nriagu J. Zinc toxicity in humans. School of Public Health, University of Michigan, Elsevier B.V. 2007.

O'Neil P. Heavy metals in soils. In: Alloway BJ, editor. Arsenic. London: Blackie Academic and Professional Arsenic; 1995. p. 105–21.

Obiri S, Yeboah PO, Osae S, Adu-Kumi S. Levels of arsenic, mercury, cadmium, copper, lead, zinc and manganese in serum and whole blood of resident adults from mining and non-mining communities in Ghana. Environ Sci Pollut Res Int. 2016;23(16):16589–97.

Oteiza PI, Keen CL, Han B, Golub MS. Aluminum accumulation and neurotoxicity in Swiss-Webster mice after long-term dietary exposure to aluminum and citrate. Metabolism. 1993;42(10):1296–300.

Park JD, Zheng W. Human exposure and health effects of inorganic and elemental mercury. J Prev Med Public Health. 2012;45(6):344–52.

Patrick L. Lead toxicity, a review of the literature. Part I: exposure, evaluation, and treatment. Altern Med Rev. 2006;11(1):2–22.

Peakall D, Burger J. Methodologies for assessing exposure to metals: speciation, bioavailability of metals, and ecological host factors. Ecotoxicol Environ Saf. 2003;56(1):110–21.

Piao F, Cheng F, Chen H, Li G, Sun X, Liu S, Yamauchi T, Yokoyama K. Effects of Zn administration on Pb toxicities in rats. Ind Health. 2007;45:546–51.

Pirpamer L, Hofer E, Gesierich B, De Guio F, Freudenberger P, Seiler S, Duering M, Jouvent E, Duchesnay E, Dichgans M, Ropele S, Schmidt R. Determinants of iron accumulation in the normal aging brain. Neurobiol Aging. 2016;43:149–55.

Pohl HR, Hansen H, Chou C-HSJ. Public health guidance values for chemical mixtures: current practice and future directions. Regul Toxicol Pharmacol. 1997;26:322–9.

Pohl HR, Roney N, Abadin HG. Metal ions affecting the neurological system. Met Ions Life Sci. 2011;8:247–62.

Prüss-Ustün A, Wolf J, Corvalán C, Bos R, Neira M. Preventing disease through healthy environments. A global assessment of the burden of disease from environmental risks. Geneva: World Health Organization (WHO); 2016.

Qato MK, Maines MD. Regulation of heme and drug metabolism activities in the brain by manganese. Biochem Biophys Res Commun. 1985;128(1):18–24.

Quintanar L. Manganese neurotoxicity: a bioinorganic chemist's perspective. Inorg Chim Acta. 2008;361:875–84.

Quintanilla-Vega B, Hernandez A, Mendoza-Figueroa T. Reduction in porphyrin excretion as a sensitive indicator of lead toxicity in primary cultures of adult rat hepatocytes. Toxicol In Vitro. 1996;10:675–83.

Rachakonda V, Pan TH, Le WD. Biomarkers of neurodegenerative disorders: how good are they? Cell Res. 2004;14(5):349–60.

Ramos-Chávez LA, Rendón-López CRR, Zepeda A, Silva-Adaya D, Del Razo LM, Gonsebatt ME. Neurological effects of inorganic arsenic exposure: altered cysteine/glutamate transport, NMDA expression and spatial memory impairment. Front Cell Neurosci. 2015;9:9–21.

Rodríguez-Barranco M, Lacasaña M, Aguilar-Garduño C, Alguacil J, Gil F, González-Alzaga B, Rojas-García A. Association of arsenic, cadmium and manganese exposure with neurodevelopment and behavioural disorders in children: a systematic review and meta-analysis. Sci Total Environ. 2013;1(454–455):562–77.

Roels HA, Bowler RM, Kim Y, Claus Henn B, Mergler D, Hoet P, Gocheva VV, Bellinger DC, Wright RO, Harris MG, Chang Y, Bouchard MF, Riojas-Rodriguez H, Menezes-Filho JA, Tellez-Rojo MM. Manganese exposure and cognitive deficits: a growing concern for manganese neurotoxicity. Neurotoxicology. 2012;33:872–80.

Roneya N, Colman J. Interaction profile for lead, manganese, zinc, and copper. Environ Toxicol Pharmacol. 2004;18:231–4.

Rosado JL, Ronquillo D, Kordas K, Rojas O, Alatorre J, Lopez P, Garcia-Vargas G, Caamaño MC, Cebrián ME, Stoltzfus RJ. Arsenic exposure and cognitive performance in Mexican schoolchildren. Environ Health Perspect. 2007;115(9):1371–5.

Rubio-Osornio M, Montes S, Heras-Romero Y, Guevara J, Rubio C, Aguilera P, Rivera-Mancia S, Floriano-Sánchez E, Monroy-Noyola A, Ríos C. Induction of ferroxidase enzymatic activity by copper reduces MPP+−evoked neurotoxicity in rats. Neurosci Res. 2013;75:250–5.

Rush T, Hjelmhaug J, Lobner D. Effects of chelators on mercury, iron, and lead neurotoxicity in cortical culture. Neurotoxicology. 2009;30:47–51.

Sahin G, Varol I, Temizer A, Benli K, Demirdamar R, Duru S. Determination of aluminum levels in the kidney, liver, and brain of mice treated with aluminum hydroxide. Biol Trace Elem Res. 1994;41(1–2):129–35.

Sánchez-Peña LC, Petrosyan P, Morales M, González NB, Gutiérrez-Ospina G, Del Razo LM, Gonsebatt ME. Arsenic species, AS3MT amount, and AS3MT gene expression in different brain regions of mouse exposed to arsenite. Environ Res. 2010;110(5):428–34.

Santamaria AB. Manganese exposure, essentiality & toxicity. Indian J Med Res. 2008;128:484–500.

Scheiber IF, Mercer JFB, Dringen R. Metabolism and functions of copper in brain. Prog Neurobiol. 2014;116:33–57.

Schneider SA. Neurodegenerations with brain iron accumulation. Parkinsonism Rel Disord. 2016;22:21–5.

Sensi SL, Jeng JM. Rethinking the excitotoxic ionic milieu: the emerging role of Zn(2+) in ischemic neuronal injury. Curr Mol Med. 2004;4(2):87–111.

Sensi SL, Yin HZ, Weiss JH. AMPA/kainate receptor-triggered Zn^{2+} entry into cortical neurons induces mitochondrial Zn^{2+} uptake and persistent mitochondrial dysfunction. Eur J Neurosci. 2000;12:3813–8.

Sethi P, Jyoti A, Hussain E, Sharma D. Curcumin attenuates aluminium-induced functional neurotoxicity in rats. Pharmacol Biochem Behav. 2009;93:31–9.

Shi LZ, Zheng W. Early lead exposure increases the leakage of the blood–cerebrospinal fluid barrier, *in vitro*. Hum Exp Toxicol. 2007;26(3):159–67.

Shukla GS, Hussain T, Chandra SV. Possible role of regional superoxide dismutase activity and lipid peroxide levels in cadmium neurotoxicity: in vivo and in vitro studies in growing rats. Life Sci. 1987;41(19):2215–21.

Shukla GS and Chandra SV. Concurrent exposure to lead, manganese, and cadmium and their distribution to various brain regions, liver, kidney, and testis of growing rats. Archives of Environ Contam Toxicol. 1987; 16(3):303–310.

Shukla A, Shukla GS, Srimal RC. Cadmium-induced alterations in blood-brain barrier permeability and its possible correlation with decreased microvessel antioxidant potential in rat. Hum Exp Toxicol. 1996;15(5):400–5.

Shy CM. Epidemiological studies of neurotoxic, reproductive, and carcinogenic effects of complex mixtures. Environ Health Perspect. 1993;101(4):183–8.

Simmons JE. Chemical mixtures: challenge for toxicology and risk assessment. Toxicology. 1995;105:11–9.

Simmons-Willis TA, Koh AS, Clarkson TW, Ballatori N. Transport of a neurotoxicant by molecular mimicry : the methylmercury-L-cysteine complex is a substrate for human L-type large neutral amino acid transporter (LAT) 1 and LAT2. Biochem J. 2002;367:239–46.

Sinczuk-Walczaki H, Szymczak M, Razniewska G, Matczak W, Szymczak W. Effects of occupational exposure to aluminium on nervous system: clinical and electroencephalographic findings. Int J Occup Med Environ Health. 2003;16(4):301–10.

Singh T, Goel RK. Neuroprotective effect of *Allium cepa* L. in aluminium chloride induced neurotoxicity. Neurotoxicology. 2015;49:1–7.

Skjørringe T, Burkhart A, Johnsen KB, Moos T. Divalent metal transporter 1 (DMT1) in the brain: implications for a role in iron transport at the blood-brain barrier, and neuronal and glial pathology. Front Mol Neurosci. 2015;8(19):1–13.

Song H, Zheng G, Liu Y, Shen X-F, Zhao Z-H, Aschner M, Luo W-J, Chen J-Y. Cellular uptake of lead in the blood-cerebrospinal fluid barrier: novel roles of Connexin 43 hemichannel and its down-regulations via Erk phosphorylation. Toxicol Appl Pharmacol. 2016;15(297):1–11.

Spurgeon DJ, Jones OAH, Dorne J-L, Svendsen C, Swain S, Stürzenbaum SR. Systems toxicology approaches for understanding the joint effects of environmental chemical mixtures. Sci Total Environ. 2010;408:3725–34.

Strak E, Ellinger I, Balthasar C, Scheinast M, Schatz J, Szattler T, Bleichert S, Saleh L, Knöfler M, Zeisler H, Hengstschläger M, Rosner M, Salzer H, Gundacker C. Mercury toxicokinetics of the healthy human term placenta involve amino acid transporters and ABC transporters. Toxicology. 2016;340:34–42.

Struys-Ponsar C, Kerkhofs A, Gauthier A, Soffié M, van den Bosch de Aguilar P. Effects of aluminum exposure on behavioral parameters in the rat. Pharmacol Biochem Behav. 1997;56(4):643–8.

Szewczyk B. Zinc homeostasis and neurodegenerative disorders. Front Aging Neurosci. 2013;5:33.

Tallkvist J, Bowlus CL, Lonnerdal B. DMT1 gene expression and cadmium absorption in human absorptive enterocytes. Toxicol Lett. 2001;122:171–7.

Tiffany-Castiglioni E, Hong S, Qian Y, Tang Y, Donnelly KC. In vitro models for assessing neurotoxicity of mixtures. Neurotoxicology. 2006;27:835–9.

Tjälve H, Henriksson J. Uptake of metals in the brain via olfactory pathways. Neurotoxicology. 1999;20(2–3):181–95.

Tougu V, Tiiman A, Palumaa P. Interactions of Zn(II) and cu(II) ions with Alzheimer's amyloid-beta peptide metal ion binding, contribution to fibrillization and toxicity. Metallomics. 2011;3:250–61.

Vazquez M, Velez D, Devesa V, Puig S. Participation of divalent cation transporter DMT1 in the uptake of inorganic mercury. Toxicology. 2015;331:119–24.

von Stackelberg K. Guzy E, Chu T, Henn BC. Mixtures, metals, genes and pathways: a systematic review. Working paper prepared for: methods for research synthesis: a cross-disciplinary workshop. Harvard Center for Risk Analysis. 2013.

Wagner GS, Tephly TR. A possible role of copper in the regulation of heme biosynthesis through ferrochelatase. Adv Exp Med Biol. 1975;58:343–54.

Walton JR. Aluminum disruption of calcium homeostasis and signal transduction resembles change that occurs in aging and Alzheimer's disease. J Alzheimers Dis. 2012;29(2):255–73.

Waalkes MP, Harvey MJ, Klaassen CD. Relative in vitro affinity of hepatic metallothionein for metals. Toxicol Lett. 1984; 20(1):33–9.

Wang G, Fowler BA. Roles of biomarkers in evaluating interactions among mixtures of lead, cadmium and arsenic. Toxicol Appl Pharmacol. 2008;233:92–9.

Wang Q, Luo W, Zheng W, Liu Y, Xu H, Zheng G, Dai Z, Zhang W, Chen Y, Chen J. Iron supplement prevents lead-induced disruption of the blood–brain barrier during rat development. Toxicol Appl Pharmacol. 2007;219(1):33–41.

Wang Q, Luo W, Zhang W, Liu M, Song H, Chen J. Involvement of DMT1 +IRE in the transport of lead in an in vitro BBB model. Toxicol In Vitro. 2011;25:991–8.

Wang L, Wang X, Zhang S, Qu G, Liu S. A protective role of heme-regulated eIF2a kinase in cadmium-induced toxicity in erythroid cells. Food Chem Toxicol. 2013;62:880–91.

Ward RJ, Zucca FA, Duyn JH, Crichton RR, Zecca L. The role of iron in brain ageing and neurodegenerative disorders. Lancet Neurol. 2014;13(10):1045–60.

Watanabe T, Hirano S. Metabolism of arsenic and its toxicological relevance. Arch Toxicol. 2013;87(6):969–79.

Weidenhamer JD, Lobunski PA, Kuepouo G, Corbin RW, Gottesfeld P. Lead exposure from aluminum cookware in Cameroon. Sci Total Environ. 2014;496:339–47.

Weiss B. Economic implications of manganese neurotoxicity. Neurotoxicology. 2006;27:362–8.

Wenting L, Ping L, Haitao J, Meng Q, Xiaofei R. Therapeutic effect of taurine against aluminum-induced impairment on learning, memory and brain neurotransmitters in rats. Neurol Sci. 2014;35(10):1579–84.

Wester RC, Maibach HI, Sedik L, Melendres J, DiZio S, Wade M. In vitro percutaneous absorption of cadmium from water and soil into human skin. Fund Appl Toxicol. 1992;19(1):1–5.

Whittaker MH, Wang G, Chen X-Q, Lipsky M, Smith D, Gwiazda R, Fowler BA. Exposure to Pb, Cd, and As mixtures potentiates the production of oxidative stress precursors: 30-day, 90-day, and 180-day drinking water studies in rats. Toxicol Appl Pharmacol. 2010;254(2):154–66.

Wills MR, Hewitt CD, Sturgill BC, et al. Long-term oral or intravenous aluminum administration in rabbits I. Renal and hepatic changes. Ann Clin Lab Sci. 1993;23(1):1–16.

Witholt R, Gwiazda RH, Smith DR. The neurobehavioral effects of subchronic manganese exposure in the presence and absence of pre-parkinsonism. Neurotoxicol Teratol. 2000;22:851–61.

Woods JS. Porphyrin metabolism as indicator of metal exposure and toxicity. In: Goyer RA, Cherian MG, editors. Handbook of experimental pharmacology. Vol. 115. Chap. 2. Toxicology of metals, biochemical aspects. Berlin: Springer; 1995. p. 19–52.

Woods JS. Altered porphyrin metabolism as a biomarker of mercury exposure and toxicity. Can J Physiol Pharmacol. 1996;74:210–5.

Woods JS, Southern MR. Studies on the etiology of trace metal-induced porphyria: effect of porphyrinogenic metals on coproporphyrinogen oxidase in rat liver and kidney. Toxicol Appl Pharmacol. 1989;97:183–90.

Woods JS, Eaton DL, Lukens CB. Studies on porphyrin metabolism in the kidney. Effects of trace metals and glutathione of renal uroporphyrinogen decarboxylase. Mol Pharmacol. 1984;26:336–41.

Woods JS, Bowers MA, Davis HA. Urinary porphyrin profiles as biomarkers of trace metal exposure and toxicity: studies on urinary porphyrin excretion patterns in rats during prolonged exposure to methyl mercury. Toxicol Appl Pharmacol. 1991;110:464–76.

Woods JS, Martin MD, Leroux BG, DeRouen TA, Bernardo MF, Luis HS, Leitão JG, Simmonds PL, Rue TC. Urinary porphyrin excretion in normal children and adolescents. Clin Chim Acta. 2009;405:104–9.

Wright RO, Baccarelli A. Metals and neurotoxicology. J Nutr. 2007;137(12):2809–13.

Yang XF, Han QG, Liu DY, Fan GY, Ma JY, Wang ZL. Microstructure and ultrastructure alterations in the pallium of immature mice exposed to cadmium. Biol Trace Elem Res. 2016;1–7.

Yasui M, Kihira T, Ota K. Calcium, magnesium and aluminum concentrations in Parkinson's disease. Neurotoxicology. 1992;13:593–600.

Yasuno T, Okamoto H, Nagai M, Kimura S, Yamamoto T, Nagano K, Furubayashi T, Yoshikawa Y, Yasui H, Katsumi H, Sakane T, Yamamoto A. The disposition and intestinal absorption of zinc in rats. Eur J Pharm Sci. 2011;44:410–5.

Yen Le TT, Vijver MG, Kinraide TB, Peijnenburg SWJGM, Hendriks AJ. Modelling metal interactions and metal toxicity to lettuce *Lactuca sativa* following mixture exposure (Cu^{2+}-Zn^{2+} and Cu^{2+}-Ag^+). Environ Pollut. 2013;176:185–92.

Yokel RA. Blood-brain barrier flux of aluminum, manganese, iron and other metals suspected to contribute to metal-induced neurodegeneration. J Alzheimers Dis. 2006;10:223–53.

Zheng W. Toxicology of choroid plexus: special reference to metal-induced neurotoxicities. Microsc Res Tech. 2001;52(1):89–103.

Zheng W, Monnot AD. Regulation of brain iron and copper homeostasis by brain barrier systems: implication in neurodegenerative diseases. Pharmacol Ther. 2012;133(2):177–88.

Zheng W, Perry DF, Nelson DL, Aposhian HV. Protection of cerebrospinal fluid against toxic metals by the choroid plexus. FASEB J. 1991;5:2188–93.

Zheng W, Aschner M, Ghersi-Egeac J-F. Brain barrier systems: a new frontier in metal neurotoxicological research. Toxicol Appl Pharmacol. 2003;192:1–11.

Zheng G, Zhang J, Xuc Y, Shen X, Song H, Jing J, Luo W, Zheng W, Chen J. Involvement of CTR1 and ATP7A in lead (Pb)-induced copper (Cu)accumulation in choroidal epithelial cells. Toxicol Lett. 2014;225:110–8.

Zhi D, Tao AIJ, Fang HJ, Sun RB, Shi Y, Wang LL, Wang Q. Influence of iron supplementation on DMT1 (IRE)-induced transport of lead by brain barrier systems in vivo. Biomed Environ Sci. 2015;28(9):651–9.

Zhou F, Chen Y, Fan G, Feng C, Dub G, Zhu G, Li Y, Jiao H, Guan L, Wang Z. Lead-induced iron overload and attenuated effects of ferroportin 1 overexpression in PC12 cells. Toxicol In Vitro. 2014;28:1339–48.

Zhu L, Ji X-J, Wang H-D, Pan H, Chen M, Lu T-J. Zinc neurotoxicity to hippocampal neurons in vitro induces ubiquitin conjugation that requires p38 activation. Brain Res. 2012;1438:1–7.

Zhu G, Fan G, Feng C, Li Y, Chen Y, Zhou F, Du G, Jiao H, Liu Z, Xiao X, Lin F, Yand J. The effect of lead exposure on brain iron homeostasis and the expression of DMT1/FP1 in the brain in developing and aged rats. Toxicol Lett. 2013;216:108–23.

Zhu H, Jia Y, Cao H, Meng F, Liu X. Biochemical and histopathological effects of subchronic oral exposure of rats to a mixture of five toxic elements. Food Chem Toxicol. 2014;71:166–75.

Methylmercury-Induced Neurotoxicity: Focus on Pro-oxidative Events and Related Consequences

Marcelo Farina and Michael Aschner

Abstract Methylmercury (MeHg) is a highly neurotoxic environmental pollutant. Even though molecular mechanisms mediating MeHg toxicity are not completely understood, several lines of evidence indicate that the neurotoxic effects resultant from MeHg exposure represent a consequence of its pro-oxidative properties. In this regard, MeHg is a soft electrophile that preferentially interacts with (and oxidize) nucleophilic groups (mainly thiols and selenols) from biomolecules, including proteins and low-molecular-weight molecules. Such interaction contributes to the occurrence of oxidative stress and impaired function of several molecules [proteins (receptors, transporters, enzymes, structural proteins), lipids (i.e., membrane constituents and intracellular messengers), and nucleic acids (i.e., DNA)], culminating in neurotoxicity.

In this chapter, an initial background on the general aspects regarding the neurotoxicology of MeHg, with a particular focus on its pro-oxidative properties and its interaction with nucleophilic thiol- and selenol-containing molecules, is provided. Even though experimental evidence indicates that symptoms (i.e., motor impairment) resultant from MeHg exposure are linked to its pro-oxidative properties, as well as to their molecular consequences (lipid peroxidation, disruption of glutamate and/or calcium homeostasis, etc.), data concerning the relationship between molecular parameters and behavioral impairment others that those related to the motor function (i.e., visual impairment, cognitive skills, etc.) are scarce. Thus, even though scientific research has provided a significant amount of knowledge concerning the mechanisms mediating MeHg-induced neurotoxicity in the last decades, the whole scenario is far from being completely understood, and further research in this area is well warranted.

Keywords Methylmercury • Pro-oxidative events • Oxidative stress • Neurotoxicity

M. Farina (✉)
Departamento de Bioquímica, Centro de Ciências Biológicas, Universidade Federal de Santa Catarina, Florianópolis, SC, Brazil
e-mail: marcelo.farina@ufsc.br

M. Aschner
Department of Molecular Pharmacology, Albert Einstein College of Medicine, Bronx, NY, USA
e-mail: Michael.Aschner@einstein.yu.edu

© Springer International Publishing AG 2017
M. Aschner and L.G. Costa (eds.), *Neurotoxicity of Metals*, Advances in Neurobiology 18, DOI 10.1007/978-3-319-60189-2_13

Abbreviations and Synonyms

GSH	Glutathione (reduced form)
H_2O_2	Hydrogen peroxide
MeHg	Methylmercury = CH_3Hg^+
-SeH	Selenol = selenohydryl
-Se$^-$ (deprotonated form of selenol)	Selenolate
-SH	Thiol = sulfhydryl
-S$^-$ (deprotonated form of thiol)	Thiolate

Methylmercury Chemistry and Toxicology: General Aspects

Mercury is a metallic element presented in liquid state at room temperature. Its atomic symbol is Hg (from *hydrargyrum*, liquid silver), and its atomic number and weight are 80 and 200.59, respectively. Hg is greatly used in industry, and its salts were considerably used for therapeutic purposes in the past (Clarkson 2002). Nowadays, Hg is still used as a preservative (thimerosal; an organic Hg compound) in vaccines such as those against hepatitis B and DPT (diphtheria, pertussis, and tetanus) especially in nondeveloped countries (Dórea 2015). However, considering that mercury and mercurials are highly toxic to humans and to the whole environment, there is a significant intent to decrease both their industrial and clinical uses (Kessler 2013).

In nature, Hg exists mainly in three chemical forms: inorganic Hg salts, elemental Hg vapor, and organic Hg compounds, such as methylmercury (MeHg) (Clarkson 2002). MeHg (CH_3Hg^+) is a pollutant ubiquitously present in environment, and its natural synthesis is consequence of the methylation of inorganic Hg, which is catalyzed by methyltransferases of aquatic microorganisms (Compeau and Bartha 1985). In the environment, MeHg is bioaccumulated through the aquatic food chain, reaching concentrations over 1 ppm in predatory fish (Hintelmann 2010).

Because most of the naturally occurring MeHg is present in the aquatic food chain (specially in predatory fish), seafood ingestion represents the most important way by which humans are exposed to MeHg. Consequently, fishing communities are highly exposed to toxic MeHg levels (Clarkson et al. 2003).

After ingested, MeHg is well absorbed by the gastrointestinal tract (approximately 90–95%) (Rahola et al. 1973). After absorption, MeHg can reach several organs, but the central nervous system (CNS) represents a preferential target for MeHg. In this regard, it is important to mention that the interaction of MeHg with thiol (-SH) groups from biomolecules has a significant role in its distribution to the different organs, as well as in its excretion and, consequently, toxicity. After interaction with L-cysteine (the unique amino acid present in proteins with a "free" thiol group), MeHg-L-cysteine complex (see Fig. 1) is taken up by cells from different tissues by molecular mimicry as a surrogate of methionine (Aschner and

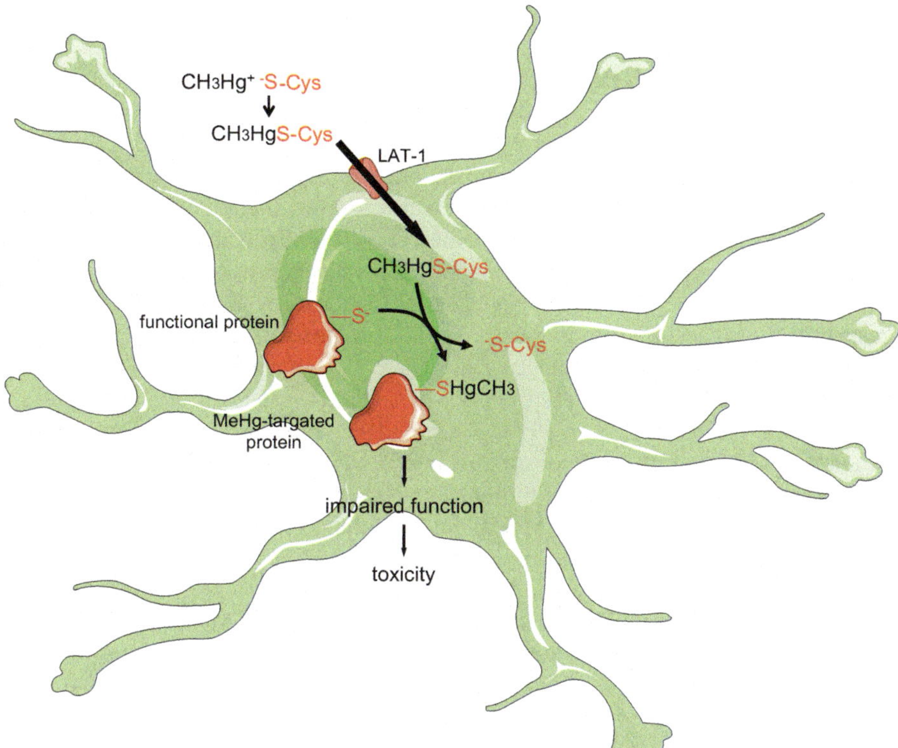

Fig. 1 MeHg-cysteine interaction. MeHg (CH$_3$Hg$^+$) interacts with the amino acid cysteine, especially with its deprotonated (thiolate) form ($^-$S-Cys). MeHg-cysteine complex (CH$_3$Hg S-Cys) is structurally similar to the amino acid methionine, thus, represents a potential substrate for the neutral amino acid transporter (LAT-1). In the intracellular environment, the occurrence of nucleophilic displacement of complexed ligand by sulfhydryl-deprotonated ligands is possible (i.e., a sulfhydryl protein), allowing for ligand exchange, resulting in the interaction of MeHg with nucleophilic biomolecules, leading to impaired function and toxic consequences. This event is not restricted to sulfhydryl-containing proteins. In fact, it can also occur (and it is more likely to occur) with selenohydryl-containing proteins

Clarkson 1988; Mokrzan et al. 1995). In this regard, it is important to mention that MeHg-L-cysteine conjugate is a substrate for the neutral amino acid transporter, LAT-1, which transports MeHg (complexed with L-cysteine) across membranes (Yin et al. 2008). Concerning the fate(s) of the MeHg-L-cysteine complex into cells, evidence shows the occurrence of nucleophilic displacement of complexed ligand by sulfhydryl-deprotonated ligands (i.e., proteins), allowing for ligand exchange (Rabenstein and Evans 1978), which can lead to the interaction with myriad of nucleophilic biomolecules, resulting in toxic consequences (Farina et al. 2009; Roos et al. 2011). Additionally, MeHg can interact with the thiol group from glutathione (GSH; discussed latter), and MeHg transport across liver canalicular membranes into bile, a major route of excretion of this toxic compound, is dependent upon intracellular GSH, as well as on the glutathione-methylmercury complex (Dutczak and Ballatori 1994).

The interaction of MeHg with nucleophilic groups from proteins is responsible, at least in part, for its toxic effects (MeHg's toxicodynamics will be further discussed in the following items). With respect to MeHg toxicokinetics, it is also important to mention that MeHg can undergo dealkylation. In fact, the carbon-mercury bound in MeHg's chemical structure can be disrupted in the biological systems, releasing the methyl group (dealkylation) and (re)generating inorganic Hg (Suda and Takahashi 1992). In this regard, the percentage (of total) of inorganic Hg in the blood, breast milk, and urine after MeHg exposure is 7%, 39%, and 73%, respectively, suggesting that inorganic Hg is an important excretable metabolite of MeHg. Of note, a high percentage of inorganic Hg (above 80%) was found in the brain of a 30-year-old individual who was exposed to MeHg at 8 years of age (22 years before) (Davis et al. 1994). This evidence indicates a high persistence of Hg in the brain after MeHg exposure. Although MeHg is well recognized as a neurotoxicant by acting at specific biomolecular sites (for a review, see Farina et al. 2011a, b), the dealkylation of MeHg into inorganic Hg likely accounts for Hg's persistence in the brain and potentially long-lasting neurological outcomes (Grandjean et al. 1997; Ninomiya et al. 2005).

Another important issue concerning the toxicokinetics of MeHg is related to the fact that it is transferred from the pregnant mother to the fetus, reaching the fetus. In this regard, an experimental study (Watanabe et al. 1999) where pregnant mice were directly exposed to MeHg detected higher levels of the metal in the fetuses' brain when compared to the dams, indicating a high transplacental transport of MeHg, as well as a great retention in the fetus brain. MeHg seems to be actively transported from the maternal to the fetal blood as its cysteine conjugate via the neutral amino acid carrier system (Kajiwara et al. 1996). In this regard, there is a huge amount of epidemiological data showing that maternal exposure to MeHg during pregnancy causes neurological deficits in their offspring (Grandjean et al. 1997; Murata et al. 2004). Interestingly, exposure to MeHg during early fetal development is linked to subtle brain injury at levels much lower than those affecting the mature brain (Grandjean and Landrigan 2014), most likely because it affects cell differentiation, migration, and synaptogenesis (Theunissen et al. 2011; Zimmer et al. 2011).

Nucleophilic Targets of MeHg

Even though Hg is a metal that does not take part in conventional reactions of electron transference, it displays pro-oxidative effects toward biomolecules; these effects are greatly resultant from its soft electrophile properties. As electrophile, MeHg is an electron-deficient species that forms covalent bonds with electron-rich nucleophiles (for a review, see LoPachin and Gavin 2016). From a toxicological point of view, the most important nucleophile groups targeted by MeHg are thiols (-SH) and selenols (−SeH). Even though these two chemical groups present relatively analogous properties (both are soft nucleophiles formed by a chalcogen bonded to a hydrogen atom), there are particular characteristics that make them dissimilar from a biological (and toxicological) perspective. The following paragraphs cover the most important chemical groups and molecules targeted by MeHg in the biological systems, as well as their oxidative consequences.

MeHg Interacts with Selenols

Firstly, it is important to mention that Hg is the softest electrophile from its periodic group or family; consequently, it will present high affinity for soft nucleophiles (for detailed aspects concerning soft and hard electrophiles/nucleophiles, see Ho 1977). Selenium is a softer nucleophile compared with sulfur, and, therefore, selenol-containing molecules are more prone to be targeted by MeHg when compared to thiol-containing ones. Moreover, selenols commonly present a lower pKa compared to thiols, and, consequently, most of selenium in selenoproteins is deprotonated ($-Se^-$; selenolate) at physiological pH, which makes it more reactive compared to the selenol form ($-SeH$). Both events are related to the great reactivity of selenol/selenolate, as well as the great number of selenoproteins targeted by MeHg; these proteins present selenol group(s) due to the presence of the amino acid selenocysteine (Lu and Holmgren 2009). As consequence, several selenoproteins have been reported as molecular targets involved in MeHg-induced neurotoxicity (Mori et al. 2007; Carvalho et al. 2008; Farina et al. 2009, 2011a, b; Wagner et al. 2010; Branco et al. 2011). Among then, the most relevant from a neurotoxicological perspective are (1) glutathione peroxidase 1, 3, and 4 (Farina et al. 2009; Franco et al. 2009; Branco et al. 2012; Zemolin et al. 2012; Usuki and Fujimura 2016), which catalyze the reduction of different types of peroxides, depending on the specific isozyme; (2) thioredoxin reductase (Wagner et al. 2010; Branco et al. 2011), the only enzyme known to catalyze the reduction of thioredoxin, thus contributing to maintain thiol redox homeostasis in proteins; (3) type II iodothyronine 5′-deiodinase (Mori et al. 2007), which is responsible for the metabolism of thyroid hormones. Even though the direct interaction of MeHg with the selenol group of these selenoproteins has been reported to be an important mechanism mediating decreased protein function (Farina et al. 2009; Branco et al. 2012), posttranscriptional (and pre-translational) events concerning mercury-selenium interaction seem to be also involved in the observed decreased activity of selenoproteins (Usuki et al. 2011; Penglase et al. 2014).

It is noteworthy that selenoproteins, such as glutathione peroxidase (Stringari et al. 2008) and deiodinases (Watanabe et al. 1999), have been reported as MeHg targets in experimental models of MeHg-induced developmental neurotoxicity. In this regard, it is important to mention that studies concerning the interaction between MeHg and dietary Se with a particular focus on the neurodevelopment are scarce; studies concerning such topic are well warranted.

MeHg Interacts with Thiols

In addition to selenols, thiols also represent important nucleophilic groups involved in MeHg-induced (neuro)toxicity. Even though a selenol-containing molecule is proner to be targeted by MeHg when compared to a thiol-containing molecule (if both are present at equimolar concentrations), it is important to mention that thiols are much more abundant than selenols in the biological systems. In fact, thiol groups can be found in low- (mainly cysteine and reduced glutathione) and high-molecular-weight proteins, whereas selenol groups are found only in a restricted group of selenoproteins (Araie and Shiraiwa 2009; Lobanov et al. 2009; Lu and Holmgren 2009). The higher nucleophilicity of selenols (compared to thiols) indicates that they represent preferential targets for MeHg *if both nucleophiles are at equimolar concentrations*. However, due to the great abundance of thiols, the *law of mass action* will favor MeHg-thiol interactions in the biological systems, which present great importance in MeHg toxicokinetics and toxicodynamics. From a toxicological perspective, the most important interaction between MeHg and thiols occurs with the amino acid cysteine, which is present in myriad of thiol proteins, as well as in the structure of low-molecular-weight antioxidant molecules, such as the tripeptide GSH (discussed latter).

Concerning the interaction between MeHg and the cysteine residue in proteins, it is important to note that such interaction changes the redox state of the protein's thiol group(s) (Kim et al. 2002). In general, the activity of sulfhydryl (thiol) proteins generally decreases when their thiol groups are oxidized. Of note, it is important to mention that several agents (i.e., reactive oxygen species) can interact with and oxidize protein thiols. In addition, protein thiol oxidation is a dynamic and generally reversible process that continuously occurs under physiological conditions (Seres et al. 1996; Mustacich and Powis 2000). Of note, even a sulfur-mercury covalent bond can be disrupted, regenerating the protein thiol group; however, the presence of a strong nucleophile is required to allow for such event (see Eq. 1).

$$\underset{\substack{\text{Protein Hg}\\\text{complex}}}{\text{Protein S Hg}} + \underset{\substack{\text{strong}\\\text{nucleophile}}}{\text{R SH}} \rightarrow \underset{\text{thiol protein}}{\text{Protein SH}} + \underset{\substack{\text{Hg nucleophile}\\\text{complex}}}{\text{R SHg}} \qquad (1)$$

With a particular emphasis to the CNS, it is important to note that oxidative modifications of thiol proteins change chemical neurotransmission due to the modulation of neurotransmitter release (LoPachin and Barber 2006) and binding to their specific synaptic receptors (Soares et al. 2003). In addition, the activity of several antioxidant proteins relies on the proper redox state of their thiol groups (i.e., thioredoxin), and, as already mentioned, MeHg is able to interact with and oxidize such thiol groups. Accordingly, MeHg modulates the functions of several sulfhydryl-containing proteins in the CNS, such as creatine kinase (Glaser et al. 2010a, b), glutathione reductase (Stringari et al. 2008), Ca^{2+}-ATPase (Freitas et al. 1996), Trx (Branco et al. 2011), choline acetyltransferase and enolase (Kung et al. 1987), etc.

As consequence, several secondary neurotoxic events are resultant from this pro-oxidative effect of MeHg (discussed in item 3).

In addition to thiol proteins, low-molecular-weight thiol molecules also represent important targets mediating MeHg-induced neurotoxicity. In this regard, it is important to mention that GSH (gamma-glutamyl-cysteinyl-glycine), the major intracellular low-molecular-weight thiol compound synthesized in all tissues, can directly interact with MeHg. The direct chemical interaction between MeHg and GSH, and its importance in mercurial toxicity, dates several decades. Such an interaction affects the deposition of MeHg in tissues (Richardson and Murphy 1975) and modifies Hg excretion in the bile of MeHg-exposed rats (Osawa and Magos 1974), indicating that this low-molecular-weight thiol compound modulates MeHg toxicity. In fact, studies on the toxicological relevance of MeHg-GSH interaction have shown that strategies that increase GSH levels are protective against MeHg-induced neurotoxicity (Kaur et al. 2011; Shanker et al. 2005). A relatively recent study (Rush et al. 2012) investigated the mechanisms of GSH-mediated attenuation of MeHg neurotoxicity in primary cortical culture. In such study, MeHg depleted GSH in neuronal, glial, and mixed cultures. The authors observed that supplementation with exogenous GSH, specifically GSH monoethyl ester (which is able to enter the cells), protected against MeHg-induced neuronal death. Of note, the authors observed that inhibition of multidrug resistance protein-1 (MRP1) potentiated MeHg neurotoxicity and increased cellular MeHg, suggesting that GSH offers neuroprotection against MeHg toxicity in a manner dependent on MRP1-mediated efflux (Rush et al. 2012). In line with the aforementioned studies, several in vitro (Franco et al. 2007; Ni et al. 2011) and in vivo (Franco et al. 2006; Stringari et al. 2008) evidences have shown that MeHg exposure causes GSH depletion. Figure 2 depicts the most important aspects concerning MeHg-GSH interaction into cells.

Because of the crucial role of GSH in maintaining redox homeostasis and detoxifying reactive oxygen species (Dringen 2005), several aspects of MeHg-induced neurotoxicity have been ascribed to GSH depletion (for a review, see Farina et al. 2011a). Taking into account the chemical interaction between MeHg and GSH, MeHg-induced GSH depletion represents a predictable phenomenon. However, intracellular GSH concentrations in the mammalian cerebrum and cerebellum are in the millimolar (mM) range. Because decreased GSH levels have been reported in the cortices (cerebral and cerebellar) of MeHg-exposed animals whose cortical mercury levels were in the low micromolar range (Franco et al. 2006; Stringari et al. 2008), one might posit that the simple interaction between both molecules is not the only cause of GSH oxidation. In fact, MeHg seems to induce the formation of reactive oxygen species (ROS) by GSH-independent mechanisms as well, leading to subsequent GSH oxidation (Franco et al. 2007; Mori et al. 2007). This event seems to be also important in terms of protein oxidation, where ROS generated from MeHg can modulate the redox state of proteins, thus affecting their function. A classical example of such phenomenon was described by who showed that MeHg induces the generation of hydrogen peroxide (a common endogenous ROS), which downregulates the activity of astrocytic glutamate transporters, culminating in excitotoxicity (Lockman et al. 2001).

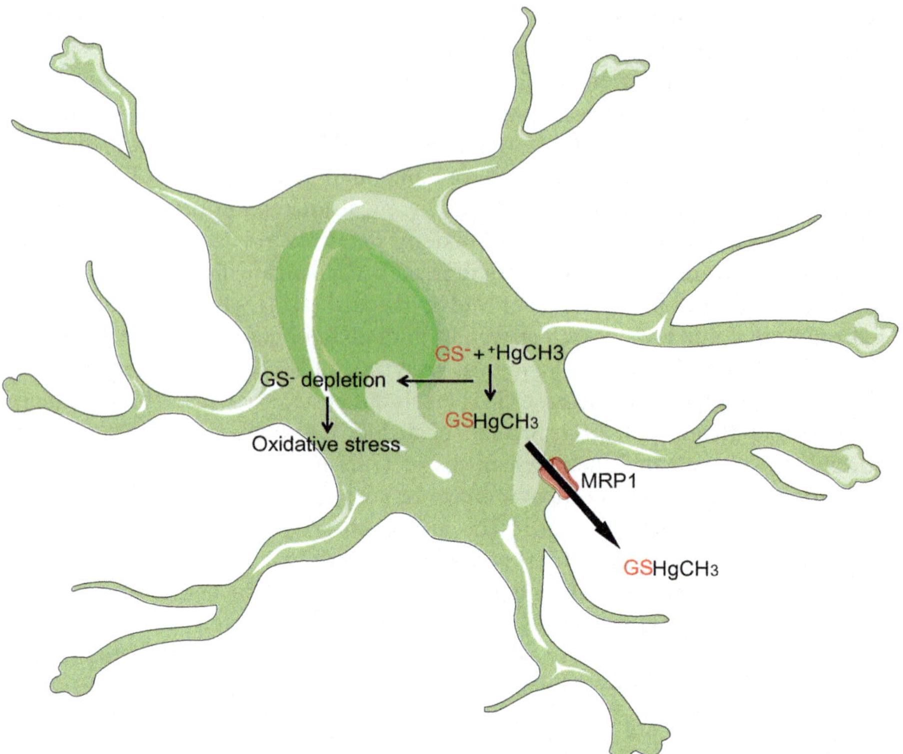

Fig. 2 MeHg-glutathione interaction. MeHg ($^+$HgCH$_3$) interacts with the reduced form of glutathione (GSH or GS$^-$) in the intracellular milieu. This interaction is responsible for decreases in the concentrations of this important thiol antioxidant, leading to oxidative stress. On the other hand, MeHg-GSH interaction leads to the formation of an excretable complex (GSHgCH$_3$), which is transported out of the cell via multidrug resistance protein-1 (MRP1)

Of note, developmental models of neurotoxicity have also pointed to GSH and enzymes involved in the GSH antioxidant system as important targets mediating MeHg neurotoxicity. It is noteworthy that the levels of GSH-related enzymes are significantly increased during the first 3 weeks after birth (early postnatal period) in the mouse brain (Khan and Black 2003). Taking into account that after birth the fetus moves from an in utero hypoxic to a relatively hyperoxic environment with an approximate fourfold elevation in oxygen concentration, the developmental changes (increase in the levels of antioxidants) have been proposed as compensatory mechanisms aimed at protecting the newborn from oxidative stress (Khan and Black 2003). In line with this, Stringari et al. (2008) showed that cerebral GSH levels significantly increased over time during the early postnatal period in mice, but gestational exposure to MeHg caused a dose-dependent inhibition of this developmental event, indicating that prenatal exposure to MeHg disrupts the postnatal development of the GSH antioxidant system in the mouse brain. In agreement, the authors observed increased lipid peroxidation in the brain of MeHg-exposed animals, reinforcing the important antioxidant role of GSH in the CNS (Dringen 2005), as well as the importance of the pro-oxidative properties of MeHg in the induction of oxidative stress.

Secondary Neurotoxic Consequences Resultant from the MeHg's Pro-oxidative Properties

As previously discussed, MeHg is able to interact with nucleophilic groups from both high- and low-molecular-weight biomolecules. This interaction leads to mis-balances in the structure and/or function of these molecules [proteins (receptors, transporters, enzymes, structural proteins), lipids (i.e., membrane constituents and intracellular messengers), and nucleic acids (i.e., DNA)], significantly contributing to MeHg's neurotoxicity. The following paragraphs cover secondary deleterious events resultant (at least partially) from the interaction of MeHg with nucleophilic groups from biomolecules, with a particular emphasis on neurotoxic events. Changes in the homeostasis of neurotransmission and antioxidant defenses are particularly discussed.

MeHg Impairs Neurotransmission: Focus on Glutamate

Taking into account the ubiquitous distribution of sulfhydryl- and selenohydryl-containing proteins in the nervous system, as well as the importance of the proper redox state in such proteins to allow for correct functioning, a *generalized* toxic effect of MeHg toward different systems should be expected. In line with this, MeHg has been reported to disrupt the homeostasis of different neurotransmission systems, such as the glutamatergic (for a review, see Aschner et al. 2007), the GABAergic (Basu et al. 2010), the dopaminergic (Daré et al., 2003), and the cholinergic (Von Burg et al. 1980; Roda et al. 2008).

With respect to the dopaminergic system, evidence shows that dopamine receptors present decreased functionality in the striatum of MeHg-exposed rats (Daré et al. 2003). Moreover, a recent study by Bridges et al. (2016) showed that MeHg-exposed fish during different developmental periods presented significant changes in dopamine concentrations in embryos, as well as in the telencephalon of adult brains, which were paralleled by significant decreases in monoamine oxidase activity in both embryos and brain tissue. Concerning the cholinergic system, the inhibitory effect of MeHg toward rat brain muscarinic receptors was reported several years ago (Von Burg et al. 1980). In agreement, developmental exposure to MeHg caused delayed MeHg exposure-related effect on M2- and M3-muscarinic receptors in the cerebellum of rats (Roda et al. 2008). MeHg also disrupts GABAergic homeostasis: in young rats chronically exposed to MeHg during the postnatal period, neurological impairment was paralleled by a significant reduction in the activity of glutamic acid decarboxylase (GAD; responsible for GABA synthesis from glutamate) in the occipital cortex, frontal cortex, and caudate-putamen (O'Kusky and McGeer 1985). The authors suggest the involvement of GABAergic neurons in MeHg-induced lesions of the cerebral cortex and neostriatum. In addition, in an in vivo experimental model with captive juvenile mink, Basu et al. (2010) showed that long-term (89 days) MeHg-exposed animals showed concentration-dependent decreases in [(3)H]-muscimol binding to GABA(A) receptors and GABA-T activity in several brain regions, with reductions as great as 94% (for GABA(A) receptor levels) and 71% (for GABA-T activity) measured in the brain stem and basal ganglia, indicating that chronic exposure to MeHg likely disrupts the transmission of GABA, the main inhibitory neurotransmitter in the mammalian nervous system.

Even though the GABAergic, dopaminergic, and cholinergic neurotransmitter systems have been pointed as potential targets for MeHg (see above), most of the available data concerning the deleterious effects of MeHg toward a neurotransmitter system is related glutamate. Glutamate is the most important excitatory neurotransmitter in the mammalian CNS, serving crucial roles on development, learning, memory, and response to injury (Fonnum 1984). Due to its direct and indirect pro-oxidative properties, MeHg increases extracellular glutamate levels, which result from both inhibition of glutamate uptake (Aschner et al. 2000; Brookes and Kristt 1989) and stimulation of its release into the synaptic cleft (Reynolds and Racz 1987), culminating in excitotoxic events (Aschner et al. 2007b). Overactivation of the NMDA subtype glutamate receptors leads to an increased Na + and Ca2+ influx, which is associated with the generation of oxidative stress and neurotoxicity (Lafon-Cazal et al. 1993). Indeed, glutamate-mediated increased intracellular Ca^{2+} concentrations leads to increased nitric oxide production (due to activation of neuronal nitric oxide synthase), as well as to mitochondrial collapse (Farina et al. 2011a). Notably, MeHg-induced Ca^{2+} and glutamate dyshomeostasis and MeHg-induced ROS generation (oxidative stress) are events that contribute independently to neurotoxicity but also represent interconnected phenomena affecting each other. Figure 3 depicts the relationship between glutamate and calcium dyshomeostasis and oxidative stress in MeHg-mediating neurotoxicity.

MeHg Induces Lipid and Nucleic Acid Oxidation

Both in vitro and in vivo studies (Andersen and Andersen 1993; Yin et al. 2007; Carvalho et al. 2007) have pointed to lipid peroxidation as an important consequence of MeHg's pro-oxidative effects. Lipid peroxidation is a free-radical-mediated chain of reactions that, once initiated, results in an oxidative deterioration of polyunsaturated lipids. The most common targets are components of biological membranes (Rosenblum et al. 1989). However, it is important to mention that the initiation and propagation of lipid peroxidations normally requires a free radical able to abstract the hydrogen atom of a lipid, as well as molecular oxygen

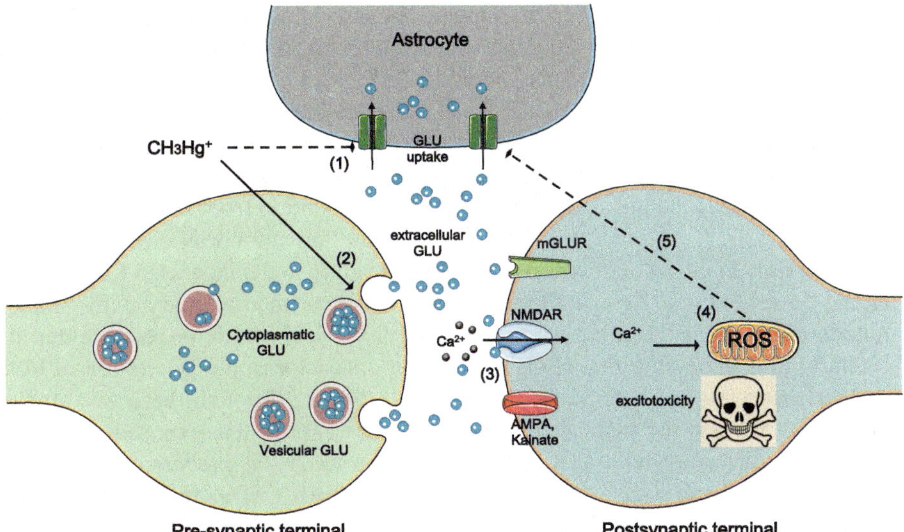

Fig. 3 Interplay between glutamate and calcium dyshomeostasis with oxidative stress in MeHg-mediating neurotoxicity. MeHg-induced inhibition of astrocyte glutamate uptake (event 1; Aschner et al. 2000) and stimulation of glutamate release (event 2; Farina et al. 2003a, b) lead to elevated extracellular glutamate levels. High levels of extracellular glutamate can overactivate its receptors, such as the ionotropic *N*-methyl D-aspartate-type glutamate receptor (event 3). Overactivation of the *N*-methyl D-aspartate-type glutamate receptor leads to an exaggerated increase in the influx of Ca^{2+} into postsynaptic neurons, resulting in excitotoxicity. Ca^{2+} can be buffered by mitochondria, leading to increased ROS generation (event 4), which contributes to decrease the activity of astrocyte glutamate transporters (event 5) *GLU* glutamate, *CH₃Hg⁺* MeHg, *ROS* reactive oxygen species, *NMDAR* N-methyl D-aspartate receptor, *mGLUR* metabotropic glutamate receptor

(Rosenblum et al. 1989). In this scenario, MeHg is unlikely to cause lipid peroxidation through direct interaction with lipids. Conversely, reactive species (i.e., H_2O_2, superoxide anion, nitric oxide) generated (or poorly detoxified) after MeHg exposure seem to be the main players in MeHg-induced lipid peroxidation.

The increased levels of H_2O_2 observed after MeHg exposure represent consequences of different phenomena, such as MeHg's inhibitory effects toward glutathione peroxidases (GPxs; Farina et al. 2009; Franco et al. 2009), which are important enzymes involved in peroxide disposal by means of glutathione (GSH). GPxs represent a family of selenoproteins whose catalytic activity (peroxide detoxification) depends on the reducing power of a selenol group located at the active site (Brigelius-Flohé 2006). Due to the extremely high affinity of MeHg for selenol groups (see above, item 2), the decreased GPx activity after MeHg exposures has been attributed to direct inhibitory events (Farina et al. 2009).

Another mechanism related to the increased H_2O_2 levels after MeHg exposure appears to be the direct hampering effect of this toxicant toward the entire GSH antioxidant system. In addition to the direct depletion of reduced GSH by MeHg, which certainly contributes to the decreased detoxification of H_2O_2 by GSH-dependent peroxidases, MeHg also impedes the physiological maturation of several enzymes involved with GSH metabolism, thus leading to increased levels of brain H_2O_2 and lipid peroxidation (Stringari et al. 2008).

In addition to the decreased H_2O_2 detoxification induced by MeHg (Stringari et al. 2008; Farina et al. 2009; Usuki et al. 2011), increased H_2O_2 generation also represents an important mechanism by which this toxicant leads to higher ROS levels. In an experimental study with isolated mitochondria from the rat cerebellum, Mori et al. (2007) observed that MeHg affects the mitochondrial electron transfer chain (mainly at the level of complexes II–III), leading to the increased formation of H_2O_2. Corroborating these findings, an in vitro experimental study with isolated mitochondria from the mouse brain showed that MeHg toxicity was blunted by catalase, thus indicating that H_2O_2 is an important factor in the generation of ROS in MeHg-exposed mitochondria (Franco et al. 2007). Although these two studies have demonstrated the increased generation of H_2O_2 in MeHg-exposed mitochondria, the actual contribution of superoxide anion (an H_2O_2 precursor) in such an event requires further investigation. While it is known that increased H_2O_2 levels represent a consequence of MeHg exposure, the precise role of this molecule in mediating MeHg-induced oxidative damage has not yet been fully determined. However, an interesting experimental study showed that catalase, which detoxifies H_2O_2, was able to abolish the inhibitory effects of MeHg on glutamate transport in cultured astrocytes (Allen et al. 2001), indicating that H_2O_2 is responsible, at least in part, for some toxic effects induced by MeHg. This notion was corroborated by a study from Franco et al. (2007), who observed that MeHg-induced H_2O_2 generation was responsible for the mitotoxic effects elicited by this compound.

As already mentioned, increased levels of reactive species other than H_2O_2 have also been observed after MeHg exposure. Particularly, increased production of superoxide anion and nitric oxide has been reported as a consequence of MeHg exposure under both in vitro and in vivo conditions. These molecules (superoxide anion and nitric oxide) seem to be responsible (at least partially) for oxidative changes in lipids because superoxide anion is a direct precursor of H_2O_2 (McCord and Fridovich 1969) and because nitric oxide can directly react with superoxide anion, generating peroxynitrite (Darley-Usmar et al. 1992), which has the ability to promote lipid peroxidation reactions (Radi et al. 1991).

The aforementioned evidences clearly show that lipid peroxidation represents an important consequence of MeHg-exposure. Protein oxidation (specifically related to the modulation of the redox state of thiols and selenols, discussed in item 2) is also pivotal in MeHg toxicity. In addition, pro-oxidative changes toward nucleic acids (specially DNA) have also reported as a consequence of MeHg exposure (Belletti et al. 2002; Glaser et al. 2014; Feng et al. 2016 10.1007/s12035-015-9643-y). Even though molecular mechanisms concerning MeHg-induced DNA oxidation are scarce (or absent) in the literature, it is reasonable to suppose that it shares mutual aspects with MeHg-induced lipid peroxidation. This is assumed taking into account that hydroxyl radical (produced by H_2O_2 in the presence of Fe^{2+}) and peroxynitrite (whose levels may also increase after MeHg exposure) have been pointed as important molecules leading to DNA oxidation (Aust and Eveleigh 1999).

Linking Molecular and Behavioral Outcomes

Even though human exposure to MeHg represents a ubiquitous phenomenon, two specific and tragic episodes – Japan (Minamata Bay) and in Iraq – were instrumental to understand several aspects concerning the toxicology of MeHg in humans (Bakir et al. 1973; Eto and Takeuchi 1977). In both episodes, the most important symptoms observed in humans highly exposed to MeHg were visual impairment, hearing impairment, cerebellar ataxia, hyperkinesia, psychiatric symptomatology, and somatosensory disturbances (numbness and paresthesia) (Bakir et al. 1973; Ekino et al. 2007). Of note, experimental studies with MeHg-exposed animals have also detected some of these aforementioned symptoms (for a review, see Farina et al. 2011b). In this scenario, one might ask whether these symptoms are necessarily linked to the pro-oxidative properties of MeHg, as well as their molecular consequences (lipid peroxidation, disruption of glutamate and/or calcium homeostasis, etc.). It is very difficult to answer this question based on epidemiological studies with humans. However, experimental evidence indicates that, at least for the case of motor impairments, the answer is *yes*.

Cerebellar ataxia (inability to coordinate movements due to cerebellar damage) is commonly observed in MeHg-exposed animals (Hirayama et al. 1985; Dietrich et al. 2005). In experimental models, cerebellar ataxia-related behaviors, which have generally been evaluated through the open field, rotarod, or beam walking tasks, have been linked to decreased cerebellar GPx activity, increased cerebellar lipid peroxidation, changes in calcium homeostasis, and changed cerebellar thiol status (Farina et al. 2005; Franco et al. 2006; Hoffman and Newland 2016). Of note, neuroprotective agents, which prevented MeHg-induced motor impairment in mice, also prevented MeHg-induced cerebellar lipid peroxidation (Farina et al. 2005), suggesting that MeHg-induced cerebellar lipid peroxidation is related to the motor impairment observed in MeHg-exposed animals.

MeHg-induced motor impairments were delayed by nimodipine, an L-type calcium channel blocker (Shen et al. 2016), suggesting that changes in calcium homeostasis are responsible, at least partially, for MeHg-induced behavioral changes (impaired motor performance). In agreement with this recent study, Sakamoto and collaborators (1996) observed that different Ca^{2+} channel blockers (flunarizine, nifedipine, nicardipine, and verapamil) prevented signs of neurotoxicity in the rats treated with MeHg.

In an experimental study using MeHg-exposed lobster cockroach *Nauphoeta cinerea*, Adedara and collaborators (2016) observed that MeHg caused locomotor deficits, which were paralleled by decreased acetylcholinesterase activity and increased dichlorofluorescein oxidation and lipid peroxidation. Co-exposure to luteolin, a polyphenolic compound with antioxidant properties, reversed the MeHg-induced locomotor deficits and enhanced the exploratory profiles of MeHg-exposed cockroaches as well as reversed the MeHg-induced acetylcholinesterase activity inhibition and decreased dichlorofluorescein oxidation and lipid peroxidation levels (Adedara et al. 2016), indicating a relationship between the pro-oxidative effects of MeHg and the behavioral consequences of MeHg-exposures.

Based on the previous-mentioned evidences, it is likely that behavioral symptoms resultant from MeHg exposure (particularly, motor impairment) are linked, at least in part, to the pro-oxidative properties of MeHg, as well as their molecular consequences. Concerning the linking between molecular and behavioral outcomes after MeHg exposure, studies on the relationship between molecular parameters and behavioral impairment others that those related to the motor function (i.e., visual impairment, cognitive skills, etc.) are scarce.

Concluding Remarks and Perspectives

As discussed in this chapter, MeHg is an electrophilic compound that interacts with sulfhydryl- and selenohydryl-containing proteins and low-molecular-weight molecules in biological systems. In the central nervous system, these molecules can act as *scavengers*, enzymes, transporters, receptors, structural components, etc. Consequently, MeHg exposure leads to several secondary deleterious events, such as impaired chemical neurotransmission, changes in the ionic homeostasis, as well as lipid, protein, and nucleic acid oxidation.

Because most of the deleterious effects caused by MeHg seem to be related to its pro-oxidative properties, antioxidant molecules have been reported as protective agents against MeHg-induced neurotoxicity (Farina et al. 2005; Wagner et al. 2010; Wormser et al. 2012; Adedara et al. 2016). Anyway, it is important to note that most data on such protective properties come from experimental studies. In fact, the current scenario of human exposure to MeHg, which is mostly linked to the long-term exposure to relatively low MeHg concentrations due to the ingestion of contaminated fish, does not favor the development of epidemiological studies concerning the potential beneficial role of antioxidants against MeHg-induced neurotoxicity. Taking into account that such kind of study is extremely important, but scarce, further research concerning the potential beneficial role of antioxidants against MeHg-induced neurotoxicity in humans is well warranted.

From a nutritional point of view, it is important to mention that both selenium and ω-3 polyunsaturated fatty acids, which represent nutritional constituents of fish, present antioxidant properties and have displayed protective effects against MeHg-induced toxicity under experimental conditions (Kaur et al. 2007, Kaur et al. 2008; Glaser et al. 2010a, b). In this regard, the development of studies comparing the potential hazardous/beneficial effects of the ingestion of different fish species, with particular focus on the ratio between MeHg and potential protective antioxidant components (i.e., selenium and ω-3 polyunsaturated fatty acids), seems to represent a promised perspective for future epidemiological studies, which could offer quantitative nutritional recommendations concerning fish consumption.

Acknowledgments The author would like to thank the colleagues/coauthors who have contributed to several studies referenced in this chapter. These studies were funded in part by grants from the Conselho Nacional de Desenvolvimento Científico e Tecnológico (CNPq), Coordenação de Aperfeiçoamento de Pessoal de Nível Superior (CAPES), and Fundação de Amparo à Pesquisa e Inovação do Estado de Santa Catarina (FAPESC).

References

Adedara IA, Rosemberg DB, Souza DO, Farombi EO, Aschner M, Rocha JB. Neuroprotection of luteolin against methylmercury-induced toxicity in lobster cockroach *Nauphoeta cinerea*. Environ Toxicol Pharmacol. 2016;42:243–51. doi:10.1016/j.etap.2016.02.001.

Allen JW, Mutkus LA, Aschner M. Methylmercury-mediated inhibition of 3H-D-aspartate transport in cultured astrocytes is reversed by the antioxidant catalase. Brain Res. 2001;902(1):92–100.

Andersen HR, Andersen O. Effects of dietary alpha-tocopherol and beta-carotene on lipid peroxidation induced by methyl mercuric chloride in mice. Pharmacol Toxicol. 1993;73(4):192–201.

Araie H, Shiraiwa Y. Selenium utilization strategy by microalgae. Molecules. 2009;14(12):4880–91. doi:10.3390/molecules14124880.

Aschner M, Clarkson TW. Uptake of methylmercury in the rat brain: effects of amino acids. Brain Res. 1988;462(1):31–9.

Aschner M, Yao CP, Allen JW, Tan KH. Methylmercury alters glutamate transport in astrocytes. Neurochem Int. 2000;37(2–3):199–206.

Aschner M, Syversen T, Souza DO, Rocha JB, Farina M. Involvement of glutamate and reactive oxygen species in methylmercury neurotoxicity. Braz J Med Biol Res. 2007;40(3):285–91.

Aust AE, Eveleigh JF. Mechanisms of DNA oxidation. Proc Soc Exp Biol Med. 1999;222(3):246–52.

Bakir F, Damluji SF, Amin-Zaki L, Murtadha M, Khalidi A, al-Rawi NY, Tikriti S, Dahahir HI, Clarkson TW, Smith JC, Doherty RA. Methylmercury poisoning in Iraq. Science. 1973;181(4096):230–41.

Basu N, Scheuhammer AM, Rouvinen-Watt K, Evans RD, Trudeau VL, Chan LH. In vitro and whole animal evidence that methylmercury disrupts GABAergic systems in discrete brain regions in captive mink. Comp Biochem Physiol C Toxicol Pharmacol. 2010;151(3):379–85. doi:10.1016/j.cbpc.2010.01.001.

Belletti S, Orlandini G, Vettori MV, Mutti A, Uggeri J, Scandroglio R, Alinovi R, Gatti R. Time course assessment of methylmercury effects on C6 glioma cells: submicromolar concentrations induce oxidative DNA damage and apoptosis. J Neurosci Res. 2002;70(5):703–11. doi:10.1002/jnr.10419.

Branco V, Canário J, Holmgren A, Carvalho C. Inhibition of the thioredoxin system in the brain and liver of zebraseabreams exposed to waterborne methylmercury. Toxicol Appl Pharmacol. 2011;251(2):95–103. doi:10.1016/j.taap.2010.12.005.

Branco V, Canário J, Lu J, Holmgren A, Carvalho C. Mercury and selenium interaction in vivo: effects on thioredoxin reductase and glutathione peroxidase. Free Radic Biol Med. 2012;52(4):781–93. doi:10.1016/j.freeradbiomed.2011.12.002.

Bridges K, Venables B, Roberts A. Effects of dietary methylmercury on the dopaminergic system of adult fathead minnows and their offspring. Environ Toxicol Chem. 2016; doi:10.1002/etc.3630.

Brigelius-Flohe R. Glutathione peroxidases and redox-regulated transcription factors. Biol Chem. 2006;387(10–11):1329–35. doi:10.1515/BC.2006.166.

Brookes N, Kristt DA. Inhibition of amino acid transport and protein synthesis by HgCl2 and methylmercury in astrocytes: selectivity and reversibility. J Neurochem. 1989;53(4):1228–37.

Carvalho MC, Franco JL, Ghizoni H, Kobus K, Nazari EM, Rocha JB, Nogueira CW, Dafre AL, Muller YM, Farina M. Effects of 2,3-dimercapto-1-propanesulfonic acid (DMPS) on methylmercury-induced locomotor deficits and cerebellar toxicity in mice. Toxicology. 2007;239(3):195–203. doi:10.1016/j.tox.2007.07.009.

Carvalho CM, Matos AI, Mateus ML, Santos AP, Batoreu MC. High-fish consumption and risk prevention: assessment of exposure to methylmercury in Portugal. J Toxicol Environ Health A. 2008;71(18):1279–88. doi:10.1080/15287390801989036.

Clarkson TW. The three modern faces of mercury. Environ Health Perspect. 2002;110(Suppl 1):11–23.

Clarkson TW, Magos L, Myers GJ. The toxicology of mercury--current exposures and clinical manifestations. N Engl J Med. 2003;349(18):1731–7. doi:10.1056/NEJMra022471.

Compeau GC, Bartha R. Sulfate-reducing bacteria: principal methylators of mercury in anoxic estuarine sediment. Appl Environ Microbiol. 1985;50(2):498–502.

Dare E, Fetissov S, Hokfelt T, Hall H, Ogren SO, Ceccatelli S. Effects of prenatal exposure to methylmercury on dopamine-mediated locomotor activity and dopamine D2 receptor binding. Naunyn Schmiedeberg's Arch Pharmacol. 2003;367(5):500–8. doi:10.1007/s00210-003-0716-5.

Darley-Usmar VM, Hogg N, O'Leary VJ, Wilson MT, Moncada S. The simultaneous generation of superoxide and nitric oxide can initiate lipid peroxidation in human low density lipoprotein. Free Radic Res Commun. 1992;17(1):9–20.

Davis LE, Kornfeld M, Mooney HS, Fiedler KJ, Haaland KY, Orrison WW, Cernichiari E, Clarkson TW. Methylmercury poisoning: long-term clinical, radiological, toxicological, and pathological studies of an affected family. Ann Neurol. 1994;35(6):680–8. doi:10.1002/ana.410350608.

Dietrich MO, Mantese CE, Anjos GD, Souza DO, Farina M. Motor impairment induced by oral exposure to methylmercury in adult mice. Environ Toxicol Pharmacol. 2005;19(1):169–75. doi:10.1016/j.etap.2004.07.004.

Dorea JG. The neurological effects of prenatal and postnatal exposure to mercury need to include ethylmercury. Chemosphere. 2015;139:667–8. doi:10.1016/j.chemosphere.2014.06.045.

Dringen R. Oxidative and antioxidative potential of brain microglial cells. Antioxid Redox Signal. 2005;7(9–10):1223–33. doi:10.1089/ars.2005.7.1223.

Dutczak WJ, Ballatori N. Transport of the glutathione-methylmercury complex across liver canalicular membranes on reduced glutathione carriers. J Biol Chem. 1994;269(13):9746–51.

Ekino S, Susa M, Ninomiya T, Imamura K, Kitamura T. Minamata disease revisited: an update on the acute and chronic manifestations of methyl mercury poisoning. J Neurol Sci. 2007;262(1–2):131–44. doi:10.1016/j.jns.2007.06.036.

Eto K, Takeuchi T. Pathological changes of human sural nerves in Minamata disease (methylmercury poisoning). Light and electron microscopic studies. Virchows Arch B Cell Pathol. 1977;23(2):109–28.

Farina M, Dahm KC, Schwalm FD, Brusque AM, Frizzo ME, Zeni G, Souza DO, Rocha JB. Methylmercury increases glutamate release from brain synaptosomes and glutamate uptake by cortical slices from suckling rat pups: modulatory effect of ebselen. Toxicol Sci. 2003a;73(1):135–40. doi:10.1093/toxsci/kfg058.

Farina M, Frizzo ME, Soares FA, Schwalm FD, Dietrich MO, Zeni G, Rocha JB, Souza DO. Ebselen protects against methylmercury-induced inhibition of glutamate uptake by cortical slices from adult mice. Toxicol Lett. 2003b;144(3):351–7.

Farina M, Franco JL, Ribas CM, Meotti FC, Missau FC, Pizzolatti MG, Dafre AL, Santos AR. Protective effects of *Polygala paniculata* extract against methylmercury-induced neurotoxicity in mice. J Pharm Pharmacol. 2005;57(11):1503–8. doi:10.1211/jpp.57.11.0017.

Farina M, Campos F, Vendrell I, Berenguer J, Barzi M, Pons S, Sunol C. Probucol increases glutathione peroxidase-1 activity and displays long-lasting protection against methylmercury toxicity in cerebellar granule cells. Toxicol Sci. 2009;112(2):416–26. doi:10.1093/toxsci/kfp219.

Farina M, Aschner M, Rocha JB. Oxidative stress in MeHg-induced neurotoxicity. Toxicol Appl Pharmacol. 2011a;256(3):405–17. doi:10.1016/j.taap.2011.05.001.

Farina M, Rocha JB, Aschner M. Mechanisms of methylmercury-induced neurotoxicity: evidence from experimental studies. Life Sci. 2011b;89(15–16):555–63. doi:10.1016/j.lfs.2011.05.019.

Feng S, Xu Z, Wang F, Yang T, Liu W, Deng Y, Xu B. Sulforaphane prevents methylmercury-induced oxidative damage and Excitotoxicity through activation of the Nrf2-ARE pathway. Mol Neurobiol. 2016; doi:10.1007/s12035-015-9643-y.

Fonnum F. Glutamate: a neurotransmitter in mammalian brain. J Neurochem. 1984;42(1):1–11.

Franco JL, Teixeira A, Meotti FC, Ribas CM, Stringari J, Garcia Pomblum SC, Moro AM, Bohrer D, Bairros AV, Dafre AL, Santos AR, Farina M. Cerebellar thiol status and motor deficit after lactational exposure to methylmercury. Environ Res. 2006;102(1):22–8. doi:10.1016/j. envres.2006.02.003.

Franco JL, Braga HC, Stringari J, Missau FC, Posser T, Mendes BG, Leal RB, Santos AR, Dafre AL, Pizzolatti MG, Farina M. Mercurial-induced hydrogen peroxide generation in mouse brain mitochondria: protective effects of quercetin. Chem Res Toxicol. 2007;20(12):1919–26. doi:10.1021/tx7002323.

Franco JL, Posser T, Dunkley PR, Dickson PW, Mattos JJ, Martins R, Bainy AC, Marques MR, Dafre AL, Farina M. Methylmercury neurotoxicity is associated with inhibition of the antioxidant enzyme glutathione peroxidase. Free Radic Biol Med. 2009;47(4):449–57. doi:10.1016/j. freeradbiomed.2009.05.013.

Freitas AJ, Rocha JB, Wolosker H, Souza DO. Effects of Hg2+ and CH3Hg+ on Ca2+ fluxes in rat brain microsomes. Brain Res. 1996;738(2):257–64.

Glaser V, Leipnitz G, Straliotto MR, Oliveira J, dos Santos V V, Wannmacher CM, de Bem AF, Rocha JB, Farina M, Latini A. Oxidative stress-mediated inhibition of brain creatine kinase activity by methylmercury. Neurotoxicology. 2010a;31(5):454–60. doi:10.1016/j.neuro.2010.05.012.

Glaser V, Nazari EM, Muller YM, Feksa L, Wannmacher CM, Rocha JB, de Bem AF, Farina M, Latini A. Effects of inorganic selenium administration in methylmercury-induced neurotoxicity in mouse cerebral cortex. Int J Dev Neurosci. 2010b;28(7):631–7. doi:10.1016/j. ijdevneu.2010.07.225.

Glaser V, Martins Rde P, Vieira AJ, Oliveira Ede M, Straliotto MR, Mukdsi JH, Torres AI, de Bem AF, Farina M, da Rocha JB, De Paul AL, Latini A. Diphenyl diselenide administration enhances cortical mitochondrial number and activity by increasing hemeoxygenase type 1 content in a methylmercury-induced neurotoxicity mouse model. Mol Cell Biochem. 2014;390(1–2):1–8. doi:10.1007/s11010-013-1870-9.

Grandjean P, Landrigan PJ. Neurobehavioural effects of developmental toxicity. Lancet Neurol. 2014;13(3):330–8. doi:10.1016/S1474-4422(13)70278-3.

Grandjean P, Weihe P, White RF, Debes F, Araki S, Yokoyama K, Murata K, Sorensen N, Dahl R, Jorgensen PJ. Cognitive deficit in 7-year-old children with prenatal exposure to methylmercury. Neurotoxicol Teratol. 1997;19(6):417–28.

Hintelmann H. Organomercurials. Their formation and pathways in the environment. Met Ions Life Sci. 2010;7:365–401. doi:10.1039/BK9781847551771-00365.

Hirayama K, Inouye M, Fujisaki T. Alteration of putative amino acid levels and morphological findings in neural tissues of methylmercury-intoxicated mice. Arch Toxicol. 1985;57(1):35–40.

Ho T-L. Hard and soft acids and bases principle in organic chemistry. 1st edn. Academic. 1977. eBook ISBN: 9780323140966. Published Date: 28th January 1977.

Hoffman DJ, Newland MC. A microstructural analysis distinguishes motor and motivational influences over voluntary running in animals chronically exposed to methylmercury and nimodipine. Neurotoxicology. 2016;54:127–39. doi:10.1016/j.neuro.2016.04.009.

Kajiwara Y, Yasutake A, Adachi T, Hirayama K. Methylmercury transport across the placenta via neutral amino acid carrier. Arch Toxicol. 1996;70(5):310–4.

Kaur P, Schulz K, Aschner M, Syversen T. Role of docosahexaenoic acid in modulating methylmercury-induced neurotoxicity. Toxicol Sci. 2007;100(2):423–32. doi:10.1093/toxsci/ kfm224.

Kaur P, Heggland I, Aschner M, Syversen T. Docosahexaenoic acid may act as a neuroprotector for methylmercury-induced neurotoxicity in primary neural cell cultures. Neurotoxicology. 2008;29(6):978–87. doi:10.1016/j.neuro.2008.06.004.

Kaur P, Aschner M, Syversen T. Biochemical factors modulating cellular neurotoxicity of methylmercury. J Toxicol. 2011;2011:721987. doi:10.1155/2011/721987.

Kessler R. The Minamata convention on mercury: a first step toward protecting future generations. Environ Health Perspect. 2013;121(10):A304–9.

Khan JY, Black SM. Developmental changes in murine brain antioxidant enzymes. Pediatr Res. 2003;54(1):77–82. doi:10.1203/01.PDR.0000065736.69214.20.

Kim JY, Park HS, Kang SI, Choi EJ, Kim IY. Redox regulation of cytosolic glycerol-3-phosphate dehydrogenase: Cys(102) is the target of the redox control and essential for the catalytic activity. Biochim Biophys Acta. 2002;1569(1–3):67–74.

Kung MP, Kostyniak P, Olson J, Malone M, Roth JA. Studies of the in vitro effect of methylmercury chloride on rat brain neurotransmitter enzymes. J Appl Toxicol. 1987;7(2):119–21.

Lafon-Cazal M, Pietri S, Culcasi M, Bockaert J. NMDA-dependent superoxide production and neurotoxicity. Nature. 1993;364(6437):535–7. doi:10.1038/364535a0.

Lobanov AV, Hatfield DL, Gladyshev VN. Eukaryotic selenoproteins and selenoproteomes. Biochim Biophys Acta. 2009;1790(11):1424–8. doi:10.1016/j.bbagen.2009.05.014.

Lockman PR, Roder KE, Allen DD. Inhibition of the rat blood-brain barrier choline transporter by manganese chloride. J Neurochem. 2001;79(3):588–94.

LoPachin RM, Barber DS. Synaptic cysteine sulfhydryl groups as targets of electrophilic neurotoxicants. Toxicol Sci. 2006;94(2):240–55. doi:10.1093/toxsci/kfl066.

LoPachin RM, Gavin T. Reactions of electrophiles with nucleophilic thiolate sites: relevance to pathophysiological mechanisms and remediation. Free Radic Res. 2016;50(2):195–205. doi:10.3109/10715762.2015.1094184.

Lu J, Holmgren A. Selenoproteins. J Biol Chem. 2009;284(2):723–7. doi:10.1074/jbc.R800045200.

McCord JM, Fridovich I. The utility of superoxide dismutase in studying free radical reactions. I. Radicals generated by the interaction of sulfite, dimethyl sulfoxide, and oxygen. J Biol Chem. 1969;244(22):6056–63.

Mokrzan EM, Kerper LE, Ballatori N, Clarkson TW. Methylmercury-thiol uptake into cultured brain capillary endothelial cells on amino acid system L. J Pharmacol Exp Ther. 1995;272(3):1277–84.

Mori K, Yoshida K, Nakagawa Y, Hoshikawa S, Ozaki H, Ito S, Watanabe C. Methylmercury inhibition of type II 5′-deiodinase activity resulting in a decrease in growth hormone production in GH3 cells. Toxicology. 2007;237(1–3):203–9. doi:10.1016/j.tox.2007.05.012.

Murata K, Weihe P, Budtz-Jorgensen E, Jorgensen PJ, Grandjean P. Delayed brainstem auditory evoked potential latencies in 14-year-old children exposed to methylmercury. J Pediatr. 2004;144(2):177–83. doi:10.1016/j.jpeds.2003.10.059.

Ni M, Li X, Yin Z, Sidoryk-Wegrzynowicz M, Jiang H, Farina M, Rocha JB, Syversen T, Aschner M. Comparative study on the response of rat primary astrocytes and microglia to methylmercury toxicity. Glia. 2011;59(5):810–20. doi:10.1002/glia.21153.

Ninomiya T, Imamura K, Kuwahata M, Kindaichi M, Susa M, Ekino S. Reappraisal of somatosensory disorders in methylmercury poisoning. Neurotoxicol Teratol. 2005;27(4):643–53. doi:10.1016/j.ntt.2005.03.008.

O'Kusky JR, McGeer EG. Methylmercury poisoning of the developing nervous system in the rat: decreased activity of glutamic acid decarboxylase in cerebral cortex and neostriatum. Brain Res. 1985;353(2):299–306.

Osawa M, Magos L. The chemical form of the methylmercury complex in the bile of the rat. Biochem Pharmacol. 1974;23(13):1903–5.

Penglase S, Hamre K, Ellingsen S. Selenium prevents downregulation of antioxidant selenoprotein genes by methylmercury. Free Radic Biol Med. 2014;75:95–104. doi:10.1016/j.freeradbiomed.2014.07.019.

Powis G, Mustacich D, Coon A. The role of the redox protein thioredoxin in cell growth and cancer. Free Radic Biol Med. 2000;29(3-4):312–22.

Rabenstein DL, Evans CA. The mobility of methylmercury in biological systems. Bioinorg Chem. 1978;8(2):107–101,104.

Radi R, Beckman JS, Bush KM, Freeman BA. Peroxynitrite-induced membrane lipid peroxidation: the cytotoxic potential of superoxide and nitric oxide. Arch Biochem Biophys. 1991;288(2):481–7.

Rahola T, Hattula T, Korolainen A, Miettinen JK. Elimination of free and protein-bound ionic mercury (20Hg2+) in man. Ann Clin Res. 1973;5(4):214–9.

Reynolds JN, Racz WJ. Effects of methylmercury on the spontaneous and potassium-evoked release of endogenous amino acids from mouse cerebellar slices. Can J Physiol Pharmacol. 1987;65(5):791–8.

Richardson RJ, Murphy SD. Effect of glutathione depletion on tissue deposition of methylmercury in rats. Toxicol Appl Pharmacol. 1975;31(3):505–19.

Roda E, Coccini T, Acerbi D, Castoldi A, Bernocchi G, Manzo L. Cerebellum cholinergic muscarinic receptor (subtype-2 and -3) and cytoarchitecture after developmental exposure to methylmercury: an immunohistochemical study in rat. J Chem Neuroanat. 2008;35(3):285–94. doi:10.1016/j.jchemneu.2008.01.003.

Roos DH, Puntel RL, Farina M, Aschner M, Bohrer D, Rocha JB, de Vargas Barbosa NB. Modulation of methylmercury uptake by methionine: prevention of mitochondrial dysfunction in rat liver slices by a mimicry mechanism. Toxicol Appl Pharmacol. 2011;252(1):28–35. doi:10.1016/j.taap.2011.01.010.

Rosenblum ER, Gavaler JS, Van Thiel DH. Lipid peroxidation: a mechanism for alcohol-induced testicular injury. Free Radic Biol Med. 1989;7(5):569–77.

Rush T, Liu X, Nowakowski AB, Petering DH, Lobner D. Glutathione-mediated neuroprotection against methylmercury neurotoxicity in cortical culture is dependent on MRP1. Neurotoxicology. 2012;33(3):476–81. doi:10.1016/j.neuro.2012.03.004.

Sakamoto M, Ikegami N, Nakano A. Protective effects of Ca2+ channel blockers against methyl mercury toxicity. Pharmacol Toxicol. 1996;78(3):193–9.

Seres T, Ravichandran V, Moriguchi T, Rokutan K, Thomas JA, Johnston RB Jr. Protein S-thiolation and dethiolation during the respiratory burst in human monocytes. A reversible post-translational modification with potential for buffering the effects of oxidant stress. J Immunol. 1996;156(5):1973–80.

Shanker G, Syversen T, Aschner JL, Aschner M. Modulatory effect of glutathione status and antioxidants on methylmercury-induced free radical formation in primary cultures of cerebral astrocytes. Brain Res Mol Brain Res. 2005;137(1–2):11–22. doi:10.1016/j.molbrainres.2005.02.006.

Shen AN, Cummings C, Hoffman D, Pope D, Arnold M, Newland MC. Aging, motor function, and sensitivity to calcium channel blockers: an investigation using chronic methylmercury exposure. Behav Brain Res. 2016;315:103–14. doi:10.1016/j.bbr.2016.07.049.

Soares FA, Farina M, Santos FW, Souza D, Rocha JB, Nogueira CW. Interaction between metals and chelating agents affects glutamate binding on brain synaptic membranes. Neurochem Res. 2003;28(12):1859–65.

Stringari J, Nunes AK, Franco JL, Bohrer D, Garcia SC, Dafre AL, Milatovic D, Souza DO, Rocha JB, Aschner M, Farina M. Prenatal methylmercury exposure hampers glutathione antioxidant system ontogenesis and causes long-lasting oxidative stress in the mouse brain. Toxicol Appl Pharmacol. 2008;227(1):147–54. doi:10.1016/j.taap.2007.10.010.

Suda I, Takahashi H. Degradation of methyl and ethyl mercury into inorganic mercury by other reactive oxygen species besides hydroxyl radical. Arch Toxicol. 1992;66(1):34–9.

Theunissen PT, Pennings JL, Robinson JF, Claessen SM, Kleinjans JC, Piersma AH. Time-response evaluation by transcriptomics of methylmercury effects on neural differentiation of murine embryonic stem cells. Toxicol Sci. 2011;122(2):437–47. doi:10.1093/toxsci/kfr134.

Usuki F, Fujimura M. Decreased plasma thiol antioxidant barrier and selenoproteins as potential biomarkers for ongoing methylmercury intoxication and an individual protective capacity. Arch Toxicol. 2016;90(4):917–26. doi:10.1007/s00204-015-1528-3.

Usuki F, Yamashita A, Fujimura M. Post-transcriptional defects of antioxidant selenoenzymes cause oxidative stress under methylmercury exposure. J Biol Chem. 2011;286(8):6641–9. doi:10.1074/jbc.M110.168872.

Von Burg R, Northington FK, Shamoo A. Methylmercury inhibition of rat brain muscarinic receptors. Toxicol Appl Pharmacol. 1980;53(2):285–92.

Wagner C, Sudati JH, Nogueira CW, Rocha JB. In vivo and in vitro inhibition of mice thioredoxin reductase by methylmercury. Biometals. 2010;23(6):1171–7. doi:10.1007/s10534-010-9367-4.

Watanabe C, Yin K, Kasanuma Y, Satoh H. In utero exposure to methylmercury and se deficiency converge on the neurobehavioral outcome in mice. Neurotoxicol Teratol. 1999;21(1):83–8.

Wormser U, Brodsky B, Milatovic D, Finkelstein Y, Farina M, Rocha JB, Aschner M. Protective effect of a novel peptide against methylmercury-induced toxicity in rat primary astrocytes. Neurotoxicology. 2012;33(4):763–8. doi:10.1016/j.neuro.2011.12.004.

Yin Z, Milatovic D, Aschner JL, Syversen T, Rocha JB, Souza DO, Sidoryk M, Albrecht J, Aschner M. Methylmercury induces oxidative injury, alterations in permeability and glutamine transport in cultured astrocytes. Brain Res. 2007;1131(1):1–10. doi:10.1016/j.brainres.2006.10.070.

Yin Z, Jiang H, Syversen T, Rocha JB, Farina M, Aschner M. The methylmercury-L-cysteine conjugate is a substrate for the L-type large neutral amino acid transporter. J Neurochem. 2008;107(4):1083–90. doi:10.1111/j.1471-4159.2008.05683.x.

Zemolin AP, Meinerz DF, de Paula MT, Mariano DO, Rocha JB, Pereira AB, Posser T, Franco JL. Evidences for a role of glutathione peroxidase 4 (GPx4) in methylmercury induced neurotoxicity in vivo. Toxicology. 2012;302(1):60–7. doi:10.1016/j.tox.2012.07.013.

Zimmer B, Schildknecht S, Kuegler PB, Tanavde V, Kadereit S, Leist M. Sensitivity of dopaminergic neuron differentiation from stem cells to chronic low-dose methylmercury exposure. Toxicol Sci. 2011;121(2):357–67. doi:10.1093/toxsci/kfr054.

Neurotoxicity of Vanadium

Hilary Afeseh Ngwa, Muhammet Ay, Huajun Jin, Vellareddy Anantharam, Arthi Kanthasamy, and Anumantha G. Kanthasamy

Abstract Vanadium (V) is a transition metal that presents in multiple oxidation states and numerous inorganic compounds and is also an ultra-trace element considered to be essential for most living organisms. Despite being one of the lightest metals, V offers high structural strength and good corrosion resistance and thus has been widely adopted for high-strength steel manufacturing. High doses of V exposure are toxic, and inhalation exposure to V adversely affects the respiratory system. The neurotoxicological properties of V are just beginning to be identified. Recent studies by our group and others demonstrate the neurotoxic potential of this metal in the nigrostriatal system and other parts of the central nervous system (CNS). The neurotoxic effects of V have been mainly attributed to its ability to induce the generation of reactive oxygen species (ROS). It is noteworthy that the neurotoxicity induced by occupational V exposure commonly occurs with co-exposure to other metals, especially manganese (Mn). This review focuses on the chemistry, pharmacology, toxicology, and neurotoxicity of V.

Keywords Vanadium • Neurotoxicity • Metals • Oxidative stress • Toxicology • Neurodegeneration • Parkinson's disease

Introduction

Vanadium (V) is a transition metal which belongs to Group VB of the periodic table, with an atomic weight of 50.9415, an atomic number of 23, and oxidizing states ranging from −1 to +5. Vanadium has many industrial uses, and its contribution to environmental contamination has been steadily growing in recent years. Vanadium is widely distributed in the earth's crust but occurs in low abundance. Vanadium is an essential trace element for normal cell growth but can be toxic when present at higher concentration. It can exist in many oxidation states with many oxyanions and

H.A. Ngwa • M. Ay • H. Jin • V. Anantharam • A. Kanthasamy • A.G. Kanthasamy (✉)
Parkinson's Disorder Research Laboratory, Iowa Center for Advanced Neurotoxicology,
Department of Biomedical Sciences, Iowa State University,
2062 College of Veterinary Medicine Building, Ames, IA 50011, USA
e-mail: akanthas@iastate.edu

© Springer International Publishing AG 2017
M. Aschner and L.G. Costa (eds.), *Neurotoxicity of Metals*, Advances
in Neurobiology 18, DOI 10.1007/978-3-319-60189-2_14

oxycations, which form in solution. The multiple oxidation states, ready hydrolysis, and polymerization confer a level of complexity to the chemistry of vanadium well above that of many metals. Vanadium dissolves in natural waters as the vanadyl ion V(IV) and the vanadate ion V(V). Both species have different nutritional and toxic properties. Studies carried out on yeast cells, for example, have demonstrated that V(V) is a strong inhibitor of the enzyme Na and K-ATPase, while V(IV) appears to be a weaker inhibitor (Patel et al. 1990).

Vanadium is among the list of essential micronutrients required in small quantities for normal metabolism (Ray et al. 2006). It has therefore been incorporated in the formulations and preparations of many multinational pharmaceutical companies (Nutrition Dynamics Inc., Texas, USA; All Nature Pharmaceuticals Inc., USA; and Ranbaxy Pvt. Ltd., Mumbai, India) along with vitamins and other essential trace elements for maintenance of normal health. Although the micronutrients lack pharmacological potencies, they assume a repair function for the essential critical molecules of the cell, such as DNA and proteins (Fenech and Ferguson 2001).

Uses of Vanadium and Its Compounds

Vanadium is used widely in industrial processes, including the production of special steels, temperature-resistant alloys, glass, pigments, and paints, for lining arc welding electrodes and as a catalyst. Its use with nonferrous metals is of particular importance in aircraft construction, atomic energy industry, and space technology (ChemIDPlus, 2016). Vanadium is preferred in the rising production of special steels and temperature-resistant alloys, namely, HSL-A, which is a high-strength, lightweight, and low-cost micro-alloyed steel, because it is one of the lightest metals with an inherent high strength. The characterization of metals in welding fumes by ICP-MS revealed that the V concentration is about 2.5% of whole transition metal content.

Vanadium is a chemical intermediate principally for V alloys and compounds. Vanadium is used (1) as a catalyst for many organic reactions; (2) as an oxidation catalyst in many industrial synthesis processes like automobile catalytic converters; (3) as a catalyst in the production of phthalic anhydride from naphthalene or 2-xylene, maleic anhydride from benzene or n-butane/butene, adipic acid from cyclohexanol/cyclohexanone, and acrylic acid from propane; (4) as a catalyst (ferrovanadium) to oxidize sulfur dioxide in sulfuric acid manufacturing; (5) as a depolarizer to manufacture yellow glass; (6) to manufacture ceramic coloring material, vanadium salts (YVO4), and pesticides; (7) in the inhibition or absorbance of UV transmission in glass; (8) as a photographic developer; (9) in dyeing textiles; and (10) in manufacturing a high capacity battery, namely, the vanadium redox battery, that uses vanadium ions in different oxidation states to store chemical potential energy.

Minor amounts of V are used to produce oxalic acid from cellulose and anthraquinone from anthracene. V is also used to lower the melting point of enamel frits for the coating of aluminum substrates. Further uses are in the making of superconductor magnets, as a corrosion inhibitor in carbon dioxide scrubbing solutions of the Benfield and related processes for the production of hydrogen from hydrocarbons, and as the cathode in primary and secondary (rechargeable) lithium batteries (ChemIDPlus 2016).

Chemistry of Vanadium

Vanadium usually occurs in aqueous solution as vanadate ions, often as polymerized isopolyvanadates, with the exact composition dependent on the protonation and condensation equilibria (Greenwood and Earnshaw 1997). Monomeric V ions are found only in very dilute solutions, since increases in the concentrations of these ions lead to polymerization, especially if the solution is acidic, reducing their bioavailability and associated toxicity (Duffus 2007). In industrial processes catalyzed by vanadium pentoxide (V_2O_5), V is involved in the oxidation of many organic compounds, forming reactive intermediates, some of which are ROS and may be carcinogenic (Hussain et al. 2003; Valko et al. 2006).

Vanadium occurs in various oxidative states with the ability to participate in reactions involving the formation of free radicals (Crans et al. 2004). V is quickly reduced to V(IV) in plasma by enzymatic (e.g., NADPH) and nonenzymatic (ascorbic acid) plasmatic antioxidants and is then transported and bound to plasma proteins. The equations below show a few such reactions which may occur inside the cell (Liochev and Fridovich 1990), forming peroxovanadyl radicals $\{V(IV) - OO\cdot\}$ and vanadyl hydroperoxide $\{V(IV) - OH\cdot\}$ (Evangelou 2002).

$$V(V) + NADPH \rightarrow V(IV) + NADP^+ + H^+.$$

$$V(IV) + O_2 \rightarrow V(V) + O_2\cdot^-.$$

$$V(V) + O_2\cdot^- \rightarrow \{V(IV) - OO\cdot\}.$$

The generated superoxide undergoes further conversion by a dismutation reaction with SOD to H_2O_2. Some studies have shown that a one-electron reduction of V(V) to V(IV), which is mediated by nonenzymatic ascorbate and phosphate, may be an important V(V) reduction pathway in vivo (Ding et al. 1994). The resulting ROS formed by V(IV) from H_2O_2 and lipid hydroperoxide through the Fenton-like reaction might be critically significant in the mechanisms of V(V)-induced cellular injury during physiological conditions (Ding et al. 1994; Zhang et al. 2001):

$$V(IV) + H_2O_2 \rightarrow V(V) + OH^- + \cdot OH.$$

Vanadium compounds even in signal transduction studies point to their ability to induce oxidative stress and mitochondrial permeability transition pore opening related to oxidative stress (Afeseh Ngwa et al. 2009; Zhao et al. 2010). Vanadium produces ROS-like hydroxyl free radicals by different ways (Cortizo et al. 2000; Gândara et al. 2005), initiating the peroxidative decomposition of cellular membrane phospholipids. This radical was also shown to damage the inner mitochondrial membrane, triggering a sequence of events leading to the loss of cell viability upon mitochondrial deenergization.

Vanadium Toxicology and Pharmacology

Vanadium has been reported in the blood, feces, and urine of workers following occupational exposure to V_2O_5 dust, demonstrating absorption as a consequence of V_2O_5 inhalation (Sjoberg 1954). Vanadium compounds released in large quantities, mainly by burning fossil fuels and also from various industrial processes, can be precipitated on the soil and drained by rain and groundwater which may be directly absorbed by plants (Pyrzyńska and Wierzbicki 2004), eventually reaching those who consume these plants. The major anthropogenic point sources of atmospheric emission are metallurgical plants (30 kg V per ton), followed by the burning of crude or residual oil and coal (0.2–2 kg V per 1000 tons and 30–300 kg V per 10^6 L). By-products containing V_2O_5 include dust, soot, boiler scale, and fly ash. The processing of V slag (about 120 g V_2O_5 per kg) is characterized by the formation of dust, with V concentrations ranging from 5 to 120 mg/m^3 (IARC 2006b). Crude oil from Venezuela is believed to have the highest V concentrations of 1400 mg/kg. Elevated levels of airborne V (4.7 mg/m^3) have been found in the breathing zone of steel industry workers (Kiviluoto et al. 1979). The toxicity of V compounds increases with its valency, making V_2O_5 the most toxic form and therefore warranting the full characterization of its toxicological properties. Studies have shown that inhaled V_2O_5 causes occupational lung diseases (bronchitis and airway fibrosis) commonly referred to as pneumoconiosis. The consequences of environmental exposure to lower levels of V_2O_5 on human health remain unclear, in part because air pollution particulates are a complex mixture of many organic and inorganic components, including a variety of metals [5].

The IARC classifies V_2O_5 as a Group 2B (possible) human carcinogen (IARC 2006a). Acute cases of V poisoning have been reported in man with sequelae of anemia, weakness, vomiting, anorexia, nausea, tinnitus, headache, dizziness, palpitations, transient coronary insufficiency, bradycardia with extra systoles, dermatitis, green discoloration of the tongue, leucopenia, leukocyte granulation, and lower cholesterol levels (Friberg et al. 1986). Epidemiological studies have reported an association between decreased birthweight and V exposure estimated from particulate matter (Jiang et al. 2016). Exposure to geogenic particulate matter (PM) comprising mineral particles has been linked to human health effects. Vanadium exposure in humans has been shown to induce motor deficits and neurobehavioral changes (Jiang et al. 2016; Li et al. 2013; Zhu et al. 2016). However, very little data exist on the neurological health effects associated with occupational dust exposure in natural settings. ICP-MS analyses of roadside dust samples revealed Al (55,090 µg/g), V (70 µg/g), Mn (511 µg/g), and Fe (21,600 µg/g). The ratio of V to Mn in inhaled dust during occupational exposures can vary from 1:1 to 1:8. People with V concentrations around 14.2 mg/L in their urine demonstrated neurobehavioral deficits, especially in visuospatial abilities and attention (Barth et al. 2002; SIMRAC 2000). Vanadium alters the viability of macrophages isolated from dogs, rabbits, and rats exposed to V_2O_5 in vitro for 20 h (Sheridan et al. 1978). The i.p. administration of V_2O_5 altered phospholipid content and induced significant increases in the levels of glucose-6-phosphate dehydrogenase and 6-phosphogluconate dehydrogenase in rats (Kacew et al. 1982). In animal studies, intranasal delivery of geogenic dust containing Mn and V (0.01–100 mg/kg dust) into adult mice via oropharyngeal aspiration induced a neuroinflammatory response (DeWitt et al. 2016; Keil et al. 2016). Very recently, Azeez et al. (2016) demonstrated that chronic postnatal V exposure in mice led to a functional deficit and region-dependent myelin damage associated with glial cell activation and proinflammatory cytokine induction. Rats experienced a significant spatial memory deficit in the Morris water maze (MWM) 3–12 months after V exposure (Folarin et al. 2016). Additional studies showed that exposure to dust containing elevated concentrations of metals can cause neuroinflammation and neurodegeneration (Calderon-Garciduenas et al. 2016; Jiang et al. 2016; Reis et al. 2014).

The cytotoxicity caused by compounds of V has been documented (Cortizo et al. 2000; Sabbioni et al. 1991). Various V compounds are known to impede the activities of ribonuclease (Lau et al. 1993), protein kinases (Bollen et al. 1990; Stankiewicz and Tracey 1995), ATPases (Sabbioni et al. 1991), and phosphatases (Tracey 2000). Some V compounds either inhibit or stimulate the activity of DNA or RNA enzymes eliciting mutagenic and genotoxic responses (Stemmler and Burrows 2001). Single-stranded DNA breaks in cells of mouse testes were observed 24 h after one intraperitoneal (i.p.) injection of V_2O_5 (5.75, 11.5, and 23 mg/kg) (Altamirano-Lozano et al. 1996), indicating an ability to cross the blood-testis barrier. Vanadium has also been reported to cross the blood-brain barrier, inducing neurochemical changes in the brain (Witkowska and Brzezinski 1979). Vanadium-containing substances alter blood levels of thyroid hormone with higher triiodothyronine plasma levels in V-treated rats (Badmaev et al. 1999; Mukherjee et al. 2004). Vanadium-containing compounds can also change the metabolism of sugars and lipids (Nakai et al. 1995). The ability of V-containing compounds to alter gene expression has generated interest among biological scientists for such compounds. In insulin receptor-overexpressing cells, greater levels of Ras, MAPK, p70s6k, and c-raf-1 have been observed following V exposure (Pandey et al. 1999). Increased levels of macrophage inflammatory protein (MIP)-2 mRNA triggered by vanadates are accompanied by increased NFκB DNA-binding activity in bronchoalveolar lavage (BAL) cells (Chong et al. 2000). Vanadate exposure has also been shown to induce gene expression of tumor necrosis factor-alpha (TNF-α), activator protein-1 (AP-1), and interleukin-8 (IL-8) (Ding et al. 1999; Jaspers et al. 1999; Ye et al. 1999).

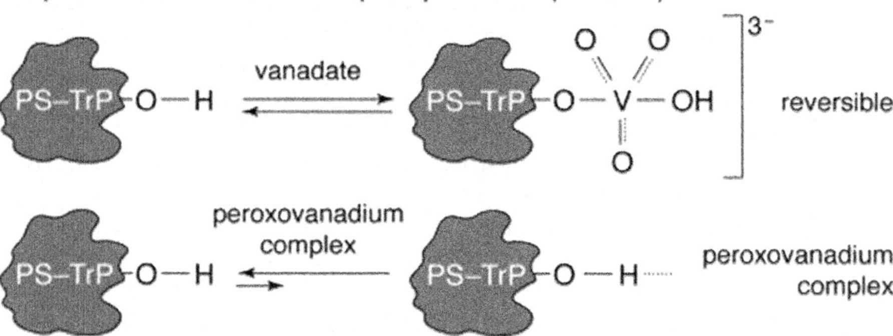

Fig. 1 A postulated mechanism of inhibiting phosphatases by vanadate and peroxovanadium complexes (Morinville et al. 1998). Vanadate acts as a transition state analogue and forms a reversible bond, thus inhibiting protein tyrosine phosphatases (PTPs). On the other hand, peroxovanadium complexes oxidize the cysteine residue in the catalytic domain of PTPs to irreversibly inhibit PTPs

Mechanistically, vanadate activates TNF-α gene promoter through NFκB (Jaspers et al. 2000; Ouellet et al. 1999).

The ability of V compounds to be potent inhibitors of protein tyrosine phosphatases (PTPs) appears to be the underlying mechanism of some of its effects (Fig. 1). Examples include neurite outgrowth in human neuroblastoma SH-SY5Y cells and the differentiation and neurite outgrowth of rat pheochromocytoma cells (PC12 cells) after treatment with sodium orthovanadate (Rogers et al. 1994). This is distinct from the kind of differentiation signaling induced by nerve growth factor (Rogers et al. 1994). However, a reduced rate of proliferation in the presence of a couple of peroxovanadium complexes has been reported in neuroblastoma NB41 and glioma C6 (Faure et al. 1995). Faure et al. demonstrated that the mechanism responsible for this is the reversible block at the G2–M transition of the cell cycle and that removal of the peroxovanadium complex restored normalcy to cell cycling.

Using apoptotic neuronal models, Farinelli and Greene (1996) have shown that substances which cause a cell-cycle block in the S, G2, and M phases don't prevent cell death, whereas substances which block cell-cycle progression before the G1–S transition prevent apoptosis. The peroxovanadium-induced cell-cycle block at G2–M therefore might be linked to apoptotic cell death.

Vanadium compounds can show antineoplastic effects in vivo (Thompson et al. 1984). In vitro, sodium orthovanadate displayed a time- and dose-dependent cytotoxicity in proliferating primary cultures and tumor cell lines, while non-proliferating cells were found to be less susceptible to vanadate-induced cytotoxicity (Cruz et al. 1995).

Since V species like vanadates and peroxovanadium complexes alter intracellular phosphorylation levels in a nonselective manner through the inhibition of protein phosphatases, it is unsurprising that they have profound effects on intracellular signaling cascades. A good example is the MAPK cascade consisting of ERK, the c-Jun N-terminal protein kinases (JNKs), and p38 (Kyriakis and Avruch 1996; Marshall 1995; Whitmarsh and Davis 1996), which are implicated in the tight regulation of some intracellular pathways and connected to both cell survival and apoptotic responses (Kummer et al. 1997; Xia et al. 1995). Interference with the MAPK signaling cascade by V compounds could explain some of the observed insulin-mimicking effects of V compounds since they can activate the MAPK signaling cascade and since the insulin-mediated activation of IRK causes the activation of ERKs and the protein kinases p70s6k and p90rsk (Pandey et al. 1995). In addition, sodium orthovanadate, vanadyl sulfate, and sodium metavanadate stimulate ERK-1, ERK-2, p70s6k and p90rsk in CHO cells (Pandey et al. 1995), while peroxovanadates activate ERK in HeLa cells (Zhao et al. 1996). Given that MAPK links to cell survival and apoptosis, the ability of V to modulate the activity of its members is possibly responsible for V-induced toxicity, although the role of the MAPKs in mediating peroxovanadium complex-induced cell death has yet to be fully elucidated (Fig. 2) (Morinville et al. 1998).

The extent of V's involvement in modulating various cell death pathways has yet to be explored. In one of the programmed cell death paradigms, extracellular ligands like TNF-α can bind to a death receptor (DR) spanning the cell membrane (Haunstetter and Izumo 1998). As previously mentioned, NFκB activity and JNK can be potentially altered by V complexes (Barbeau et al. 1997; Gopalbhai and Meloche 1998), with a real potential to modify cell death signal sites or routes (Morinville et al. 1998). The inhibition of PTPs can affect the transduction signals arising from DRs. Apoptotic signals from DRs involve caspase activation. The precise regulation of each caspase has yet to be fully elucidated even though it is known that these caspases are activated by cleavage at a specific aspartate residue. Also, peroxovanadium complexes can activate caspases (Morinville et al. 1998).

Vanadium compounds might not only cause apoptotic cell death but also death through necrosis. Despite evidence suggesting that the mechanism underlying V toxicity is independent of its ability to inhibit PTPs, the relationship between V exposure and necrosis needs to be extensively probed as very little is known of the players involved in this distinct necrotic cell death (Morinville et al. 1998).

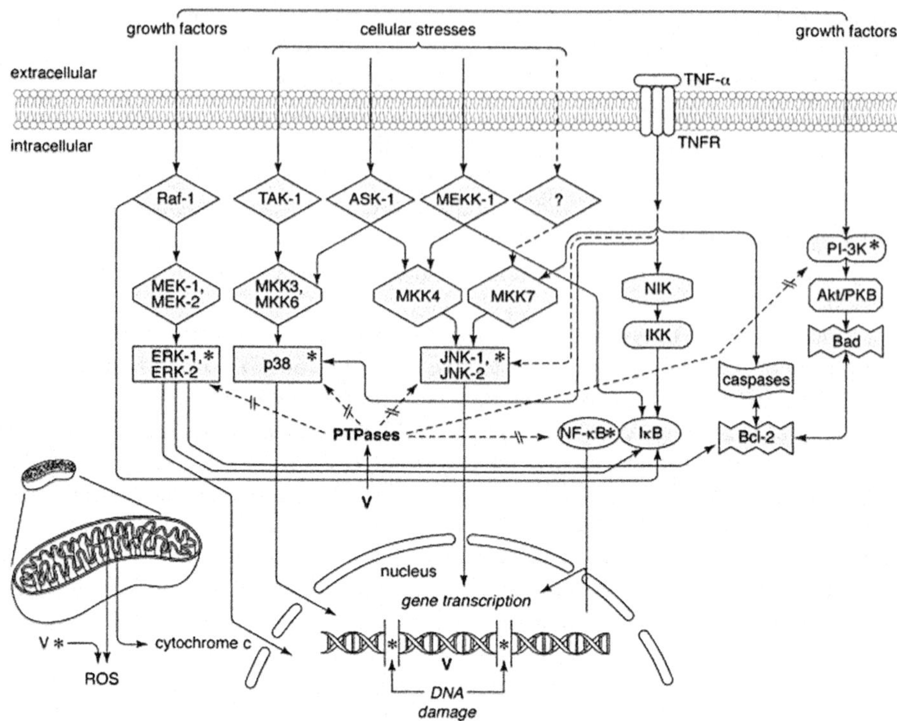

Fig. 2 A simplified diagram of the apoptotic signaling pathways modified by vanadium (V) compounds (Morinville et al. 1998). Vanadium has been shown to modulate multiple signaling pathways including the MAPK and NFκB signaling cascades that contribute to apoptotic cell death. The proteins that can be modulated by V are marked with an *asterisk*. *Dotted lines* indicate putative connections

Vanadium-Induced Neurotoxicity and its Relevance to Neurodegeneration

Parkinson's disease (PD) is one of the most common neurodegenerative disorders, affecting over one million Americans and about 2% worldwide of those over the age of 60 years (Bergareche et al. 2004; Elbaz et al. 2002). Since most PD cases are sporadic, enormous interest has emerged in understanding the role of environmental factors in various Parkinsonian disorders (Hanna et al. 1999; Kumar et al. 2004). Many occupational exposures have been linked to the etiology and progression of PD, including farming, steel/alloy manufacturing, mining, wood/pulp processing, carpentry, planer milling, cleaning, forestry/logging, orchard farming, as well as certain occupations comprising body and fender repair, working in oil and gas fields, auto painting, and railroad and auto mechanics. These occupational exposures are often related to environmental exposure to certain metals, fuel oil, pesticides, herbicides, well water, magnetic fields, and rural living (Fall et al. 1999;

Gorell et al. 2004; Jankovic 2005; Noonan et al. 2002). Some reports have suggested that welders may be at a higher risk of developing Parkinsonism and welding is a risk factor for PD pathogenesis (Racette et al. 2001, 2005). In a mortality study of occupational information reported on US death certificates, a significantly higher proportion of PD fatality cases correlated with likely exposure to Mn through welding-related jobs (Park et al. Park et al. 2005a, b). On the other hand, some occupational studies did not find any statistically significant association between PD and welding (Gorell et al. 1999, 2004).

Sundin (1998) estimated that more than one million people were employed as welders worldwide, and millions around the world are exposed to welding aerosols on a daily basis (Antonini et al. 2009a; Antonini et al. 2009b). The American welding association (http://www.aws.org/w/a/research/outlook.html), McInerny et al. 2009) expects that the number of welders will continue to grow to meet the increasing demand for steel and metal products around the world as developing countries continue to modernize. In 1991, it was also reported that more than three million people performed welding as a part of their work at least intermittently (Sferlazza and Beckett 1991). Welding fumes comprise a complex mixture of gases with fine and ultrafine particles of different metals and their oxides, which form during welding by metal vaporization and oxidation (McNeilly et al. 2004; Yu et al. 2000). The fumes from welding have also been found to contain silicates and fluorides of metals like chromium, Mn, V, titanium, molybdenum, cobalt, nickel, copper, and zinc (Sanderson 1968). These fumes, which also produce gases such as hydrogen fluoride, carbon monoxide, nitrogen oxide, fluorine, and ozone, can adversely affect the health of welders as well as the health of those in the immediate area (American society of safety engineers, http://www.asse.org/practicespecialties/articles/weldingfumes.php). The exact nature of the welding fumes is largely dependent on the composition of the electrode, the filler wire, and the type of welding being performed (Antonini et al. 1996; Sferlazza and Beckett 1991). More than 90% of V goes into steelmaking, and the dominant market driver of V production in recent years has been the rapid growth in worldwide steel production (Bunting 2006).

Individuals exposed to V have manifested neurological symptoms like tremor and CNS depression (Done 1979). Inhaled V_2O_5 has induced significant dopaminergic neuronal loss in the substantia nigra of mice, accompanied by morphological alterations of striatal medium spiny neurons (Avila-Costa et al. 2004). The same group also reported in their animal studies that V inhalation produced necrotic-like cell death, a loss of dendritic spines, and notorious alterations in the neuropile, resulting in the impairment of spatial memory as evaluated by the Morris water maze (Avila-Costa et al. 2004, 2006).

Our laboratory recently probed the cell death signaling mechanisms leading to the loss of dopaminergic neuronal cells following exposure to V (Afeseh Ngwa et al. 2009). Vanadium (V_2O_5) was found to be neurotoxic to rat dopaminergic neuronal (N27) cells, with an EC_{50} of 37 μM. ICP-MS analysis determined that a time-dependent uptake of V into the cells accompanied the neurotoxic effects. Also, the metal transporter proteins transferrin (Tf) and divalent metal transporter 1 (DMT1) were upregulated. We further showed that V exposure generated up to a threefold increase in ROS, which was accompanied by the release of mitochondrial cytochrome c into the cytoplasm with consequential activation of initiator caspase-9 and activator caspase-3. Interestingly, we also observed that V exposure further induced the caspase-mediated proteolytic cleavage of a pro-apoptotic kinase, protein kinase C delta (PKCδ), resulting in persistently increased kinase activity. Co-treating V with the pan-caspase inhibitor Z-VAD-FMK significantly blocked V-induced PKCδ proteolytic activation and increases in DNA fragmentation, hence supporting the role of caspase-mediated PKCδ cleavage in V-induced neurotoxicity. Importantly, V was also highly neurotoxic to murine primary mesencephalic dopaminergic neurons.

In another animal model study (Ngwa et al. 2014), we examined the neurotoxic effects of V on the olfactory bulb since anosmia is considered an early symptom of neurological diseases, including PD. C57 black mice were exposed intranasally to an environmentally relevant exposure dose of 182 µg V_2O_5 three times a week for up to 1 month. Behavioral, neurochemical, and histological studies were performed following the intranasal exposure. When compared to controls, the treatment group experienced dramatic decreases in olfactory bulb weights, tyrosine hydroxylase levels, as well as dopamine and DOPAC levels. The severe neurotoxic effect of V in the olfactory system had a neuroinflammatory component, as evidenced by the accumulation of astroglia in the glomerular layer of the olfactory bulb where dopaminergic neurons were degenerating. Neurobiological changes in response to intranasal V exposure were severe enough to be manifested at the behavioral level as impaired olfaction and significant locomotor deficits. These results suggest exposure to V is toxic to dopaminergic neurons and impairs olfaction in mouse models. However, more evidence is needed to prove a cause-and-effect relationship between PD and V exposure.

Conclusions

This review has covered the evidence supporting the idea that V and its compounds may interfere with various cellular functions including neuronal functions, leading to changes through the generation of ROS and interactions with protein tyrosine phosphatases (PTP) that affect cell signaling pathways, which may in turn produce or inhibit cell death depending on V's oxidation state and the type of V compound. Much research on V and its compounds has been on its respiratory effects, as well as some on its effects on the kidney and liver, whereas comparatively little has been done on its possible neurotoxic effects. Vanadium and its compounds, often in synergy with other neurotoxic compounds like Mn that co-occur in occupational fumes, are likely neurotoxic. Manganese, which has been linked with Parkinson-like symptoms, has garnered almost all the attention for its association with welding fumes and neurotoxicity, while V and its compounds have thus far been largely neglected. Based on this review, much more work is warranted to explore how mixed metals, and their individual components like V, potentiate the neurotoxic effects caused by welding fumes.

Acknowledgment This chapter was supported by National Institutes of Health Grants ES10586 and ES26892. The W. Eugene and Linda Lloyd Endowed Chair for AGK is also acknowledged. We thank Gary Zenitsky for assistance in preparing this chapter.

References

Afeseh Ngwa H, Kanthasamy A, Anantharam V, Song C, Witte T, Houk R, Kanthasamy AG. Vanadium induces dopaminergic neurotoxicity via protein kinase Cdelta dependent oxidative signaling mechanisms: relevance to etiopathogenesis of Parkinson's disease. Toxicol Appl Pharmacol. 2009;240(2):273–85.

Altamirano-Lozano M, Alvarez-Barrera L, Basurto-Alcantara F, Valverde M, Rojas E. Reprotoxic and genotoxic studies of vanadium pentoxide in male mice. Teratog Carcinog Mutagen. 1996;16(1):7–17.

Antonini JM, Murthy GGK, Rogers RA, Albert R, Ulrich GD, Brain JD. Pneumotoxicity and pulmonary clearance of different welding fumes after Intratracheal instillation in the rat. Toxicol Appl Pharmacol. 1996;140(1):188–99.

Antonini JM, Roberts JR, Stone S, Chen BT, Schwegler-Berry D, Frazer DG. Short-term inhalation exposure to mild steel welding fume had no effect on lung inflammation and injury but did alter defense responses to bacteria in rats. Inhal Toxicol. 2009a;21(3):182–92.

Antonini JM, Sriram K, Benkovic SA, Roberts JR, Stone S, Chen BT, Schwegler-Berry D, Jefferson AM, Billig BK, Felton CM, Hammer MA, Ma F, Frazer DG, O'Callaghan JP, Miller DB. Mild steel welding fume causes manganese accumulation and subtle neuroinflammatory changes but not overt neuronal damage in discrete brain regions of rats after short-term inhalation exposure. Neurotoxicology. 2009b;30(6):915–25.

Avila-Costa MR, Flores EM, Colin-Barenque L, Ordoñez JL, Gutiérrez AL, Niño-Cabrera HG, Mussali-Galante P, Fortoul TI. Nigrostriatal modifications after vanadium inhalation: an Immunocytochemical and cytological approach. Neurochem Res. 2004;29(7):1365–9.

Avila-Costa MR, Fortoul TI, Niño-Cabrera G, Colín-Barenque L, Bizarro-Nevares P, Gutiérrez-Valdez AL, Ordóñez-Librado JL, Rodríguez-Lara V, Mussali-Galante P, Díaz-Bech P, Anaya-Martínez V. Hippocampal cell alterations induced by the inhalation of vanadium pentoxide (V2O5) promote memory deterioration. Neurotoxicology. 2006;27(6):1007–12.

Azeez IA, Olopade F, Laperchia C, Andrioli A, Scambi I, Onwuka SK, Bentivoglio M, Olopade JO. Regional myelin and axon damage and Neuroinflammation in the adult mouse brain after long-term postnatal vanadium exposure. J Neuropathol Exp Neurol. 2016;75(9):843–54.

Badmaev V, Prakash S, Majeed M. Vanadium: a review of its potential role in the fight against diabetes. J Altern Complement Med. 1999;5(3):273–91.

Barbeau B, Bernier R, Dumais N, Briand G, Olivier M, Faure R, Posner BI, Tremblay M. Activation of HIV-1 long terminal repeat transcription and virus replication via NF-kappaB-dependent and -independent pathways by potent phosphotyrosine phosphatase inhibitors, the peroxovanadium compounds. J Biol Chem. 1997;272(20):12968–77.

Barth A, Schaffer AW, Konnaris C, Blauensteiner R, Winker R, Osterode W, Rudiger HW. Neurobehavioral effects of vanadium. J Toxicol Environ Health A. 2002;65(9):677–83.

Bergareche A, De la Puente E, López deMunain A, Sarasqueta C, de Arce A, Poza JJ, Martí-Massó JF. Prevalence of Parkinson's disease and other types of Parkinsonism. J Neurol. 2004;251(3):340–5.

Bollen M, Miralpeix M, Ventura F, Toth B, Bartrons R, Stalmans W. Oral administration of vanadate to streptozotocin-diabetic rats restores the glucose-induced activation of liver glycogen synthase. Biochem J. 1990;267(1):269–71.

Bunting RM. Vanadium: how market developments affect the titanium industry. Strategic minerals corporation, Titanium 2006, International Titanium Association Conference. San Diego, California. 2006.

Calderon-Garciduenas L, Leray E, Heydarpour P, Torres-Jardon R, Reis J. Air pollution, a rising environmental risk factor for cognition, neuroinflammation and neurodegeneration: the clinical impact on children and beyond. Rev Neurol. 2016;172(1):69–80.

ChemIDPlus. Hazardous Substance Database. 2016.

Chong IW, Lin SR, Hwang JJ, Huang MS, Wang TH, Tsai MS, Hou JJ, Paulauskis JD. Expression and regulation of macrophage inflammatory protein-2 gene by vanadium in mouse macrophages. Inflammation. 2000;24(2):127–39.

Cortizo AMA, Bruzzone L, Molinuevo S, Etcheverry SB. A possible role of oxidative stress in the vanadium-induced cytotoxicity in the MC3T3E1 osteoblast and UMR106 osteosarcoma cell lines. Toxicology. 2000;147(2):89–99.

Crans DC, Smee JJ, Gaidamauskas E, Yang L. The chemistry and biochemistry of vanadium and the biological activities exerted by vanadium compounds. Chem Rev. 2004;104(2):849–902.

Cruz TF, Morgan A, Min W. In vitro and in vivo antineoplastic effects of orthovanadate. Mol Cell Biochem. 1995;153(1-2):161–6.

DeWitt J, Buck B, Goossens D, Hu Q, Chow R, David W, Young S, Teng Y, Leetham-Spencer M, Murphy L, Pollard J, McLaurin B, Gerads R, Keil D. Health effects following subacute exposure to geogenic dusts from arsenic-rich sediment at the Nellis dunes recreation area, Las Vegas, NV. Toxicol Appl Pharmacol. 2016;304:79–89.

Ding M, Gannett PM, Rojanasakul Y, Liu K, Shi X. One-electron reduction of vanadate by ascorbate and related free radical generation at physiological pH. J Inorg Biochem. 1994;55(2):101–12.

Ding M, Li JJ, Leonard SS, Ye JP, Shi X, Colburn NH, Castranova V, Vallyathan V. Vanadate-induced activation of activator protein-1: role of reactive oxygen species. Carcinogenesis. 1999;20(4):663–8.

Done AK. Of metals and chelation. AK: Done; 1979. p. 186–9.

Duffus JH. Carcinogenicity classification of vanadium pentoxide and inorganic vanadium compounds, the NTP study of carcinogenicity of inhaled vanadium pentoxide, and vanadium chemistry. Regul Toxicol Pharmacol. 2007;47(1):110–4.

Elbaz A, Bower JH, Maraganore DM, McDonnell SK, Peterson BJ, Ahlskog JE, Schaid DJ, Rocca WA. Risk tables for parkinsonism and Parkinson's disease. J Clin Epidemiol. 2002;55(1):25–31.

Evangelou AM. Vanadium in cancer treatment. Crit Rev Oncol Hematol. 2002;42(3):249–65.

Fall PA, Fredrikson M, Axelson O, Granerus AK. Nutritional and occupational factors influencing the risk of Parkinson's disease: a case-control study in southeastern Sweden. Mov Disord. 1999;14(1):28–37.

Farinelli SE, Greene LA. Cell cycle blockers mimosine, ciclopirox, and deferoxamine prevent the death of PC12 cells and postmitotic sympathetic neurons after removal of trophic support. J Neurosci. 1996;16(3):1150–62.

Faure R, Vincent M, Dufour M, Shaver A, Posner BI. Arrest at the G2/M transition of the cell cycle by protein-tyrosine phosphatase inhibition: studies on a neuronal and a glial cell line. J Cell Biochem. 1995;59(3):389–401.

Fenech M, Ferguson LR. Vitamins/minerals and genomic stability in humans. Mutat Res/ Fundament Mol Mech Mutagen. 2001;475(1–2):1–6.

Folarin O, Olopade F, Onwuka S, Olopade J. Memory deficit recovery after chronic vanadium exposure in mice. Oxidative Med Cell Longev. 2016;2016:4860582.

Friberg L, Nordberg GF, Kessler E, Vouk VB. Handbook of the toxicology of metals. New York: Elsevier Science Publishers BV; 1986.

Gândara RMC, Soares SS, Martins H, Gutiérrez-Merino C, Aureliano M. Vanadate oligomers: in vivo effects in hepatic vanadium accumulation and stress markers. J Inorg Biochem. 2005;99(5):1238–44.

Gopalbhai K, Meloche S. Repression of mitogen-activated protein kinases ERK1/ERK2 activity by a protein tyrosine phosphatase in rat fibroblasts transformed by upstream oncoproteins. J Cell Physiol. 1998;174(1):35–47.

Gorell JM, Rybicki BA, Cole Johnson C, Peterson EL. Occupational metal exposures and the risk of Parkinson's disease. Neuroepidemiology. 1999;18(6):303–8.

Gorell JM, Peterson EL, Rybicki BA, Johnson CC. Multiple risk factors for Parkinson's disease. J Neurol Sci. 2004;217(2):169–74.

Greenwood NN, Earnshaw A. Chemistry of the elements (2nd edition): Butterworth-Heinemann; 1997.

Hanna PA, Jankovic J, Kirkpatrick JB. Multiple system atrophy: the putative causative role of environmental toxins. Arch Neurol. 1999;56(1):90–4.

Haunstetter A, Izumo S. Apoptosis: basic mechanisms and implications for cardiovascular disease. Circ Res. 1998;82(11):1111–29.

Hussain SP, Hofseth LJ, Harris CC. Radical causes of cancer. Nat Rev Cancer. 2003;3(4):276–85.

IARC. Cobalt in hard metals and cobalt sulfate, gallium arsenide, indium phosphide and vanadium pentoxide, IARC monographs on the evaluation of carcinogenic risks to humans. Lyon: International Agency for Research on Cancer; 2006a. p. 227–92.

IARC. Cobalt in hard metals and cobalt sulfate, gallium arsenide, indium phosphide and vanadium pentoxide, IARC monographs on the evaluation of carcinogenic risks to humans. International Agency for Research on Cancer, Lyon, France; 2006b. p. 227–92.

Jankovic J. Searching for a relationship between manganese and welding and Parkinson's disease. Neurology. 2005;64(12):2021–8.

Jaspers I, Samet JM, Reed W. Arsenite exposure of cultured airway epithelial cells activates kappaB-dependent interleukin-8 gene expression in the absence of nuclear factor-kappaB nuclear translocation. J Biol Chem. 1999;274(43):31025–33.

Jaspers I, Samet JM, Erzurum S, Reed W. Vanadium-induced kappaB-dependent transcription depends upon peroxide-induced activation of the p38 mitogen-activated protein kinase. Am J Respir Cell Mol Biol. 2000;23(1):95–102.

Jiang M, Li Y, Zhang B, Zhou A, Zheng T, Qian Z, Du X, Zhou Y, Pan X, Hu J, Wu C, Peng Y, Liu W, Zhang C, Xia W, Xu S. A nested case-control study of prenatal vanadium exposure and low birthweight. Hum Reprod. 2016;31(9):2135–41.

Kacew S, Parulekar MR, Merali Z. Effects of parenteral vanadium administration on pulmonary metabolism of rats. Toxicol Lett. 1982;11(1):119–24.

Keil D, Buck B, Goossens D, Teng Y, Leetham M, Murphy L, Pollard J, Eggers M, McLaurin B, Gerads R, DeWitt J. Immunotoxicological and neurotoxicological profile of health effects following subacute exposure to geogenic dust from sand dunes at the Nellis dunes recreation area, Las Vegas, NV. Toxicol Appl Pharmacol. 2016;291:1–12.

Kiviluoto M, Pyy L, Pakarinen A. Serum and urinary vanadium of vanadium-exposed workers. Scand J Work Environ Health. 1979;5(4):362–7.

Kumar A, Calne SM, Schulzer M, Mak E, Wszolek Z, Van Netten C, Tsui JK, Stoessl AJ, Calne DB. Clustering of Parkinson disease: shared cause or coincidence? Arch Neurol. 2004;61(7):1057–60.

Kummer JL, Rao PK, Heidenreich KA. Apoptosis induced by withdrawal of trophic factors is mediated by p38 mitogen-activated protein kinase. J Biol Chem. 1997;272(33):20490–4.

Kyriakis JM, Avruch J. Sounding the alarm: protein kinase cascades activated by stress and inflammation. J Biol Chem. 1996;271(40):24313–6.

Lau JY, Qian KP, Wu PC, Davis GL. Ribonucleotide vanadyl complexes inhibit polymerase chain reaction. Nucleic Acids Res. 1993;21(11):2777.

Li H, Zhou D, Zhang Q, Feng C, Zheng W, He K, Lan Y. Vanadium exposure-induced neurobehavioral alterations among Chinese workers. Neurotoxicology. 2013;36:49–54.

Liochev SI, Fridovich I. Vanadate-stimulated oxidation of NAD(P)H in the presence of biological membranes and other sources of O2−. Arch Biochem Biophys. 1990;279(1):1–7.

Marshall CJ. Specificity of receptor tyrosine kinase signaling: transient versus sustained extracellular signal-regulated kinase activation. Cell. 1995;80(2):179–85.

McInerny SC, Brown AL, Smith DW. Region-specific changes in mitochondrial D-loop in aged rat CNS. Mech Ageing Dev. 2009;130(5):343–9.

McNeilly JD, Heal MR, Beverland IJ, Howe A, Gibson MD, Hibbs LR, MacNee W, Donaldson K. Soluble transition metals cause the pro-inflammatory effects of welding fumes in vitro. Toxicol Appl Pharmacol. 2004;196(1):95–107.

Morinville A, Maysinger D, Shaver A. From Vanadis to Atropos: vanadium compounds as pharmacological tools in cell death signalling. Trends Pharmacol Sci. 1998;19(11):452–60.

Mukherjee B, Patra B, Mahapatra S, Banerjee P, Tiwari A, Chatterjee M. Vanadium—an element of atypical biological significance. Toxicol Lett. 2004;150(2):135–43.

Nakai M, Watanabe H, Fujiwara C, Kakegawa H, Satoh T, Takada J, Matsushita R, Sakurai H. Mechanism on insulin-like action of vanadyl sulfate: studies on interaction between rat adipocytes and vanadium compounds. Biol Pharm Bull. 1995;18(5):719–25.

Ngwa HA, Kanthasamy A, Jin H, Anantharam V, Kanthasamy AG. Vanadium exposure induces olfactory dysfunction in an animal model of metal neurotoxicity. Neurotoxicology. 2014;43:73–81.

Noonan CW, Reif JS, Yost M, Touchstone J. Occupational exposure to magnetic fields in case-referent studies of neurodegenerative diseases. Scand J Work Environ Health. 2002;28(1):42–8.

Ouellet M, Barbeau B, Tremblay MJ. p56(lck), ZAP-70, SLP-76, and calcium-regulated effectors are involved in NF-kappaB activation by bisperoxovanadium phosphotyrosyl phosphatase inhibitors in human T cells. J Biol Chem. 1999;274(49):35029–36.

Pandey SK, Chiasson JL, Srivastava AK. Vanadium salts stimulate mitogen-activated protein (MAP) kinases and ribosomal S6 kinases. Mol Cell Biochem. 1995;153(1-2):69–78.

Pandey SK, Theberge JF, Bernier M, Srivastava AK. Phosphatidylinositol 3-kinase requirement in activation of the ras/C-raf-1/MEK/ERK and p70(s6k) signaling cascade by the insulinomimetic agent vanadyl sulfate. Biochemistry. 1999;38(44):14667–75.

Park J, Yoo CI, Sim CS, Kim HK, Kim JW, Jeon BS, Kim KR, Bang OY, Lee WY, Yi Y, Jung KY, Chung SE, Kim Y. Occupations and Parkinson's disease: a multi-center case-control study in South Korea. Neurotoxicology. 2005a;26(1):99–105.

Park RM, Schulte PA, Bowman JD, Walker JT, Bondy SC, Yost MG, Touchstone JA, Dosemeci M. Potential occupational risks for neurodegenerative diseases. Am J Ind Med. 2005b;48(1):63–77.

Patel B, Henderson GE, Haswell SJ, Grzeskowiak R. Speciation of vanadium present in a model yeast system. Analyst. 1990;115(8):1063–6.

Pyrzyńska K, Wierzbicki T. Determination of vanadium species in environmental samples. Talanta. 2004;64(4):823–9.

Racette BA, McGee-Minnich L, Moerlein SM, Mink JW, Videen TO, Perlmutter JS. Welding-related parkinsonism: clinical features, treatment, and pathophysiology. Neurology. 2001;56(1):8–13.

Racette BA, Tabbal SD, Jennings D, Good L, Perlmutter JS, Evanoff B. Prevalence of parkinsonism and relationship to exposure in a large sample of Alabama welders. Neurology. 2005;64(2):230–5.

Ray RS, Rana B, Swami B, Venu V, Chatterjee M. Vanadium mediated apoptosis and cell cycle arrest in MCF7 cell line. Chem Biol Interact. 2006;163(3):239–47.

Reis AP, Patinha C, Noack Y, Robert S, Dias AC. Assessing human exposure to aluminium, chromium and vanadium through outdoor dust ingestion in the Bassin Minier de Provence, France. Environ Geochem Health. 2014;36(2):303–17.

Rogers MV, Buensuceso C, Montague F, Mahadevan L. Vanadate stimulates differentiation and neurite outgrowth in rat pheochromocytoma PC12 cells and neurite extension in human neuroblastoma SH-SY5Y cells. Neuroscience. 1994;60(2):479–94.

Sabbioni E, Pozzi G, Pintar A, Casella L, Garattini S. Cellular retention, cytotoxicity and morphological transformation by vanadium(IV) and vanadium(V) in BALB/3T3 cell lines. Carcinogenesis. 1991;12(1):47–52.

Sanderson JT. Hazards of the arc-air gouging process. Ann Occup Hyg. 1968;11(2):123–33.

Sferlazza SJ, Beckett WS. The respiratory health of welders. Am Rev Respir Dis. 1991;143(5 Pt 1):1134–48.

Sheridan CJ, Pfleger RC, McClellan RO. Cytotoxicity of vanadium pentoxide on pulmonary alveolar macrophages from dog, rabbit, and rat: effect on viability and effect on lipid metabolism. Ann Resp Inhalation Toxicol; 1978. p. 294–8.

SIMRAC. Hazardous metals in mining processing plants in South Africa. The risk of occupational exposure Mine Health and Safety Council (Safety in Mines Research Advisory Committee Report). 2000.

Sjoberg S-G. Vanadium bronchitis from cleaning oil-fired boilers. Occup Med. 1954;4(1):31.

Stankiewicz PJ, Tracey AS. Stimulation of enzyme activity by oxovanadium complexes. Met Ions Biol Syst. 1995;31:249–85.

Stemmler AJ, Burrows CJ. Guanine versus deoxyribose damage in DNA oxidation mediated by vanadium(IV) and vanadium(V) complexes. J Biol Inorg Chem. 2001;6(1):100–6.

Sundin DS. National occupational exposure survey database 1981–1983. 1998.

Thompson HJ, Chasteen ND, Meeker LD. Dietary vanadyl(IV) sulfate inhibits chemically-induced mammary carcinogenesis. Carcinogenesis. 1984;5(6):849–51.

Tracey AS. Hydroxamido vanadates: aqueous chemistry and function in protein tyrosine phosphatases and cell cultures. J Inorg Biochem. 2000;80(1–2):11–6.

Valko M, Rhodes CJ, Moncol J, Izakovic M, Mazur M. Free radicals, metals and antioxidants in oxidative stress-induced cancer. Chem Biol Interact. 2006;160(1):1–40.

Whitmarsh AJ, Davis RJ. Transcription factor AP-1 regulation by mitogen-activated protein kinase signal transduction pathways. J Mol Med. 1996;74(10):589–607.

Witkowska D, Brzezinski J. Alteration of brain noradrenaline, dopamine and 5-hydroxytryptamine levels during vanadium poisoning. Pol J Pharmacol Pharm. 1979;31(4):393–8.

Xia Z, Dickens M, Raingeaud J, Davis RJ, Greenberg ME. Opposing effects of ERK and JNK-p38 MAP kinases on apoptosis. Science. 1995;270(5240):1326–31.

Ye J, Ding M, Zhang X, Rojanasakul Y, Nedospasov S, Vallyathan V, Castranova V, Shi X. Induction of TNFalpha in macrophages by vanadate is dependent on activation of transcription factor NF-kappaB and free radical reactions. Mol Cell Biochem. 1999;198(1-2):193–200.

Yu IJ, Kim KJ, Chang HK, Song KS, Han KT, Han JH, Maeng SH, Chung YH, Park SH, Chung KH, Han JS, Chung HK. Pattern of deposition of stainless steel welding fume particles inhaled into the respiratory systems of Sprague–Dawley rats exposed to a novel welding fume generating system. Toxicol Lett. 2000;116(1–2):103–11.

Zhang Z, Huang C, Li J, Leonard SS, Lanciotti R, Butterworth L, Shi X. Vanadate-induced cell growth regulation and the role of reactive oxygen species. Arch Biochem Biophys. 2001;392(2):311–20.

Zhao Z, Tan Z, Diltz CD, You M, Fischer EH. Activation of mitogen-activated protein (MAP) kinase pathway by pervanadate, a potent inhibitor of tyrosine phosphatases. J Biol Chem. 1996;271(36):22251–5.

Zhao Y, Ye L, Liu H, Xia Q, Zhang Y, Yang X, Wang K. Vanadium compounds induced mitochondria permeability transition pore (PTP) opening related to oxidative stress. J Inorg Biochem. 2010;104(4):371–8.

Zhu CW, Liu YX, Huang CJ, Gao W, Hu GL, Li J, Zhang Q, Lan YJ. Effect of vanadium exposure on neurobehavioral function in workers. Zhonghua lao dong wei sheng zhi ye bing za zhi = Zhonghua laodong weisheng zhiyebing zazhi Chin J Indust Hyg Occup Dis. 2016;34(2):103–6.

Neurotoxicity of Zinc

Deborah R. Morris and Cathy W. Levenson

Abstract Zinc-induced neurotoxicity has been shown to play a role in neuronal damage and death associated with traumatic brain injury, stroke, seizures, and neurodegenerative diseases. During normal firing of "zinc-ergic" neurons, vesicular free zinc is released into the synaptic cleft where it modulates a number of postsynaptic neuronal receptors. However, excess zinc, released after injury or disease, leads to excitotoxic neuronal death. The mechanisms of zinc-mediated neurotoxicity appear to include not only neuronal signaling but also regulation of mitochondrial function and energy production, as well as other mechanisms such as aggregation of amyloid beta peptides in Alzheimer's disease. However, recent data have raised questions about some of our long-standing assumptions about the mechanisms of zinc in neurotoxicity. Thus, this review explores the most recent published findings and highlights the current mechanistic controversies.

Keywords Zinc • Abeta • Excitotoxicity • Neurotoxicity

Introduction

Most neuronal zinc is protein-bound. There are, however, small pools of free zinc located in presynaptic vesicles of glutamatergic neurons as well as other, less well understood, intracellular non-vesicular zinc pools. While there has been a significant amount of work on protective roles of zinc in specific clinical situations such as traumatic brain injury (Young et al. 1996; Cope et al. 2011; 2016), there is also a great deal of data showing that the release of excess free zinc can cause acute

D.R. Morris
Department of Biomedical Sciences, The Florida State University College of Medicine, Tallahassee, FL 32306-4300, USA

C.W. Levenson (✉)
Department of Biomedical Sciences, The Florida State University College of Medicine, Tallahassee, FL 32306-4300, USA

Program in Neuroscience, The Florida State University College of Medicine, Tallahassee, FL 32306-4300, USA
e-mail: Cathy.Levenson@med.fsu.edu

© Springer International Publishing AG 2017
M. Aschner and L.G. Costa (eds.), *Neurotoxicity of Metals*, Advances in Neurobiology 18, DOI 10.1007/978-3-319-60189-2_15

neuronal damage and death. Zinc-induced neurotoxicity has been associated with a number of brain disorders and injury including ischemic brain injury (stroke), traumatic brain injury, and seizures. Under these conditions, excessive free zinc is released into the synaptic cleft. Upon release, zinc modulates a number of postsynaptic neuronal receptors, with excess zinc leading to neuronal death (Inoue et al. 2015). While the exact mechanisms of zinc toxicity during these pathological conditions are not fully understood, there is evidence that the influx and accumulation of excess zinc causes excitotoxicity, generates oxidative stress, and impairs neuronal energy production (Morris and Levenson 2012). In this review, we will explore the most recent studies examining the mechanisms of zinc-induced toxicity.

Zinc Toxicity and Neuronal Signaling

There are a number of important postsynaptic receptors, such as the voltage-gated calcium channels (VGCC), N-methyl-D-aspartate receptors (NMDAR), and α-amino-3-hydroxy-5-methyl-4-isoxazolepropionate receptors (AMPAR), that are not only modulated by zinc but also permeable to zinc ions (Inoue et al. 2015). NMDA receptors, for example, have a high-affinity zinc-binding site that can bind synaptically released zinc at nanomolar concentrations. Zinc is thus responsible for fine-tuning the activity of these important glutamate receptors. Synaptic zinc has also been shown to diffuse outside of the synaptic cleft where it can inhibit extrasynaptic NMDA receptors (Anderson et al. 2015). The excess release of free ionic zinc can not only regulate the amount and type of NMDA receptor expression (Wang et al. 2015) but acts as an intermediate in the cascade of biochemical events including calcium dysregulation, production of reactive oxygen species, mitochondrial disruption, and excitotoxicity leading to postsynaptic neuronal damage and death (Granzotto and Sensi 2015). This cascade has been shown to be important in a wide variety of clinical scenarios such as trauma, epilepsy, and stroke. Additionally, there is now evidence of mutations in NMDA receptors with altered zinc affinities that may have implications for neurodevelopment leading to a variety of developmental disorders including childhood epilepsy and cognitive deficits (Serraz et al. 2016).

Zinc is also an allosteric modulator of gamma-aminobutyric acid A receptors (GABA$_A$R) and glycine receptors (GlyR). GABA$_A$R and GlyR are pentameric ionotropic receptors and ligand-gated chloride ion channels that mediate inhibitory neurotransmission. Because the balance between inhibitory and excitatory synaptic transmission is needed for normal brain function, these receptors play an important role in preventing neurotoxicity. Under normal conditions, zinc inhibits GABA receptors. Zinc has a biphasic effect on Gly receptors such that low concentrations of zinc potentiate GlyR function and high concentrations inhibit GlyR function (Kuenzel et al. 2016). Using a recombinant in vitro model exposed to the ligands GABA or glycine and increasing concentrations of zinc (10 nM–1 mM), zinc concentrations of >100 μM were required for zinc inhibition of GABA$_A$R and GlyR

(Kuenzel et al. 2016). Reports have also suggested that $GABA_A$ receptors without the γ subunit are the most sensitive to zinc inhibition, while receptors with the γ subunit are the least sensitive to zinc inhibition (Smart et al. 1991; Hosie et al. 2003). This suggests that the subunit combinations of $GABA_A$R and potentially GlyR are important for the sensitivity of zinc inhibition. Further examination of these receptors could give further insight into the mechanisms underlying neurotoxicity of zinc.

Mitochondrial Function and Energy Production

It has been known for quite some time that excess zinc can inhibit cellular energy production leading to neuronal death. A wide variety of mechanistic hypotheses have been proposed to explain the role of excess zinc and mitochondrial dysfunction including zinc-mediated alterations in Kreb's cycle intermediates, electron transport chain components, mitochondrial calcium (Pivovarova et al. 2014), and essential cofactors (e.g., NAD^+), many of which have been thoroughly reviewed (Dineley et al. 2003; Floriańczyk and Trojanowski 2009). The following represents the most recent data on the role of excess zinc in cellular energy regulation and production.

AMPK Recently, new information on a possible role for AMP-activated protein kinase (AMPK) in zinc-mediated neuronal death has appeared. AMPK is an energy-sensing protein needed for cellular energy homeostasis. During zinc-induced neuronal death, the upstream kinase liver kinase B1 (LKB1) activates AMPK which in turn induces the pro-apoptotic protein Bim. This leads to caspase-3 activation and zinc-induced apoptosis. Additionally, inhibition of AMPK has a neuroprotective effect by reducing zinc-induced neuronal death (Eom et al. 2016). Thus, blocking this LKB1-AMPK-Bim signaling cascade could have potential therapeutic efficacy in ischemic and acute brain injury by preventing zinc-induced neuronal damage.

NAD^+ Depletion It is generally accepted that excess zinc reduces levels of nicotinamide-adenine dinucleotide (NAD^+) (Sheline et al. 2010; Cai et al. 2006). More recently, treatment of primary cortical cultures with 40 μM zinc for 12 h and 24 h reduced NAD^+ levels to approximately $28 \pm 12\%$ and $14 \pm 6\%$ of controls, respectively (Kim et al. 2016a). Treatment with ethyl pyruvate (EP), an anti-inflammatory and anti-oxidative agent, during zinc-induced toxicity not only replenished NAD^+ levels but also chelated intracellular zinc, resulting in decreased neuronal cell death (Kim et al. 2016a). In animal models of stroke and traumatic brain injury, EP has shown to have neuroprotective qualities (Shi et al. 2015; Turkmen et al. 2016). NAD^+ replacement and zinc chelation therapy could underlie the effect of EP in these disorders.

ERK Activation and Signaling The RAS/MEK/ERK signaling pathway also appears to be involved in zinc-induced mitochondrial dysfunction (He and Aizenman 2010). Under normal conditions, Ras activates the kinases MEK1/MEK2 resulting in the phosphorylation and activation of ERK1/ERK2. ERK1/ERK2 is then translocated to the nucleus where it activates transcription factors such as Elk1. The most recent analysis of the role of excess zinc in the process suggests that ERK1/ERK2 is required for the zinc-mediated mitochondrial hyperpolarization that leads to neuronal death (He and Aizenman 2010).

An additional role for ERK activation in zinc-induced neurotoxicity involves the striatal-enriched protein tyrosine phosphatase (STEP). Excess zinc results in the hyperphosphorylation of the membrane-associated STEP isoform, $STEP_{61}$. At the same time, exogenous zinc activates brain-derived neurotrophic factor (BDNF) and its receptor TrkB resulting in protein kinase A (PKA) activity. Together $STEP_{61}$ and PKA lead to ERK2 phosphorylation and activation (Poddar et al. 2016), which has been associated with mitochondrial dysfunction and neuronal death.

Hormone Regulated Mechanisms of Zinc Toxicity

Angiotensin II Angiotensin II is best known for its role in vasoconstriction. However, an early finding that the angiotensin system may be involved in neuroprotection led to the hypothesis that angiotensin modulated zinc-induced neuronal toxicity (Park et al. 2013). Using mouse cortical cell cultures, they showed that high levels of exogenously applied zinc (300 μM for 15 min) induced cell death, while a combined treatment with zinc and angiotensin II significantly increased NADPH oxidase, ROS levels, and cell death only in neurons, with no effect on astrocytes. While the exact mechanisms are not known, these actions appear to be through the activity of the angiotensin II type 2 receptor (AT2R) and not the type 1 receptor (AT1R) (Park et al. 2013). Preclinical studies involving animal models of both traumatic brain injury (Villapol et al. 2015) and ischemia (Panahpour et al. 2014) have demonstrated that blocking AT1R can provide neuroprotective effects. Given that Park et al. (2013) found that AT2R is involved in zinc-induced oxidative stress, these data suggest that more than one mechanism involving the hormone angiotensin may be involved in zinc neurotoxicity.

Stress Hormones There is a link between the stress response and neuronal zinc toxicity. Using an in vitro cell model of dopaminergic neurons, pretreatment with urocortin, a member of the corticotrophin-releasing factor neuropeptide family, did not affect cell death but increased the zinc permeable TRPM7 channel expression. However, urocortin-pretreated cells followed by incubation with extracellular zinc resulted in increased cell death via zinc influx into the cell through the TRPM7 channel, ultimately resulting in increased ROS levels and cell death (Kim et al. 2016b). Interestingly, the anesthetic lidocaine has been shown to not only inhibit the TRPM7 channel but also reduces TRPM7-medicted zinc toxicity (Leng et al. 2015).

Zinc-Mediated Aβ Aggregation and Neurotoxicity

Alzheimer's disease (AD) is an all-too-common neurodegenerative disorder, characterized by progressive loss of cortical and hippocampal neurons resulting in cognitive decline. There are a large number of factors including age, genetics, epigenetics, and a variety of modifiable environmental risk factors that appear to play a role in the development and progression of AD (Uchoa et al. 2016; Klein et al. 2016; Bellou et al. 2016). The pathological hallmarks of this disorder include hyperphosphorylation of the tau protein, leading to neurofibrillary tangles that disrupt neuronal function, and the development of extraneuronal AD plaques. These plaques are rich in aggregates of the Aβ peptide formed by the cleavage of amyloid precursor protein (APP) by the enzymes β- and ϒ-secretase, producing Aβ peptides that range from 39 to 42 amino acids in length (Greenough et al. 2013). While the role of zinc and these peptides in the etiology of AD are discussed in detail in other chapters in this volume, this section will focus on the most recent data related to the specific role of zinc interactions with the long form of Aβ ($A\beta_{42}$) and our expanding understanding of the role of zinc in neurotoxicity.

Zinc and Extracellular Aβ Aggregation The etiology and progression of AD has long been linked to the dysregulation of metals in the brain such as copper, iron, and particularly the essential trace element zinc (Bush 2013). This hypothesis is supported by findings that brain zinc concentrations in AD patients are significantly higher (>1000 µM) than age-matched controls (350 µM) (Lovelle et al. 1998). Because Aβ has long been known to form aggregates that precipitate in the presence of high concentrations of zinc in vitro, as reviewed in Greenough et al. (2013), a prevailing hypothesis has been that abnormally high zinc concentrations in the brain could lead to zinc-mediated Aβ aggregation and plaque formation. This model, however, was soon shown to be overly simplistic. Recent work has revealed significant debates over the role of zinc in Aβ-mediated neurotoxicity.

Historically, the possible interactions between zinc and $A\beta_{42}$ leading to aggregation of Aβ peptides have received a great deal of attention. However, the most recent data on the kinetics of Aβ fiber formation show that zinc can facilitate the aggregation of both $A\beta_{42}$ and $A\beta_{40}$. Interestingly, even very low concentrations of ionic zinc (Zn^{2+}) have been shown to result in extracellular aggregation of Aβ (Matheou et al. 2016). Additional evidence for the role of zinc in $A\beta_{40}$ aggregation came from work showing that Zn^{2+} coordinates with specific amino acid residues of $A\beta_{40}$ resulting in a shift from $A\beta_{40}$ monomers to dimers and oligomers. Zinc was shown to be responsible for stabilization of the potentially toxic oligomers while inhibiting the formation of less toxic fibrils (Xu et al. 2015; Abelein et al. 2015).

Even more recently, attention has focused on the N-terminal domain of the Aβ peptide ($A\beta_{1-16}$), a region of the peptide known to have a metal-binding domain (Mezentsev et al. 2016). Specifically, the AD-associated variant containing an isoaspartate at position 7 (isoAβ) and a phosphorylated serine at position 8 (pS8-Aβ) has now been shown to coordinate with the Aβ1–16 region of native Aβ through the metal-binding domain. The resulting zinc-dependent dimers (isoAβ – Zn – Aβ_{1-16}) then serve as seeds for the initiation of Aβ toxic aggregation (Mezentsev et al. 2016). Other work narrowed the key amino acid residues in the N-terminal region to Aβ_{6-14} and showed a specific role for the histidine at position 6, as well as the segment between amino acids 11 and 14 in the zinc-mediated interactions between Aβ molecules (Istrate et al. 2016).

Despite the significant progress that the above work represents, there are still contradictory data, likely due, at least in part, to the use of different methods to measure aggregation (Sharma et al. 2013). Thus, Sharma and colleagues sought to use a variety of methods including thioflavin (ThT) staining to detect amyloid aggregation, native gel electrophoresis and Western blotting, as well a transmission electron microscopy (TEM) to evaluate the effect of zinc ions on Aβ aggregation and neuronal toxicity. Their work confirms that the neurotoxic molecules are derived from Aβ_{42}, not Aβ_{40} peptides. Aβ_{42} monomers (that spontaneously form soluble oligomers) and preformed oligomers reduced the survival of cultured Neuro-2A cells by approximately 40% and 55%, respectively. They also confirmed that longer fibrils had little neurotoxicity with an 86% survival rate. The addition of zinc ions to Neuro-2A cells also resulted in in very little cell death, with an approximately 85% survival rate. What was surprising, however, was the finding that the addition of zinc ions combined with Aβ_{42} also did not significantly reduce cell survival. In contrast, the combination of copper (Cu^{2+}) and Aβ_{42} resulted in significant cytotoxicity, leading to the conclusion that copper is responsible for the formation of neurotoxic oligomers of Aβ_{42}, while sub-stoichiometric levels of zinc participate in the formation and stabilization of nontoxic, insoluble, amorphous aggregates of Aβ_{42} (Sharma et al. 2013).

Subsequent work similarly found that very low zinc concentrations (nM) reduced the self-affinity of Aβ molecules and the formation of Aβ–Aβ dimers that are more toxic than monomers or fibrils (Hane et al. 2016). These findings, while in contrast to much of the currently accepted data on the role of zinc and Aβ formation, are not without precedent (Garai et al. 2007). Clearly, these controversial, but very important, studies need to be addressed with future work to understand the full role of zinc in Aβ oligomer formation, neurotoxicity, and Alzheimer's disease.

Zinc and Intracellular Aβ Aggregation While most of the attention has been directed at the role of zinc in extracellular Aβ aggregation, plaque formation, and neurotoxicity, there is now a growing body of literature suggesting a neurotoxic function for intracellular zinc and Aβ oligomers. For example, lysosomal zinc has been implicated in the accumulation of Aβ oligomers in Chinese hamster ovary cells expressing the amyloid precursor protein/mutant presenilin 1 (APP/mPS1) gene. Treatment of cells with the zinc ionophore clioquinol not only reduced the intracellular accumulation of Aβ_{42} in these cells but protected them from lysosomal dilation and autophagy stimulated by the drug chloroquine (Seo et al. 2015). The authors suggest that manipulation of lysosomal zinc may represent a novel strategy for clearing Aβ from intracellular compartments.

In addition to lysosomal zinc stores that appear to modulate Aβ, there is also evidence of intraneuronal Aβ that is localized to the nucleus (Khmeleva et al. 2016). It appears that zinc increases the binding of DNA and RNA to aggregates of Aβ$_{42}$ and to amorphous aggregates, forming complexes that may contribute to neuronal toxicity (Khmeleva et al. 2016). These intracellular interactions appear to be particularly relevant in the medial performant pathways-dentate gyrus granule cell synapses of the hippocampus. This region of the brain not only has high concentrations of Zn^{2+} but is also especially vulnerable to damage in AD (Takeda and Tamano 2016).

Calcium and Zinc Interactions A recent review highlighted the possible interactions between calcium and zinc cations leading to Aβ aggregation and tau phosphorylation in AD (Sensi 2014). There are a number of reasons this relationship should not be ignored. First, when zinc is released from glutamateric synapses in the hippocampus, it modulates the activity of postsynaptic voltage-gated calcium channels (VGCC) and the glutamate receptors NMDAR and AMPAR, all of which function as calcium channels (Corona et al. 2011). A recent review highlighted the fact that glial cells also have functional NMDA receptors and hypothesized a role for zinc and Aβ in neuronal-glial communication (Hancock et al. 2014). Additionally, newly published data show that zinc activates a metabotropic Gq-coupled Zn^{2+}-sensing receptor known as mZnR/GPR39 (Abramovitch-Dahan et al. 2016). This receptor is expressed in brain regions including the frontal cortex, amygdala, and hippocampus (Khan 2016). Under normal conditions, synaptically released zinc activates mZnR/GPR39 and induces calcium signaling. However, in the presence of Aβ, zinc ions appear to be sequestered, resulting in the inhibition of mZnR/GPR39 and calcium signaling (Abramovitch-Dahan et al. 2016). Given the importance of calcium signaling in neuronal function, the role of zinc and Aβ and the possible implications for neurotoxicity clearly requires more attention and investigation.

Conclusions

While the presence of free zinc in neurons and the role of excess zinc in neuronal damage and death has been known for almost three decades, the mechanisms responsible for zinc-mediated neurotoxicity are still being explored and debated. Recent controversies, particularly surrounding the role of mitochondrial zinc, calcium, and Aβ aggregation, have all highlighted not only the importance of zinc in the brain but also the clear need for more research that will impact our understanding and treatment of brain injury, stroke, epilepsy, and Alzheimer's disease.

References

Abelein A, Gräslund A, Danielsson J. Zinc as chaperone-mimicking agent for retardation of amyloid β peptide fibril formation. Proc Natl Acad Sci U S A. 2015;112:5407–12.

Abramovitch-Dahan C, Asraf H, Bogdanovic M et al. Amyloid β attenuates metabotropic zinc sensing receptor, mZnR/GPR39, dependent Ca2+ , ERK1/2 and Clusterin signaling in neurons. J Neurochem. 2016. doi:10.1111/jnc.13760.

Anderson CT, Radford RJ, Zastrow ML, et al. Modulation of extrasynaptic NMDA receptors by synaptic and tonic zinc. Proc Natl Acad Sci U S A. 2015;112:E2705–14.

Bellou V, Belbasis L, Tzoulaki I, et al. Systematic evaluation of the associations between environmental risk factors and dementia: an umbrella review of systematic reviews and meta-analyses. Alzheimers Dement pii. 2016;S1552-5260(16):32853–9.

Bush AI. The metal theory of Alzheimer's disease. J Alzheimers Dis. 2013;33(Suppl 1):S277–81.

Cai AL, Zipfel GJ, Sheline CT. Zinc neurotoxicity is dependent on intracellular NAD levels and the sirtuin pathway. Eur J Neurosci. 2006;24:2169–76.

Cope EC, Morris DR, Scrimgeour AG, et al. Zinc supplementation provides behavioral resiliency in a rat model of traumatic brain injury. Physiol Behav. 2011;104:942–7.

Cope EC, Morris DR, Gower-Winter SD, et al. Effect of zinc supplementation on neuronal precursor proliferation in the rat hippocampus after traumatic brain injury. Exp Neurol. 2016;279:96–103.

Corona C, Pensalfini A, Frazzini V, et al. New therapeutic targets in Alzheimer's disease: brain deregulation of calcium and zinc. Cell Death Dis. 2011;2:e176.

Dineley KE, Votyakova TV, Reynolds IJ. Zinc inhibition of cellular energy production: implications for mitochondria and neurodegeneration. J Neurochem. 2003;85:563–70.

Eom JW, Lee JM, Koh JY, et al. AMP-activated protein kinase contributes to zinc-induced neuronal death via activation by LKB1 and induction of Bim in mouse cortical cultures. Mol Brain. 2016;9:14.

Floriańczyk B, Trojanowski T. Inhibition of respiratory processes by overabundance of zinc in neuronal cells. Folia Neuropathol. 2009;47:234–9.

Garai K, Sahoo B, Kaushalya SK, et al. Zinc lowers amyloid-beta toxicity by selectively precipitating aggregation intermediates. Biochemistry. 2007;46:10655–63.

Granzotto A, Sensi SL. Intracellular zinc is a critical intermediate in the excitotoxic cascade. Neurobiol Dis. 2015;81:25–37.

Greenough MA, Camakaris J, Bush AI. Metal dyshomeostasis and oxidative stress in Alzheimer's disease. Neurochem Int. 2013;62:540–55.

Hancock SM, Finkelstein DI, Adlard PA. Glia and zinc in ageing and Alzheimer's disease: a mechanism for cognitive decline? Front Aging Neurosci. 2014;6:137.

Hane FT, Hayes R, Lee BY, et al. Effect of copper and zinc on the single molecule self-affinity of Alzheimer's amyloid-β peptides. PLoS One. 2016;11:e0147488.

He K, Aizenman E. ERK signaling leads to mitochondrial dysfunction in extracellular zinc-induced neurotoxicity. J Neurochem. 2010;114:452–61.

Hosie AM, Dunne EL, Harvey RJ, et al. Zinc-mediated inhibition of GABA(a) receptors: discrete binding sites underlie subtype specificity. Nat Neurosci. 2003;6:362–9.

Inoue K, O'Bryant Z, Xiong ZG. Zinc-permeable ion channels: effects on intracellular zinc dynamics and potential physiological/pathophysiological significance. Curr Med Chem. 2015;22:1248–57.

Istrate AN, Kozin SA, Zhokhov SS, et al. Interplay of histidine residues of the Alzheimer's disease Aβ peptide governs its Zn-induced oligomerization. Sci Rep. 2016;6:21734.

Khan MZ. A possible significant role of zinc and GPR39 zinc sensing receptor in Alzheimer disease and epilepsy. Biomed Pharmacother. 2016;79:263–72.

Khmeleva SA, Radko SP, Kozin SA, et al. Zinc-mediated binding of nucleic acids to amyloid-β aggregates: role of histidine residues. J Alzheimers Dis. 2016;54:809–19.

Kim SW, Lee HK, Kim HJ, et al. Neuroprotective effect of ethyl pyruvate against Zn(2+) toxicity via NAD replenishment and direct Zn(2+) chelation. Neuropharmacology. 2016a;105:411–9.

Kim Y, Oh HG, Cho YY, et al. Stress hormone potentiates Zn(2+)-induced neurotoxicity via TRPM7 channel in dopaminergic neuron. Biochem Biophys Res Commun. 2016b;470:362–7.

Klein HU, Bennett DA, De Jager PL. The epigenome in Alzheimer's disease: current state and approaches for a new path to gene discovery and understanding disease mechanism. Acta Neuropathol. 2016;132:503–14.

Kuenzel K, Friedrich O, Gilbert DF. A recombinant human pluripotent stem cell line stably expressing halide-sensitive YFP-I152L for GABAAR and GlyR-targeted high-throughput drug screening and toxicity testing. Front Mol Neurosci. 2016;9:51.

Leng TD, Lin J, Sun HW, et al. Local anesthetic lidocaine inhibits TRPM7 current and TRPM7-mediated zinc toxicity. CNS Neurosci Ther. 2015;21:32–9.

Lovelle MA, Robertson JD, Teesdale WJ, et al. Copper, iron and zinc in Alzheimer's disease senile plaques. J. Neurol Sci. 1998;158:47–52.

Matheou CJ, Younan ND, Viles JH. The rapid exchange of zinc(2+) enables trace levels to profoundly influence amyloid-β Misfolding and dominates assembly outcomes in cu(2+)/Zn(2+) mixtures. J Mol Biol. 2016;428:2832–46.

Mezentsev YV, Medvedev AE, Kechko OI, et al. Zinc-induced heterodimer formation between metal-binding domains of intact and naturally modified amyloid-beta species: implication to amyloid seeding in Alzheimer's disease? J Biomol Struct Dyn. 2016;21:1–10.

Morris DR, Levenson CW. Ion channels and zinc: mechanisms of neurotoxicity and neurodegeneration. J Toxicol. 2012;2012:785647.

Panahpour H, Nekooeian AA, Dehghani GA. Blockade of central angiotensin II AT1 receptor protects the brain from ischemia/reperfusion injury in normotensive rats. Iran J Med Sci. 2014;39:536–42.

Park MH, Kim HN, Lim JS, et al. Angiotensin II potentiates zinc-induced cortical neuronal death by acting on angiotensin II type 2 receptor. Mol Brain. 2013;6:50.

Pivovarova NB, Stanika RI, Kazanina G, et al. The interactive roles of zinc and calcium in mitochondrial dysfunction and neurodegeneration. J Neurochem. 2014;128:592–602.

Poddar R, Rajagopal S, Shuttleworth CW, et al. Zn2+−dependent activation of the Trk signaling pathway induces phosphorylation of the brain-enriched tyrosine phosphatase STEP: MOLECULAR BASIS FOR ZN2+−INDUCED ERK MAPK ACTIVATION. J Biol Chem. 2016;291:813–25.

Sensi S. Metal homeostasis in dementia. Free Radic Biol Med. 2014;75(Suppl 1):S9.

Seo BR, Lee SJ, Cho KS, et al. The zinc ionophore clioquinol reverses autophagy arrest in chloroquine-treated ARPE-19 cells and in APP/mutant presenilin-1-transfected Chinese hamster ovary cells. Neurobiol Aging. 2015;36:3228–38.

Serraz B, Grand T, Paoletti P. Altered zinc sensitivity of NMDA receptors harboring clinically-relevant mutations. Neuropharmacology. 2016;109:196–204.

Sharma AK, Pavlova ST, Kim J, et al. The effect of cu(2+) and Zn(2+) on the Aβ42 peptide aggregation and cellular toxicity. Metallomics. 2013;5:1529–36.

Sheline CT, Cai AL, Zhu J, et al. Serum or target deprivation-induced neuronal death causes oxidative neuronal accumulation of Zn2+ and loss of NAD+. Eur J Neurosci. 2010;32:894–904.

Shi H, Wang HL, Pu HJ, et al. Ethyl pyruvate protects against blood-brain barrier damage and improves long-term neurological outcomes in a rat model of traumatic brain injury. CNS Neurosci Ther. 2015;21:374–84.

Smart TG, Moss SJ, Xie X, et al. GABAA receptors are differentially sensitive to zinc: dependence on subunit composition. Br J Pharmacol. 1991;103:1837–9.

Takeda A, Tamano H. Innervation from the entorhinal cortex to the dentate gyrus and the vulnerability to Zn2. J trace Elem med Biol 2016. pii:S0946-672X(16)30076-1.

Turkmen S, Cekic Gonenc O, Karaca Y, et al. The effect of ethyl pyruvate and N-acetylcysteine on ischemia-reperfusion injury in an experimental model of ischemic stroke. Am J Emerg Med. 2016;34:1804–7.

Uchoa MF, Moser VA, Pike CJ. Interactions between inflammation, sex steroids, and Alzheimer's disease risk factors. Front Neuroendocrinol pii. 2016;S0091-3022(16):30039–5.

Villapol S, Balarezo MG, Affram K, et al. Neurorestoration after traumatic brain injury through angiotensin II receptor blockage. Brain. 2015;138:3299–315.

Wang G, Yu X, Wang D, et al. Altered levels of zinc and N-methyl-D-aspartic acid receptor underlying multiple organ dysfunctions after severe trauma. Med Sci Monit. 2015;21:2613–20.

Xu L, Shan S, Chen Y, et al. Coupling of zinc-binding and secondary structure in Nonfibrillar Aβ40 peptide Oligomerization. J Chem Inf Model. 2015;55:1218–30.

Young B, Ott L, Kasarskis E, et al. Zinc supplementation is associated with improved neurologic recovery rate and visceral protein levels of patients with severe closed head injury. J Neurotrauma. 1996;13:25–34.

Neurotoxicity of Copper

Felix Bulcke, Ralf Dringen, and Ivo Florin Scheiber

Abstract Copper is an essential trace metal that is required for several important biological processes, however, an excess of copper can be toxic to cells. Therefore, systemic and cellular copper homeostasis is tightly regulated, but dysregulation of copper homeostasis may occur in disease states, resulting either in copper deficiency or copper overload and toxicity. This chapter will give an overview on the biological roles of copper and of the mechanisms involved in copper uptake, storage, and distribution. In addition, we will describe potential mechanisms of the cellular toxicity of copper and copper oxide nanoparticles. Finally, we will summarize the current knowledge on the connection of copper toxicity with neurodegenerative diseases.

Keywords Copper • Nanoparticles • Neurotoxicity • Neurodegenerative disease • Oxidative stress • Brain

Introduction

Copper represents the third most abundant essential transition metal in humans (Lewińska-Preis et al. 2011). After the liver, the brain is the organ containing the highest copper content (Szerdahelyi and Kása 1986). In its function as a cofactor and/or as structural component for several enzymes, copper participates in many physiological pathways, including energy metabolism, antioxidative defense and iron metabolism (Scheiber et al. 2014). Furthermore, copper has been linked to important biological processes including angiogenesis, response to hypoxia and neuromodulation (Scheiber et al. 2014). However, excess of cellular copper above the needs is deleterious. Given the requirement for copper on the one hand and the potential toxicity of copper on the other hand, cells have evolved mechanisms to

F. Bulcke • R. Dringen • I.F. Scheiber (✉)
Center for Biomolecular Interactions Bremen, Faculty 2 (Biology/Chemistry), University of Bremen, Bremen, Germany

Center for Environmental Research and Sustainable Technology, Bremen, Germany
e-mail: bulcke@uni-bremen.de; ralf.dringen@uni-bremen.de; ifscheiber@gmail.com

© Springer International Publishing AG 2017
M. Aschner and L.G. Costa (eds.), *Neurotoxicity of Metals*, Advances in Neurobiology 18, DOI 10.1007/978-3-319-60189-2_16

maintain cellular copper concentrations in a proper range. However, in genetic copper dyshomeostasis and in neurodegenerative diseases, these homeostatic mechanisms may fail and as a consequence copper deficiency or copper overload may occur. Following a brief overview on copper homeostasis and the essentiality of copper, this chapter will review the potential mechanisms of copper toxicity and list the neurologic diseases that have been connected to noxious effects of copper. In addition, we will discuss the toxicity of copper nanoparticles.

Brain Copper Content and Spatial Distribution

Total brain copper content has been estimated to be 3.1 µg g^{-1} wet weight in humans (Lech and Sadlik 2007), 5.5 µg g^{-1} wet weight in mice (Waggoner et al. 2000), and 1.0 µg g^{-1} wet weight in rat (Olusola et al. 2004). However, the brain is a heterogeneous organ with anatomically and physiologically different regions which vary in their specific copper contents (Davies et al. 2012; Krebs et al. 2014; Ramos et al. 2014). In humans, by far the highest copper contents are found in locus coeruleus and substantia nigra (Warren et al. 1960; Davies et al. 2012; Krebs et al. 2014), two structures which are rich in neuromelanin, but also areas within the hippocampus are strongly enriched in copper (Dobrowolska et al. 2008). While the copper concentration of the cerebrospinal fluid (CSF) in humans and rodents ranges between 0.2 and 0.5 µM (Stuerenburg 2000; Forte et al. 2004; Strozyk et al. 2009; Fu et al. 2015), the extracellular copper concentration in brain tissue may be higher. At least for the synaptic cleft, copper concentrations of up to 250 µM have been reported (Kardos et al. 1989; Hopt et al. 2003).

Brain copper content and distribution change during development, with age and in neurodegenerative diseases. An increase in copper content with age has been reported for rodents (Maynard et al. 2002; Tarohda et al. 2004; Wang et al. 2010; Fu et al. 2015) and cattle (Zatta et al. 2008), whereas no significant alteration with age was observed for most human brain regions (Loeffler et al. 1996; Davies et al. 2012; Ramos et al. 2014). The copper content in brains of Wilson's disease (WD) patients was shown to be almost eight times that of control brains, with homogeneous copper accumulation in all brain regions (Litwin et al. 2013). Such a nonselective increase of copper throughout the brain was also observed in the ATP7B null mice, a rodent model of Wilson's disease (Boaru et al. 2014). Brain copper contents of Menkes disease (MD) patients (Nooijen et al. 1981; Willemse et al. 1982) and mouse models of MD (Camakaris et al. 1979; Lenartowicz et al. 2015) were found to be lowered to values down to 20% of those found for controls. The amyloid plaques in Alzheimer's disease (AD) brain are strongly enriched in copper (Lovell et al. 1998), while cerebral cortex, frontal cortex, amygdala, and hippocampus were shown to be decreased by up to 50% in copper content (Deibel et al. 1996; Akatsu et al. 2012; James et al. 2012; Rembach et al. 2013). In Parkinson's disease (PD) and incidental Lewy body disease, a reduction by about 50% in copper content of substantia nigra and locus coeruleus has been reported (Ayton et al. 2013; Davies et al. 2014). Substantial lower copper levels have also been observed in hippocampal tissue from patients with mesial temporal lobe epilepsy associated with hippocampal sclerosis (Ristić et al. 2014) and in brains of scrapie-infected mice (Thackray et al. 2002), whereas an increase in copper was shown for the striatum of Huntington's disease (HD) patients (Dexter et al. 1992) and in iron-rich areas of the dentate nucleus of patients suffering from Friedreich's ataxia and spinocerebellar ataxia type 3 (Koeppen et al. 2012).

Copper Homeostasis

Cellular Copper Homeostasis

Many components of the cellular copper homeostasis machinery have been described at the molecular level (Fig. 1). The copper transport receptor (Ctr) 1 is considered as the major entry pathway for copper into mammalian cells (Lee et al. 2002a, b), but other copper uptake systems have also been reported (Lee et al. 2002b; Moriya et al. 2008; Kidane et al. 2012). Further evidence for such alternative transport mechanism was provided by data from cell-specific Ctr1 knockout mice (Nose et al. 2006; Kim et al. 2009). The copper transporter Ctr2 (Bertinato et al. 2008), the divalent metal transporter (DMT) 1 (Arredondo et al. 2003; Espinoza et al. 2012; Monnot et al. 2012; Lin et al. 2015), and anion transporters (Alda and Garay 1990; Zimnicka et al. 2011) have been discussed as possible candidate proteins mediating this alternative transport mechanism (Fig. 1). The accumulation of copper in the cytosol bears the risk of copper toxicity. However, under physiological conditions, the concentration of free copper within the cell is kept very low at around 10^{-18} M (Rae et al. 1999). This low concentration of free copper is maintained by efficient binding of copper to metallothioneins (MTs) and ligands of low molecular mass such as glutathione (GSH) (Scheiber et al. 2014). In addition, mitochondria are likely to contribute to the cellular copper buffering capacity (Cobine et al. 2004; Maxfield et al. 2004; Leary et al. 2009). A group of specialized proteins, termed copper chaperones, shuttle copper to copper-dependent enzymes and to organelles (Fig. 1), thereby protecting it from being scavenged by MTs or GSH. Atox1 transfers Cu^+ to the N-terminal metal-binding domains of the copper-transporting P-type ATPases ATP7A and ATP7B; the copper chaperone for superoxide dismutase (CCS) facilitates the insertion of copper into superoxide dismutase (SOD) 1, while Cox17, Sco1, Sco2, and Cox11 participate in the insertion of copper ions into mammalian cytochrome c oxidase (Robinson and Winge 2010). In addition, a yet to be identified copper ligand aids in the transport of copper into the mitochondrial matrix (Cobine et al. 2004; Vest et al. 2013). Cellular copper export in mammals relies on the function of two proteins, ATP7A and ATP7B (Fig. 1). These proteins belong to the protein family of P1B-type ATPases that use the energy of ATP hydrolysis to transport heavy metals across cellular membranes

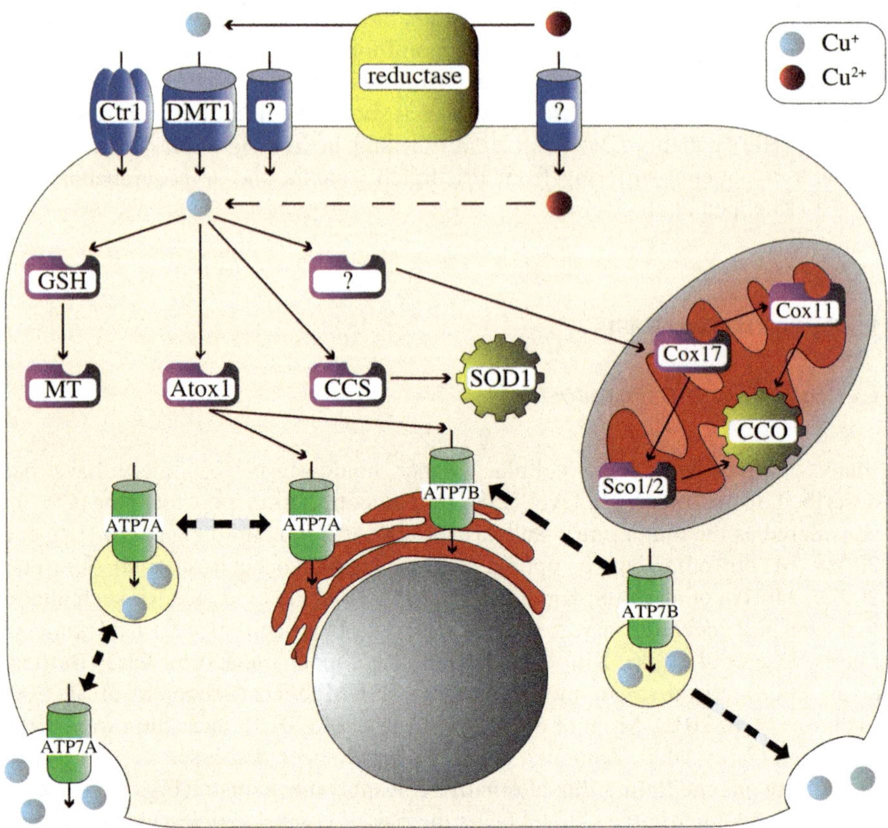

Fig. 1 Mechanism of cellular copper transport. Copper enters cells via the copper transporter Ctr1, DMT 1, and/or yet-unknown transporters. A cuprireductase provides Cu+, the preferred substrate for Ctr1 and DMT1. In cells, accumulated copper is sequestered by GSH or stored in metallothioneins (MT). Copper chaperones shuttle copper to its specific cellular targets. CCS provides copper to SOD1. A yet-unknown copper ligand aids in the transport of copper into the mitochondrial matrix and Cox17, Sco1/Sco2 and Cox11 participate in the insertion of copper into cytochrome c oxidase (CCO). Atox1 delivers copper to the copper-transporting P-type ATPases ATP7A and ATP7B that shuttle copper to the secretory pathway for subsequent incorporation into copper-dependent enzymes and mediate cellular copper efflux

(Arguello et al. 2007). In addition to their critical function in the efflux of cellular copper, ATP7A and ATP7B shuttle copper to the secretory pathway for incorporation into copper-dependent enzymes such as tyrosinase, peptidylglycine-amidating monooxygenase (PAM), dopamine β-monooxygenase (DβM), lysyl oxidase (LOX), and ceruloplasmin (Cp) (Scheiber et al. 2014). In the brain, ATP7A is further required for the release of copper from hippocampal neurons upon NMDA activation (Schlief et al. 2005).

Systemic Copper Homeostasis

Overall balance of systemic copper in the body is achieved by regulation of the rate of uptake of copper in the small intestine and efflux of copper from the liver in the bile (Scheiber et al. 2013). Most dietary copper is absorbed in the small intestine (Linder and Hazegh-Azam 1996), and Ctr1 has been shown to be essential for this process as mice with intestinal-specific knockout of Ctr1 exhibited severe copper deficiency and death by 3 weeks of age due to intestinal block of copper absorption (Nose et al. 2006). While it is clear that Ctr1 is required for copper to be bioavailable (Nose et al. 2006), its function in apical copper entry is still under controversial debate. In most studies, Ctr1 was observed to be localized to the apical surface (Kuo et al. 2006; Nose et al. 2010), but Zimnicka et al. (2007) reported that Ctr1 is located at the basolateral membrane in the enterocytes. Furthermore, enterocytes deficient in Ctr1 hyperaccumulated copper (Nose et al. 2006), suggesting the contribution of other transporters in the transport of copper across the brush border of the intestinal epithelial cells. Indeed, DMT1 (Arredondo et al. 2003; Espinoza et al. 2012) and anion transporters (Zimnicka et al. 2011) have been implicated in this process. The copper efflux protein ATP7A is responsible for the transport of copper across the basolateral surface of intestinal epithelia cells into portal circulation (Scheiber et al. 2013). Increasing dietary copper causes ATP7A in intestinal enterocytes to traffic from the *trans*-Golgi network (TGN) to sub-basolateral membrane vesicles that periodically fuse with the plasma membrane to release copper into the basolateral milieu (Monty et al. 2005; Nyasae et al. 2007). ATP7B is the transporter responsible for efflux of copper from the liver into the bile, the principle pathway for removing excess copper from the body (Scheiber et al. 2013). Excess copper in the hepatocyte stimulates trafficking of this protein from the TGN to vesicles close to the apical membrane of the hepatocyte that abuts the biliary canaliculus (Cater et al. 2006), thus increasing the capacity of rapid copper sequestration from the cytosol and allowing for subsequent excretion of excess copper via exocytosis.

Brain Copper Homeostasis

Brain copper homeostasis is regulated by the brain barrier systems, i.e., the blood-brain barrier (BBB) and blood-CSF barrier (BCB). The main route for copper entry into the brain parenchyma appears to be the BBB (Fig. 2), requiring the combined action of Ctr1 and ATP7A (Choi and Zheng 2009; Monnot et al. 2011; Zheng and Monnot 2012; Fu et al. 2014). Ctr1 is strongly expressed in brain capillary endothelial cells (Kuo et al. 2006) and has been proposed to locate on the luminal side of these cells (Kaler 2011) making it an ideal candidate in regulating copper uptake from the blood. Copper levels in brains of Ctr1-heterozygous knockout mice are reduced to about 50% of that of wild-type animals (Lee et al. 2001) confirming the fundamental role for Ctr1 in the transport of copper across the BBB into the brain.

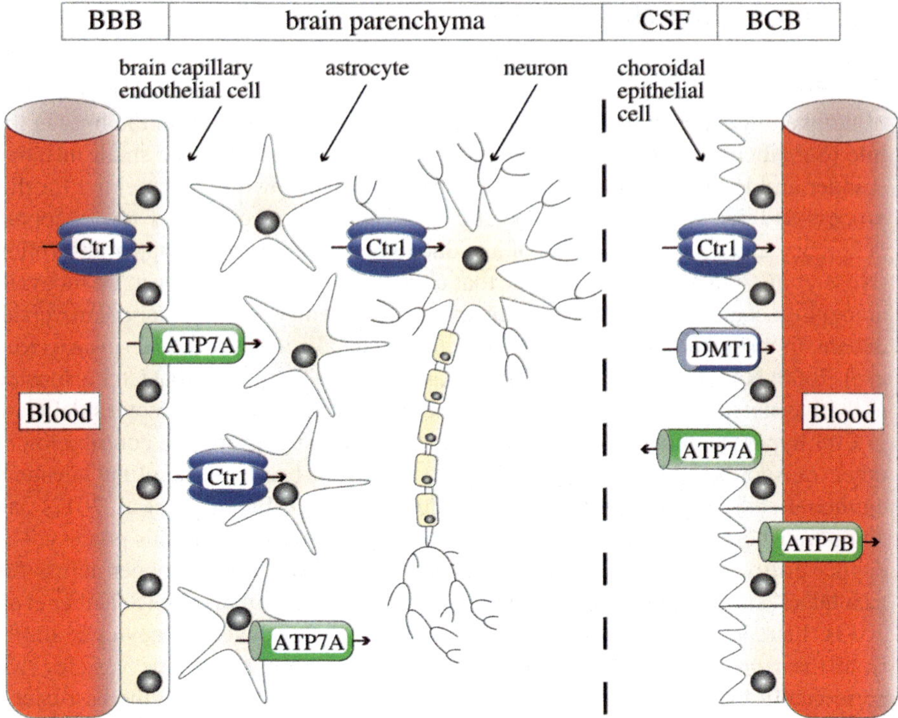

Fig. 2 Brain copper homeostasis. The blood-brain barrier (BBB) appears to be the main route for copper entry into the brain. Brain capillary endothelial cells take up copper from the blood via Ctr1. These cells release copper via ATP7A into the brain parenchyma and copper is subsequently taken up into astrocytes, neurons and other brain cells, most likely predominantly via Ctr1. At least astrocytes release via ATP7A excess of copper into the CSF. The choroid plexus functions in extracting copper from the CSF. Copper taken up via Ctr1 and/or DMT1 into choroidal epithelial cells that constitute the blood-CSF barrier (BCB) is either released into the blood via ATP7B or stored for potential release by ATP7A back into the CSF

The requirement of ATP7A in copper export from brain capillary endothelial cells has been demonstrated in a cell culture model for these cells (Qian et al. 1998) and dysfunction of ATP7A results in hyperaccumulation of copper in brain capillaries of mouse models of Menkes disease (Kodama 1993; Yoshimura et al. 1995). ATP7A mRNA levels in the BBB were found to be about 13 times higher than ATP7B mRNA levels, supporting a predominant role for ATP7A in copper export from brain capillary endothelial cells into brain parenchyma (Fu et al. 2014). Although the transport of copper from blood circulation into the choroid plexus (CP) is faster than into cerebral capillaries, further transport of copper from the CP into the CSF is very slow, virtually prohibiting the passage of copper from blood to CSF (Choi and Zheng 2009; Fu et al. 2014). Moreover, in vitro and in vivo data demonstrated that the direction of BCB in transporting copper is from the CSF to blood (Fig. 2), providing evidence that the BCB's role in CNS copper homeostasis is to remove copper from the CSF (Monnot et al. 2011). However, the situation might be different in the developing brain for which the BCB has been hypothesized to be the primary route of copper entry (Donsante et al. 2010). Using a choroidal cell model, it was shown that both Ctr1 and DMT1 mediate copper accumulation by choroidal epithelial cells (Monnot et al. 2012; Zheng et al. 2012) although Ctr1 appears to play a much more significant role in transporting Cu into the cells than does DMT1 (Zheng et al. 2012). Both transporters are enriched at the apical membrane of epithelial cells of the CP (Kuo et al. 2006; Wang et al. 2008; Davies et al. 2012; Zheng and Monnot 2012) consistent with the proposed function of the CP in extracting copper from the CSF. In contrast to the BBB, ATP7B mRNA is more abundantly expressed in choroidal epithelial cells than ATP7A. However, data from siRNA knockdown experiments indicates that both Cu-transporting ATPases, ATP7A and ATP7B, contribute to copper transport across the BCB (Fu et al. 2014). Furthermore, upon copper incubation of rat choroid plexus tissue, ATP7B was shown to traffic from a perinuclear location toward the basolateral membrane, whereas ATP7A translocated toward the apical microvilli, suggesting that while ATP7B is responsible for release of copper into the blood, ATP7A is responsible for the efflux of copper from choroidal epithelial cells into the CSF (Fu et al. 2014). Such trafficking behavior of ATP7A and ATP7B in choroidal epithelial cells has been previously hypothesized by Kaler (Kaler 2011) but contrasts the situation reported for other polarized cells (Monty et al. 2005; Llanos et al. 2008; Michalczyk et al. 2008) and to the localization of ATP7A and ATP7B reported for human epithelial cells of the CP (Davies et al. 2012).

Essentiality of Copper

By virtue of its function as cofactor and/or structural component in a number of important enzymes, copper is essential for a variety of biological pathways (Scheiber et al. 2014). The final step of the electron transfer in the mitochondrial respiratory chain, the oxidation of reduced cytochrome c by dioxygen, is catalyzed by cytochrome c oxidase, a member of the superfamily of heme-copper-containing oxidases (Ferguson-Miller and Babcock 1996). The copper-dependent SODs 1 and 3 contribute to the antioxidative defense by catalyzing the dismutation of superoxide to oxygen and hydrogen peroxide (Perry et al. 2010). The multi-copper oxidase Cp plays an important role in iron homeostasis and thus links copper and iron metabolism (Healy and Tipton 2007). Lysyl oxidase has a crucial role in the formation, maturation, and stabilization of connective tissues by catalyzing the cross-linking of elastin and collagen (Lucero and Kagan 2006). Both DβM and PAM belong to a small class of copper proteins found exclusively in mammals (Klinman 2006). DβM catalyzes the final step in noradrenaline synthesis, the oxidative hydroxylation of dopamine to noradrenaline, and thus plays an important role in the catecholamine metabolism (Timmers et al. 2004). PAM exclusively catalyzes the C-terminal α-amidation of propeptides, a posttranslational modification essential for the bioactivity of diverse physiological regulators, including peptide hormones, neurotransmitters, and growth factors (Bousquet-Moore et al. 2010b). Tyrosinase is the key enzyme in the biogenesis of melanin pigments. Among others, tyrosinase catalyzes the hydroxylation of L-tyrosine to L-3,4-dihydroxyphenylalanine (L-DOPA), the rate-limiting step in the biosynthesis of melanins and dopamine, and its subsequent oxidation to DOPA quinone (Olivares and Solano 2009). Primary and secondary copper amine oxidases regulate biogenic amine levels by catalyzing their oxidative deamination (Klinman 1996).

In addition to its requirement for enzymes, a growing body of evidence indicates a role for copper in biological processes such as coagulation (Wakabayashi et al. 2001), angiogenesis (Urso and Maffia 2015), response to hypoxia (Feng et al. 2009), nonclassical secretion (Prudovsky et al. 2008), and neuromodulation (Gaier et al. 2013). Synaptosomes and primary hippocampal neurons have been shown to release copper following depolarization (Kardos et al. 1989; Hopt et al. 2003; Schlief et al. 2005) in concentrations sufficient to modulate excitatory and inhibitory neurotransmission (Gaier et al. 2013; Scheiber et al. 2014). Several of the neuromodulatory functions of copper appear to be directly linked to interactions of copper with receptors, but copper may exert its neuromodulatory functions also by altering intracellular signaling pathways (Gaier et al. 2013; Scheiber et al. 2014). The exact role of copper in synaptic physiology remains to be elucidated (for review, see Gaier et al. 2013). However, synaptically released endogenous copper and exogenously applied copper protect primary hippocampal neurons against NMDA-mediated excitotoxic cell death (Schlief and Gitlin 2006) in a process that involves the cellular prion protein (Gasperini et al. 2015). While an inhibitory effect of copper on long-term potentiation (LTP) has been demonstrated using hippocampus slices that had been exposed to exogenous copper (Doreulee et al. 1997; Salazar-Weber and Smith 2011) and hippocampal slices of rats that had been fed a high-copper diet (Goldschmith et al. 2005; Leiva et al. 2009), copper has been shown to be required for amygdalar LTP (Gaier et al. 2014a, b).

The essentiality of copper is best illustrated by MD, a rare, X-linked recessive disorder caused by genetic defects in the copper-transporting ATPase ATP7A that manifests with clinical symptoms, including severe progressive neurological degeneration, increased seizure frequency, connective tissue abnormalities, muscular hypotonia, hypothermia, and abnormalities of the skin and hair (Kaler 2011; Kodama et al. 2011). As ATP7A is required for the transport of copper across the basolateral surface of intestinal epithelia cells into portal circulation, loss of function of ATP7A leads to failure of copper absorption in the intestine and hence to a systemic copper deficiency (Kodama et al. 2011). Treatment with parental copper can improve neurological outcomes when initiated in the neonatal period and the BBB is immature, but proves ineffective when initiated at later age due to the essential role of ATP7A for copper transport across the BBB (Kaler 2011; Kodama et al. 2011). Many of the clinical symptoms of MD can be ascribed to a decrease in the activities of secreted copper-dependent enzymes that rely on the function of ATP7A to receive their copper (Kaler 2011; Kodama et al. 2011). Decreased PAM activity and the subsequent lack of α-amidated peptides are thought to contribute to the neurodevelopmental delay and increased seizure frequency associated with MD (Bousquet-Moore et al. 2010a; Kaler 2011). Partial deficiency of DβM accounts for the elevated dopamine to noradrenaline ratio in plasma and CSF of MD patients (Kaler 1998). Hypopigmentation of the skin and hair is a consequence from reduced tyrosinase activity and lowered LOX activity is responsible for bone and connective tissue abnormalities (Kaler 2011; Kodama et al. 2011). However, low CCO activity as a consequence of impaired transport of copper into the brain is likely to be the major cause of the severe neurodegeneration associated with MD (Kaler 2013; Scheiber et al. 2014). In support of this view, a mouse model (Atp7a^{Nes}) in which the Atp7a gene was selective deleted in neural cells showed normal to slightly elevated brain copper levels and no signs of Menkes-like degenerative neuropathology and early mortality (Hodgkinson et al. 2015). Nevertheless, ATP7A has been shown to have a critical role in the availability of an NMDA receptor-dependent releasable pool of copper in primary hippocampal neurons (Schlief et al. 2005), which has been shown to protect these cells against NMDA-mediated excitotoxic cell death (Schlief et al. 2006). Failure of this copper-dependent neuroprotective pathway in MD may contribute to the extensive neurodegeneration seen in this fatal disease (Schlief et al. 2006; Schlief and Gitlin 2006; Hodgkinson et al. 2015).

Toxicity of Copper

Copper toxicity in individuals without genetic susceptibility is rare (de Romaña et al. 2011). Acute copper toxicity has been described for individuals that accidentally or with suicidal intention ingested high doses of copper (Franchitto et al. 2008). For copper doses up to 1 gram, gastrointestinal symptoms predominate. Ingestion of higher copper doses may result in nausea, vomiting, headache, diarrhea, hemolytic anemia, gastrointestinal hemorrhage, liver and kidney failure and even death may occur (Franchitto et al. 2008). Chronic copper toxicity is a feature of WD, Indian childhood cirrhosis, and idiopathic chronic toxicosis that originate from genetic defects affecting copper metabolism (Scheiber et al. 2013). In addition, copper may contribute as a noxious metal to the pathology of neurodegenerative disorders, including AD, PD, and HD (Scheiber et al. 2014).

Mechanisms of Copper Toxicity

Oxidative Mechanisms

Copper toxicity is in large part a consequence of the redox activity of copper. Copper can easily cycle between the reduced Cu(I) and the oxidized Cu(II) oxidation state, allowing it to facilitate redox reactions and to coordinate a large variety of ligands (Liu et al. 2014). This feature is utilized by most of the copper-dependent enzymes that employ copper as a cofactor in fundamental redox reactions (Liu et al. 2014). However, the redox nature that makes copper biologically useful also renders it potentially toxic. Redox cycling of copper in the presence of superoxide or reducing agents such as ascorbic acid or GSH may catalyze the generation of highly reactive hydroxyl radicals from hydrogen peroxide via the Haber-Weiss cycle (Gunther et al. 1995). The hydroxyl radical, being the most powerful oxidizing radical likely to arise in biological systems, is capable of initiating oxidative damage by abstracting the hydrogen from an amino-bearing carbon to form a carbon-centered protein radical or from an unsaturated fatty acid to form a lipid radical and by inducing DNA strand breaks and oxidation of bases (Gaetke et al. 2014). In addition, copper ions are capable of accelerating lipid peroxidation by splitting lipid hydroperoxides in a reaction analogous to the Fenton reaction, giving alkoxyl and peroxyl radicals thereby propagating the chain reaction (Halliwell 2006).

Mitochondria are major targets for copper-induced oxidative damage. Ultrastructural changes of liver mitochondria in WD patients; in the Long-Evans Cinnamon (LEC) rat, a rat model of WD; and in rats with dietary copper overload (Sokol et al. 1990; Zischka et al. 2011; Fanni et al. 2014) are accompanied by functional impairment of enzymes of mitochondrial respiration (Sokol et al. 1993; Gu et al. 2000; Zischka et al. 2011). Altered activities of respiratory chain enzyme complexes similar to that found in the liver have been observed in brain tissue of ATP7B$^{-/-}$ mice (Sauer et al. 2011). Treatment of cultured hepatocytes mixed neuronal/glial cultures or neuroblastoma cultures with copper was shown to inhibit mitochondrial pyruvate and α-ketoglutarate dehydrogenase complexes, which was attributed to mitochondrial ROS formation (Sheline and Choi 2004; Arciello et al. 2005). As markers of lipid peroxidation are elevated in hepatocyte mitochondria of WD patients, animal models of WD and rats with dietary copper overload, oxidative membrane damage is likely to contribute to the mitochondrial alterations observed under these copper-overload conditions (Sokol et al. 1990, 1994; Zischka et al. 2011). In addition, increased levels of phosphatidic acid and phosphatidyl hydroxyl acetone have been observed in liver mitochondria of ATP7B$^{-/-}$ mice (Yurkova et al. 2011), indicative of ROS-mediated fragmentation of mitochondrial cardiolipin (Yurkova et al. 2008). Cardiolipin is a phospholipid crucial for integrity and function of the mitochondrial inner membrane and oxidation of cardiolipin has been shown to impair oxidative phosphorylation and to cause induction of apoptosis (Hauck and Bernlohr 2016). The induction of the mitochondrial permeability transition as a consequence of copper-mediated oxidative stress was observed in primary hepatocytes (Roy et al. 2009) and primary astrocytes, but not in primary neurons (Reddy et al. 2008). Mitochondrial permeability transition results in increased permeability of the inner mitochondrial membrane leading to cell death via apoptosis and/or necrosis (Javadov and Kuznetsov 2013).

Extensive genome damage is a common feature of metal-overload conditions, including many neurological disorders, in particular base modifications and strand breaks (Hegde et al. 2011; Mitra et al. 2014). The induction of oxidative DNA damage by copper and various copper complexes has been demonstrated in vitro with isolated DNA (Sagripanti and Kraemer 1989; Tkeshelashvili et al. 1991) and cultured mammalian cell cultures (Ma et al. 1998; Alimba et al. 2016) as well as in vivo (Prá et al. 2008; Georgieva et al. 2013). Copper is thought to exert its genotoxic effect via a site-specific mechanism that involves the generation of singlet oxygen and/or hydroxyl radicals bound to or in close proximity of high-affinity copper-binding sites on double-stranded DNA rather than via the generation of free hydroxyl radicals (Sagripanti and Kraemer 1989; Tkeshelashvili et al. 1991; Frelon et al. 2003). Facilitation of autoxidation of catecholamines such as adrenaline, L-DOPA, dopamine, and 6-hydroxydopamine by copper results not only in an increased production of superoxide (Halliwell 2006), but complexes resulting from catecholamine oxidation products and copper also oxidatively damage DNA (Lévay et al. 1997; Spencer et al. 2011). This observation has been used to explain the selective copper neurotoxicity in neurodegenerative diseases, in particular PD (Spencer et al. 2011).

The oxidative DNA damage exerted by copper and/or copper-induced oxidative stress may lead to activation of the tumor suppressor protein p53 (Phatak and Muller 2015) which in turn can trigger apoptosis by transcriptionally activating or repressing the expression of a panel of pro- and antiapoptotic proteins or by direct action at the mitochondria (Wang et al. 2014). Indeed, elevated p53 mRNA and protein levels and nuclear translocation of p53 have been shown in liver cells and neurons upon copper exposure (Strand et al. 1998; Narayanan et al. 2001; VanLandingham et al. 2002). A supporting role of p53 in copper-induced cell death has been demonstrated for neurons and liver cells deficient or mutated in p53 which are more resistant to the toxic effect of copper (Strand et al. 1998; VanLandingham et al. 2002). The induction of apoptosis in hepatocytes in response to copper has further been shown to involve the activation of the endogenous CD95 system (Strand et al. 1998), a downstream effector of p53-dependent apoptosis (Haupt et al. 2003), and the activation of acid sphingomyelinase and subsequent release of ceramide (Lang et al. 2007) by copper-induced ROS. As the induction of apoptosis via the CD95 system in hepatocytes has been shown to require the activation of acid sphingomyelinase in vivo (Kirschnek et al. 2000), copper may stimulate acid sphingomyelinase in these cells at least in part through the endogenous CD95 system (Lang et al. 2007). However, in erythrocytes, copper induced phosphatidylserine exposure and death via leukocyte-secreted acid sphingomyelinase, suggesting that ceramide might also be involved in CD95-independent pathways leading to hepatocyte and erythrocyte death after copper treatment (Lang et al. 2007).

Binding to Biomolecules

Although copper toxicity is ascribed in large part as a consequence to copper-induced oxidative stress, direct binding of copper to proteins should be considered. In this regard, copper has been shown to bind to the X-linked inhibitor of apoptosis (XIAP), an antiapoptotic protein that directly binds to and inhibits specific caspases, thereby inducing a conformational change in the protein as well as a decrease in its half-life (Mufti et al. 2006). These two changes make the cell more susceptible to apoptotic stimuli and may contribute to the pathophysiology of copper toxicosis syndromes (Mufti et al. 2007). In addition, copper may nonspecifically bind to thiol and amino groups in proteins unrelated to copper metabolism, thereby altering protein structure and modifying their biological functions (Letelier et al. 2005). Binding of copper has been shown to inhibit enzymatic activities of the cytochrome P450 oxidative system, GSH transferases, and lactate dehydrogenase (Letelier et al. 2005, 2006; Pamp et al. 2005). Noncompetitive inhibition of Na^+/K^+-ATPase from rat brain synaptic plasma membranes (Vasić et al. 1999; Krstić et al. 2005; Nedeljković and Horvat 2007) and rabbit kidney (Li et al. 1996) by copper most likely occurs via binding of copper to protein sulfhydryl groups (Vujisić et al. 2004). Na^+/K^+-ATPase is concentrated in the synaptic membranes where it mediates potassium uptake and sodium release, which are required to restore ionic equilibria after the passage of nervous impulse (de Lores Arnaiz and Ordieres 2014). Consequently, inhibition of Na^+/K^+-ATPase will lead to diverse alterations of neuronal behavior (de Lores Arnaiz and Ordieres 2014). Copper binding to proteins involved in DNA repair may contribute to copper-induced DNA damage. Copper has been shown to inhibit the activities of the DNA glycosylases NEIL1 and NEIL2 by forming stable complexes with these proteins (Hegde et al. 2010) and to inhibit both phosphatase and kinase activities of the enzyme polynucleotide kinase 3′-phosphatase (PNKP) that is responsible for preparing nicked DNA for ligation (Whiteside et al. 2010). Copper has further been shown to strongly inhibit DNA-binding affinity of the DNA nick-sensor poly(ADP-ribose)polymerase-1 (PARP-1) and H_2O_2-induced poly(ADP-ribosyl)ation in HeLa S3 cells (Schwerdtle et al. 2007). As binding to DNA lesions and the activity of PARP-1 depends on three zinc finger domains (Eustermann et al. 2011), copper may exert its inhibitory effect by displacing zinc and/or by oxidation of the cysteines complexing zinc in these zinc finger structure (Schwerdtle et al. 2007).

Alteration of gene expression and metabolic pathways may also contribute to copper toxicity. Utilizing the ATB7B$^{-/-}$ mice, an animal model for WD, Huster et al. (2007) provided evidence that despite significant copper accumulation, copper-mediated oxidative stress does not play a major role at early stages of the disease. Instead, in presymptomatic ATB7B$^{-/-}$ mice, copper overload was shown to have a distinct and selective effect on liver gene expression and metabolism: Accumulated copper selectively upregulated the molecular machinery associated with cell cycle and chromatin structure and downregulated lipid metabolism (Huster et al. 2007). In fact, transcripts of genes involved in lipid metabolism remain significantly down-regulated in ATP7B$^{-/-}$ mice liver at all stages of WD (Ralle et al. 2010). Transcripts of enzymes involved in key steps of cholesterol biosynthesis were found to be most affected and accompanied by a marked decrease in liver cholesterol and VLDL cholesterol in serum (Huster et al. 2007; Ralle et al. 2010). Furthermore, severe dysregulation of sterol metabolism was observed in brains of ATP7B$^{-/-}$ mice (Sauer et al. 2011). The mechanism through which copper induces its effects on gene expression is not yet fully revealed. However, analysis of downregulated signaling pathways revealed a significant involvement of specific nuclear receptors (Burkhead et al. 2011). Indeed, NR3C1/glucocorticoid receptor (GR) and NR1H4/farnesoid X receptor (FXR), two key nuclear receptors with functions in lipid metabolism, are less abundant in nuclei of ATP7B$^{-/-}$ hepatocytes (Wilmarth et al. 2012). Also nuclear receptor target gene expression and activity are impaired in HepG2 cells treated with copper, livers from ATP7B$^{-/-}$ mice, and hepatic autopsy samples of WD patients (Wooton-Kee et al. 2015). Recent evidence suggests that copper directly decreases nuclear receptor function by competing with zinc for occupancy of the DNA-binding zinc finger domains (Wooton-Kee et al. 2015). The selective effects of copper on gene expression may be explained by differences in zinc finger coordination among different zinc-containing transcription factors that may result in a spectrum of susceptibility to copper interaction with the zinc finger proteins (Wooton-Kee et al. 2015).

Increasing evidence suggests a neuromodulatory function of copper (Gaier et al. 2013; Scheiber et al. 2014). Several of the neuromodulatory functions of copper appear to be linked to its effects on voltage-gated ion channels and synaptic receptors, but copper may exert its neuromodulatory functions also by altering intracellular signaling pathways in neurons (Gaier et al. 2013; Scheiber et al. 2014). Thus, copper neurotoxicity may be in part a consequence of excess copper adversely affecting synaptic transmission and functions.

Neurotoxicity of Copper

A number of neurodegenerative disorders have been connected with disturbances in copper homeostasis in brain (Rivera-Mancia et al. 2010; Scheiber et al. 2014; Bandmann et al. 2015). Here we will only shortly mention the main characteristics of the disorders and will focus more on the evidence presented so far on the roles that copper deprivation or copper excess may play in the pathology of the diseases.

Neurologic Wilson Disease

WD is a rare, inherited autosomal recessive disease of copper metabolism that originates from a genetic defect in the copper-transporting ATPase ATP7B. Impaired ATP7B function in WD results in failure of biliary copper secretion, leading to copper accumulation in the liver, brain and other tissues as well as in failure of loading of Cp with copper (Dusek et al. 2015). The majority of patients with WD present either predominantly hepatic or neuropsychiatric symptoms, the latter occurring in up to 50% of WD patients (Das and Ray 2006). Neurologic symptoms in WD are manifold and include dysarthria, tremor, Parkinsonism, dystonia, ataxia, chorea and cognitive impairments (Lorincz 2010). Ventricular dilatation and generalized atrophy are common neuropathological abnormalities in the WD brain (Meenakshi-Sundaram et al. 2008). Macroscopic structural changes are most consistently observed in the basal ganglia, particularly in the dorsal striatum, but have also been reported for the thalamus, brainstem, and frontal cortex (Brewer and

Yuzbasiyan-Gurkan 1992; Meenakshi-Sundaram et al. 2008). Involvement of the white matter has been considered to be present in at least 10% of cases (Mikol et al. 2005). Copper toxicity is considered as primary cause of the brain damage associated with WD, although other factors, such as decreased Cp oxidase activity and subsequent disturbance of iron metabolism, may also contribute (Dusek et al. 2015). Copper content in brains of WD patients is strongly increased in all brain regions (Litwin et al. 2013) and a fair degree of correlation between the severity of neurodegeneration and cerebral copper content has been reported (Horoupian et al. 1988).

The occurrence of abnormal astrocytes, i.e. Alzheimer type I and II cells and Opalski cells, already in early stages of the disease is a typical neuropathological feature of WD (Mossakowski et al. 1970; Bertrand et al. 2001; Das and Ray 2006). Astrocytes, localized in the brain between neurons and capillary endothelial cells, are considered the first parenchymal cells to encounter metals crossing the BBB (Scheiber and Dringen 2013) and abnormal astrocytes in WD stain strongly for MT and copper (Bertrand et al. 2001; Mikol et al. 2005), suggesting that astrocytes accumulate excess copper, in order to protect neurons from copper toxicity. Such a neuroprotective function of astrocytes has been reported for cultured brain cells (Brown 2004) and is supported by data from the North Ronaldsay sheep, an animal model for copper toxicosis, where an elevated brain copper content was accompanied by increased expression of MT and copper accumulation in astrocytes (Haywood et al. 2008). However, during the course of WD, the storage capacity of astrocytes is likely to get exhausted, leading to astrocyte damage as well as to an increase in extracellular copper in the brain parenchyma. Thus, both impairments of astrocyte functions that are required for normal neuronal function (Parpura et al. 2012) and exposure of neurons to excess copper should be considered to contribute to neuronal death in WD.

Alzheimer Disease

AD is the most common neurodegenerative disease in humans with most of the cases representing the late-onset form that is sporadic with no obvious implication of genetic factors (Prakash et al. 2016). The disease is characterized by a progressive decline and ultimately loss of memory and multiple other cognitive functions along with psychiatric disturbances (Castellani et al. 2010). Aside from age, other risk factors include family history of dementia and genetic and environmental factors (Castellani et al. 2010). The major pathological hallmarks of AD are the presence of extracellular senile plaques, primarily composed of amyloid-β (Aβ) peptides of 40 and 42 residues, and intracellular neurofibrillary tangles, primarily constituted of hyperphosphorylated tau protein (Ballard et al. 2011).

Strong evidence implicates a dyshomeostasis of copper in the etiology of AD, but controversy exists regarding the role of copper in the pathogenic process. While some evidence supports a detrimental role of copper in AD, other studies suggest the opposite. In support of the former, Aβ peptides bind copper with high affinity, and the senile plaques are strongly enriched in copper (Eskici and Axelsen 2012).

Copper has been shown to precipitate Aβ peptides in vitro, and it has been suggested that copper triggers the formation of senile plaques (Roberts et al. 2012). However, with increasing copper:Aβ ratios, the aggregation pathway changes, and the aggregating peptide is diverted into soluble oligomeric forms that are thought to be the most neurotoxic Aβ species (Eskici and Axelsen 2012; Matheou et al. 2015). Although the precise mechanisms by which oligomeric Aβ species exert their toxic effects are unknown, copper may exacerbated the toxicity of such Aβ oligomers through the formation of ROS, as Aβ can mediate the reduction of Cu^{2+} to Cu^+ (Roberts et al. 2012), by increasing the specific inhibition of cytochrome c oxidase (Crouch et al. 2005) or by enhancing microglial activation (Yu et al. 2015). Moreover, copper has been implicated in tau pathology associated with AD, by stimulating the phosphorylation and aggregation of tau and by enhancing the toxicity of tau aggregates (Du et al. 2014; Voss et al. 2014).

On the contrary, lower copper contents in affected brain regions of AD patients (Loeffler et al. 1996) and mouse models for AD (Bayer et al. 2003) as compared to controls rather argue for a copper deficit contributing to the neurodegeneration in AD. Copper supplementation and administration of Cu(gtsm) as copper source improved the survival and cognitive functions in mouse models of AD (Bayer et al. 2003; Crouch et al. 2009). However, intake of copper had no effect on cognition in patients with mild AD (Kessler et al. 2008). Mechanistically, copper deficiency may exacerbate disease progression by influencing amyloid precursor protein processing and Aβ metabolism (Cater et al. 2008). In addition, copper deficiency may impair the activity of copper-dependent enzymes. In this regard, low activities of cytochrome c oxidase (Maurer et al. 2000) and SOD1 (Marcus et al. 1998) have been reported for the AD brain.

Parkinson Disease

PD is the second most common neurodegenerative disease in humans, with the majority of cases representing idiopathic PD (Thomas and Flint Beal 2007). PD is characterized by a complex motor disorder known as Parkinsonism that manifests with resting tremor, bradykinesia, rigidity and postural instability (Thomas and Flint Beal 2007). The pathological hallmarks of the disease are the loss of neuromelanin-containing dopaminergic neurons in the substantia nigra pars compacta and the presence of α-synuclein aggregates, named Lewy bodies (Thomas and Flint Beal 2007). The precise mechanisms underlying α-synuclein aggregation and nigral cell loss are unknown. Among others, oxidative stress, mitochondrial dysfunction, inflammation and dyshomeostasis of metals have been suggested to contribute to the pathogenesis of PD (Jomova et al. 2010).

The role of copper in PD is controversial, as some evidence points to a noxious role of copper in the pathology of PD, while other studies suggest a deficiency of copper in PD. Thus, copper has been demonstrated to bind to both soluble and membrane-bound α-synucleins with high affinity, to accelerate aggregation of soluble α-synuclein (Uversky et al. 2001), and a copper-binding oligomer of α-synuclein has been discussed as neurotoxic form of α-synuclein (Brown 2010). However, while the total copper content in brains of PD patients does not differ strongly from healthy controls, copper levels are substantial lower in substantia nigra of PD patients (Loeffler et al. 1996; Ayton et al. 2013; Davies et al. 2014). This reduction in the copper content of the substantia nigra in PD has been discussed to result in the impairment of copper-dependent pathways, thereby contributing to the pathogenesis of PD (Double 2012; Ayton et al. 2013; Davies et al. 2014). In support of this view, copper supplementation (Alcaraz-Zubeldia et al. 2001, 2009) and the use of the BBB-permeable copper complex Cu(II)atsm (Hung et al. 2012) have been shown to be neuroprotective in animal models of PD, whereas copper chelation was not (Youdim et al. 2007).

Huntington's Disease

HD is a rare autosomal-dominant, progressive neurodegenerative disease characterized by motor, cognitive, and psychiatric abnormalities (Anderson 2011). HD is caused by polyglutamine expansion at the N-terminus of the huntingtin protein (McFarland and Cha 2011) that finally leads to brain atrophy, predominantly in the striatum and the cerebral cortex (Anderson 2011). Aggregation of the mutant huntingtin protein, oxidative stress, impaired energy metabolism, loss of neurotrophic support and transcriptional dysregulation have been discussed to contribute to development and progression of HD, but the exact pathogenic mechanism remains unknown. Accumulation of copper in the HD brain has been hypothesized to foster disease progression by promoting aggregation of the mutant huntingtin protein (Fox et al. 2007; Hands et al. 2010; Xiao et al. 2013). Further supporting a potential role of copper in disease progression, treatment with copper chelators, dietary copper reduction and genetic manipulation of copper transporters delayed disease progression in animal models for HD (Nguyen et al. 2005; Tallaksen-Greene et al. 2009; Cherny et al. 2012; Xiao et al. 2013).

Autism Spectrum Disorders

Autism spectrum disorders (ASD) are a group of neurodevelopmental disorders, including autistic disorder and Asperger syndrome, that are characterized by pervasive difficulties since early childhood across reciprocal social communication and restricted, repetitive interests and behaviors (Murphy et al. 2016). The etiology of ASD is currently unknown but is likely to be multifactorial encompassing both genetic and environmental factors (Murphy et al. 2016). There is some evidence for an alteration of copper homeostasis in ASDs. Homozygous deletions of the COMMD1 gene have been linked to autism (Levy et al. 2011), which loss of function results in copper overload in hepatic cell lines and is the cause of copper toxicosis in Bedlington Terriers (Fedoseienko et al. 2014). Hair and nail samples of autistic children contain significant elevated levels of copper when compared to healthy controls and the levels of copper correlated positively with the severity of autism (Lakshmi Priya and Geetha 2011; Russo and de Vito 2011). Excess copper has further been shown to affect a pathway at the glutamatergic synapses associated with autism (Baecker et al. 2014).

Neurotoxicity of Copper Nanoparticles

Nanoparticles are usually defined as objects with at least two dimensions in the nanoscale (Borm et al. 2006). Due to their small size and their relative high surface, compared to the bulk material, they provide various interesting material properties. The chemical and physical properties of nanoparticles do not only depend on their size but also on their composition, shape, surface area, catalytic activity, and surface modifications (Kettler et al. 2014; Amin et al. 2015). Due to the huge variety of these materials, nanoparticles gained a lot of interest from industry and the scientific community over the last decades (Borm et al. 2006; Cupaioli et al. 2014).

The cheap price and the special features of copper oxide nanoparticles (CuO-NPs) led to an increased interest from the industry toward this material (Yurderi et al. 2015). However, despite their high application potential, there are various disadvantages of this material. The biocidal activity of CuO-NPs is a double-edged feature. On the one hand, CuO-NPs are effectively used in wood preservatives, anti-fouling paint, water filters, sterile surface coatings or textiles and bandages (Almeida et al. 2007; Ben-Sasson et al. 2014; Dankovich and Smith 2014). On the other hand, the biocidal activity of CuO-NPs could be unintentionally harmful to the human health and the environment (Karlsson et al. 2008).

It is important to elucidate the uptake and distribution of CuO-NPs in the body to understand the toxic mechanisms of CuO-NPs. Several studies report that nanoparticles are able to enter the body by different routes but inhalation is the most probable uptake route for nanoparticles, whereas the skin is hardly penetrated (Oberdörster et al. 2004; Borm et al. 2006; Kimura et al. 2012). Nanoparticles are able to enter the brain upon inhalation either directly by translocation over the nerve endings of the olfactory bulb or indirectly after uptake into the blood stream and crossing of the BBB (Kreyling et al. 2002; Oberdörster et al. 2004; Sharma and Sharma 2012). Especially for the occupational exposure scenario, it has to be considered that high amounts of Cu-containing NPs can unintentionally be released from electric motors or during welding (Szymczak et al. 2007). The majority of airborne copper is present as fine particles and nanoparticles. A recent study identified such airborne copper as source for poor motor neuron performance and altered basal ganglia in school kids, demonstrating the impact of nano-particular copper on the brain (Pujol et al. 2016).

The high toxic potential of CuO-NPs was demonstrated by in vitro studies on lung cell lines (Kim et al. 2013; Ivask et al. 2015). This high toxicity of CuO-NPs was confirmed by in vivo inhalation and injection studies on rats and mice (Chen et al. 2006; Liao and Liu 2012; Privalova et al. 2014; Jing et al. 2015). Hereby, one particular inhalation study reported the high toxicity of CuO-NPs in comparison to the less toxic micrometer-sized copper oxide particles (Yokohira et al. 2008). In vivo studies have also shown that CuO-NPs can accumulate in the brain and have a high capacity to alter brain functionality (An et al. 2012; Privalova et al. 2014). The animals treated with CuO-NPs suffered severe cognitive impairments and damage of the BBB (An et al. 2012; Sharma and Sharma 2012). Wistar rats treated with CuO-NPs showed a decrease in learning and memory abilities as well as an impaired hippocampal LTP (An et al. 2012) which may involve the reported effects of CuO-NPs on neuronal potassium and sodium channels (Xu et al. 2009; Liu et al. 2011).

Several studies have evaluated the toxicity of CuO-NPs on brain cells including neurons (Li et al. 2007; Chen et al. 2008; Xu et al. 2009; Prabhu et al. 2010; Liu et al. 2011; Perreault et al. 2012) and astrocytes (Bulcke et al. 2014; Bulcke and Dringen 2014; Bulcke and Dringen 2016; Joshi et al. 2016). In contrast to iron oxide nanoparticles (Petters et al. 2014), CuO-NPs have a high toxic potential on primary cultured astrocytes (Bulcke and Dringen 2014) and alter in sub-toxic concentrations their glucose and glutathione metabolism and induce the synthesis of MTs (Bulcke and Dringen 2014; Bulcke and Dringen 2016). CuO-NP application leads to substantial cellular copper accumulation. CuO-NPs are likely to enter astrocytes by endocytotic mechanisms (Bulcke and Dringen 2016), but also extracellular liberation of copper ions has been suggested to be involved in the copper accumulation observed in glial cells after exposure to CuO-NPs (Joshi et al. 2016). The consequence of an exposure of cells to CuO-NPs is most likely mediated by an increase in cytosolic copper concentration which is caused by accumulation of copper liberated from particles rather than adverse particle effects (Bulcke and Dringen 2016). Thus, the reported toxicity of CuO-NPs to brain cells is most likely mediated by accelerated ROS production and oxidative damage (Bulcke et al. 2014).

Conclusions

Copper is an essential trace element which is involved in a large variety of different cellular functions. However, as copper in excess leads to accelerated formation of ROS and inactivation of cellular enzymes, the availability of copper is tightly regulated both on the systemic and cellular level. Both excess of copper and copper deprivation have severe adverse consequences on cells and organism as clearly shown by the different types of neurodegenerative disorders which have been connected with disturbances in copper homeostasis. The dilemma that sufficient amounts of copper have to be available but that an excess of copper has to be prevented makes therapeutic approaches to correct disturbances of copper homeostasis in neurological disorders a challenging task.

References

Akatsu H, Hori A, Yamamoto T, et al. Transition metal abnormalities in progressive dementias. Biometals. 2012;25:337–50. doi:10.1007/s10534-011-9504-8.

Alcaraz-Zubeldia M, Rojas P, Boll C, Ríos C. Neuroprotective effect of acute and chronic administration of copper (II) sulfate against MPP$^+$ neurotoxicity in mice. Neurochem Res. 2001;26:59–64. doi:10.1023/A:1007680616056.

Alcaraz-Zubeldia M, Boll-Woehrlen MC, Montes-Lopez S, et al. Copper sulfate prevents tyrosine hydroxylase reduced activity and motor deficits in a Parkinson's disease model in mice. Rev Investig Clin. 2009;61:405–11.

Alda JO, Garay R. Chloride (or bicarbonate)-dependent copper uptake through the anion exchanger in human red blood cells. Am J Phys. 1990;259:C570–6.

Alimba CG, Dhillon V, Bakare AA, Fenech M. Genotoxicity and cytotoxicity of chromium, copper, manganese and lead, and their mixture in WIL2-NS human B lymphoblastoid cells is enhanced by folate depletion. Mutat Res Genet Toxicol Environ Mutagen. 2016;798-799:35–47. doi:10.1016/j.mrgentox.2016.02.002.

Almeida E, Diamantino TC, de Sousa O. Marine paints: the particular case of antifouling paints. Prog Org Coatings. 2007;59:2–20. doi:10.1016/j.porgcoat.2007.01.017.

Amin ML, Joo JY, Yi DK, An SSA. Surface modification and local orientations of surface molecules in nanotherapeutics. J Control Release. 2015;207:131–42. doi:10.1016/j.jconrel.2015.04.017.

An L, Liu S, Yang Z, Zhang T. Cognitive impairment in rats induced by nano-CuO and its possible mechanisms. Toxicol Lett. 2012;213:220–7. doi:10.1016/j.toxlet.2012.07.007.

Anderson KE. Huntington's disease. In:Handbook of Clinical Neurology. New York: Wiley; 2011. p. 15–24.

Arciello M, Rotilio G, Rossi L. Copper-dependent toxicity in SH-SY5Y neuroblastoma cells involves mitochondrial damage. Biochem Biophys Res Commun. 2005;327:454–9. doi:10.1016/j.bbrc.2004.12.022.

Arguello JM, Eren E, Gonzalez-Guerrero M. The structure and function of heavy metal transport P1B-ATPases. Biometals. 2007;20:233–48. doi:10.1007/s10534-006-9055-6.

Arredondo M, Muñoz P, Mura C, Nùñez M. DMT1, a physiologically relevant apical Cu^{1+} transporter of intestinal cells. Am J Physiol Cell Physiol. 2003;284:C1525–30. doi:10.1152/ajpcell.00480.2002.

Ayton S, Lei P, Duce JA, et al. Ceruloplasmin dysfunction and therapeutic potential for Parkinson disease. Ann Neurol. 2013;73:554–9. doi:10.1002/ana.23817.

Baecker T, Mangus K, Pfaender S, et al. Loss of COMMD1 and copper overload disrupt zinc homeostasis and influence an autism-associated pathway at glutamatergic synapses. Biometals. 2014;27:715–30. doi:10.1007/s10534-014-9764-1.

Ballard C, Gauthier S, Corbett A, et al. Alzheimer's disease. Lancet. 2011;377:1019–31. doi:10.1016/S0140-6736(10)61349-9.

Bandmann O, Weiss KH, Kaler SG. Wilson's disease and other neurological copper disorders. Lancet Neurol. 2015;14:103–13. doi:10.1016/S1474-4422(14)70190-5.

Bayer TA, Schäfer S, Simons A, et al. Dietary cu stabilizes brain superoxide dismutase 1 activity and reduces amyloid Abeta production in APP23 transgenic mice. Proc Natl Acad Sci U S A. 2003;100:14187–92. doi:10.1073/pnas.2332818100.

Ben-Sasson M, Zodrow KR, Genggeng Q, et al. Surface functionalization of thin-film composite membranes with copper nanoparticles for antimicrobial surface properties. Environ Sci Technol. 2014;48:384–93. doi:10.1021/es404232s.

Bertinato J, Swist E, Plouffe LJ, et al. Ctr2 is partially localized to the plasma membrane and stimulates copper uptake in COS-7 cells. Biochem J. 2008;409:731–40. doi:10.1042/BJ20071025.

Bertrand E, Lewandowska E, Szpak M, et al. Neuropathological analysis of pathological forms of astroglia in Wilson's disease. Folia Neuropathol. 2001;39:73–9.

Boaru SG, Merle U, Uerlings R, et al. Simultaneous monitoring of cerebral metal accumulation in an experimental model of Wilson's disease by laser ablation inductively coupled plasma mass spectrometry. BMC Neurosci. 2014;15:1–13. doi:10.1186/1471-2202-15-98.

Borm PJA, Robbins D, Haubold S, et al. The potential risks of nanomaterials: a review carried out for ECETOC. Part Fibre Toxicol. 2006;3:11–46. doi:10.1186/1743-8977-3-11.

Bousquet-Moore D, Mains RE, Eipper BA. Peptidylglycine α-amidating monooxygenase and copper: a gene-nutrient interaction critical to nervous system function. J Neurosci Res. 2010a;88:2535–45. doi: 10.1002/jnr.22404.

Bousquet-Moore D, Prohaska JR, Nillni EA, et al. Interactions of peptide amidation and copper: novel biomarkers and mechanisms of neural dysfunction. Neurobiol Dis. 2010b;37:130–40. doi:10.1016/j.nbd.2009.09.016.

Brewer GJ, Yuzbasiyan-Gurkan V. Wilson disease. Medicine (Baltimore). 1992;71:139–64.

Brown DR. Role of the prion protein in copper turnover in astrocytes. Neurobiol Dis. 2004;15:534–43. doi:10.1016/j.nbd.2003.11.009.

Brown DR. Oligomeric alpha-synuclein and its role in neuronal death. IUBMB Life. 2010;62:334–9.

Bulcke F, Dringen R. Copper oxide nanoparticles stimulate glycolytic flux and increase the cellular contents of glutathione and metallothioneins in cultured astrocytes. Neurochem Res. 2014;40:15–26. doi:10.1007/s11064-014-1458-0.

Bulcke F, Dringen R. Handling of copper and copper oxide nanoparticles by astrocytes. Neurochem Res. 2016;41:33–43. doi:10.1007/s11064-015-1688-9.

Bulcke F, Thiel K, Dringen R. Uptake and toxicity of copper oxide nanoparticles in cultured primary brain astrocytes. Nanotoxicology. 2014;8:775–85. doi:10.3109/17435390.2013.829591.

Burkhead JL, Gray LW, Lutsenko S. Systems biology approach to Wilson's disease. Biometals. 2011;24:455–66. doi:10.1007/s10534-011-9430-9.

Camakaris J, Mann JR, Danks DM. Copper metabolism in mottled mouse mutants: copper concentrations in tissues during development. Biochem J. 1979;180:597–604.

Castellani RJ, Rolston RK, Smith MA. Alzheimer disease. Dis Mon. 2010;56:484–546. doi:10.1016/j.disamonth.2010.06.001.

Cater MA, La Fontaine S, Shield K, et al. ATP7B mediates vesicular sequestration of copper: insight into biliary copper excretion. Gastroenterology. 2006;130:493–506. doi:10.1053/j.gastro.2005.10.054.

Cater MA, KT MI, Li Q-X, et al. Intracellular copper deficiency increases amyloid-beta secretion by diverse mechanisms. Biochem J. 2008;412:141–52. doi:10.1042/BJ20080103.

Chen Z, Meng H, Xing G, et al. Acute toxicological effects of copper nanoparticles in vivo. Toxicol Lett. 2006;163:109–20. doi:10.1016/j.toxlet.2005.10.003.

Chen J, Zhu J, Cho H-H, et al. Differential cytotoxicity of metal oxide nanoparticles. J Exp Nanosci. 2008;3:321–8. doi:10.1080/17458080802235765.

Cherny RA, Ayton S, Finkelstein DI, et al. PBT2 reduces toxicity in a C. elegans model of polyQ aggregation and extends lifespan, reduces striatal atrophy and improves motor performance in the R6/2 mouse model of Huntington's disease. J Huntingtons Dis. 2012;1:211–9. doi:10.3233/JHD-120029.

Choi BS, Zheng W. Copper transport to the brain by the blood-brain barrier and blood-CSF barrier. Brain Res. 2009;1248:14–21. doi:10.1016/j.brainres.2008.10.056.

Cobine PA, Ojeda LD, Rigby KM, Winge DR. Yeast contain a non-proteinaceous pool of copper in the mitochondrial matrix. J Biol Chem. 2004;279:14447–55. doi:10.1074/jbc.M312693200.

Crouch PJ, Blake R, Duce JA, et al. Copper-dependent inhibition of human cytochrome c oxidase by a dimeric conformer of amyloid-beta1-42. J Neurosci. 2005;25:672–679. doi:25/3/672 [pii]\r10.1523/JNEUROSCI.4276-04.2005.

Crouch PJ, Wai L, Adlard PA, et al. Increasing Cu bioavailability inhibits Aβ oligomers and tau phosphorylation. Proc Natl Acad Sci U S A. 2009;106:381–6. doi:10.1073/pnas.0809057106.

Cupaioli FA, Zucca FA, Boraschi D, Zecca L. Engineered nanoparticles. How brain friendly is this new guest? Prog Neurobiol. 2014;119-120:20–38.

Dankovich TA, Smith JA. Incorporation of copper nanoparticles into paper for point-of-use water purification. Water Res. 2014;63:245–51. doi:10.1016/j.watres.2014.06.022.

Das SK, Ray K. Wilson's disease: an update. Nat Clin Pract Neurol. 2006;2:482–93. doi:10.1038/ncpneuro0291.

Davies KM, Hare DJ, Cottam V, et al. Localization of copper and copper transporters in the human brain. Metallomics. 2012;5:43–51. doi:10.1039/c2mt20151h.

Davies KM, Bohic S, Carmona A, et al. Copper pathology in vulnerable brain regions in Parkinson's disease. Neurobiol Aging. 2014;35:858–66. doi:10.1016/j.neurobiolaging.2013.09.034.

de Lores Arnaiz GR, Ordieres MGL. Brain Na$^+$, K$^+$−ATPase activity in aging and disease. Int J Biomed Sci. 2014;10:85–102.

de Romaña DL, Olivares M, Uauy R, Araya M. Risks and benefits of copper in light of new insights of copper homeostasis. J Trace Elem Med Biol. 2011;25:3–13. doi:10.1016/j.jtemb.2010.11.004.

Deibel MA, Ehmann WD, Markesbery WR. Copper, iron, and zinc imbalances in severely degenerated brain regions in Alzheimer's disease: possible relation to oxidative stress. J Neurol Sci. 1996;143:137–42. doi:10.1016/S0022-510X(96)00203-1.

Dexter DT, Jenner P, Schapira AH, Marsden CD. Alterations in levels of iron, ferritin, and other trace metals in neurodegenerative diseases affecting the basal ganglia. Ann Neurol. 1992;32(Suppl):S94–100.

Dobrowolska J, Dehnhardt M, Matusch A, et al. Quantitative imaging of zinc, copper and lead in three distinct regions of the human brain by laser ablation inductively coupled plasma mass spectrometry. Talanta. 2008;74:717–23. doi:10.1016/j.talanta.2007.06.051.

Donsante A, Johnson P, Jansen LA, Kaler SG. Somatic mosaicism in Menkes disease suggests choroid plexus-mediated copper transport to the developing brain. Am J Med Genet Part A. 2010;152(A):2529–34. doi:10.1002/ajmg.a.33632.

Doreulee N, Yanovsky Y, Haas HL. Suppression of long-term potentiation in hippocampal slices by copper. Hippocampus. 1997;7:666–9. doi:10.1002/(SICI)1098-1063(1997)7:6<666::AID-HIPO8>3.0.CO;2-C.

Double KL. Neuronal vulnerability in Parkinson's disease. Parkinsonism Relat Disord. 2012;18:S52–4. doi:10.1016/S1353-8020(11)70018-9.

Du X, Zheng Y, Wang Z, et al. Inhibitory act of selenoprotein P on Cu$^+$/Cu^{2+}-induced tau aggregation and neurotoxicity. Inorg Chem. 2014;53:11221–30. doi:10.1021/ic501788v.

Dusek P, Litwin T, Czlonkowska A. Wilson disease and other neurodegenerations with metal accumulations. Neurol Clin. 2015;33:175–204. doi:10.1016/j.ncl.2014.09.006.

Eskici G, Axelsen PH. Copper and oxidative stress in the pathogenesis of Alzheimer's disease. Biochemistry. 2012;51:6289–311. doi:10.1021/bi3006169.

Espinoza A, Le Blanc S, Olivares M, et al. Iron, copper, and zinc transport: inhibition of divalent metal transporter 1 (DMT1) and human copper transporter 1 (hCTR1) by shRNA. Biol Trace Elem Res. 2012;146:281–6. doi:10.1007/s12011-011-9243-2.

Eustermann S, Videler H, Yang JC, et al. The DNA-binding domain of human PARP-1 interacts with DNA single-strand breaks as a monomer through its second zinc finger. J Mol Biol. 2011;407:149–70. doi:10.1016/j.jmb.2011.01.034.

Fanni D, Fanos V, Gerosa C, et al. Effects of iron and copper overload on the human liver: an ultrastructural study. Curr Med Chem. 2014;21:3768–74.

Fedoseienko A, Bartuzi P, Van de Sluis B. Functional understanding of the versatile protein copper metabolism MURR1 domain 1 (COMMD1) in copper homeostasis. Ann N Y Acad Sci. 2014;1314:6–14. doi:10.1111/nyas.12353.

Feng W, Ye F, Xue W, et al. Copper regulation of hypoxia-inducible factor-1 activity. Mol Pharmacol. 2009;75:174–82. doi:10.1124/mol.108.051516.

Ferguson-Miller S, Babcock GT. Heme/copper terminal oxidases. Chem Rev. 1996;96:2889–907. doi:10.1021/cr950051s.

Forte G, Bocca B, Senofonte O, et al. Trace and major elements in whole blood, serum, cerebrospinal fluid and urine of patients with Parkinson's disease. J Neural Transm. 2004;111:1031–40. doi:10.1007/s00702-004-0124-0.

Fox JH, Kama JA, Lieberman G, et al. Mechanisms of copper ion mediated Huntington's disease progression. PLoS One. 2007;2:e334. doi:10.1371/journal.pone.0000334.

Franchitto N, Gandia-Mailly P, Georges B, et al. Acute copper sulphate poisoning: a case report and literature review. Resuscitation. 2008;78:92–6. doi:10.1016/j.resuscitation.2008.02.017.

Frelon S, Douki T, Favier A, Cadet J. Hydroxyl radical is not the main reactive species involved in the degradation of DNA bases by copper in the presence of hydrogen peroxide. Chem Res Toxicol. 2003;16:191–7. doi:10.1021/tx025650q.

Fu X, Zhang Y, Jiang W, et al. Regulation of copper transport crossing brain barrier systems by CU-ATPases: effect of manganese exposure. Toxicol Sci. 2014;139:432–51. doi:10.1093/toxsci/kfu048.

Fu S, Jiang W, Zheng W. Age-dependent increase of brain copper levels and expressions of copper regulatory proteins in the subventricular zone and choroid plexus. Front Mol Neurosci. 2015;8:1–10. doi:10.3389/fnmol.2015.00022.

Gaetke LM, Chow-Johnson HS, Chow CK. Copper: toxicological relevance and mechanisms. Arch Toxicol. 2014;88:1929–38.

Gaier ED, Eipper BA, Mains RE. Copper signaling in the mammalian nervous system: synaptic effects. J Neurosci Res. 2013;91:2–19.

Gaier ED, Eipper BA, Mains RE. Pam heterozygous mice reveal essential role for Cu in amygdalar behavioral and synaptic function. Ann N Y Acad Sci. 2014a;1314:15–23. doi:10.1111/nyas.12378.

Gaier ED, Rodriguiz RM, Zhou J, et al. In vivo and in vitro analyses of amygdalar function reveal a role for copper. J Neurophysiol. 2014b;111:1927–39. doi:10.1152/jn.00631.2013.

Gasperini L, Meneghetti E, Pastore B, et al. Prion protein and copper cooperatively protect neurons by modulating NMDA receptor through S-nitrosylation. Antioxid Redox Signal. 2015;22:772–84. doi:10.1089/ars.2014.6032.

Georgieva S, Popov B, Petrov V. Genotoxic effects of copper sulfate in rabbits. Arch Biol Sci. 2013;65:963–7. doi:10.2298/ABS1303963G.

Goldschmith A, Infante C, Leiva J, et al. Interference of chronically ingested copper in long-term potentiation (LTP) of rat hippocampus. Brain Res. 2005;1056:176–82. doi:10.1016/j.brainres.2005.07.030.

Gu M, Cooper JM, Butler P, et al. Oxidative-phosphorylation defects in liver of patients with Wilson's disease. Lancet. 2000;356:469–74. doi:10.1016/S0140-6736(00)02556-3.

Gunther MR, Hanna PM, Mason RP, Cohen MS. Hydroxyl radical formation from cuprous ion and hydrogen peroxide: a spin-trapping study. Arch Biochem Biophys. 1995;316:515–22. doi:10.1006/abbi.1995.1068.

Halliwell B. Oxidative stress and neurodegeneration: where are we now? J Neurochem. 2006;97:1634–58. doi:10.1111/j.1471-4159.2006.03907.x.

Hands SL, Mason R, Sajjad MU, et al. Metallothioneins and copper metabolism are candidate therapeutic targets in Huntington's disease. Biochem Soc Trans. 2010;38:552–558. doi:BST0380552 [pii]\r10.1042/BST0380552.

Hauck AK, Bernlohr DA. Oxidative stress and lipotoxicity. J Lipid Res. 2016:1–37. doi:10.1194/jlr.R066597.

Haupt S, Berger M, Goldberg Z, Haupt Y. Apoptosis – the p53 network. J Cell Sci. 2003;116:4077–85. doi:10.1242/jcs.00739.

Haywood S, Paris J, Ryvar R, Botteron C. Brain copper elevation and neurological changes in North Ronaldsay sheep: a model for neurodegenerative disease? J Comp Pathol. 2008;139:252–5. doi:10.1016/j.jcpa.2008.06.008.

Healy J, Tipton K. Ceruloplasmin and what it might do. J Neural Transm. 2007;114:777–81. doi:10.1007/s00702-007-0687-7.

Hegde ML, Hegde PM, Holthauzen LMF, et al. Specific inhibition of NEIL-initiated repair of oxidized base damage in human genome by copper and iron: potential etiological linkage to neurodegenerative diseases. J Biol Chem. 2010;285:28812–25. doi:10.1074/jbc.M110.126664.

Hegde ML, Hegde PM, Rao KS, Mitra S. Oxidative genome damage and its repair in neurodegenerative diseases: function of transition metals as a double-edged sword. J Alzheimers Dis. 2011;24:183–98.

Hodgkinson VL, Zhu S, Wang Y, et al. Autonomous requirements of the Menkes disease protein in the nervous system. Am J Physiol Cell Physiol. 2015;309:C660–8. doi:10.1152/ajpcell.00130.2015.

Hopt A, Korte S, Fink H, et al. Methods for studying synaptosomal copper release. J Neurosci Methods. 2003;128:159–72. doi:10.1016/S0165-0270(03)00173-0.

Horoupian D, Sternlieb I, Scheinberg I. Neuropathological findings in penicillamine-treated patients with Wilson's disease. Clin Neuropathol. 1988;7:62–7.

Hung LW, Villemagne VL, Cheng L, et al. The hypoxia imaging agent CuII(atsm) is neuroprotective and improves motor and cognitive functions in multiple animal models of Parkinson's disease. J Exp Med. 2012;209:837–54. doi:10.1084/jem.20112285.

Huster D, Purnat TD, Burkhead JL, et al. High copper selectively alters lipid metabolism and cell cycle machinery in the mouse model of Wilson disease. J Biol Chem. 2007;282:8343–55. doi:10.1074/jbc.M607496200.

Ivask A, Titma T, Visnapuu M, et al. Toxicity of 11 metal oxide nanoparticles to three mammalian cell types in vitro. Curr Top Med Chem. 2015;15:1914–29. doi:10.2174/1568026615666150506150109.

James SA, Volitakis I, Adlard PA, et al. Elevated labile Cu is associated with oxidative pathology in Alzheimer disease. Free Radic Biol Med. 2012;52:298–302. doi: 10.1016/j.freeradbiomed.2011.10.446.

Javadov S, Kuznetsov A. Mitochondrial permeability transition and cell death: the role of cyclophilin D. Front Physiol. 2013; doi:10.3389/fphys.2013.00076.

Jing X, Park JH, Peters TM, Thorne PS. Toxicity of copper oxide nanoparticles in lung epithelial cells exposed at the air-liquid interface compared with in vivo assessment. Toxicol Vitr. 2015;29:502–11. doi:10.1016/j.tiv.2014.12.023.

Jomova K, Vondrakova D, Lawson M, Valko M. Metals, oxidative stress and neurodegenerative disorders. Mol Cell Biochem. 2010;345:91–104.

Joshi A, Rastedt W, Faber K, et al. Uptake and toxicity of copper oxide nanoparticles in C6 glioma cells. Neurochem Res 2016;41:3004–19.doi: 10.1007/s11064-016-2020-z .

Kaler SG. Diagnosis and therapy of Menkes syndrome, a genetic form of copper deficiency. Am J Clin Nutr. 1998;67:1029S–34S.

Kaler SG. ATP7A-related copper transport diseases-emerging concepts and future trends. Nat Rev Neurol. 2011;7:15–29. doi:10.1038/nrneurol.2010.180.

Kaler SG. Inborn errors of copper metabolism. Handb Clin Neurol. 2013;113:1745–54. doi:10.1016/B978-0-444-59565-2.00045-9.

Kardos J, Kovacs I, Hajos F, et al. Nerve endings from rat brain tissue release copper upon depolarization. A possible role in regulating neuronal excitability. Neurosci Lett. 1989;103:139–44. doi:10.1016/0304-3940(89)90565-X.

Karlsson HL, Cronholm P, Gustafsson J, Möller L. Copper oxide nanoparticles are highly toxic: a comparison between metal oxide nanoparticles and carbon nanotubes. Chem Res Toxicol. 2008;21:1726–32. doi:10.1021/tx800064j.

Kessler H, Bayer TA, Bach D, et al. Intake of copper has no effect on cognition in patients with mild Alzheimer's disease: a pilot phase 2 clinical trial. J Neural Transm. 2008;115:1181–7. doi:10.1007/s00702-008-0080-1.

Kettler K, Veltman K, van de Meent D, et al. Cellular uptake of nanoparticles as determined by particle properties, experimental conditions, and cell type. Environ Toxicol Chem. 2014;33:481–92. doi:10.1002/etc.2470.

Kidane TZ, Farhad R, Lee KJ, et al. Uptake of copper from plasma proteins in cells where expression of CTR1 has been modulated. Biometals. 2012;25:697–709. doi:10.1007/s10534-012-9528-8.

Kim H, Son H-Y, Bailey SM, Lee J. Deletion of hepatic Ctr1 reveals its function in copper acquisition and compensatory mechanisms for copper homeostasis. Am J Physiol Gastrointest Liver Physiol. 2009;296:G356–64. doi:10.1152/ajpgi.90632.2008.

Kim JS, Peters TM, O'Shaughnessy PT, et al. Validation of an in vitro exposure system for toxicity assessment of air-delivered nanomaterials. Toxicol Vitr. 2013;27:164–73. doi:10.1016/j.tiv.2012.08.030.

Kimura E, Kawano Y, Todo H, et al. Measurement of skin permeation/penetration of nanoparticles for their safety evaluation. Biol Pharm Bull. 2012;35:1476–86. doi:10.1248/bpb.b12-00103.

Kirschnek S, Paris F, Weller M, et al. CD95-mediated apoptosis in vivo involves acid sphingomyelinase. J Biol Chem. 2000;275:27316–23. doi:10.1074/jbc.M002957200.

Klinman JP. Mechanisms whereby mononuclear copper proteins functionalize organic substrates. Chem Rev. 1996;96:2541–62. doi:10.1021/cr950047g.

Klinman JP. The copper-enzyme family of dopamine beta-monooxygenase and peptidylglycine alpha-hydroxylating monooxygenase: resolving the chemical pathway for substrate hydroxylation. J Biol Chem. 2006;281:3013–6.

Kodama H. Recent developments in Menkes disease. J Inherit Metab Dis. 1993;16:791–9. doi:10.1007/BF00711911.

Kodama H, Fujisawa C, Bhadhprasit W. Pathology, clinical features and treatments of congenital copper metabolic disorders – focus on neurologic aspects. Brain and Development. 2011;33:243–51. doi:10.1016/j.braindev.2010.10.021.

Koeppen AH, Ramirez RL, Yu D, et al. Friedreich's ataxia causes redistribution of iron, copper, and zinc in the dentate nucleus. Cerebellum. 2012;11:845–60. doi:10.1007/s12311-012-0383-5.

Krebs N, Langkammer C, Goessler W, et al. Assessment of trace elements in human brain using inductively coupled plasma mass spectrometry. J Trace Elem Med Biol. 2014;28:1–7. doi:10.1016/j.jtemb.2013.09.006.

Kreyling WG, Semmler M, Erbe F, et al. Translocation of ultrafine insoluble iridium particles fromm lung epithelium to extrapulmonary organs is size dependent but very low. J Toxicol Environ Heal Part A. 2002;65:1513–30. doi:10.1080/00984100290071649.

Krstić DZ, Krinulović K, Vasić VM. Inhibition of Na^+/K^+-ATPase and Mg^{2+}-ATPase by metal ions and prevention and recovery of inhibited activities by chelators. J Enzyme Inhib Med Chem. 2005;20:469–76. doi:10.1080/14756360500213280.

Kuo Y-M, Gybina AA, Pyatskowit JW, et al. Copper transport protein (Ctr1) levels in mice are tissue specific and dependent on copper status. J Nutr. 2006;136:21–6. doi: 136/1/21 [pii]

Lakshmi Priya MD, Geetha A. Level of trace elements (copper, zinc, magnesium and selenium) and toxic elements (lead and mercury) in the hair and nail of children with autism. Biol Trace Elem Res. 2011;142:148–58. doi:10.1007/s12011-010-8766-2.

Lang PA, Schenck M, Nicolay JP, et al. Liver cell death and anemia in Wilson disease involve acid sphingomyelinase and ceramide. Nat Med. 2007;13:164–70. doi:10.1038/nm1539.

Leary SC, Winge DR, Cobine PA. "Pulling the plug" on cellular copper: the role of mitochondria in copper export. Biochim Biophys Acta, Mol Cell Res. 2009;1793:146–53. doi:10.1016/j.bbamcr.2008.05.002.

Lech T, Sadlik JK. Copper concentration in body tissues and fluids in normal subjects of southern Poland. Biol Trace Elem Res. 2007;118:10–5. doi:10.1007/s12011-007-0014-z.

Lee J, Prohaska JR, Thiele DJ. Essential role for mammalian copper transporter Ctr1 in copper homeostasis and embryonic development. Proc Natl Acad Sci U S A. 2001;98:6842–7. doi:10.1073/pnas.111058698.

Lee J, Pena MMO, Nose Y, Thiele DJ. Biochemical characterization of the human copper transporter Ctr1. J Biol Chem. 2002a;277:4380–7. doi:10.1074/jbc.M104728200.

Lee J, Petris MJ, Thiele DJ. Characterization of mouse embryonic cells deficient in the Ctr1 high affinity copper transporter: identification of a Ctr1-independent copper transport system. J Biol Chem. 2002b;277:40253–9. doi:10.1074/jbc.M208002200.

Leiva J, Palestini M, Infante C, et al. Copper suppresses hippocampus LTP in the rat, but does not alter learning or memory in the morris water maze. Brain Res. 2009;1256:69–75. doi:10.1016/j.brainres.2008.12.041.

Lenartowicz M, Krzeptowski W, Lipiński P, et al. Mottled mice and non-mammalian models of Menkes disease. Front Mol Neurosci. 2015;8:1–18. doi:10.3389/fnmol.2015.00072.

Letelier ME, Lepe AM, Faúndez M, et al. Possible mechanisms underlying copper-induced damage in biological membranes leading to cellular toxicity. Chem Biol Interact. 2005;151:71–82. doi:10.1016/j.cbi.2004.12.004.

Letelier ME, Martinez M, Gonzalez-Lira V, et al. Inhibition of cytosolic glutathione S-transferase activity from rat liver by copper. Chem Biol Interact. 2006;164:39–48. doi:10.1016/j.cbi.2006.08.013.

Lévay G, Ye Q, Bodell WJ. Formation of DNA adducts and oxidative base damage by copper mediated oxidation of dopamine and 6-hydroxydopamine. Exp Neurol. 1997;146:570–4. doi:10.1006/exnr.1997.6560.

Levy D, Ronemus M, Yamrom B, et al. Rare de novo and transmitted copy-numbervariation in autistic spectrum disorders. Neuron. 2011;70:886–97. doi:10.1016/j.neuron.2011.05.015.

Lewińska-Preis L, Jabłońska M, Fabiańska MJ, Kita A. Bioelements and mineral matter in human livers from the highly industrialized region of the upper Silesia Coal Basin (Poland). Environ Geochem Health. 2011;33:595–611. doi:10.1007/s10653-011-9373-7.

Li J, Lock RAC, Klaren PHM, et al. Kinetics of Cu^{2+} inhibition of Na^+/K^+-ATPase. Toxicol Lett. 1996;87:31–8. doi:10.1016/0378-4274(96)03696-X.

Li F, Zhou X, Zhu J, et al. High content image analysis for human H4 neuroglioma cells exposed to CuO nanoparticles. BMC Biotechnol. 2007;7:66. doi:10.1186/1472-6750-7-66.

Liao M, Liu H. Gene expression profiling of nephrotoxicity from copper nanoparticles in rats after repeated oral administration. Environ Toxicol Pharmacol. 2012;34:67–80. doi:10.1016/j.etap.2011.05.014.

Lin C, Zhang Z, Wang T, et al. Copper uptake by DMT1: a compensatory mechanism for CTR1 deficiency in human umbilical vein endothelial cells. Metallomics. 2015;7:1285–9. doi:10.1039/c5mt00097a.

Linder MC, Hazegh-Azam M. Copper biochemistry and molecular biology. Am J Clin Nutr. 1996;63:797S–811S.

Litwin T, Gromadzka G, Szpak GM, et al. Brain metal accumulation in Wilson's disease. J Neurol Sci. 2013;329:55–8. doi:10.1016/j.jns.2013.03.021.

Liu Z, Liu S, Ren G, et al. Nano-CuO inhibited voltage-gated sodium current of hippocampal CA1 neurons via reactive oxygen species but independent from G-proteins pathway. J Appl Toxicol. 2011;31:439–45. doi:10.1002/jat.1611.

Liu J, Chakraborty S, Hosseinzadeh P, et al. Metalloproteins containing cytochrome, iron-sulfur, or copper redox centers. Chem Rev. 2014;114:4366–9.

Llanos RM, Michalczyk AA, Freestone DJ, et al. Copper transport during lactation in transgenic mice expressing the human ATP7A protein. Biochem Biophys Res Commun. 2008;372:613–7. doi:10.1016/j.bbrc.2008.05.123.

Loeffler DA, LeWitt PA, Juneau PL, et al. Increased regional brain concentrations of ceruloplasmin in neurodegenerative disorders. Brain Res. 1996;738:265–74. doi:10.1016/S0006-8993(96)00782-2.

Lorincz MT. Neurologic Wilson's disease. Ann N Y Acad Sci. 2010;1184:173–87. doi:10.1111/j.1749-6632.2009.05109.x.

Lovell MA, Robertson JD, Teesdale WJ, et al. Copper, iron and zinc in Alzheimer's disease senile plaques. J Neurol Sci. 1998;158:47–52. doi:10.1016/S0022-510X(98)00092-6.

Lucero HA, Kagan HM. Lysyl oxidase: an oxidative enzyme and effector of cell function. Cell Mol Life Sci. 2006;63:2304–16. doi:10.1007/s00018-006-6149-9.

Ma Y, Cao L, Kawabata T, et al. Cupric nitrilotriacetate induces oxidative DNA damage and apoptosis in human leukemia HL-60 cells. Free Radic Biol Med. 1998;25:568–75.

Marcus DL, Thomas C, Rodriguez C, et al. Increased peroxidation and reduced antioxidant enzyme activity in Alzheimer's disease. Exp Neurol. 1998;150:40–4. doi:10.1006/exnr.1997.6750.

Matheou CJ, Younan ND, Viles JH. Cu^{2+} accentuates distinct misfolding of $A\beta_{1-40}$ and $A\beta_{1-42}$ peptides, and potentiates membrane disruption. Biochem J. 2015;466:233–42. doi:10.1042/BJ20141168.

Maurer I, Zierz S, Möller HJ. A selective defect of cytochrome c oxidase is present in brain of Alzheimer disease patients. Neurobiol Aging. 2000;21:455–62. doi:10.1016/S0197-4580(00)00112-3.

Maxfield AB, Heaton DN, Winge DR. Cox17 is functional when tethered to the mitochondrial inner membrane. J Biol Chem. 2004;279:5072–80. doi:10.1074/jbc.M311772200.

Maynard CJ, Cappai R, Volitakis I, et al. Overexpression of Alzheimer's disease amyloid-β opposes the age-dependent elevations of brain copper and iron. J Biol Chem. 2002;277:44670–6. doi:10.1074/jbc.M204379200.

McFarland KN, Cha J-HJ. Molecular biology of Huntington's disease. In: Handbook of clinical neurology. 2011. pp 25–81.

Meenakshi-Sundaram S, Mahadevan A, Taly AB, et al. Wilson's disease: a clinico-neuropathological autopsy study. J Clin Neurosci. 2008;15:409–17. doi:10.1016/j.jocn.2006.07.017.

Michalczyk A, Bastow E, Greenough M, et al. ATP7B expression in human breast epithelial cells is mediated by lactational hormones. J Histochem Cytochem. 2008;56:389–99. doi:10.1369/jhc.7A7300.2008.

Mikol J, Vital C, Wassef M, et al. Extensive cortico-subcortical lesions in Wilson's disease: Clinico-pathological study of two cases. Acta Neuropathol. 2005;110:451–8. doi:10.1007/s00401-005-1061-1.

Mitra J, Guerrero EN, Hegde PM, et al. New perspectives on oxidized genome damage and repair inhibition by pro-oxidant metals in neurological diseases. Biomol Ther. 2014;4:678–703. doi:10.3390/biom4030678.

Monnot AD, Behl M, Ho S, Zheng W. Regulation of brain copper homeostasis by the brain barrier systems: effects of Fe-overload and Fe-deficiency. Toxicol Appl Pharmacol. 2011;256:249–57. doi:10.1016/j.taap.2011.02.003.

Monnot AD, Zheng G, Zheng W. Mechanism of copper transport at the blood-cerebrospinal fluid barrier: influence of iron deficiency in an in vitro model. Exp Biol Med (Maywood). 2012;237:327–33. doi:10.1258/ebm.2011.011170.

Monty J-FF, Llanos RM, Mercer JFB, et al. Copper exposure induces trafficking of the Menkes protein in intestinal epithelium of ATP7A transgenic mice. Biochem J. 2005;135:2762–766. doi:135/12/2762 [pii].

Moriya M, Ho Y-H, Grana A, et al. Copper is taken up efficiently from albumin and α_2-macroglobulin by cultured human cells by more than one mechanism. Am J Phys Cell Phys. 2008;295:C708–21. doi: 10.1152/ajpcell.00029.2008.

Mossakowski MJ, Renkawek K, Kraśnicka Z, et al. Morphology and histochemistry of Wilsonian and hepatogenic gliopathy in tissue culture. Acta Neuropathol. 1970;16:1–16. doi:10.1007/BF00686958.

Mufti AR, Burstein E, Csomos RA, et al. XIAP is a copper binding protein deregulated in Wilson's disease and other copper toxicosis disorders. Mol Cell. 2006;21:775–85. doi:10.1016/j.molcel.2006.01.033.

Mufti AR, Burstein E, Duckett CS. XIAP: cell death regulation meets copper homeostasis. Arch Biochem Biophys. 2007;463:168–74. doi:10.1016/j.abb.2007.01.033.

Murphy CM, Wilson CE, Robertson DM, et al. Autism spectrum disorder in adults: diagnosis, management, and health services development. Neuropsychiatr Dis Treat. 2016;12:1669–86. doi:10.2147/NDT.S65455.

Narayanan VS, Fitch CA, Levenson CW. Tumor suppressor protein p53 mRNA and subcellular localization are altered by changes in cellular copper in human Hep G2 cells. J Nutr. 2001;131:1427–32.

Nedeljković N, Horvat A. One-step bioluminescence ATPase assay for the evaluation of neurotoxic effects of metal ions. Monatshefte fur Chemie. 2007;138:253–60. doi:10.1007/s00706-007-0595-4.

Nguyen T, Hamby A, Massa SM. Clioquinol down-regulates mutant huntingtin expression in vitro and mitigates pathology in a Huntington's disease mouse model. Proc Natl Acad Sci U S A. 2005;102:11840–5. doi:10.1073/pnas.0502177102.

Nooijen JL, De Groot CJ, Van den Hamer CJ, et al. Trace element studies in three patients and a fetus with Menkes' disease. Effect of copper therapy. Pediatr Res. 1981;15:284–9.

Nose Y, Kim BE, Thiele DJ. Ctr1 drives intestinal copper absorption and is essential for growth, iron metabolism, and neonatal cardiac function. Cell Metab. 2006;4:235–44. doi:10.1016/j.cmet.2006.08.009.

Nose Y, Wood LK, Kim BE, et al. Ctr1 is an apical copper transporter in mammalian intestinal epithelial cells in vivo that is controlled at the level of protein stability. J Biol Chem. 2010;285:32385–92. doi:10.1074/jbc.M110.143826.

Nyasae L, Bustos R, Braiterman L, et al. Dynamics of endogenous ATP7A (Menkes protein) in intestinal epithelial cells: copper-dependent redistribution between two intracellular sites. Am J Physiol Gastrointest Liver Physiol. 2007;292:G1181–94. doi:10.1152/ajpgi.00472.2006.

Oberdörster G, Sharp Z, Atudorei V, et al. Translocation of inhaled ultrafine particles to the brain. Inhal Toxicol. 2004;16:437–45. doi:10.1080/08958370490439597.

Olivares C, Solano F. New insights into the active site structure and catalytic mechanism of tyrosinase and its related proteins. Pigment Cell Melanoma Res. 2009;22:750–60. doi:10.1111/j.1755-148X.2009.00636.x.

Olusola A, Obodozie O, Nssien M, et al. Concentrations of copper, iron, and zinc in the major organs of the wistar albino and wild black rats: a comparative study. Biol Trace Elem Res. 2004;98:265–74. doi:10.1385/BTER:98:3:265.

Pamp K, Bramey T, Kirsch M, et al. NAD(H) enhances the Cu(II)-mediated inactivation of lactate dehydrogenase by increasing the accessibility of sulfhydryl groups. Free Radic Res. 2005;39:31–40. doi:10.1080/10715760400023671.

Parpura V, Heneka MT, Montana V, et al. Glial cells in (patho)physiology. J Neurochem. 2012;121:4–27. doi:10.1111/j.1471-4159.2012.07664.x.

Perreault F, Pedroso Melegari S, Henning da Costa C, et al. Genotoxic effects of copper oxide nanoparticles in Neuro 2A cell cultures. Sci Total Environ. 2012;441:117–24. doi:10.1016/j.scitotenv.2012.09.065.

Perry JJP, Shin DS, Getzoff ED, Tainer JA. The structural biochemistry of the superoxide dismutases. Biochim Biophys Acta Proteins Proteomics. 2010;1804:245–62. doi:10.1016/j.bbapap.2009.11.004.

Petters C, Irrsack E, Koch M, Dringen R. Uptake and metabolism of iron oxide nanoparticles in brain cells. Neurochem Res. 2014;39:1648–60.

Phatak VM, Muller PAJ. Metal toxicity and the p53 protein: an intimate relationship. Toxicol Res. 2015;4:576–91. doi:10.1039/C4TX00117F.

Prá D, Franke SIR, Giulian R, et al. Genotoxicity and mutagenicity of iron and copper in mice. Biometals. 2008;21:289–97. doi:10.1007/s10534-007-9118-3.

Prabhu BM, Ali SF, Murdock RC, et al. Copper nanoparticles exert size and concentration dependent toxicity on somatosensory neurons of rat. Nanotoxicology. 2010;4:150–60. doi:10.3109/17435390903337693.

Prakash A, Dhaliwal GK, Kumar P, Majeed ABA. Brain biometals and Alzheimer's disease – boon or bane? Int J Neurosci. 2016;7454:1–34. doi:10.3109/00207454.2016.1174118.

Privalova LI, Katsnelson BA, Loginova NV, et al. Subchronic toxicity of copper oxide nanoparticles and its attenuation with the help of a combination of bioprotectors. Int J Mol Sci. 2014;15:12379–406. doi:10.3390/ijms150712379.

Prudovsky I, Tarantini F, Landriscina M, et al. Secretion without Golgi. J Cell Biochem. 2008;103:1327–43. doi:10.1002/jcb.21513.

Pujol J, Fenoll R, Macià D, et al. Airborne copper exposure in school environments associated with poorer motor performance and altered basal ganglia. Brain Behav. 2016;e00467. doi:10.1002/brb3.467.

Qian Y, Tiffany-castiglioni E, Welsh J, Harris ED. Copper efflux from murine microvascular cells requires expression of the Menkes Cu-ATPase. J Nutr. 1998;128:1276–82.

Rae TD, Schmidt PJ, Pufahl RA, et al. Undetectable intracellular free copper: the requirement of a copper chaperone for superoxide dismutase. Science. 1999;284:805–8. doi:10.1126/science.284.5415.805.

Ralle M, Huster D, Vogt S, et al. Wilson disease at a single cell level: intracellular copper trafficking activates compartment-specific responses in hepatocytes. J Biol Chem. 2010;285:30875–83. doi:10.1074/jbc.M110.114447.

Ramos P, Santos A, Pinto NR, et al. Anatomical region differences and age-related changes in copper, zinc, and manganese levels in the human brain. Biol Trace Elem Res. 2014;161:190–201. doi:10.1007/s12011-014-0093-6.

Reddy PVB, Rao KVR, Norenberg MD. The mitochondrial permeability transition, and oxidative and nitrosative stress in the mechanism of copper toxicity in cultured neurons and astrocytes. Lab Investig. 2008;88:816–30. doi:10.1038/labinvest.2008.49.

Rembach A, Hare DJ, Lind M, et al. Decreased copper in Alzheimer's disease brain is predominantly in the soluble extractable fraction. Int J Alzheimers Dis. 2013;2013(1–2) doi: 10.1155/2013/623241.

Ristić AJ, Sokić D, Baščarević V, et al. Metals and electrolytes in sclerotic hippocampi in patients with drug-resistant mesial temporal lobe epilepsy. Epilepsia. 2014;55:e34–7. doi:10.1111/epi.12593.

Rivera-Mancia S, Perez-Neri I, Rios C, et al. The transition metals copper and iron in neurodegenerative diseases. Chem Biol Interact. 2010;186:184–99. doi:10.1016/j.cbi.2010.04.010.

Roberts BR, Ryan TM, Bush AI, et al. The role of metallobiology and amyloid-β peptides in Alzheimer's disease. J Neurochem. 2012;120:149–66. doi: 10.1111/j.1471-4159.2011.07500.x .

Robinson NJ, Winge DR. Copper metallochaperones. Annu Rev Biochem. 2010;79:537–62. doi:10.1146/annurev-biochem-030409-143539.

Roy DN, Mandal S, Sen G, Biswas T. Superoxide anion mediated mitochondrial dysfunction leads to hepatocyte apoptosis preferentially in the periportal region during copper toxicity in rats. Chem Biol Interact. 2009;182:136–47. doi:10.1016/j.cbi.2009.08.014.

Russo AJ, de Vito R. Analysis of copper and zinc plasma concentration and the efficacy of zinc therapy in individuals with Asperger's syndrome, pervasive developmental disorder not otherwise specified (PDD-NOS) and autism. Biomark Insights. 2011;6:127–33. doi:10.4137/BMI.S7286.

Sagripanti JL, Kraemer KH. Site-specific oxidative DNA damage at polyguanosines produced by copper plus hydrogen peroxide. J Biol Chem. 1989;264:1729–34.

Salazar-Weber NL, Smith JP. Copper inhibits NMDA receptor-independent LTP and modulates the paired-pulse ratio after LTP in mouse hippocampal slices. Int J Alzheimers Dis. 2011;2011:864753. doi:10.4061/2011/864753.

Sauer SW, Merle U, Opp S, et al. Severe dysfunction of respiratory chain and cholesterol metabolism in Atp7b$^{-/-}$ mice as a model for Wilson disease. Biochim Biophys Acta Mol basis Dis. 1812;2011:1607–15. doi: 10.1016/j.bbadis.2011.08.011.

Scheiber IF, Dringen R. Astrocyte functions in the copper homeostasis of the brain. Neurochem Int. 2013;62:556–65. doi:10.1016/j.neuint.2012.08.017.

Scheiber I, Dringen R, Mercer JFB. Copper: effects of deficiency and overload. Met Ions Life Sci. 2013;13:359–87. doi:10.1007/978-94-007-7500-8-11.

Scheiber IF, Mercer JFB, Dringen R. Metabolism and functions of copper in brain. Prog Neurobiol. 2014;116:33–57. doi:10.1016/j.pneurobio.2014.01.002.

Schlief ML, Gitlin JD. Copper homeostasis in the CNS: a novel link between the NMDA receptor and copper homeostasis in the hippocampus. Mol Neurobiol. 2006;33:81–90. doi:10.1385/MN:33:2:81.

Schlief ML, Craig AM, Gitlin JD. NMDA receptor activation mediates copper homeostasis in hippocampal neurons. J Neurosci. 2005;25:239–46. doi:10.1523/JNEUROSCI.3699-04.2005.

Schlief ML, West T, Craig AM, et al. Role of the Menkes copper-transporting ATPase in NMDA receptor-mediated neuronal toxicity. Proc Natl Acad Sci. 2006;103:14919–24. doi:10.1073/pnas.0605390103.

Schwerdtle T, Hamann I, Jahnke G, et al. Impact of copper on the induction and repair of oxidative DNA damage, poly(ADP-ribosyl)ation and PARP-1 activity. Mol Nutr Food Res. 2007;51:201–10. doi:10.1002/mnfr.200600107.

Sharma HS, Sharma A. Neurotoxicity of engineered nanoparticles from metals. CNS Neurol Disord Drug Targets. 2012;11:65–80. doi:10.2174/187152712799960817.

Sheline CT, Choi DW. Cu^{2+} toxicity inhibition of mitochondrial dehydrogenases in vitro and in vivo. Ann Neurol. 2004;55:645–53. doi: 10.1002/ana.20047.

Sokol RJ, Devereaux M, Mierau GW, et al. Oxidant injury to hepatic mitochondrial lipids in rats with dietary copper overload. Modif Vitam E Def Gastroenterol. 1990;99:1061–71.

Sokol RJ, Devereaux MW, O'Brien K, et al. Abnormal hepatic mitochondrial respiration and cytochrome C oxidase activity in rats with long-term copper overload. Gastroenterology. 1993;105:178–87.

Sokol RJ, Twedt D, McKim JM Jr, et al. Oxidant injury to hepatic mitochondria in patients with Wilson's disease and Bedlington terriers with copper toxicosis. Gastroenterology. 1994;107:1788–98.

Spencer WA, Jeyabalan J, Kichambre S, Gupta RC. Oxidatively generated DNA damage after Cu(II) catalysis of dopamine and related catecholamine neurotransmitters and neurotoxins: role of reactive oxygen species. Free Radic Biol Med. 2011;50:139–47. doi:10.1016/j.freeradbiomed.2010.10.693.

Strand S, Hofmann WJ, Grambihler A, et al. Hepatic failure and liver cell damage in acute Wilson's disease involve CD95 (APO-1/Fas) mediated apoptosis. Nat Med. 1998;4:588–93. doi:10.1038/nm0598-588.

Strozyk D, Launer LJ, Adlard PA, et al. Zinc and copper modulate Alzheimer Aβ levels in human cerebrospinal fluid. Neurobiol Aging. 2009;30:1069–77. doi: 10.1016/j.neurobiolaging.2007.10.012.

Stuerenburg HJ. CSF copper concentrations, blood-brain barrier function, and ceruloplasmin synthesis during the treatment of Wilson's disease. J Neural Transm. 2000;107:321–9. doi:10.1007/s007020050026.

Szerdahelyi P, Kása P. Histochemical demonstration of copper in normal rat brain and spinal cord. Histochem Cell Biol. 1986;85:341–7.

Szymczak W, Menzel N, Keck L. Emission of ultrafine copper particles by universal motors controlled by phase angle modulation. J Aerosol Sci. 2007;38:520–31. doi:10.1016/j.jaerosci.2007.03.002.

Tallaksen-Greene SJ, Janiszewska A, Benton K, et al. Evaluation of tetrathiomolybdate in the R6/2 model of Huntington disease. Neurosci Lett. 2009;452:60–2. doi:10.1016/j.neulet.2009.01.040.

Tarohda T, Yamamoto M, Amamo R. Regional distribution of manganese, iron, copper, and zinc in the rat brain during development. Anal Bioanal Chem. 2004;380:240–6. doi:10.1007/s00216-004-2697-8.

Thackray AM, Knight R, Haswell SJ, et al. Metal imbalance and compromised antioxidant function are early changes in prion disease. Biochem J. 2002;362:253–8. doi:10.1042/0264-6021:3620253.

Thomas B, Flint Beal M. Parkinson's disease. Hum Mol Genet. 2007;16:R183–94. doi:10.1093/hmg/ddm159.

Timmers HJLM, Deinum J, Wevers RA, JWM L. Congenital dopamine-β-hydroxylase deficiency in humans. Ann N Y Acad Sci. 2004;1018:520–3. doi: 10.1196/annals.1296.064.

Tkeshelashvili LK, McBride T, Spence K, Loeb LA. Mutation spectrum of copper-induced DNA damage. J Biol Chem. 1991;266:6401–6.

Urso E, Maffia M. Behind the link between copper and angiogenesis: established mechanisms and an overview on the role of vascular copper transport systems. J Vasc Res. 2015;52:172–96. doi:10.1159/000438485.

Uversky VN, Li J, Fink AL. Metal-triggered structural transformations, aggregation, and fibrillation of human alpha-synuclein: a possible molecular link between parkinson's disease and heavy metal exposure. J Biol Chem. 2001;276:44284–96. doi:10.1074/jbc.M105343200.

VanLandingham JW, Fitch CA, Levenson CW. Zinc inhibits the nuclear translocation of the tumor suppressor protein p53 and protects cultured human neurons from copper-induced neurotoxicity. NeuroMolecular Med. 2002;1:171–82. doi:10.1385/NMM:1:3:171.

Vasić V, Jovanović D, Krstić D, et al. Prevention and recovery of $CuSO_4$-induced inhibition of Na^+/K^+-ATPase and Mg^{2+}-ATPase in rat brain synaptosomes by EDTA. Toxicol Lett. 1999;110:95–104. doi: 10.1016/S0378-4274(99)00144-7.

Vest KE, Leary SC, Winge DR, Cobine PA. Copper import into the mitochondrial matrix in Saccharomyces cerevisiae is mediated by Pic2, a mitochondrial carrier family protein. J Biol Chem. 2013;288:23884–92. doi:10.1074/jbc.M113.470674.

Voss K, Harris C, Ralle M, et al. Modulation of tau phosphorylation by environmental copper. Transl Neurodegener. 2014;3:24. doi:10.1186/2047-9158-3-24.

Vujisić L, Krstić D, Krinulović K, Vasić V. The influence of transition and heavy metal ions on ATP-ases activity in rat synaptic plasma membranes. J Serbian Chem Soc. 2004;69:541–7. doi:10.2298/JSC0407541V.

Waggoner DJ, Drisaldi B, Bartnikas TB, et al. Brain copper content and cuproenzyme activity do not vary with prion protein expression level. J Biol Chem. 2000;275:7455–8. doi:10.1074/jbc.275.11.7455.

Wakabayashi H, Koszelak ME, Mastri M, Fay PJ. Metal ion-independent association of factor VIII subunits and the roles of calcium and copper ions for cofactor activity and inter-subunit affinity. Biochemistry. 2001;40:10293–300. doi:10.1021/bi010353q.

Wang X, Li GJ, Zheng W. Efflux of iron from the cerebrospinal fluid to the blood at the blood-CSF barrier: effect of manganese exposure. Exp Biol Med (Maywood). 2008;233:1561–71. doi:10.3181/0803-RM-104.

Wang L-M, Becker JS, Wu Q, et al. Bioimaging of copper alterations in the aging mouse brain by autoradiography, laser ablation inductively coupled plasma mass spectrometry and immunohistochemistry. Metallomics. 2010;2:348–53. doi:10.1039/c003875j.

Wang DB, Kinoshita C, Kinoshita Y, Morrison RS. P53 and mitochondrial function in neurons. Biochim Biophys Acta Mol basis Dis. 2014;1842:1186–97. doi:10.1016/j.bbadis.2013.12.015.

Warren PJ, Earl CJ, Thompson RHS. The distribution of copper in human brain. Brain. 1960;83:709–17. doi:10.1093/brain/83.4.709.

Whiteside JR, Box CL, McMillan TJ, Allinson SL. Cadmium and copper inhibit both DNA repair activities of polynucleotide kinase. DNA Repair (Amst). 2010;9:83–9. doi:10.1016/j.dnarep.2009.11.004.

Willemse J, Van den Hamer CJ, Prins HW, Jonker PL. Menkes' kinky hair disease. I. Comparison of classical and unusual clinical and biochemical features in two patients. Brain and Development. 1982;4:105–14.

Wilmarth PA, Short KK, Fiehn O, et al. A systems approach implicates nuclear receptor targeting in the $Atp7b^{-/-}$ mouse model of Wilson's disease. Metallomics. 2012;4:660–8. doi: 10.1039/c2mt20017a.

Wooton-Kee CR, Jain AK, Wagner M, et al. Elevated copper impairs hepatic nuclear receptor function in Wilson's disease. J Clin Invest. 2015;125:3449–60. doi:10.1172/JCI78991.

Xiao G, Fan Q, Wang X, Zhou B. Huntington disease arises from a combinatory toxicity of polyglutamine and copper binding. Proc Natl Acad Sci U S A. 2013;110:14995–5000. doi:10.1073/pnas.1308535110.

Xu LJ, Zhao JX, Zhang T, et al. In vitro study on influence of nano particles of CuO on CA1 pyramidal neurons of rat hippocampus potassium currents. Environ Toxicol. 2009;24:211–217. doi:10.1002/Tox.20418.

Yokohira M, Kuno T, Yamakawa K, et al. Lung toxicity of 16 fine particles on intratracheal instillation in a bioassay model using f344 male rats. Toxicol Pathol. 2008;36:620–31. doi:10.1177/0192623308318214.

Yoshimura N, Kida K, Usutani S. Histochemical localization of copper in various organs of brindled mice after copper therapy. Pathol Int. 1995;45:10–8.

Youdim MBH, Grünblatt E, Mandel S. The copper chelator, D-penicillamine, does not attenuate MPTP induced dopamine depletion in mice. J Neural Transm. 2007;114:205–9. doi:10.1007/s00702-006-0499-1.

Yu F, Gong P, Hu Z, et al. Cu(II) enhances the effect of Alzheimer's amyloid-β peptide on microglial activation. J Neuroinflammation. 2015;12:122. doi:10.1186/s12974-015-0343-3.

Yurderi M, Bulut A, Ertas IE, et al. Supported copper-copper oxide nanoparticles as active, stable and low-cost catalyst in the methanolysis of ammonia-borane for chemical hydrogen storage. Appl Catal B Environ. 2015;165:169–75. doi:10.1016/j.apcatb.2014.10.011.

Yurkova IL, Stuckert F, Kisel MA, et al. Formation of phosphatidic acid in stressed mitochondria. Arch Biochem Biophys. 2008;480:17–26. doi:10.1016/j.abb.2008.09.007.

Yurkova IL, Arnhold J, Fitzl G, Huster D. Fragmentation of mitochondrial cardiolipin by copper ions in the Atp7b $^{-/-}$ mouse model of Wilson's disease. Chem Phys Lipids. 2011;164:393–400. doi: 10.1016/j.chemphyslip.2011.05.006.

Zatta P, Drago D, Zambenedetti P, et al. Accumulation of copper and other metal ions, and metallothionein I/II expression in the bovine brain as a function of aging. J Chem Neuroanat. 2008;36:1–5. doi:10.1016/j.jchemneu.2008.02.008.

Zheng W, Monnot AD. Regulation of brain iron and copper homeostasis by brain barrier systems: implication in neurodegenerative diseases. Pharmacol Ther. 2012;133:177–88. doi:10.1016/j.pharmthera.2011.10.006.

Zheng G, Chen J, Zheng W. Relative contribution of CTR1 and DMT1 in copper transport by the blood-CSF barrier: implication in manganese-induced neurotoxicity. Toxicol Appl Pharmacol. 2012;260:285–93. doi:10.1016/j.taap.2012.03.006.

Zimnicka AM, Maryon EB, Kaplan JH. Human copper transporter hCTR1 mediates basolateral uptake of copper into enterocytes: implications for copper homeostasis. J Biol Chem. 2007;282:26471–80. doi:10.1074/jbc.M702653200.

Zimnicka AM, Ivy K, Kaplan JH. Acquisition of dietary copper: a role for anion transporters in intestinal apical copper uptake. Am J Physiol Cell Physiol. 2011;300:C588–99. doi:10.1152/ajpcell.00054.2010.

Zischka H, Lichtmannegger J, Schmitt S, et al. Liver mitochondrial membrane crosslinking and destruction in a rat model of Wilson disease. J Clin Invest. 2011;121:1508–18. doi:10.1172/JCI45401.

Thallium Toxicity: General Issues, Neurological Symptoms, and Neurotoxic Mechanisms

Laura Osorio-Rico, Abel Santamaria, and Sonia Galván-Arzate

Abstract Thallium (Tl⁺) is a ubiquitous natural trace metal considered as the most toxic among heavy metals. The ionic ratio of Tl^+ is similar to that of potassium (K^+), therefore accounting for the replacement of the latter during enzymatic reactions. The principal organelle damaged after Tl^+ exposure is mitochondria. Studies on the mechanisms of Tl^+ include intrinsic pathways altered and changes in antiapoptotic and proapoptotic proteins, cytochrome c, and caspases. Oxidative damage pathways increase after Tl^+ exposure to produce reactive oxygen species (ROS), changes in physical properties of the cell membrane caused by lipid peroxidation, and concomitant activation of antioxidant mechanisms. These processes are likely to account for the neurotoxic effects of the metal. In humans, Tl^+ is absorbed through the skin and mucous membranes and then is widely distributed throughout the body to be accumulated in bones, renal medulla, liver, and the Central Nervous System. Given the growing relevance of Tl^+ intoxication, in recent years there is a notorious increase in the number of reports attending Tl^+ pollution in different countries. In this sense, the neurological symptoms produced by Tl^+ and its neurotoxic effects are gaining attention as they represent a serious health problem all over the world. Through this review, we present an update to general information about Tl^+ toxicity, making emphasis on some recent data about Tl^+ neurotoxicity, as a field requiring attention at the clinical and preclinical levels.

Keywords Thallium • Pollution • Metal • Human health • Neurotoxicity

L. Osorio-Rico • S. Galván-Arzate (✉)
Departamento de Neuroquímica, Instituto Nacional de Neurología y Neurocirugía, Insurgentes Sur 3877, Mexico City 14269, Mexico
e-mail: sonia_galvan@yahoo.com

A. Santamaria
Laboratorio de Aminoácidos Excitadores, Instituto Nacional de Neurología y Neurocirugía, Insurgentes Sur 3877, Mexico City 14269, Mexico

© Springer International Publishing AG 2017
M. Aschner and L.G. Costa (eds.), *Neurotoxicity of Metals*, Advances in Neurobiology 18, DOI 10.1007/978-3-319-60189-2_17

345

Introduction

Thallium (Tl^+) is a toxic heavy metal that was accidentally discovered in 1861 by Sir William Crookes by burning the dust from a sulfuric acid industrial plant. He observed a bright green spectral band that quickly disappeared. The new element was named "thallium" (after thallos meaning young shoot) (Galván-Arzate and Santamaría 1998).

Tl^+ poisoning is one of the most complex and serious toxic patterns known to man. The symptomatology is usually nonspecific due to the multi-organ involvement. The first symptoms of Tl^+ poisoning include fever, gastrointestinal alterations, and neurological symptoms such as delirium, convulsions, and coma. Symptoms may appear rapidly, but during acute toxicity, they are replaced by a gradual development and expression of mild gastrointestinal disturbances, polyneuritis, encephalopathy, tachycardia, skin eruptions, stomatitis, atrophic changes of the skin, Mee's lines, and skin hyperesthesia (mainly in the soles of the feet). Development of psychotic behavior with hallucinations and dementia has also been reported in advanced stages of intoxication. In human beings, the most characteristic sign of Tl^+ toxicity is alopecia, which usually appears 15–20 days after intoxication (Saddique and Peterson 1983).

Tl^+ is a nonessential heavy metal exhibiting environmental and occupational threats, as well as therapeutic hazards, because of its use in medicine. It is found in two oxidation states, thallous (Tl^+) and thallic (Tl^{3+}) salts, both of which are considered highly toxic to humans, as well as domestic and wild organisms. Many Tl^+ compounds are colorless, odorless, and tasteless, and these characteristics, combined with the high level of toxicity of Tl^+ compounds, have led to their use as a poisoning agent. Because of its similarity to potassium (K^+) ions, plants and animals readily absorb Tl^+ through the skin and digestive and respiratory systems. In mammals, it can cross the placental, blood-brain, and gonadal barriers (Rodríguez-Mercado and Altamirano-Lozano 2013).

Clinical and Industrial Use of Tl^+

In the past, Tl^+ salts were used for treatment against syphilis, as well as to reduce night sweats in patients suffering from tuberculosis and malaria. Nowadays, Tl^{201} is employed as a tracker in radio-medicine studies. Since 1920, Tl^+ was used as poison for rodents and insects, and this practice was sustained unfortunately in some countries until the 1980s. Tl^+ salts have also been utilized as depilatory agents. Other uses of this metal include the manufacture of imitation jewelry, fabrication of low-temperature thermometers, ceramic semiconductor material, scintillation counters for radioactivity quantification, and optical lenses (Peter and Viraraghavan 2005). Nowadays, Tl^+ is restricted to industrial purposes only in few countries. Still, the environmental and health problems derived from its use persist.

Pollution by Tl⁺

As above mentioned, industrial uses of Tl^+ include electronic devices, as well as glass and Tl^+-containing catalysts, all of them patented for industrial processes (Ponton et al. 2016). Since 1975, Tl^+ was considered as a potential agent contributing to environmental pollution when discharged in wastes from mines and ore-processing and coal-burning plants (Zitko 1975; Gómez-González et al. 2015).

Peter and Viraraghavan (2005) proposed Tl^+ as a toxic metal contributing to public health problems. Studies in China have demonstrated that Tl^+ produces specific endemic diseases as a result of natural geochemical processes and anthropogenic activities. These problems are due to Tl^+ contamination in local drinking water and vegetables surrounding the Tl^+-rich sulfide mineralized areas (Li et al. 2012). The main factors facilitating the release of Tl^+ from sulfide minerals and rocks are its mobility in most aqueous environments, and its ability to be dispersed easily during oxidation of Tl^+-bearing sulfides. High rainfall precipitation (around 1000 mm per year), warm temperature, and hydrological conditions aid Tl^+ dispersion from the source points (i.e., the mine wastes containing high contents of sulfide minerals) to the downstream areas (Xiao et al. 2012; Zhuang and Gao 2015). Other environmental factors are known to contribute to Tl + poisoning; for example, in estuaries of South West England, Tl^+ is accumulated by macroalgae and deposit-feeding invertebrates (Tuner et al. 2013).

Other countries such as Brazil (Alves et al. 2014), Italy (Malandrino et al. 2016), Czech Republic (Loula et al. 2016), provinces of Flanders (Govarts et al. 2016), and Canada (Ponton et al. 2016) have suffered from important human health issues associated with exposure to Tl^+ in water of rivers.

In Area of Doñana, Spain, 29 shrews were studied, and their livers and kidneys were dissected. They showed high concentrations of Tl^+ that were dependent on the organ studied and gender. These subjects were collected from a protected area with great amount of acidic waters and sludge from pyrite mines (Sánchez-Chardi 2007).

Whatever the case, it seems clear that Tl^+ emerges as an important environmental pollutant causing severe clinical alterations in subjects exposed to the metal; therefore, the characterization of those potential sources of Tl^+ for its release is an important health issue to prevent communal exposures.

Human Poisoning and Neurological Symptoms

Features of Tl^+ poisoning depend on the dose, route of administration, individual susceptibility, and onset of treatment. Tl^+ intoxication in human beings is principally accidental, through its ingestion from Tl^+-containing rodenticides or its direct consumption for homicidal or suicidal purposes. For instance, two patients were acutely intoxicated with Tl^+ in water, and they exhibited the following CNS manifestations: confusion, disorientation, and hallucinations followed by anxiety, depression, lack of attention, as well as memory and verbal fluency impairment.

Noteworthy, the elevated concentrations of Tl^+ in urine (14,520 µg/L), blood (2056 µg/L), and drinking water (3124 mg/L) found in this study also produced lesions and degeneration in the striatum of one of the patients (Tsai et al. 2006).

In another case, three patients were intoxicated by contamination of heroin with Tl^+ (content in urine 200–300 µg/dL). The main symptoms exhibited by them included generalized weakness, insomnia, loss of vision, and alopecia (Afshari et al. 2012). In China, 30 patients with endemic thallotoxicosis showed several clinical manifestations similar to those previously described for other cases (Li et al. 2012). Another study (100 cases) in Iran showed that long-term opioid abuse contaminated with the metal led to Tl^+ intoxication; the mean urinary Tl^+ level in these subjects was close to 21 µg/L, and their symptoms comprised ataxia, tremor, insomnia, neuropathy, sweeting, scalp hair loss, nausea, and vomiting (Ghaderi et al. 2015).

Macro- and Micro-distribution

Tl^+ (30 mg/kg; 4 h) administered to rats produced toxic effects and altered the morphology and function of kidney. The magnitude of distribution and damage was followed by the ileum, stomach, and liver (Leung and Ooi 2000). This evidence suggests that regions with high metabolic rate and elevated content of biochemical substrates for oxidative activity, such as the kidney, liver, and brain, could be more vulnerable to Tl^+ toxicity. In the brain, Tl^+ distribution exhibited a differential pattern over time. For instance, after 5 min, tissue was characterized by very low Tl^+ contents in white matter fiber tracts and a highly heterogeneous pattern of the metal distribution in different nuclei, layers, and cell types in different brain regions. Uptake was higher in the glomerular layer of the olfactory bulbs, with varying intensities in different glomeruli, in neurons, in layer II/III and V of cingulate cortical areas, in the lateral habenula, in mammillary bodies in the central nucleus of the inferior colliculus, in the oculomotor nucleus, and in many brainstem and facial nerve nuclei. High Tl^+ uptake was also present in interneurons from layer IV and pyramidal cells of layer V in the cerebral cortex. At 24 h, Tl^+ distribution was remarkably different from that of 5 min. Such differences in distribution can obey to different mechanisms of clearance (Wanger et al. 2012). Previous studies of our group support a pattern of differential distribution throughout the brain regions (Ríos et al. 1989; Galván-Arzate et al. 2000).

Neurotoxic Mechanisms of Tl^+

Probably one of the most important mechanisms for Tl^+ toxicity lies in its ability to interfere with energy production by inhibiting the Na^+/K^+-ATPase. Tl^+-induced brain damage is associated with an increased oxidative stress via induction of lipid peroxidation (del Carmen Puga Molina and Verstraeten 2008), supporting the concept that reactive oxygen species (ROS) production plays a pivotal role for its toxic pattern. An increased generation of ROS, and the concomitant disruption of cellular energy production from impaired mitochondria, can block cell cycle progression, which in turn leads to cell apoptosis. Tl^+ also inhibits DNA replication, leading to cell cycle arrest and death. The induction of cell cycle arrest is associated with upregulation of the CDK (cyclin-dependent kinases) inhibitor p21. In addition, apoptosis is associated with the elevation of proapoptotic proteins such as Apaf and Bad and downregulation of antiapoptotic proteins such as Bcl-2 and Bcl-x_L. Tl^+ also produces a decreased cell viability in C6 glioma cell, coursing with cell cycle progression at G_2/M phase and CDK2 protein, as well as increased expression of p53 and p21 (Chia et al. 2005).

Another recent potential mechanism described for this metal involves excitotoxicity. MK-801 is a potent, noncompetitive glutamate receptor antagonist that blocks N-methyl-D-aspartate receptors (NMDAr) and the toxic events elicited by their overactivation in models of ischemia and neurodegeneration. Since Tl^+ competes with K^+ for Na^+/K^+-ATPases, blocking their activity, the metal might exert its neurotoxic effects in part through secondary excitotoxicity, involving membrane depolarization and further overactivation of NMDAr, as evidenced by the protective effect exerted by MK-801 on Tl^+-induced alterations in motor activity and oxidative damage to lipids in different rat brain regions (Osorio-Rico et al. 2015). Current experiments are actually in progress to confirm this hypothesis.

In turn, Tl^+-induced apoptosis (at doses ranging 100–500 μM) involves swelling in mitochondria and opening of the membrane transition pore (MTP) in Jurkat cells (Bragadin et al. 2003). In turn, MTP opening induces a decrease in the liver mitochondrial potential and an increase in states III and IV of the mitochondrial respiratory chain (Korotkov et al. 2008; Korotkov 2009). Succinate dehydrogenase (complex II) activity, ATP levels, and ATP/ADP ratio are also decreased by Tl^+ in liver mitochondria (Eskandari et al. 2015), and this could be taking place also in the CNS. Moreover, in pheochromocytoma cells (PC12), Tl^+ produces apoptosis by altering intrinsic pathways in a manner dependent of caspase 9 and caspase 3 activation (Hanzel and Verstraeten 2009).

Central and Peripheral Effects of Tl^+ on Enzymes

Tl^+ produces histopathological damage in liver and kidney when administered at high doses (30 mg/kg for 4 days) in an acute scheme. In serum, aspartate aminotransferase (AST) and alanine aminotransferase (ALT) activities were increased in a dose-dependent manner after administration of 30, 60, and 120 mg/kg of Tl^+ for 16 h (Leung and Ooi 2000). Heme oxygenase and ALA synthase were increased, whereas NADPH cytochrome P-450 was decreased after Tl^+ exposure (50, 100, or 200 mg/kg) (Woods et al. 1984).

In vivo studies have demonstrated that glutathione (GSH) concentrations and glutathione peroxidase (GPx) were not decreased by Tl^+ in the renal cortex, medulla, or brain regions (24 h after administration of the metal). This evidence suggests that Tl^+ toxicity takes time to express markers of oxidative damage, a consideration that is supported by the fact that in the brain, only the striatum exhibited a decrease in

GHS at 7 days after the Tl^+ administration (Appenroth and Winnefeld 1999; Osorio-Rico et al. 2015; Galván-Arzate et al. 2005). In contrast, in vitro exposure to Tl^+ decreased the content of GSH and GPx enzyme activity (Hanzel et al. 2005), emphasizing the role of oxidative stress as part of the toxic pattern of this metal.

Changes in Locomotor Activity

There are only few reports describing the effects of Tl^+ on motor and behavioral skills. The metal decreased endpoints of locomotor activity in an open-field test at different doses (8, 16, and 32 mg/kg) at 24 h and 7 days after its administration (Osorio-Rico et al. 2015; Galván-Arzate et al. 2005). While early changes in motor activity could be due to immediate alteration in the function of neurochemical systems, late changes could be more related with lack of motor regulation after brain tissue has been damaged, which is the case of the striatum. More detailed studies are needed to support this suggestion.

Therapy Against Tl^+ Intoxication

Unfortunately, there are no controlled trials for treatment of Tl^+-poisoned patients. Then, the controversial reports are useless. Literature is mostly represented by a limited number of toxicological studies in animals and case reports offering very limited data. Strong evidence points against the use of traditional metal chelators, such as dimercaprol (British anti-Lewisite) and penicillamine, especially since the latter may cause redistribution of Tl^+ into the CNS. Likewise, forced K^+ diuresis appears harmful. The use of single- or multiple-dose activated charcoal is supported by in vitro binding experiments and some animal data, and therefore, charcoal hemoperfusion may be a useful adjunct. Multiple animal studies provide evidence for enhanced elimination and improved survival with Prussian blue. Unfortunately, despite the fact that many patients have been treated with Prussian blue, the data presented are insufficient to comment definitively on its efficacy. However, Prussian blue's safety profile is superior to that of other proposed therapies (and it should be considered the drug of choice for treatment of acute Tl^+ poisoning). Public health efforts should focus on greater restrictions on access to, and use of Tl^+ salts (Hoffman 2003).

Huang et al. (2014) reported that treating severe Tl^+ poisoning requires lowering of its blood levels as soon as possible. These authors reported the case of a patient with supralethal blood levels of Tl^+ who was treated successfully using combined hemoperfusion (HP) and continuous veno-venous hemofiltration (CVVH). Three rounds of HP alone decreased blood Tl^+ levels by 20.2%, 34.8%, and 32.2%, while each of the five subsequent rounds of CVVH reduced Tl^+ blood levels by 63.5%, 64.2%, 42.1%, 18.6%, and 22.6%. The reversal of symptoms and prevention of lasting neurological damage indicates that combining HP, CVVH, and 2,3-dimercapto propane-1-sulfonate constitutes a suitable therapy comprising neuroprotective agents, along with supportive therapy that can be considered as successful to treat cases of severe Tl$^+$ poisoning. Nonetheless, additional efforts require finding more effective therapies for Tl$^+$ intoxication.

Conclusion

Tl$^+$, similar to other toxicants, represents a complex threat to mankind. Knowledge about the toxic mechanism exerted by this element is crucial for the design of strategies for professional health care. Early recognition of the clinical characteristics of poisoning is also important to initiate appropriate therapy and minimizing deceases. In addition, since the toxic pattern exerted by Tl$^+$ favors alterations in the CNS at the biochemical and molecular levels, more preclinical and clinical investigations associated with neurophysiological featuring will be helpful for the early identification of Tl$^+$ intoxication and

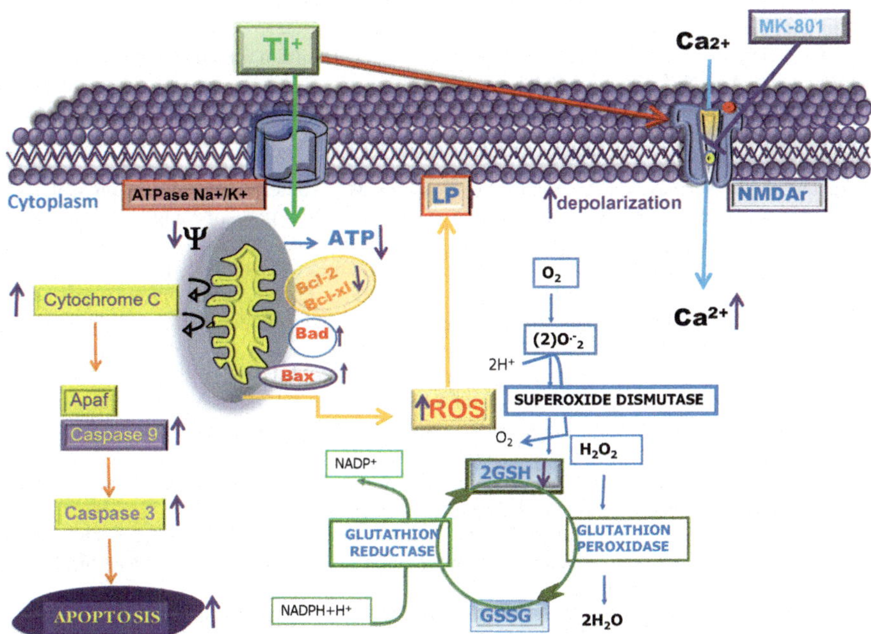

Fig. 1 Summarized mechanisms proposed for Tl$^+$ toxicity. Tl$^+$ induces direct blockade of Na$^+$/K$^+$-ATPase; mitochondrial swelling; dysfunction of complexes I, II, and IV of the electron transport chain (ETC); ATP depletion; reactive oxygen species (ROS) formation; decrease of antioxidant enzyme activity; and oxidative stress evidenced by lipid peroxidation (LP). Through these mechanisms, this metal could induce membrane depolarization and further ion channel activation, as well as voltage-gated N-methyl-D-aspartate receptor (NMDAr) activation. Also in mitochondria, Tl$^+$ can alter the balance of Bad-Bax/Bcl2 proteins, increasing cytochrome C release, activating PAF1, inducing apoptosome formation, and activating caspases 9 and 3, thus leading to apoptotic cell death

treatment. In the meantime, the major mechanisms involved in Tl$^+$ neurotoxicity involve mitochondrial impairment, energy depletion, Na$^+$/K$^+$-ATPase dysfunction, ROS formation and oxidative stress, and proapoptotic signaling, with a potential risk to involve secondary excitotoxicity. These events are summarized in Fig. 1.

Disclosure of Conflict of Interest Authors declare no conflict of interest.

References

Afshari R, Mégarbane B, Zavar A. Thallium poisoning: one additional and unexpected risk of heroin abuse. Clinical Toxicol. (Phila) 2012;50:791–2.

Alves RI, Sampaio CF, Nadal M, Schumacher M, Domingo JL, Segura-Muñoz SY. Metal concentrations in surface water and sediments from Pardo River, Brazil: human health risks. Environ Res. 2014;133:149–55.

Appenroth D, Winnefeld K. Is thallium-induced nephrotoxicity in rats connected with riboflavin and/or GSH?-reconsideration of hypotheses on the mechanism of thallium toxicity. J Appl Toxicol. 1999;19:61–6.

Bragadin M, Toninello A, Bindoli A, Rigobello MP, Canton M. Thallium induces apoptosis in Jurkat cells. Ann N Y Acad Sci. 2003;1010:283–91.

Chia CF, Chen SC, Chen CS, Shih CM, Lee HM, Wu CH. Thallium acetate induces C6 glioma cell apoptosis. Ann N Y Acad Sci. 2005;1042:523–30.

del Carmen Puga Molina L, Verstraeten SV. Thallium (III)-mediated changes in membrane physical properties and lipid oxidation affect cardiolipin-cytochrome c interactions. Biochim Biophys Acta. 2008;1778:2157–64.

Eskandari MR, Mashayekhi V, Aslani M, Hosseini MJ. Toxicity of thallium on isolated rat liver mitochondria: the role of oxidative stress and MPT pore opening. Environ Toxicol. 2015;30:232–41.

Galván-Arzate S, Santamaría A. Thallium toxicity. Toxicol Lett. 1998;99:1–13.

Galván-Arzate S, Martínez A, Medina E, Santamaría A, Ríos C. Subchronic administration of sublethal doses of thallium to rats: effects on distribution and lipid peroxidation in brain regions. Toxicol Lett. 2000;116(1–2):37–43.

Galván-Arzate S, Pedraza-Chaverri J, Medina-Campos ON, Maldonado PD, Vázquez-Román B, Ríos C, Santamaria A. Delayed effects of thallium in the rat brain: regional changes in lipid peroxidation and behavioral markers, but moderate alterations in antioxidants, after a single administration. Food Chem Toxicol. 2005;43:1037–45.

Ghaderi A, Vahdati-Mashhadian N, Oghabian Z, Moradi V, Afshari R, Mehrpour O. Thallium exists in opioid poisoned patients. DARU 2015;23:39–42.

Gómez-González MA, García-Guinea J, Laborda F, Garrido F. Thallium occurrence and partitioning in soils and sediments affected by mining activities in Madrid Province (Spain). Sci Total Environ. 2015;536:268–78.

Govarts E, Remy S, Bruckers L, Den Hond E, Sioen I, Nelen V, Baeyens W, Nawrot T, Loots I, Van Larebeke N, Schoeters G. Combined effects of prenatal exposures to environmental chemicals on birth weight. Int J Environ Res Public Health. 2016;13:495.

Hanzel CE, Villaverde MS, Verstraeten SV. Glutathione metabolism is impaired in vitro by thallium (III) hydroxide. Toxicology. 2005;207:501–10.

Hanzel C, Verstraeten S. Tl (I) and Tl (III) activate both mitochondrial and extrinsic pathways of apoptosis in rat pheochromocytoma (PC12) cells. Toxicol Appl Pharmacol. 2009;236:59–70.

Hoffman RS. Thallium toxicity and the role of Prussian blue in therapy. Toxicol Rev. 2003;22:29–40.

Huang C, Zhang X, Li G, Jiang Y, Wang Q, Tian R. A case of severe thallium poisoning successfully treated with hemoperfusion and continuous veno-venous hemofiltration. Hum Exp Toxicol. 2014;33:554–8.

Korotkov SM. Effects of Tl^+ on ion permeability, membrane potential and respiration of isolated rat liver mitochondria. J Bioenerg Biomembr. 2009;41:277–87.

Korotkov SM, Emel'yanova LV, Yagodina OV. Inorganic phosphate stimulates the toxic effects of Tl^+ in rat liver mitochondria. J Biochem Mol Toxicol. 2008;22:148–57.

Leung KM, Ooi VE. Studies on thallium toxicity, its tissue distribution and histopathological effects in rats. Chemosphere. 2000;41:155–9.

Li S, Xiao T, Zheng B. Medical geology of arsenic, selenium and thallium in China. Sci Total Environ. 2012;421-422:31–40.

Loula M, Kaňa A, Vosmanská M, Koplík R, Mestek O. Transfer of thallium from rape seed to rape oil is negligible and oil is fit for human consumption. Food Addit Contam Part A Chem Anal Control Expo Risk Assess. 2016;33:668–73.

Malandrino P, Russo M, Ronchi A, Minoia C, Cataldo D, Regalbuto C, Giordano C, Attard M, Squatrito S, Trimarchi F, Vigneri R. Increased thyroid cancer incidence in basaltic volcanic area is associated with non-anthropogenic pollution and bicontamination. Endocrine. 2016;53:471–9.

Osorio-Rico L, Villeda-Hernández J, Santamaría A, Königsberg M, Galván-Arzate S. The N-methyl-D-aspartate receptor antagonist MK-801 prevents thallium-induced behavioral and biochemical alterations in the rat brain. Int J Toxicol. 2015;34:505–13.

Peter AL, Viraraghavan T. Thallium: a review of public health and environmental concerns. Environ Int. 2005;31:493–501.

Ponton DE, Caron A, Hare L, Campbell PG. Hepatic oxidative stress and metal subcellular partitioning are affected by selenium exposure in wild yellow perch (*Perca flavescens*). Environ Pollut. 2016;214:608–17.

Ríos C, Galván-Arzate S, Tapia R. Brain regional thallium distribution in rats acutely intoxicated with Tl2SO4. Arch Toxicol. 1989;63(1):34–7.

Rodríguez-Mercado JJ, Altamirano-Lozano MA. 2013. Genetic toxicology of thallium: a review. Drug Chem Toxicol. 2013;36:369–83.

Saddique A, Peterson CD. Thallium poisoning: a review. Vet Hum Toxicol. 1983;25:16–22.

Sánchez-Chardi A. Tissue, age, and sex distribution of thallium in shrews from Doñana, a protected area in SW Spain. Sci Total Environ. 2007;383:237–40.

Tsai YT, Huang CC, Kuo HC, Wang HM, Shen WS, Shih TS, Chu NS. Central nervous system effects in acute thallium poisoning. Neurotoxicology. 2006;27:291–5.

Turner A, Turner D, Braungardt C. Biomonitoring of thallium availability in two estuaries of southwest England. Mar Pollut Bull. 2013;69:172–7.

Wanger T, Scheich H, Ohl FW, Goldschmidt J. The use of thallium diethyldithiocarbamate for mapping CNS potassium metabolism and neuronal activity: Tl^+-redistribution, Tl^+-kinetics and Tl^+-equilibrium distribution. J Neurochem. 2012;122:106–14.

Woods JS, Fowler BA, Eaton DL. Studies on the mechanisms of thallium-mediated inhibition of hepatic mixed function oxidase activity. Correlation with inhibition of NADPH-cytochrome c (P-450) reductase. Biochem Pharmacol. 1984;33:571–6.

Xiao T, Yang F, Li S, Zheng B, Ning Z. Thallium pollution in China: a geo-environmental perspective. Sci Total Environ. 2012;1:51–8.

Zhuang W, Gao X. Distribution, enrichment and sources of thallium in the surface sediments of the southwestern coastal Laizhou Bay, Bohai Sea. Mar Pollut Bull. 2015;96:502–7.

Zitko V. Toxicity and pollution potential of thallium. Sci. Total Environ. 1975;4:185–92.

Neurodegeneration Induced by Metals in *Caenorhabditis elegans*

Felix Antunes Soares, Daiandra Almeida Fagundez, and Daiana Silva Avila

Abstract Metals are a component of a variety of ecosystems and organisms. They can generally be divided into essential and nonessential metals. The essential metals are involved in physiological processes once the deficiency of these metals has been associated with diseases. Although iron, manganese, copper, and zinc are important for life, it has been evidenced that they are also involved in neuronal damage in many neurodegenerative disorders. Nonessential metals, which are metals without physiological functions, are present in trace or higher levels in living organisms. Occupational, environmental, or deliberate exposures to lead, mercury, aluminum, and cadmium are clearly correlated with the increase of toxicity and varied kinds of pathological situations. Actually, the field of neurotoxicology needs to satisfy two opposing demands: the testing of a growing list of chemicals and resource limitations and ethical concerns associated with testing using traditional mammalian species. Toxicological assays using alternative animal models may relieve some of this pressure by allowing testing of more compounds while reducing expenses and using fewer mammals. The nervous system is by far the more complex system in *C. elegans*. Almost a third of their cells are neurons (302 neurons versus 959 cells in adult hermaphrodite). It initially underwent extensive development as a model organism in order to study the nervous system, and its neuronal lineage and the complete wiring diagram of its nervous system are stereotyped and fully described. The neurotransmission systems are phylogenetically conserved from nematodes to vertebrates, which allows for findings from *C. elegans* to be extrapolated and further confirmed in vertebrate systems. Different strains of *C. elegans* offer a new perspective on neurodegenerative processes. Some genes have been found to be related to neurodegeneration induced by metals. Studying these interactions may be an effective tool to slow neuronal loss and deterioration.

Keywords Lead • Mercury • Aluminum • Cadmium • Iron • Manganese • Copper • Zinc • Neurodegenerative diseases

F.A. Soares (✉)
Departamento de Bioquimica e Biologia Molecular, Universidade Federal de Santa Maria, Santa Maria, Rio Grande do Sul 97105 900, Brazil
e-mail: felix@ufsm.br

D.A. Fagundez • D.S. Avila (✉)
Universidade Federal do Pampa, Uruguaiana, Rio Grande do Sul 97508-000, Brazil
e-mail: avilads1@gmail.com

© Springer International Publishing AG 2017
M. Aschner and L.G. Costa (eds.), *Neurotoxicity of Metals*, Advances in Neurobiology 18, DOI 10.1007/978-3-319-60189-2_18

Introduction

Metals are a component of a variety of ecosystems and organisms. They can generally be divided in essential and nonessential metals. The essential metals are involved in different kinds of physiological processes once the deficiency of these metals has been associated with various diseases. Although metals are important elements for life, they usually are required in trace amounts, and the excessive quantities of metal levels can accumulate in various organs. For this reason, the elevated levels of metals may induce various pathological events. Although essential metals are important for life, it has been evidenced that they are also involved in neuronal damage in many neurodegenerative disorders.

Some metals are essential for life processes, but they are also culpable for several degenerative mechanisms. Due to the widespread existence of metals in our environment from both natural and anthropogenic sources, understanding the mechanisms that they act in the organisms must be well studied. Many organisms have evolved cellular detoxification systems including glutathione (GSH), metallothioneins (MTs), pumps and transporters, and heat shock proteins (HSPs) to regulate intracellular metal levels and avoid some toxic effect that could be caused by essential or nonessential metals.

Iron (Fe) has been related to several neurodegenerative disorders which can result from both Fe accumulation in specific brain regions and defects in its metabolism and/or homeostasis.

Mn neurotoxicity has been associated to dopamine (DA) oxidation, mitochondrial dysfunction, and astrocytosis, thus leading to a syndrome named manganism, which resembles Parkinson's disease (Benedetto et al. 2009).

Recent studies have reported oxidative damage due to copper (Cu) in various tissues, including brain tissue, which is important to study because the vulnerability of the brain to oxidative stress and disturbances in the copper homeostasis have been connected with neurodegenerative disorders (Scheiber et al. 2014).

Excessive zinc (Zn) is neurotoxic and causes neurodegeneration following transient global ischemia and plays a crucial role in the pathogenesis of vascular-type dementia (VD). In addition, increasing evidence suggests that the etiology of Alzheimer's Disease (AD) may involve disruptions of zinc homeostasis, and oxidative stress, thus facilitating reactive oxygen species production is an early and sustained event in AD disease progression.

The presence of toxic amounts of metals in the environment may originate from some human activities and a diversity of natural processes. The unquestionable increase levels of contaminants in the environment are a huge concern to human health and for the environment. Nonessential metals, which are metals without physiological functions, are present in trace or higher levels in living organisms. Occupational, environmental, or deliberate exposures to this kind of metals are clearly correlated with the increase of toxicity and varied kinds of pathological situations.

Lead (Pb^{2+}) has been known for centuries to be a neurotoxin, able to cross the blood-brain barrier more readily in children than in adults, and impair ontogenetic development of brain –adversely affecting cognitive ability and producing behavioral abnormalities.

Mercury (Hg), like other toxic metals, may enter motor neurons as a possible pathogenic factor in amyotrophic lateral sclerosis and other neurodegenerative disorders (Pamphlett and Kum Jew 2013). The molecular mechanisms mediating methylmercury (MeHg)-induced neurotoxicity and neurodegeneration are better known when compared with other Hg forms (organic and ethylmercury).

Aluminum (Al^{3+}) is the most widely distributed metal in the environment. Al was used as a phosphate-binding gel in patients with chronic renal failure, and the first Al neurotoxicity cases were reported in these patients (Alfrey et al. 1976). Al has neurotoxic effects, and it has long been implicated in the pathogenesis of AD and other neurodegenerative diseases (Campbell 2002; Gupta et al. 2005).

Cadmium (Cd) can enter into the brain parenchyma and neurons (Antonio et al. 2003) causing neurological alterations in humans and animal models (Lukawski et al. 2005), leading to lower attention, hypernociception, olfactory dysfunction, and memory deficits. Additionally, there are in vitro studies showing the neurotoxicity of Cd on cell culture of neurons and glial cells (Lopez et al. 2006).

In recent years, neurodegenerative diseases have become an important worldwide health issue. Neurodegeneration is characterized by cell death and/or loss of structure and/or function of neurons. Many neurodegenerative diseases including Parkinson's disease (PD) and Alzheimer's disease (AD) are the result of neurodegenerative processes. Owing to the prevalence, morbidity, and mortality, as well as social, ethical, and personal burden of neurodegenerative disorders, considerable effort has been directed toward the identification of a rational strategy to treat these devastating brain pathological conditions. A substantial challenge is to discern phenomena that may represent causes from those that may represent effects. The neurodegenerative process is associated with many events, of which each one may correspond to a typical feature of a specific disease.

There was a time when non-mammals were thought to be far from ideal models for the study of biomedical sciences because they are phylogenically too distant from humans. However, it has now become abundantly clear that some nonmammals are not only convenient models but also are endowed with physiological and pharmacological properties common to humans. Thus, several species have become very popular alternative organisms and are being used extensively as complementary/alternative animal models.

The identification of the molecular components and mechanisms of neurodegenerative diseases has often been inhibited by the complexities of the vertebrate brain and the difficulties of modeling the diseases in cell cultures. The recent advances in genetic technologies and the high sequence similarity between human and invertebrate genomes allow for the dissection of the molecular pathways involved in neurological diseases using model organisms. The field of neurotoxicology needs to satisfy two opposing demands: the testing of a growing list of chemicals and resource limitations and ethical concerns associated with testing using traditional mammalian species. National and international government agencies have defined a need to reduce, refine, or replace mammalian species in toxicological testing with alternative testing methods and nonmammalian models. Toxicological assays using alternative animal models may relieve some of this pressure by allowing testing of more compounds while reducing expense and using fewer mammals.

Caenorhabditis elegans as a Model for Metal-Induced Neurodegeneration

The nervous system is by far the more complex system in *C. elegans*. Almost a third of their cells are neurons (302 neurons versus 959 cells in adult hermaphrodite). It initially underwent extensive development as a model organism in order to study the nervous system (Brenner 1974), and its neuronal lineage and the complete wiring diagram of its nervous system are stereotyped and fully described (Sulston 1983; Sulston et al. 1983; White et al. 1986). The structure of the nervous system has been described in unprecedented detail by electron microscopic reconstruction (White et al. 1986). The high-resolution images obtained with electron microscopy allowed White and colleagues to identify all the synapses, about 890 electrical junctions, 1410 neuromuscular junctions, and 6393 chemical synapses, using the same neurotransmitter systems (cholinergic, gamma-aminobutyric acid [GABA]ergic, glutamatergic, dopaminergic [DAergic], and serotonergic) that are expressed in vertebrates (Chen et al. 2006). The neurotransmission systems are phylogenetically conserved from nematodes to vertebrates, which allows for findings from *C. elegans* to be extrapolated and further confirmed in vertebrate systems.

In addition, several genes involved in neurotransmission have been identified in *C. elegans*. As a result, knockout mutants can be generated and the synaptic transmission can be evaluated. For instance, worms lacking the tyrosine hydroxylase gene (*cat-1*), responsible for the biosynthesis of DA (Lints and Emmons 1999), present defective response to bacterial mechanosensation, i.e., they do not slow their basal movement in response to food presence, as worms with normal DA signaling do (Sawin et al. 2000). Individual neurons can be ablated by laser, and behavioral evaluations can be performed to obtain the phenotype related to that neuronal loss (Avery and Horvitz 1989; Bourgeois and Ben-Yakar 2007). Remarkably, transgenic worms can be generated by the fusion of the green fluorescent protein (GFP) to a reporter gene of interest, thus allowing in vivo imaging of any desired neuron (Chalfie et al. 1994).

As a result of the extensive study of *C. elegans* nervous system and the advantages of using mutants and transgenics, the nematode offers unique perspectives as a neurotoxicological model, which use has been recently increasing in the toxicology community. Studies examining metals as neurotoxicants address a vast array of outcomes including measuring endpoints on behavioral, structural, signaling, and molecular levels. Motor and mechanosensory functions of glutamatergic neurons are evaluated by measuring the pharyngeal pumping rate and the response to touch. Mechanosensory functions of DAergic and serotoninergic neurons are appraised by observing the ability of worms to slow down when they encounter food. Neurodegeneration of specific neurons induced by metal exposure can be visualized, and the mechanism of neurotoxicity can be determined by biochemical assays or by using mutants. Of note, the use of whole-organism assays allows the study of a functional multicellular unit, such as a dopaminergic synapse, instead of a single cell, which is of relevance to the extrapolation of the findings from worms to mammals.

Metal-Induced Neurodegeneration

Heavy metal pollution is of serious concern, along with the development of human production activities and industrial and agricultural waste water emissions (Jiang et al. 2016). Metals are persistent environmental contaminants; consequently the occurrence of unbalance in metal metabolism in the brain was described by a vast literature as being associated with neurodegenerative disease. Some of the most compelling evidence for the contribution of metals into neurodegeneration comes from studies of postmortem tissue implicating metal accumulation in the areas of the brain coincident with cell death in patients with confirmed neurodegenerative disease (Martinez-Finley et al. 2011; Berg and Youdim 2006; Chen et al. 2013). It is not certain, however, whether metal presence is the cause or consequence of the disease. The mechanism by which metals produce neurodegenerative damage is metal and dose dependent; however, they share common mechanisms including free radical production, protein aggregation, bioenergetic dysfunction, calcium dysregulation and metal transport alteration (Chen et al. 2016; Farina et al. 2013; Gaeta and Hider 2005) and most likely a combination of these factors ultimately triggers the neurodegenerative process. For instance, magnetic resonance imaging and postmortem studies determined, for instance, that Fe is accumulated in brain areas responsible for different neurodegenerative diseases: cortex for Huntington's disease (Rosas et al. 2012); caudate nucleus, globus pallidus, and putamen in progressive supranuclear palsy (Boelmans et al. 2012); and basal ganglia in multiple sclerosis (Grimaud et al. 1995). Moreover, recent evidences have found direct links between divalent metal transporter (DMT) and Parkinson's disease (PD), as the postmortem brains of PD patients show upregulation of DMT1 protein in the substantia nigra pars compacta (SNpc) (Salazar et al. 2008).

Neurodegeneration is characterized by the cell death or loss of structure and/or function of neurons. Remarkably, many neurodegenerative diseases including PD and Alzheimer's disease (AD) are the result of neurodegenerative processes induced by essential and nonessential metals. Neurodegenerative diseases have become more prevalent and of great epidemiological importance. These diseases are characterized by progressive accumulation of proteins aggregates and disruption of the proteostasis in neuron cells, resulting in progressive degeneration and consequent debilitating conditions. Neurons normally do not reproduce or replace themselves, so when they become damaged or die, they cannot be replaced by the organism.

Remarkably, neurological symptoms manifest when certain percentage of neurons are degenerated and the available therapeutic options cannot revert the damage. Because of that, understanding the mechanisms that underlie metal induced neurotoxicity may contribute to the development of effective therapies.

Essential Metals Neurotoxicity

Iron (Fe^{2+}) is an important metal to the organism homeostasis and exists abundantly in the environment. Iron participates in many cellular functions, is essential for normal neural development and physiology, and plays a fundamental role as a component of mitochondrial respiratory chain complexes (Farina et al. 2013). However, if inappropriately managed, the transition metal can cause toxic effects in many organisms (Hu et al. 2008). All the absorbed iron is bound to storage or transporting proteins and the levels of intra and extracellular free iron are very low. Indeed, even low levels of free Fe can cause toxic effects in different types of cells (Hu et al. 2008).

As in mammals, Fe^{2+} is essential to worms, as it is necessary for the cytochromes of the respiratory chain complexes and for P450 metabolism enzymes. Genes involved in Fe and energy homeostasis in vertebrates are conserved in the nematode. These include aconitase, ferritin, divalent metal transporter-1 (DMT-1), frataxin, and Fe sulfur cluster assembly proteins. The Fe regulating protein-1 (IRP-1) homologue (ACO-1) of *C. elegans* has aconitase activity and is posttranslationally regulated by Fe.

Fe overload in worms causes phenotypic and behavioral defects as well as alteration of the resistance to oxidative stress, characterized by reduced life span, body size, generation time, brood size, head thrash, and body bend frequencies, as well as chemotaxis plasticity (Hu et al. 2008; Valentini et al. 2012). Several of these defects (body bend frequency and life span) were transferred from Fe-exposed *C. elegans* to their progeny (Hu et al. 2008).

Some disorders have been related to genetic causes and called neurodegeneration with brain iron accumulation (NBIA), two of which, aceruloplasminemia and neuroferritinopathy, are caused by mutations in genes directly involved in iron metabolic pathway and others, such as pantothenate kinase 2-, phospholipase-A2-, and fatty acid 2-hydroxylase-associated neurodegeneration (Dusek et al. 2012). The observation that pharmacological agents with Fe chelation capacity prevent neuronal death induced by parkinsonian toxins (Zhu et al. 2007) highlights the pivotal role of Fe in PD neuronal death. Notably, as the brain ages, Fe accumulates in regions that are affected by Alzheimer's disease, Parkinson's disease, or Huntington's disease (Bartzokis et al. 1997; Schipper 2012; Pfefferbaum et al. 2009).

Using *C. elegans*, Klang et al. linked Fe aging to protein insolubility, and other research has associated Fe and the neurodegeneration in diseases (Klang et al. 2014). The adverse effects of Fe on locomotive behavior suggest that Fe might be involved in disruption of synaptic function between neurons and muscle cells. In *C. elegans* models of Aß toxicity, Fe was shown to possess high affinity for Aß.

Aß accumulation in the Aß-expressing strain CL2006 resulted in Fe homeostasis disruption. In addition to increasing Fe content, Aß has also been shown to increase ROS generation (Wan et al. 2011).

In a recent study, Fagundez et al. found that *C. elegans* acutely treated with Fe^{2+} depicted altered DAergic neurons, which was associated with oxidative stress, decreased locomotor activity, and reduction in egg laying and longevity (Fagundez et al. 2015). This is a plausible consequence to evidences that show that PD brains exhibit increased total Fe concentration (Gotz 2006; Friedman et al. 2009; Norfray et al. 1988). An increase in iron as seen in PD brains may increase DA synthesis, causing excess DA to be released into the cytoplasm, which may lead to increased ROS production (Chege and McColl 2014).

It is known that $Fe^{2+/3+}$accumulation causes free radical damage through the Fenton reaction (Fraga and Oteiza 2002; Aisen et al. 2001), as Fe^{2+} reduces hydrogen peroxide to the highly cytotoxic hydroxyl radical (OH•) (Fig. 1). Mitochondrion is the main site of superoxide production and an important site of Fe metabolism. Hence, the continuous Fe influx renders this organelle susceptible to the oxidative effects of Fe. In mitochondria, Fe can trigger different cell death pathways and lead to ferroptosis, a form of cell death very different from necrosis or apoptosis (Dixon et al. 2012), which may explain the neurotoxicological effects of this metal.

Manganese (Mn) is an essential ubiquitous metal ion required for normal growth, development, and cellular homeostasis (Erikson et al. 2005) and is one of the most abundant naturally occurring elements in the earth's crust. Mn exists in various chemical forms, oxidation states (Mn^{2+}, Mn^{3+}, Mn^{4+}, Mn^{6+}, Mn^{7+}), salts (sulfate,

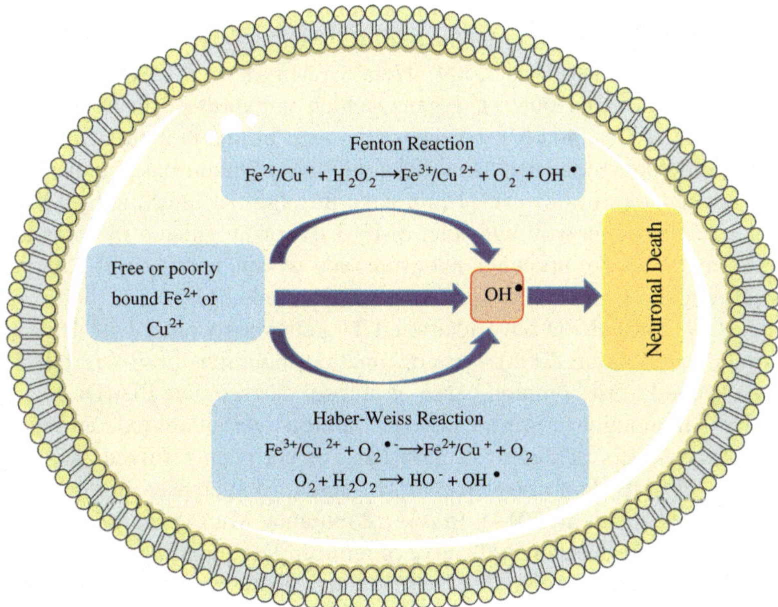

Fig. 1 Scheme of Fenton reaction and Haber-Weis reaction triggered by Fe or Cu

chloride, and gluconate), and chelates (aspartate, fumarate, succinate) (Farina et al. 2013). In humans and animals, Mn plays an important role in the development and functioning of the brain as a cofactor of several enzymes involved in neurotransmitter synthesis and metabolism (Takeda 2003).

Manganese (Mn) depicts vital functions in *C. elegans*, as it participates as cofactor of important enzymes as Mn-SOD. It has been demonstrated that Mn supplementation in the worm's growth medium (up to 1 mM) increases mean life span and fertility and causes thermal stress resistance (Lin et al. 2006). On the other hand, Mn toxicity in *C. elegans* has been associated with increased ROS formation and glutathione production, altered mitochondria membrane potential, and DA neuronal death (Benedetto et al. 2010; Settivari et al. 2009). Therefore, maintaining Mn homeostasis is very important.

Short-term exposure of young worms (L1 for 30 min) is enough to cause distinctive puncta and discontinuous GFP signal of neuronal processes of the DAergic CEP mechanosensory neurons, in a dose-dependent manner (Benedetto et al. 2010). On the other hand, no serotoninergic, cholinergic, GABAergic, or glutamatergic neurons were affected in this experimental design. In addition, the same study revealed that the lack of dopamine transporter-1 (DAT-1) depicted hypersensitivity to Mn, thus indicating that lack of clearance of DA and its extracellular increased levels at the synaptic cleft facilitates its reaction with the metal and generating reactive oxidative species (Sistrunk et al. 2007). Corroborating to that, Mn toxicity was increased by administration of exogenous DA at 10 mM (Benedetto et al. 2010).

The control of Mn uptake is tightly regulated in eukaryotes. Many transporters are involved in this regulation, being one of the most important the divalent metal transporter (DMT-1) (Au et al. 2008). In *C. elegans*, three isoforms of DMT-1, named SMF-1, SMF-2, and SMF-3, are distributed in different regions of the worm (Au et al. 2009; Settivari et al. 2009). Notably, Mn induced DAergic neurotoxicity is reduced by knocking out *smf-1* gene, which is expressed in DAergic neurons (Settivari et al. 2009), and Mn levels are reduced in this mutant (Au et al. 2009). This finding demonstrates that Mn uptake into DAergic neurons is in part related to the neurodegeneration; however, Mn uptake through the intestine by SMF-3 is of particular relevance as well. Of note, SMF-3 is downregulated in order to reduce excessive metal uptake upon Mn exposure, and worms lacking *smf-3* depict lower Mn levels and hyper-resistance to this metal (Au et al. 2009).

Studies of familial PD have identified 11 genes associated with heritable PD, including *dj-1* (Bonifati 2005). Loss of function mutations in *dj-1 (PARK7)* represents the second most common cause of autosomal recessive PD (De Marco et al. 2010). DJ-1 is thought to protect DAergic neurons via an antioxidant mechanism, but the precise basis of this protection has not yet been resolved. A preliminary study in *C. elegans* demonstrated that Mn uptake is increased in *djr-1.1* deletion mutants (Brinkhaus et al. 2014). In *djr-1.2* mutants, Mn exposure increased dauer movement, which is a strong indicative of reduced DA signaling (Chen et al. 2015).

PD is a neurodegenerative disease characterized by accumulation of misfolded α-synuclein, which aggregates and forms Lewy bodies. Moreover, Mn has been shown to stimulate the aggregation of α-synuclein in vitro and potentially exacerbate other neurodegenerative diseases such as amyotrophic lateral sclerosis (ALS) and prion disease (Bowman et al. 2011; Santner and Uversky 2010). In this context, it has been investigated in vivo the role of Mn on protein aggregation disorders. Mn treatment activates the ER-unfolded protein response, severely exacerbates toxicity in a disease model of protein misfolding (polyglutamine), and increases aggregate insolubility (Angeli et al. 2014), which is an important clue to the development of PD-like syndrome in humans. In contrast, worms expressing α-synuclein in a *pdr-1* and *djr-1.1* background are not more sensitive to Mn as it would be expected (Bornhorst et al. 2014). Actually, wild-type α-synuclein is protected from Mn-induced neurotoxicity. Although some evidences indicate its neurotoxic effects, α-synuclein has been implicated in neuroprotection and can play multiple roles in metal homeostasis.

In summary, studies on Mn-induced neurotoxicity in *C. elegans* brought new targets to combat Mn toxicity, as BLI-3 (Benedetto et al. 2010), lipocalin-related protein LPR-5 (Rudgalvyte et al. 2016), TRT-1 (a catalytic subunit of telomerase) (Ijomone et al. 2016), and others to be further evaluated (Parmalee et al. 2015).

Copper (Cu) is an indispensable element for all organisms that have an oxidative metabolism. Copper is after iron and zinc the third most abundant essential transition metal in human liver (Lewinska-Preis et al. 2011). The brain concentrates heavy metals including copper for metabolic use (Bush 2000). Cu is of great importance for the normal development and function of the brain.

Although essential to *C. elegans*, it has been described that copper (Cu) exposure induces significant behavioral defects (Williams and Dusenbery 1990; Anderson et al. 2004; Jiang et al. 2016; Zhang et al. 2010). Indeed, even with the cessation of exposure (post 20 h), worms still exhibited reduced locomotion (Anderson et al. 2004) Notably, it was confirmed that low and high Cu concentrations (2.5 and 200 μM) cause GABAergic neuron degeneration, especially AVL, RMEs, and RIS neurons, as verified by GFP labeling (Du and Wang 2009). Reduction in GABAergic input has been related to locomotor deficits in *C. elegans* (Jorgensen 2005).

C. elegans avoids toxic chemicals by reversing their movement, this behavior is mediated by amphid sensory neurons, particularly the ASH neurons (Hilliard et al. 2002; Bargmann et al. 1990; Sambongi et al. 1999). Sensory modulation is essential for animal sensations, behaviors, and survival (Guo et al. 2015). Studies have been demonstrating that $CuSO_4$ (Cu^{2+}) is a potent chemical repellent to *C. elegans*. Mutants that have structural defects in ciliated neurons (*che-2* and *osm-3*) as well as worms with three laser-operated neurons (ADL, ASE, and ASH) showed no avoidance behavior from Cu^{2+} (Sambongi et al. 1999; Wang et al. 2015; Esposito et al. 2010), thus corroborating to the effects of this metal in sensory neurons.

It is known that amyloid precursor protein (APP) contains a Cu-binding domain (CuBD) localized between amino acids 135 and 156 (APP135–APP156), which can reduce Cu^{2+} to Cu^{1+} in vitro (Multhaup et al. 1996). Cerpa et al. (2004) and White et al. (2002) demonstrated that the worm homologue, APL-1 CuBD, has protective properties against Cu^{2+} neurotoxicity. In agreement, in vivo studies demonstrated that even though exposure to Cu increases wild-type Aβ aggregation in worms and accelerated their paralysis, animals show decreased sensitivity to toxic $CuCl_2$ expo-

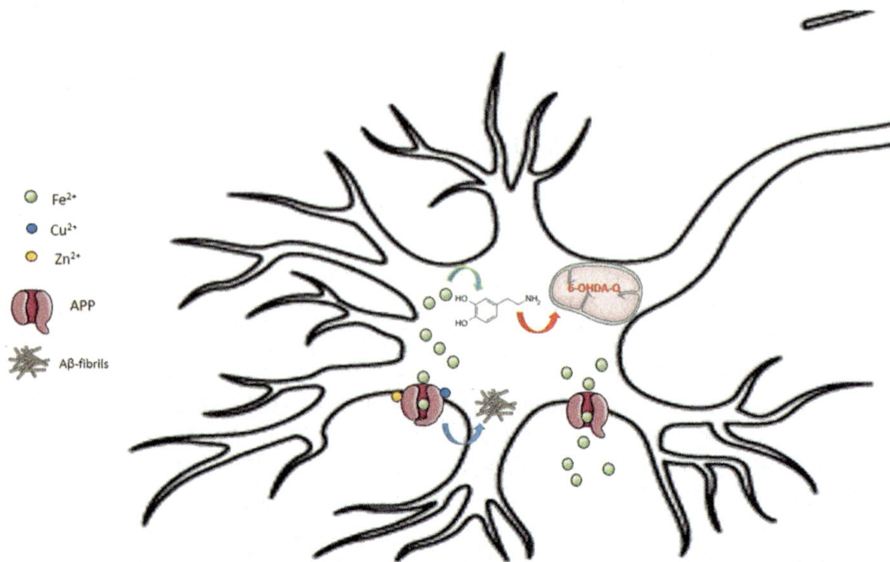

Fig. 2 Proposed mechanism of loss of control of Fe export from APP induced by Zn and Cu. APP exports excessive Fe from the neurons; however Cu and Zn may bind to the E2 domain of APP, thus causing conformational changes and impairing its function. Malfunction of APP may contribute to Aβ fibril formation and aggregation. Increased Fe levels may oxidize dopamine, generating by-products such as 6-hydrohydopamine quinone, which increases oxidative stress in the mitochondria of the neurons

sures (150–450 mM) compared to control worms (Luo et al. 2011; Minniti et al. 2009). These results illuminate a complex and dynamic relationship between Cu homeostasis and the role of Aβ in *C. elegans*, with metal-induced changes on the aggregation state of Aβ being coupled with protection against Cu^{2+} toxicity from the aggregates themselves.

APP is an integrated protein present in many cells and notably at the synapses of neurons, being one of its functions Fe export from the cells (Duce et al. 2010). It has been demonstrated that APP transgenic mice have imbalanced homeostasis of divalent metals such as Cu and Zn. Zn or Cu can bind to E2 domain of APP, thus inhibiting Fe export from neurons (Dahms et al. 2012). Increased Fe levels inside the neurons can lead to the aforementioned neurodegenerative effects of Fe, besides APP cleavage and formation of Aβ peptides (Fig. 2).

Zinc (Zn) is an essential trace metal that participates in numerous biological processes, including enzymatic function, protein structure, and cell signaling pathways. As the other essential metals, both excess and deficiency of zinc can lead to detrimental effects on development and metabolism, resulting in abnormalities and diseases.

Zinc is also essential to *C. elegans* development and reproduction (Dietrich et al. 2016). In *C. elegans*, families of the CDFs (cation diffusion facilitators), ZIPs (Zrt- and Irt-like proteins), and MTs (metallothioneins) are involved in Zn metabolism. Deletion of MT-1 and MT-2 results in increased Zn accumulation, while *mtl-1* knockout worms show heightened sensitivity to increased Zn level. The Zn source is the peptone, a component of NGM, as well as the agar. *E. coli* acquires Zn from the NGM, and worms acquire the metal from bacteria ingestion. Little is known about Zn neurotoxicity in *C. elegans*, and the preliminary studies point out reduction in body bends, head trashes' frequency, and feeding starting at 5 mg/L, which is a sublethal concentration (Dhawan et al. 1999; Jiang et al. 2016). Related to that, Zn inhibits acetylcholine esterase activity (AChE), with an EC50 of 2.76 mg/L.

Moreover, Wang et al. (Wang et al. 2007) described that Zn exposure caused locomotion deficits that were transferred to the progeny. In addition, chemotaxis plasticity defects were observed in Zn-exposed worms and their progeny. Notably, authors described the appearance of uncoordinated worms following Zn exposure, which indicates that besides reducing worms' movement, the metal changes the characteristics of the movement. By using thermotaxis and associative learning endpoints, Zhang et al. observed that Zn exposure (50 μM) reduced worms' response in both assays (Zhang et al. 2010). Altogether, these data suggest that Zn may cause neuronal alterations; however it still needs to be further confirmed which neurotransmissions systems are damaged by this metal.

Nonessential Metals

Toxicological metals, such as mercury, cadmium, lead, and aluminum, have no known normal biological function and are detrimental to any organism when absorbed (Duce 2010). These metals can also induce oxidative toxicity but more likely work by binding to proteins and interfering with metal transport and protein function (Wright and Baccarelli 2007). When absorbed in high doses, they usually damage specific organs such as the kidneys, bone, and brain and may also be implicated in the pathogenesis of neurodegenerative diseases, particularly Alzheimer's or Parkinson's disease (Poujois et al. 2016). Neurodegenerative diseases are mediated or triggered by an increase in free radical production, protein aggregation, mitochondrial dysfunction, calcium unbalance, and metal transport alteration, and these actions initiate a cascade of events which finally lead to neurodegeneration and cell death (Gaeta and Hider 2005; Levenson 2005; Price 1999). As a result of specific reactions, different chemical forms of mercury are present, such as elemental mercury (Hg^0) and inorganic (Hg^{2+} and Hg^+) and organic mercury compounds. The toxic properties and target organs of Hg are dependent upon its chemical speciation. Organomercurials with a short aliphatic chain (methyl or ethyl) are the most harmful compounds, and they may cause irreversible damage to the nervous system (Sanfeliu et al. 2003). The major source of methylmercury (CH_3Hg^+ or MeHg) exposure is the consumption of some predator fish, and, therefore, if fish is a dietary component, mercury intake is practically inevitable. MeHg has a higher entrance rate into the central nervous system when compared to inorganic forms, rendering it an important neurotoxicant agent (Debes et al. 2006). The occupational exposure to Hg (mainly Hg^0) is caused by its use in gold mining and in industry (Lubick 2010; Neghab et al. 2012). Hg^0 is also of toxicological relevance but less important for neurodegenerative process.

MeHg interacts with and oxidizes nucleophilic groups of several biomolecules; sulfhydryl (thiol/thiolate; -SH/-S−) groups are important and relevant targets of MeHg in the biological systems (Farina et al. 2013). MeHg can modify the oxidation state of the −SH groups on proteins by direct interaction with thiols, as well as indirect mechanisms, modulating their functions (Kim et al. 2002). MeHg is known to affect a variety of neuronal activities including generation of ROS, increased calcium influx, dopamine metabolism, neural stem cell differentiation, DNA damage, and mitochondrial dysfunction (Petroni et al. 2012; Sadiq et al. 2012; Tamm et al. 2008; Tiernan et al. 2015). MeHg poisoning is characterized by severe neurological deficits due to brain lesions and disruptions of neurotransmitter systems (Aschner and Syversen 2005). In addition, MeHg exposure has been shown to increase β-amyloid in the hippocampus and decrease it in the cerebrospinal fluid, both hallmarks of Alzheimer disease (Kim et al. 2014). An interesting aspect of MeHg neurotoxicology is its preferential affinity for specific structures of the central nervous system; commonly, the cerebral and cerebellar cortices are the regions more severely affected (Eto et al. 2010).

An interesting work of McElwee and colleagues demonstrate that mercurials depicted a clearly different mechanism of toxicity. *C. elegans* exposure to the $HgCl_2$ and MeHgCl caused different effects on gene expression, and different genes were important in the cellular response to the two mercurials (McElwee et al. 2013). $HgCl_2$ was the most toxic among the inorganic metals, with severe toxicity on associative learning behavior, thermotaxis, and locomotion behavior in nematodes. $HgCl_2$ at concentrations of 10 and 50 μM also induced moderately but significantly higher associative learning behavior, higher body bend, or thermotaxis to cultivation temperature. Since the locomotion behavior defects reflect the possible dysfunction of the nervous systems, authors suggest that $HgCl_2$ could cause some neuronal disturb in *C. elegans* (Wang and Xing 2008; Zhang et al. 2010). Du and Wang demonstrated that *C. elegans* GABAergic system could be affected by exposure to high concentrations (75 μM and 200 μM) of $HgCl_2$, by causing clear axonal degeneration and neuronal loss in nerve cords (Du and Wang 2009).

However, MeHg was more toxic to *C. elegans* than $HgCl_2$ when assessing feeding, movement, and reproduction, all of which requiring proper neuromuscular activity. MeHg exposure resulted in increased steady-state levels of the stress response genes at lower concentrations than $HgCl_2$ (McElwee and Freedman 2011). In addition, MeHg causes severe toxic effects such as decreased survival, developmental delay, and decreased pharyngeal pumping in worms (Helmcke et al. 2009). MeHg causes oxidative stress in *C. elegans* as indicated by the disruption in glutathione levels as well as the increase expression of HSP and GST (Helmcke and Aschner 2010). The data from studies of Helmcke and Aschner suggest that *gst-4* contributes to the response to MeHg exposure but that knockdown of this gene does not affect lethality. Metallothionein knockouts displayed increased lethality upon exposure to MeHg. Increases in *mtl-1* expression following acute MeHg exposure was observed in acute exposure at L1 stage but no change following chronic exposure (Helmcke and Aschner 2010). Van Duyn and colleagues demonstrated that worms exposed chronically to 1 μM of MeHg depicted a dopaminergic neuronal degeneration of 30% under reduction in *skn-1* gene expression. Notably, no degeneration was observed in wild-type animals, indicating that SKN-1 is an important pathway in *C. elegans* to inhibit MeHg-induced dopaminergic neuronal degeneration (Vanduyn et al. 2010). The dependency of SKN-1 was demonstrated in early-life MeHg exposure, once it led to decreases in DA-mediated behavior seen at 72 h which positively correlated to dopaminergic degeneration at 96 h (adult stage). In this work, Martinez-Finley and colleagues (Martinez-Finley et al. 2013) demonstrate the presence of irregularities in DAergic neurons 96 h following 30 min exposure (at L1 stage) to 20 µM MeHg. Taken together, these results indicate that exposure to MeHg confers dopaminergic neurodegeneration in *C. elegans* model, and it could occur later in life. Furthermore, knockdown of *skn-1* amplified MeHg's effect.

Lead (Pb^{+2}) is a nonessential heavy metal and a ubiquitously present pollutant in the ecosystem. Pb^{+2} is a neurotoxicant agent that could be considered as the main global environmental health danger, which can predominantly affect populations in undeveloped countries and in urban centers (Oteiza et al. 2004). In humans, inhalation and oral ingestion are the major routes of Pb exposure. Pb accumulates in different brain regions, and lipid oxidation was found in the parietal cortex, striatum, hippocampus, thalamus, and cerebellum in rats exposed to Pb during gestation and until postnatal day 45 (Villeda-Hernandez et al. 2001). The mechanisms of lead neurotoxicity are complex and still not fully understood. One of the basic mechanisms proposed for Pb-induced neuronal toxicity is its substitution for calcium in intracellular signal transduction. In particular, acute and chronic exposure to lead would predominantly affect two specific protein complexes: protein kinase C and the *N*-methyl-D-aspartate subtype of glutamate receptor (Marchetti 2003). Pb can promote the formation of protein aggregates at nanomolar concentrations (Basha et al. 2005) as well as cause the accumulation of intracellular β-amyloid protein and increase dense-core plaques in the primate model (Wu et al. 2008). Pb alters several cell signals, in particular those sensitive to calcium, affecting second messengers that subsequently modulate transcription factors and ultimately gene expression (Oteiza et al. 2004). The main target of Pb-induced toxicity is the nervous system, and children are particularly sensitive to Pb intoxication due to a higher rate of gastrointestinal absorption, decreased excretion, a high access of Pb to the brain, and the vulnerability of the developing central nervous system to this metal (Lidsky and Schneider 2003).

Pb-exposed *C. elegans* show a variety of alterations in some parameters such as life span, development, locomotion, learning, and memory behaviors (Ye et al. 2008; Zhang et al. 2010). In a study concerning acute toxicity, Roh and cols observed that LD50 for different metals tested in *C. elegans* after 24 h of exposure was as follows: Pb > As > Cr > Cd. The calculated LD50 for Pb was 34 mg/L (Roh et al. 2006). Another study demonstrated that younger (L1–L3) larvae show more sensitivity to Pb-induced neurotoxicity regarding neuronal survival and synaptic function than L4 larvae and young adult nematodes (Xing et al. 2009). Du and Wang (2009) studied the effect of Pb in the GABAergic neurons and demonstrated the neurodegeneration, and abnormal structures can be formed in these motor neurons after Pb exposure. Exposure to Pb at 2.5 μM could also induce noticeable neuronal loss, and exposure to 75 μM and/or 200 μM significantly reduced the relative size and fluorescent intensities of AVL, RMEs, and RIS neurons in *C. elegans*. As mercury, lead depicted high toxicity in associative learning behavior, thermotaxis, and locomotion behavior in this nematode. A low concentration of 2.5 μM of Pb resulted in significant decrease of associative learning behavior and higher concentrations as 50 μM also induced moderately but significantly higher body bend or thermotaxis (Zhang et al. 2010). Ye and colleagues (2010) observed that expression of MTL-1 and MTL-2 did not rescue the neurobehavioral toxicity induced by exposure to 200 mM of Pb in *mtl-1* and *mtl-2* mutants. However, the overexpression of MTL-1 and MTL-2 at the L2 larval stage significantly suppressed the toxicity on locomotion behavior caused by Pb exposure. In contrast, expression of proteins not related with stress response and adaptive response, such as MOD-5 and EAT-4, did not influence the toxicity on behaviors induced by Pb.

Aluminum (Al^{3+}) is the third most abundant element and the most abundant metal in the earth's crust. Its toxicity is directly linked to its bioavailability. Although Al is one of the most common elements in the biosphere, the amounts taken up into living cells are extremely small and are exceptionally difficult to measure accurately. Even when the bioavailability of this metal is low due to the formation of Al-silicate complexes, certain environmental and industrial factors could raise Al availability. Human exposure to Al^{3+} occurs through a number of mechanisms including soil and fertilizers, cookware, and water from purification systems, as well as pharmaceutical and cosmetic preparations (Verstraeten et al. 2008). In biological fluids, this trivalent cation is rarely present as an ion because it complexes extensively with biologically available ligands such as phosphate, hydroxide, and citrate (Duce and Bush 2010). The exact function of Al in animals remains unknown. Al is highly reactive with carbon and oxygen, making it toxic to living organisms. Al^{3+} can increase the production of reactive species and produce oxidative stress on its own and synergistically with Cu and Fe (Walton 2013). Al can stabilize the intracellular $Fe2^+$ by preventing its oxidation but enhances Fe-initiated oxidative damage (Exley 2006). Al and Fe, but neither copper nor zinc, are key to the precipitation of beta-sheets of Abeta_1–42 in senile plaque cores in AD (Exley 2006). Al^{+3} has long been implicated in the pathogenesis of neurodegenerative diseases (Gupta et al. 2005) since it is associated with the abnormal aggregation of Aβ (Domingo 2006), tau aggregation (Mizoroki et al. 2007), and cellular dysfunction (Bharathi et al. 2008). The contributions of Al^{3+} to AD pathogenesis and neuropathology involving APP and Aβ metabolism, formation and growth of tau pathology, and neuron-to-neuron spreading of Al inducing the progression of AD have been reviewed recently (Walton 2013). In order to protect the brain from the noxious effects of Al, there is an active efflux of the metal at the blood-brain barrier mediated by a monocarboxylate transporter (Yokel et al. 2001). However, exposure to high amounts of Al or an increased blood Al concentration due to a decreased renal functionality can lead to brain Al accumulation. However, the results of many of these reports do not address confounding variables such as genetic backgrounds that may predispose an organism to the susceptibility of Al-induced neurological damage.

A study evaluating chronic exposure to Al demonstrated that it can be toxic to *C. elegans*. Page and collaborators (Page et al. 2012) demonstrate that Al can induce changes in growth, development, lifespan, and fertility. Al also induces changes in elemental composition of whole worms. The exposure occurred during development, as lifespan was unaffected by Al exposure during adulthood. In addition lower levels of Al slowed *C. elegans* development and reduced hermaphrodite self-fertility and adult body size. A significant developmental delay was observed even when Al exposure was restricted to embryogenesis period of *C. elegans*. Ye et al. (Ye et al. 2008) demonstrated that Al exposure (2.5 and 75 μmol/L) caused a significantly decrease in memory functions at least 7 h post exposure. Furthermore, they observed that posttreatment with vitamin E could recover the memory defects in worms exposed to 75 μmol/ L Al. In addition, exposure to Al (2.5 and 75 μmol /L) and posttreatment with vitamin E (100 and 200 lg/mL) did not affect the body bend behavior. VanDuyn and collegues (VanDuyn et al. 2013) described a novel model for Al^{3+} toxicity and have shown that the *C. elegans* transporter SMF-3 plays a significant role in modulating Al^{3+}-induced dopaminergic neuron degeneration through the intracellular sequestration of Al^{3+}. In addition they demonstrated that SMF-3 expression is sensitive to Al^{3+}, and the PD-associated proteins α-synuclein, Nrf2/SKN-1, and Apaf1/CED-4 modulate Al^{3+}-associated dopaminergic neuron cell death.

Cadmium (Cd^{2+}) is a nonessential transition heavy metal and an environmental pollutant that has been classified as a category 1 human carcinogen (IARC 1993). The main pathways of exposition in humans include diet and smoke (EFSA 2009). Cd exposure is directly associated with teratogenic and mutagenic problems (WHO 1996). Workers as miners, welders, smokers, and workers in battery production are at risk of high Cd^{2+} occupational exposure (Wang and Du 2013). The occupational Cd^{2+} exposure can be correlated with lung cancer and other cancers such as the prostate, renal, liver, hematopoietic system, urinary bladder, pancreatic, testis, and stomach (Waalkes 2000). In cells, Cd^{2+} can induce oxidative stress, suppress gene expression, and inhibit DNA damage repair and apoptosis (Bishak et al. 2015). Chronic exposure to Cd may severely interfere with normal function of the nervous system, and infants and children are more susceptible than adults. The neurotoxic effects of Cd^{2+} were complex and could be associated with both biochemical changes of the cell and functional changes of central nervous system, indicating that Cd neurotoxic effects play a role in the systemic toxic effects of the exposure to Cd^{2+}, particularly the long-term exposure (Wang and Du 2013). Cd^{2+} impairs cell viability and disturbs MAPKs pathways (Rigon et al. 2008), induced mitophagy in brain tissue by ROS production (Wei et al. 2015), and induces oligodendrocyte progenitors cell death mainly by apoptosis (Hossain et al. 2009). Some studies indicate that Cd could be a possible etiological factor of neurodegenerative diseases, including Alzheimer and Parkinson diseases. Jiang et al. (2007b) in an early study found that Cd^{2+} accelerates self-aggregation of Alzheimer's tau peptide, and it has been reported in a case study that a 64-year-old man developed PD symptoms 3 months after acute exposure to Cd^{2+} (Okuda et al. 1997).

In *C. elegans*, Cd^{2+} has been shown to alter behavior; decrease growth, life span, and reproduction; and to affect feeding and movement (Boyd et al. 2010; Chen et al. 2013; Hoss et al. 2011). Low Cd exposure in phytochelation synthase-1 (*pcs-1*) RNAi worms resulted smaller, necrotic and sterile worms which had a shorter life span. Following higher concentrations of Cd, *pcs-1* worms arrested at L2–L4 stage presented necrotic cells and died (Vatamaniuk et al. 2001). 290 genes were identified that are differentially expressed following a 4- or 24-h exposure to cadmium. Several of these genes are known to be involved in metal detoxification, including *mtl-1, mtl-2, cdr-1,* and *ttm-1* (Cui et al. 2007). Cd but not Cu or Zn was able to influence temporal transcription response in a concentration-dependent manner. Cd accumulation found to be highest in *mtl-2* and double-mutant strains (Bofill et al. 2009; Swain et al. 2004; Zeitoun-Ghandour et al. 2010). A study of Gonzalez-Hunt (Gonzalez-Hunt et al. 2014) depicted that exposure to mitochondrial genotoxins, as Cd^{2+}, during early development may predispose to dopaminergic neurodegeneration later in life. Cd exposure was also shown to cause GABAergic neurodegeneration in worms. At low Cd concentration, neuronal loss was observed, while at high Cd concentration, axonal degeneration and neuronal loss, as well as reduced size of AVL, RMEs, and RIS neurons, were noted in fluorescently labeled GABAergic neurons (Du and Wang 2009).

Uranium (U) is present in the environment as a result of natural deposits and release by human applications. Experimental studies show that after exposure, uranium can reach the brain and lead to neurobehavioral impairments, including increased locomotor activity, perturbation of the sleep-wake cycle, decreased memory, and increased anxiety (Dinocourt et al. 2015). There are concerns that U exposure may also result in neurologic sequelae, particularly since it readily crosses the blood-brain barrier (BBB), accumulates in specific brain regions, and decreases neurocognitive performance, as observed in Gulf War veterans (Fitsanakis et al. 2006; McDiarmid et al. 2009; Dobson et al. 2006). Accordingly, rats exposed to 4% enriched U for 1.5 months through drinking water accumulated twice as much U in some key areas such as the hippocampus, hypothalamus, and adrenals than did control rats. The U accumulation was correlated with a 38% increase in paradoxical sleep, a reduction of spatial working memory, and an increase in anxiety-like behavior (Houpert et al. 2005). In addition, exposure for 1.5 months to depleted U did not induce these effects, suggesting that the radiological activity induces the primary events of these effects of uranium. However, U depicted a low cytotoxicity in primary rat cortical neuron cultures upon exposure to uranyl acetate until cultures are exposed to 100 µM. Furthermore, no significant changes in F2-prostanes and thiol metabolite levels were observed, and only minimal changes in total adenosine nucleotides (ATP + ADP + AMP) were detected.

In earlier studies, (Jiang et al. 2007a, 2009) used *C. elegans* as an in vivo model to determinate the U accumulation and the capacity to produce neurodegeneration. In these studies, U demonstrates an ability to accumulate in a dose-dependent manner at the same time increase the toxicity to the nematode. However, when 1 mM of U was used, the nematode did not present any signal of neurodegeneration in a pan-neuronal and a dopamine-specific GFP-tagged strains. In the same set of experiments, accumulation studies further indicated that MT1 appears to be the protein form that is associated with the uptake of U in *C. elegans*. *C. elegans mtl-1* knockout mutants displayed increased cellular accumulation of U (Jiang et al. 2009). Uranium exposure in different generations can cause differential effects of development plasticity, selection pressures, and evolutionary responses. Exposure at P0 generation showed that *C. elegans* individuals were smaller, slower, and less fertile and developed slower than untreated worms. Across generations, phenotypic changes can vary in amplitude and in direction, in a dose-dependent manner (Dutilleul et al. 2013).

Metal Nanoparticle Neurotoxicity

Nanotechnology is rapidly developing and increasingly playing important roles in various fields, particularly medicine and chemistry. Nanoparticles (NPs) are structures of 100–500 nm, varying in composition, size, shape, and surface properties. They may be presented as polymeric, metal/metal oxide NPs, or quantum nanodots, for instance. The surface of nanomaterials can be modulated according to their application such as for drug delivery, the biocompatibility of the nanomaterials can be modified, and their cell specific targeting ability can also be enhanced by attaching them with targeting ligand (Subbiah et al. 2010; Thanh and Green 2010). In vivo, NPs have to avoid nonspecific interactions with plasma proteins and must contain their colloidal stability under physiological conditions, especially in a wide range of pH (Thanh and Green 2010). In an attempt to reduce NP toxicity, polymeric coatings have been used in order to increase water solubility, to reduce toxicity, and to direct site-specific metal delivery (Subbiah et al. 2010; Thanh and Green 2010). Toxicological evaluation of these nanomaterials has become necessary as there is a growing concern on the short- and long-term effects following exposure.

One of the most tested NPs is AgNPs, which are commonly found in consumer products and were proven to have anti-HIV properties (Elechiguerra et al. 2005); however their action on neurons is still uncertain. In worms, sublethal concentrations of AgNPs may cause adverse neurological responses. Contreras et al. (Contreras et al. 2014) found that exposure to 100 mg AgNP/L reduced the flex, amplitude, wavelength, and velocity of the body bend of exposed worms, which was worsened in the progeny following multiple generation exposures (four generations). Different NP sizes (2.5 or 10 nm) caused different effects on worms' life span and reproduction; however the motility was reduced equally following exposure to all different sized AgNPs. Of note, worms' locomotion is regulated by GABAergic, cholinergic, and dopaminergic neurons; then alteration in these parameters may indicate neuronal damage (Rand 2007; Jorgensen 2005). Notably, these AgNPs were coated with thiolated polyethylene glycol (mPEG-SH) polymer to render them water soluble.

Long-term early onset exposure to cadmium telluride (CdTe) quantum dots (QD-0.1 and 1 μg/L) caused abnormal foraging behavior, which is related to altered function of the motor neurons (Zhao et al. 2015). In accordance, there was a decreased fluorescence of motor neuron cell bodies, indicating alteration in their development. Furthermore, authors demonstrated that these QDs crossed the intestinal barrier and reached RME neurons, which are GABAergic motor neurons. In addition, this prolonged exposure increased defecation cycle length, an indication of alterations in other GABAergic neurons, AVL and DVB. Indeed, CdTe QDs also caused decreased fluorescence in these neurons as well.

Another study with CdTe QDs coated with 3-mercaptopropionic acid demonstrated that exposed worms depicted behavioral defects, including alterations in body bending, head trashing, pharyngeal pumping, and defecation cycle. Impaired learning and memory were also affected by these QDs. Of particular importance, CdTe QDs altered the expression of genes related to glutamatergic, serotoninergic, and dopaminergic neurotransmission, thus indicating that behavioral alterations are associated with neuronal modifications (Wu et al. 2015).

Al_2O_3 NPs have been used in industry and biomedical applications; however their toxicity can be very high. These NPs were able to alter worms' locomotor behavior due to alterations in glutamate, serotonin, and dopamine neurotransmitter systems (Li et al. 2012, 2013). The study demonstrated some molecular targets for Al2O3 NP neurotoxicity: non-NMDA glutamate receptors GLR-2 and GLR-6, ionotropic serotonin receptor MOD-1, and D1-like dopamine receptor DOP-1.

NPs caused severe deficits in gut development, defecation behavior, lethality, locomotion, growth, reproduction, ROS production, and changes in gene expression (Rui et al. 2013; Zhao et al. 2014). Notably, locomotor activity in *C. elegans* is a good parameter to evaluate neurotoxicity. In view of that, many studies with metal NPs use this endpoint to characterize their toxicity level. For instance, the behavioral toxicity of TiO_2, ZnO, and SiO_2 NPs of 30 nm size in a prolonged exposure in L1 worms has been evaluated and compared. Authors described that worms reduced significantly head trashes and body bends following exposure to the three NPs and that toxicity order was $ZnO > TiO_2 > SiO_2$ (Wu et al. 2013). Notably, N-acetyl cysteine reverted locomotor alterations caused by these NPs, demonstrating that oxidative stress plays an important role in these metals NP-induced neurotoxicity.

In view of that, the same research group identified that *sod-2, sod-3,mtl-2*, and *hsp-16.48* mutants exhibited a more severe decrease in both head thrash and body bend than that of wild-type N2 nematodes (Wu et al. 2014). This indicates that the lack of SOD isoforms, metallothioneins, and heat shock proteins, which are proteins involved in oxidative stress protection and metal elimination, renders worms more susceptible to metal/metal oxide NPs. Corroborating to that, a study of Ma and cols (Ma et al. 2009) demonstrated that exposure to ZnO NPs caused reduced movement speed that was associated to increased *mtl-2*:GFP expression. On the other hand, Jung et al. used a multi-endpoint high-throughput assay to compare nano-Ag, TiO_2, CeO_2, and SiO_2 with higher sizes, in a chronic exposure paradigm using L4 worms; however they did not find significant alterations in worms' speed (Jung et al. 2015). In agreement, Khare et al. reported that TiO_2 and ZnO NPs sized <100 nm were much less toxic to exposed worms, in contrast to <25 nm NPs (Khare et al. 2011). Moreover, ZnO NPs sized 21 nm decreased in a more significative manner head trashes and body bends in comparison to 35 and 65 nm ZnO (Khare et al. 2015). These different findings indicate that the size of the NPs and the age of the worms are important variables to take into account. Notably, smaller NPs are generally more neurotoxic, probably due to easier absorption and uptake from the neurons. In addition, young worms are more susceptible to neuronal damage, especially when exposed chronically. For instance, worms exposed to Fe-NPs from L4 stage for 24 h depicted slightly lower behavioral defects compared to L1 worms treated until they reached adulthood (Wu et al. 2012).

Concluding Remarks/Perspectives

This chapter revised the effects of essential and nonessential metals in the nervous system of a nematode. By using *C. elegans* as animal model, molecular targets involved in metal-induced neuronal degeneration have been discovered as well as new relationships between metals and molecular targets. Various genetic strains of *C. elegans* – mutants and GFP tagged – offer a unique perspective on neurodegenerative processes and their etiology. A collection of genes have been found to be related to neurodegeneration induced by metals, especially SKN-1, MTL (1 and 2 isoforms), and different HSPs (Fig. 3). Targeting these interactions may be an effective approach to modify the vulnerability of these neurons and to thus slow neuronal loss and clinical deterioration.

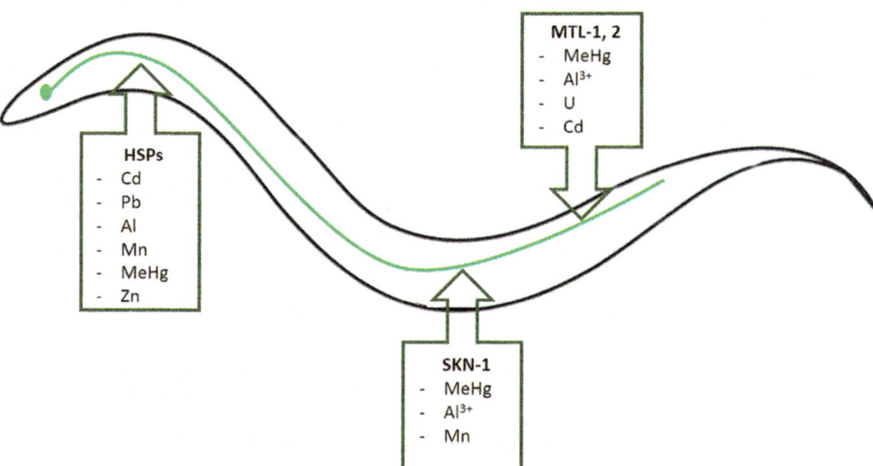

Fig. 3 Putative molecular targets that protect from metal-induced neurodegeneration

C. elegans is also an excellent tool for genetic analysis and manipulations. The availability of mutants and green fluorescent protein (GFP) tagging makes it easy to explore a wide range of chemicals and their effects. Several effects in response to exposure to metals, especially those involving gene expression and behavior, have been reported using the nematode as a model.

References

Aisen P, Enns C, Wessling-Resnick M. Chemistry and biology of eukaryotic iron metabolism. Int J Biochem Cell Biol. 2001;33(10):940–59.

Alfrey AC, LeGendre GR, Kaehny WD. The dialysis encephalopathy syndrome. Possible aluminum intoxication. N Engl J Med. 1976;294(4):184–8. doi:10.1056/NEJM197601222940402.

Anderson GL, Cole RD, Williams PL. Assessing behavioral toxicity with Caenorhabditis elegans. Environ Toxicol Chem. 2004;23(5):1235–40.

Angeli S, Barhydt T, Jacobs R, Killilea DW, Lithgow GJ, Andersen JK. Manganese disturbs metal and protein homeostasis in Caenorhabditis elegans. Metallomics : integrated biometal science. 2014;6(10):1816–23. doi:10.1039/c4mt00168k.

Antonio MT, Corredor L, Leret ML. Study of the activity of several brain enzymes like markers of the neurotoxicity induced by perinatal exposure to lead and/or cadmium. Toxicol Lett. 2003;143(3):331–40.

Aschner M, Syversen T. Methylmercury: recent advances in the understanding of its neurotoxicity. Ther Drug Monit. 2005;27(3):278–83.

Au C, Benedetto A, Aschner M. Manganese transport in eukaryotes: the role of DMT1. Neurotoxicology. 2008;29(4):569–76. doi:10.1016/j.neuro.2008.04.022.

Au C, Benedetto A, Anderson J, Labrousse A, Erikson K, Ewbank JJ, Aschner Mn.d.. SMF-1, SMF-2 and SMF-3 DMT1 orthologues regulate and are regulated differentially by manganese levels in C. elegans. PloS one. 2009;4(11):e7792. doi:10.1371/journal.pone.0007792.

Avery L, Horvitz HR. Pharyngeal pumping continues after laser killing of the pharyngeal nervous system of C. elegans. Neuron. 1989;3(4):473–85.

Bargmann CI, Thomas JH, Horvitz HR. Chemosensory cell function in the behavior and development of Caenorhabditis elegans. Cold Spring Harb Symp Quant Biol. 1990;55:529–38.

Bartzokis G, Beckson M, Hance DB, Marx P, Foster JA, Marder SR. MR evaluation of age-related increase of brain iron in young adult and older normal males. Magn Reson Imaging. 1997;15(1):29–35.

Basha MR, Wei W, Bakheet SA, Benitez N, Siddiqi HK, Ge YW, Lahiri DK, Zawia NH. The fetal basis of amyloidogenesis: exposure to lead and latent overexpression of amyloid precursor protein and beta-amyloid in the aging brain. J Neurosci. 2005;25(4):823–9. doi:10.1523/JNEUROSCI.4335-04.2005.

Benedetto A, Au C, Aschner M. Manganese-induced dopaminergic neurodegeneration: insights into mechanisms and genetics shared with Parkinson's disease. Chem Rev. 2009;109(10): 4862–84. doi:10.1021/cr800536y.

Benedetto A, Au C, Avila DS, Milatovic D, Aschner M. Extracellular dopamine potentiates mn-induced oxidative stress, lifespan reduction, and dopaminergic neurodegeneration in a BLI-3-dependent manner in Caenorhabditis elegans. PLoS Genet. 2010;6(8) doi:10.1371/journal.pgen.1001084.

Berg D, Youdim MB. Role of iron in neurodegenerative disorders. Top Magn Resonan Imag TMRI. 2006;17(1):5–17. doi:10.1097/01.rmr.0000245461.90406.ad.

Bharathi VP, Govindaraju M, Palanisamy AP, Sambamurti K, Rao KS. Molecular toxicity of aluminium in relation to neurodegeneration. Indian J Med Res. 2008;128(4):545–56.

Bishak YK, Payahoo L, Osatdrahimi A, Nourazarian A. Mechanisms of cadmium carcinogenicity in the gastrointestinal tract. Asian Pac J Cancer Prev. 2015;16(1):9–21.

Boelmans K, Holst B, Hackius M, Finsterbusch J, Gerloff C, Fiehler J, Munchau A. Brain iron deposition fingerprints in Parkinson's disease and progressive supranuclear palsy. Mov Disord. 2012;27(3):421–7. doi:10.1002/mds.24926.

Bofill R, Orihuela R, Romagosa M, Domenech J, Atrian S, Capdevila M. Caenorhabditis elegans Metallothionein isoform specificity--metal binding abilities and the role of histidine in CeMT1 and CeMT2. FEBS J. 2009;276(23):7040–56. doi:10.1111/j.1742-4658.2009.07417.x.

Bonifati V. Genetics of Parkinson's disease. Minerva Med. 2005;96(3):175–86.

Bornhorst J, Chakraborty S, Meyer S, Lohren H, Brinkhaus SG, Knight AL, Caldwell KA, Caldwell GA, Karst U, Schwerdtle T, Bowman A, Aschner M. The effects of pdr1, djr1.1 and pink1 loss in manganese-induced toxicity and the role of alpha-synuclein in C. elegans. Metallom Integr Biometal Sci. 2014;6(3):476–90. doi:10.1039/c3mt00325f.

Bourgeois F, Ben-Yakar A. Femtosecond laser nanoaxotomy properties and their effect on axonal recovery in C. elegans. Opt Express. 2007;15(14):8521–31.

Bowman AB, Kwakye GF, Herrero Hernandez E, Aschner M. Role of manganese in neurodegenerative diseases. J Trace Elem Med Biol. 2011;25(4):191–203. doi:10.1016/j.jtemb.2011.08.144.

Boyd WA, Smith MV, Kissling GE, Freedman JH. Medium- and high-throughput screening of neurotoxicants using C. elegans. Neurotoxicol Teratol. 2010;32(1):68–73. doi:10.1016/j.ntt.2008.12.004.

Brenner S. The genetics of Caenorhabditis elegans. Genetics. 1974;77(1):71–94.

Brinkhaus SG, Bornhorst J, Chakraborty S, Wehe CA, Niehaus R, Reifschneider O, Aschner M, Karst U. Elemental bioimaging of manganese uptake in C. elegans. Metallom Integr Biometal Sci. 2014;6(3):617–21. doi:10.1039/c3mt00334e.

Bush AI. Metals and neuroscience. Curr Opin Chem Biol. 2000;4(2):184–91.

Campbell A. n.d. The potential role of aluminium in Alzheimer's disease. Nephrol Dial Transplant. (2002;17(Suppl 2):17–20.

Cerpa WF, Barria MI, Chacon MA, Suazo M, Gonzalez M, Opazo C, Bush AI, Inestrosa NC. The N-terminal copper-binding domain of the amyloid precursor protein protects against Cu2+ neurotoxicity in vivo. FASEB J. 2004;18(14):1701–3. doi:10.1096/fj.03-1349fje.

Chalfie M, Tu Y, Euskirchen G, Ward WW, Prasher DC. Green fluorescent protein as a marker for gene expression. Science. 1994;263(5148):802–5.

Chege PM, McColl G. Caenorhabditis elegans: a model to investigate oxidative stress and metal dyshomeostasis in Parkinson's disease. Front Aging Neurosci. 2014;6:89. doi:10.3389/fnagi.2014.00089.

Chen BL, Hall DH, Chklovskii DB. Wiring optimization can relate neuronal structure and function. Proc Natl Acad Sci U S A. 2006;103(12):4723–8. doi:10.1073/pnas.0506806103.

Chen P, Martinez-Finley EJ, Bornhorst J, Chakraborty S, Aschner Mn.d.-a. Metal-induced neurodegeneration in C. elegans. Front Aging Neurosci. 2013;5:18. doi:10.3389/fnagi.2013.00018.

Chen P, DeWitt MR, Bornhorst J, Soares FA, Mukhopadhyay S, Bowman AB, Aschner M. Age- and manganese-dependent modulation of dopaminergic phenotypes in a C. elegans DJ-1 genetic model of Parkinson's disease. Metallom Integr Biometal Sci. 2015;7(2):289–98. doi:10.1039/c4mt00292j.

Chen P, Miah MR, Aschner M. n.d.-b Metals and Neurodegeneration. F1000Research 5. 2016. doi:10.12688/f1000research.7431.1

Contreras EQ, Puppala HL, Escalera G, Zhong W, Colvin VL. Size-dependent impacts of silver nanoparticles on the lifespan, fertility, growth, and locomotion of Caenorhabditis elegans. Environ Toxicol Chem. 2014;33(12):2716–23. doi:10.1002/etc.2705.

Cui Y, McBride SJ, Boyd WA, Alper S, Freedman JH. Toxicogenomic analysis of Caenorhabditis elegans reveals novel genes and pathways involved in the resistance to cadmium toxicity. Genome Biol. 2007;8(6):R122. doi:10.1186/gb-2007-8-6-r122.

Dahms SO, Konnig I, Roeser D, Guhrs KH, Mayer MC, Kaden D, Multhaup G, Than ME. Metal binding dictates conformation and function of the amyloid precursor protein (APP) E2 domain. J Mol Biol. 2012;416(3):438–52. doi:10.1016/j.jmb.2011.12.057.

De Marco EV, Annesi G, Tarantino P, Nicoletti G, Civitelli D, Messina D, Annesi F, Arabia G, Salsone M, Condino F, Novellino F, Provenzano G, Rocca FE, Colica C, Morelli M, Scornaienchi V, Greco V, Giofre L, Quattrone A. DJ-1 is a Parkinson's disease susceptibility gene in southern Italy. Clin Genet. 2010;77(2):183–8. doi:10.1111/j.1399-0004.2009.01310.x.

Debes F, Budtz-Jorgensen E, Weihe P, White RF, Grandjean P. Impact of prenatal methylmercury exposure on neurobehavioral function at age 14 years. Neurotoxicol Teratol. 2006;28(5):536–47. doi:10.1016/j.ntt.2006.02.005.

Dhawan R, Dusenbery DB, Williams PL. Comparison of lethality, reproduction, and behavior as toxicological endpoints in the nematode Caenorhabditis elegans. J Toxicol Environ Health A. 1999;58(7):451–62.

Dietrich N, Tan CH, Cubillas C, Earley BJ, Kornfeld K. Insights into zinc and cadmium biology in the nematode Caenorhabditis elegans. Arch Biochem Biophys. 2016; doi:10.1016/j.abb.2016.05.021.

Dinocourt C, Legrand M, Dublineau I, Lestaevel P. The neurotoxicology of uranium. Toxicology. 2015;337:58–71. doi:10.1016/j.tox.2015.08.004.

Dixon SJ, Lemberg KM, Lamprecht MR, Skouta R, Zaitsev EM, Gleason CE, Patel DN, Bauer AJ, Cantley AM, Yang WS, Morrison B 3rd, Stockwell BR. Ferroptosis: an iron-dependent form of nonapoptotic cell death. Cell. 2012;149(5):1060–72. doi:10.1016/j.cell.2012.03.042.

Dobson AW, Lack AK, Erikson KM, Aschner M. Depleted uranium is not toxic to rat brain endothelial (RBE4) cells. Biol Trace Elem Res. 2006;110(1):61–72. doi:10.1385/BTER:110:1:61.

Domingo JL. Aluminum and other metals in Alzheimer's disease: a review of potential therapy with chelating agents. J Alzheimer's Dis JAD. 2006;10(2–3):331–41.

Du M, Wang D. The neurotoxic effects of heavy metal exposure on GABAergic nervous system in nematode Caenorhabditis elegans. Environ Toxicol Pharmacol. 2009;27(3):314–20. doi:10.1016/j.etap.2008.11.011.

Duce JA, Bush AI. Biological metals and Alzheimer's disease: implications for therapeutics and diagnostics. Prog Neurobiol. 2010;92(1):1–18. doi:10.1016/j.pneurobio.2010.04.003.

Duce JA, Tsatsanis A, Cater MA, James SA, Robb E, Wikhe K, Leong SL, Perez K, Johanssen T, Greenough MA, Cho HH, Galatis D, Moir RD, Masters CL, McLean C, Tanzi RE, Cappai R, Barnham KJ, Ciccotosto GD, Rogers JT, Bush AI. Iron-export ferroxidase activity of beta-amyloid precursor protein is inhibited by zinc in Alzheimer's disease. Cell. 2010;142(6):857–67. doi:10.1016/j.cell.2010.08.014.

Dusek P, Jankovic J, Le W. Iron dysregulation in movement disorders. Neurobiol Dis. 2012;46(1):1–18. doi:10.1016/j.nbd.2011.12.054.

Dutilleul M, Lemaire L, Reale D, Lecomte C, Galas S, Bonzom JM. Rapid phenotypic changes in Caenorhabditis elegans under uranium exposure. Ecotoxicology. 2013;22(5):862–8. doi:10.1007/s10646-013-1090-9.

EFSA. Cadmium in food. EFSA J. 2009;53:1–139.

Elechiguerra JL, Burt JL, Morones JR, Camacho-Bragado A, Gao X, Lara HH, Yacaman MJ. Interaction of silver nanoparticles with HIV-1. J Nanobiotechnol. 2005;3:6. doi:10.1186/1477-3155-3-6.

Erikson KM, John CE, Jones SR, Aschner M. Manganese accumulation in striatum of mice exposed to toxic doses is dependent upon a functional dopamine transporter. Environ Toxicol Pharmacol. 2005;20(3):390–4. doi:10.1016/j.etap.2005.03.009.

Esposito G, Amoroso MR, Bergamasco C, Di Schiavi E, Bazzicalupo P. The G protein regulators EGL-10 and EAT-16, the Gialpha GOA-1 and the G(q)alpha EGL-30 modulate the response of the C. elegans ASH polymodal nociceptive sensory neurons to repellents. BMC Biol. 2010;8:138. doi:10.1186/1741-7007-8-138.

Eto K, Marumoto M, Takeya M. The pathology of methylmercury poisoning (Minamata disease): the 50th anniversary of Japanese Society of Neuropathology. Neuropathology. 2010;30(5):471–9. doi:10.1111/j.1440-1789.2010.01119.x.

Exley C. Aluminium and iron, but neither copper nor zinc, are key to the precipitation of beta-sheets of Abeta_{42} in senile plaque cores in Alzheimer's disease. J Alzheimer's Dis JAD. 2006;10(2–3):173–7.

Fagundez DA, Camara DF, Salgueiro WG, Noremberg S, Luiz Puntel R, Piccoli JE, Garcia SC, da Rocha JBT, Aschner M, Avila DS. Behavioral and dopaminergic damage induced by acute iron toxicity in Caenorhabditis elegans. Toxicol Res. 2015;4(4):878–84. doi:10.1039/C4TX00120F.

Farina M, Avila DS, da Rocha JB, Aschner M. Metals, oxidative stress and neurodegeneration: a focus on iron, manganese and mercury. Neurochem Int. 2013;62(5):575–94. doi:10.1016/j. neuint.2012.12.006.

Fitsanakis VA, Erikson KM, Garcia SJ, Evje L, Syversen T, Aschner M. Brain accumulation of depleted uranium in rats following 3- or 6-month treatment with implanted depleted uranium pellets. Biol Trace Elem Res. 2006;111(1–3):185–97. doi:10.1385/BTER:111:1:185.

Fraga CG, Oteiza PI. Iron toxicity and antioxidant nutrients. Toxicology. 2002;180(1):23–32.

Friedman A, Galazka-Friedman J, Koziorowski D. Iron as a cause of Parkinson disease - a myth or a well established hypothesis? Parkinsonism Relat Disord. 2009;15(Suppl 3):S212–4. doi:10.1016/S1353-8020(09)70817-X.

Gaeta A, Hider RC. The crucial role of metal ions in neurodegeneration: the basis for a promising therapeutic strategy. Br J Pharmacol. 2005;146(8):1041–59. doi:10.1038/sj.bjp.0706416.

Gonzalez-Hunt CP, Leung MC, Bodhicharla RK, McKeever MG, Arrant AE, Margillo KM, Ryde IT, Cyr DD, Kosmaczewski SG, Hammarlund M, Meyer JN. Exposure to mitochondrial genotoxins and dopaminergic neurodegeneration in Caenorhabditis elegans. PLoS One. 2014;9(12):e114459. doi:10.1371/journal.pone.0114459.

Gotz W. History of treatment of Parkinson disease. Pharm Unserer Zeit. 2006;35(3):190–7.

Grimaud J, Millar J, Thorpe JW, Moseley IF, McDonald WI, Miller DH. Signal intensity on MRI of basal ganglia in multiple sclerosis. J Neurol Neurosurg Psychiatry. 1995;59(3):306–8.

Guo M, Wu TH, Song YX, Ge MH, Su CM, Niu WP, Li LL, Xu ZJ, Ge CL, Al-Mhanawi MT, Wu SP, Wu ZX. Reciprocal inhibition between sensory ASH and ASI neurons modulates nociception and avoidance in Caenorhabditis elegans. Nat Commun. 2015;6:5655. doi:10.1038/ ncomms6655.

Gupta VB, Anitha S, Hegde ML, Zecca L, Garruto RM, Ravid R, Shankar SK, Stein R, Shanmugavelu P, Jagannatha Rao KS. Aluminium in Alzheimer's disease: are we still at a crossroad? Cellul Mol Life Sci CMLS. 2005;62(2):143–58. doi:10.1007/s00018-004-4317-3.

Helmcke KJ, Aschner M. Hormetic effect of methylmercury on Caenorhabditis elegans. Toxicol Appl Pharmacol. 2010;248(2):156–64. doi:10.1016/j.taap.2010.07.023.

Helmcke KJ, Syversen T, Miller DM 3rd, Aschner M. Characterization of the effects of methylmercury on Caenorhabditis elegans. Toxicol Appl Pharmacol. 2009;240(2):265–72. doi:10.1016/j. taap.2009.03.013.

Hilliard MA, Bargmann CI, Bazzicalupo P. C. elegans Responds to chemical repellents by integrating sensory inputs from the head and the tail. Current biology : CB. 2002;12(9):730–4.

Hoss S, Schlottmann K, Traunspurger W. Toxicity of ingested cadmium to the nematode Caenorhabditis elegans. Environ Sci Technol. 2011;45(23):10219–25. doi:10.1021/es2027136.

Hossain S, Liu HN, Nguyen M, Shore G, Almazan G. Cadmium exposure induces mitochondriadependent apoptosis in oligodendrocytes. Neurotoxicology. 2009;30(4):544–54. doi:10.1016/j. neuro.2009.06.001.

Houpert P, Lestaevel P, Bussy C, Paquet F, Gourmelon P. Enriched but not depleted uranium affects central nervous system in long-term exposed rat. Neurotoxicology. 2005;26(6):1015–20. doi:10.1016/j.neuro.2005.05.005.

Hu YO, Wang Y, Ye BP, Wang DY. Phenotypic and behavioral defects induced by iron exposure can be transferred to progeny in Caenorhabditis elegans. Biomedical Environ Sci BES. 2008;21(6):467–73. doi:10.1016/S0895-3988(09)60004-0.

IARC. n.d. Cadmium and certain cadmium compounds in IARC monographs on the EValuation the carcinogenic risk of chemicals to humans. Beryllium cadmium, mercury and exposures in the glass manufacturing industry, vol 58. World Health Organization, Lyon. 1993.

Ijomone OM, Miah MR, Peres TV, Nwoha PU, Aschner M. Null allele mutants of trt-1, the catalytic subunit of telomerase in Caenorhabditis elegans, are less sensitive to Mn-induced toxicity and DAergic degeneration. Neurotoxicology. 2016;57:54–60. doi:10.1016/j.neuro.2016.08.016.

Jiang GC, Tidwell K, McLaughlin BA, Cai J, Gupta RC, Milatovic D, Nass R, Aschner M. Neurotoxic potential of depleted uranium effects in primary cortical neuron cultures and in Caenorhabditis elegans. Toxicol Sci. 2007a;99(2):553–65. doi:10.1093/toxsci/kfm171.

Jiang LF, Yao TM, Zhu ZL, Wang C, Ji LN. Impacts of cd(II) on the conformation and self-aggregation of Alzheimer's tau fragment corresponding to the third repeat of microtubule-binding domain. Biochim Biophys Acta. 2007b;1774(11):1414–21. doi:10.1016/j.bbapap.2007.08.014.

Jiang GC, Hughes S, Sturzenbaum SR, Evje L, Syversen T, Aschner M. Caenorhabditis elegans Metallothioneins protect against toxicity induced by depleted uranium. Toxicol Sci. 2009;111(2):345–54. doi:10.1093/toxsci/kfp161.

Jiang Y, Chen J, Wu Y, Wang Q, Li H. Sublethal toxicity endpoints of heavy metals to the nematode Caenorhabditis elegans. PLoS One. 2016;11(1):e0148014. doi:10.1371/journal.pone.0148014.

Jorgensen EM. n.d. Gaba. WormBook : the online review of C elegans biology. 2005;1–13. doi:10.1895/wormbook.1.14.1.

Jung SK, Qu X, Aleman-Meza B, Wang T, Riepe C, Liu Z, Li Q, Zhong W. Multi-endpoint, high-throughput study of nanomaterial toxicity in Caenorhabditis elegans. Environ Sci Technol. 2015;49(4):2477–85. doi:10.1021/es5056462.

Khare P, Sonane M, Pandey R, Ali S, Gupta KC, Satish A. Adverse effects of TiO2 and ZnO nanoparticles in soil nematode, Caenorhabditis elegans. J Biomed Nanotechnol. 2011;7(1):116–7.

Khare P, Sonane M, Nagar Y, Moin N, Ali S, Gupta KC, Satish A. Size dependent toxicity of zinc oxide nano-particles in soil nematode Caenorhabditis elegans. Nanotoxicology. 2015;9(4):423–32. doi:10.3109/17435390.2014.940403.

Kim SO, Merchant K, Nudelman R, Beyer WF Jr, Keng T, DeAngelo J, Hausladen A, Stamler JS. OxyR: a molecular code for redox-related signaling. Cell. 2002;109(3):383–96.

Kim DK, Park JD, Choi BS. Mercury-induced amyloid-beta (Abeta) accumulation in the brain is mediated by disruption of Abeta transport. J Toxicol Sci. 2014;39(4):625–35.

Klang IM, Schilling B, Sorensen DJ, Sahu AK, Kapahi P, Andersen JK, Swoboda P, Killilea DW, Gibson BW, Lithgow GJ. n.d. Iron promotes protein insolubility and aging in C. elegans. Aging. 2014;6(11):975–91. doi:10.18632/aging.100689.

Levenson CW. Trace metal regulation of neuronal apoptosis: from genes to behavior. Physiol Behav. 2005;86(3):399–406. doi:10.1016/j.physbeh.2005.08.010.

Lewinska-Preis L, Jablonska M, Fabianska MJ, Kita A. Bioelements and mineral matter in human livers from the highly industrialized region of the upper Silesia Coal Basin (Poland). Environ Geochem Health. 2011;33(6):595–611. doi:10.1007/s10653-011-9373-7.

Li Y, Yu S, Wu Q, Tang M, Pu Y, Wang D. Chronic Al2O3-nanoparticle exposure causes neurotoxic effects on locomotion behaviors by inducing severe ROS production and disruption of ROS defense mechanisms in nematode Caenorhabditis elegans. J Hazard Mater. 2012;219-220:221–30. doi:10.1016/j.jhazmat.2012.03.083.

Li Y, Yu S, Wu Q, Tang M, Wang D. Transmissions of serotonin, dopamine, and glutamate are required for the formation of neurotoxicity from Al2O3-NPs in nematode Caenorhabditis elegans. Nanotoxicology. 2013;7(5):1004–13. doi:10.3109/17435390.2012.689884.

Lidsky TI, Schneider JS. Lead neurotoxicity in children: basic mechanisms and clinical correlates. Brain J Neurol. 2003;126(Pt 1):5–19.

Lin YT, Hoang H, Hsieh SI, Rangel N, Foster AL, Sampayo JN, Lithgow GJ, Srinivasan C. Manganous ion supplementation accelerates wild type development, enhances stress resistance, and rescues the life span of a short-lived Caenorhabditis elegans mutant. Free Radic Biol Med. 2006;40(7):1185–93. doi:10.1016/j.freeradbiomed.2005.11.007.

Lints R, Emmons SW. Patterning of dopaminergic neurotransmitter identity among Caenorhabditis elegans ray sensory neurons by a TGFbeta family signaling pathway and a Hox gene. Development. 1999;126(24):5819–31.

Lopez E, Arce C, Oset-Gasque MJ, Canadas S, Gonzalez MP. Cadmium induces reactive oxygen species generation and lipid peroxidation in cortical neurons in culture. Free Radic Biol Med. 2006;40(6):940–51. doi:10.1016/j.freeradbiomed.2005.10.062.

Lubick N. Mercury alters immune system response in artisanal gold miners. Environ Health Perspect. 2010;118(6):A243. doi:10.1289/ehp.118-a243.

Lukawski K, Nieradko B, Sieklucka-Dziuba M. Effects of cadmium on memory processes in mice exposed to transient cerebral oligemia. Neurotoxicol Teratol. 2005;27(4):575–84. doi:10.1016/j.ntt.2005.05.009.

Luo Y, Zhang J, Liu N, Luo Y, Zhao B. Copper ions influence the toxicity of beta-amyloid(1-42) in a concentration-dependent manner in a Caenorhabditis elegans model of Alzheimer's disease. Sci China Life Sci. 2011;54(6):527–34. doi:10.1007/s11427-011-4180-z.

Ma H, Bertsch PM, Glenn TC, Kabengi NJ, Williams PL. Toxicity of manufactured zinc oxide nanoparticles in the nematode Caenorhabditis elegans. Environ Toxicol Chem. 2009;28(6):1324–30. doi:10.1897/08-262.1.

Marchetti C. Molecular targets of lead in brain neurotoxicity. Neurotox Res. 2003;5(3):221–36.

Martinez-Finley EJ, Avila DS, Chakraborty S, Aschner M. Insights from Caenorhabditis elegans on the role of metals in neurodegenerative diseases. Metallom Integr Biometal Sci. 2011;3(3):271–9. doi:10.1039/c0mt00064g.

Martinez-Finley EJ, Chakraborty S, Slaughter JC, Aschner M. Early-life exposure to methylmercury in wildtype and pdr-1/parkin knockout C. elegans. Neurochem Res. 2013;38(8):1543–52. doi:10.1007/s11064-013-1054-8.

McDiarmid MA, Engelhardt SM, Dorsey CD, Oliver M, Gucer P, Wilson PD, Kane R, Cernich A, Kaup B, Anderson L, Hoover D, Brown L, Albertini R, Gudi R, Squibb KS. Surveillance results of depleted uranium-exposed gulf war I veterans: sixteen years of follow-up. J Toxicol Environ Health A. 2009;72(1):14–29. doi:10.1080/15287390802445400.

McElwee MK, Freedman JH. Comparative toxicology of mercurials in Caenorhabditis elegans. Environ Toxicol Chem. 2011;30(9):2135–41. doi:10.1002/etc.603.

McElwee MK, Ho LA, Chou JW, Smith MV, Freedman JH. Comparative toxicogenomic responses of mercuric and methyl-mercury. BMC Genomics. 2013;14:698. doi:10.1186/1471-2164-14-698.

Minniti AN, Rebolledo DL, Grez PM, Fadic R, Aldunate R, Volitakis I, Cherny RA, Opazo C, Masters C, Bush AI, Inestrosa NC. Intracellular amyloid formation in muscle cells of Abeta-transgenic Caenorhabditis elegans: determinants and physiological role in copper detoxification. Mol Neurodegener. 2009;4:2. doi:10.1186/1750-1326-4-2.

Mizoroki T, Meshitsuka S, Maeda S, Murayama M, Sahara N, Takashima A. Aluminum induces tau aggregation in vitro but not in vivo. J Alzheimer's Dis JAD. 2007;11(4):419–27.

Multhaup G, Schlicksupp A, Hesse L, Beher D, Ruppert T, Masters CL, Beyreuther K. The amyloid precursor protein of Alzheimer's disease in the reduction of copper(II) to copper(I). Science. 1996;271(5254):1406–9.

Neghab M, Norouzi MA, Choobineh A, Kardaniyan MR, Zadeh JH. Health effects associated with long-term occupational exposure of employees of a chlor-alkali plant to mercury. Int J Occup Saf Ergonom JOSE. 2012;18(1):97–106. doi:10.1080/10803548.2012.11076920.

Norfray JF, Chiaradonna NL, Heiser WJ, Song SH, Manyam BV, Devleschoward AB, Eastwood LM. Brain iron in patients with Parkinson disease: MR visualization using gradient modification. AJNR Am J Neuroradiol. 1988;9(2):237–40.

Okuda B, Iwamoto Y, Tachibana H, Sugita M. Parkinsonism after acute cadmium poisoning. Clin Neurol Neurosurg. 1997;99(4):263–5.

Oteiza PI, Mackenzie GG, Verstraeten SV. Metals in neurodegeneration: involvement of oxidants and oxidant-sensitive transcription factors. Mol Asp Med. 2004;25(1–2):103–15. doi:10.1016/j.mam.2004.02.012.

Page KE, White KN, McCrohan CR, Killilea DW, Lithgow GJ. Aluminium exposure disrupts elemental homeostasis in Caenorhabditis elegans. Metallom Integr Biometal Sci. 2012;4(5):512–22. doi:10.1039/c2mt00146b.

Pamphlett R, Kum Jew S. Uptake of inorganic mercury by human locus ceruleus and cortico-motor neurons: implications for amyotrophic lateral sclerosis. Acta Neuropathol Commun. 2013;1:13. doi:10.1186/2051-5960-1-13.

Parmalee NL, Maqbool SB, Ye B, Calder B, Bowman AB, Aschner M. n.d. RNASeq in C. elegans following manganese exposure. Curr Protocols Toxicol. 2015;65:11 20 11–17. doi:10.1002/0471140856.tx1120s65.

Petroni D, Tsai J, Agrawal K, Mondal D, George W. Low-dose methylmercury-induced oxidative stress, cytotoxicity, and tau-hyperphosphorylation in human neuroblastoma (SH-SY5Y) cells. Environ Toxicol. 2012;27(9):549–55. doi:10.1002/tox.20672.

Pfefferbaum A, Adalsteinsson E, Rohlfing T, Sullivan EV. MRI estimates of brain iron concentration in normal aging: comparison of field-dependent (FDRI) and phase (SWI) methods. NeuroImage. 2009;47(2):493–500. doi:10.1016/j.neuroimage.2009.05.006.

Poujois A, Devedjian JC, Moreau C, Devos D, Chaine P, Woimant F, Duce JA. Bioavailable trace metals in neurological diseases. Curr Treat Options Neurol. 2016;18(10):46. doi:10.1007/s11940-016-0426-1.

Price DL. New order from neurological disorders. Nature. 1999;399(6738 Suppl):A3–5.

Rand JB. Acetylcholine. WormBook. 2007:1–21. doi:10.1895/wormbook.1.131.1.

Rigon AP, Cordova FM, Oliveira CS, Posser T, Costa AP, Silva IG, Santos DA, Rossi FM, Rocha JB, Leal RB. Neurotoxicity of cadmium on immature hippocampus and a neuroprotective role for p38 MAPK. Neurotoxicology. 2008;29(4):727–34. doi:10.1016/j.neuro.2008.04.017.

Roh JY, Lee J, Choi J. Assessment of stress-related gene expression in the heavy metal-exposed nematode Caenorhabditis elegans: a potential biomarker for metal-induced toxicity monitoring and environmental risk assessment. Environ Toxicol Chem. 2006;25(11):2946–56.

Rosas HD, Chen YI, Doros G, Salat DH, Chen NK, Kwong KK, Bush A, Fox J, Hersch SM. Alterations in brain transition metals in Huntington disease: an evolving and intricate story. Arch Neurol. 2012;69(7):887–93. doi:10.1001/archneurol.2011.2945.

Rudgalvyte M, Peltonen J, Lakso M, Nass R, Wong G. RNA-Seq reveals acute manganese exposure increases endoplasmic reticulum related and Lipocalin mRNAs in Caenorhabditis elegans. J Biochem Mol Toxicol. 2016;30(2):97–105. doi:10.1002/jbt.21768.

Rui Q, Zhao Y, Wu Q, Tang M, Wang D. Biosafety assessment of titanium dioxide nanoparticles in acutely exposed nematode Caenorhabditis elegans with mutations of genes required for oxidative stress or stress response. Chemosphere. 2013;93(10):2289–96. doi:10.1016/j.chemosphere.2013.08.007.

Sadiq S, Ghazala Z, Chowdhury A, Busselberg D. Metal toxicity at the synapse: presynaptic, postsynaptic, and long-term effects. J Toxicol. 2012;2012:132671. doi:10.1155/2012/132671.

Salazar J, Mena N, Hunot S, Prigent A, Alvarez-Fischer D, Arredondo M, Duyckaerts C, Sazdovitch V, Zhao L, Garrick LM, Nunez MT, Garrick MD, Raisman-Vozari R, Hirsch EC. Divalent metal transporter 1 (DMT1) contributes to neurodegeneration in animal models of Parkinson's disease. Proc Natl Acad Sci U S A. 2008;105(47):18578–83. doi:10.1073/pnas.0804373105.

Sambongi Y, Nagae T, Liu Y, Yoshimizu T, Takeda K, Wada Y, Futai M. Sensing of cadmium and copper ions by externally exposed ADL, ASE, and ASH neurons elicits avoidance response in Caenorhabditis elegans. Neuroreport. 1999;10(4):753–7.

Sanfeliu C, Sebastia J, Cristofol R, Rodriguez-Farre E. Neurotoxicity of organomercurial compounds. Neurotox Res. 2003;5(4):283–305.

Santner A, Uversky VN. Metalloproteomics and metal toxicology of alpha-synuclein. Metall Integr Biometal Sci. 2010;2(6):378–92. doi:10.1039/b926659c.

Sawin ER, Ranganathan R, Horvitz HR. C. elegans Locomotory rate is modulated by the environment through a dopaminergic pathway and by experience through a serotonergic pathway. Neuron. 2000;26(3):619–31.

Scheiber IF, Mercer JF, Dringen R. Metabolism and functions of copper in brain. Prog Neurobiol. 2014;116:33–57. doi:10.1016/j.pneurobio.2014.01.002.

Schipper HM. Neurodegeneration with brain iron accumulation - clinical syndromes and neuroimaging. Biochim Biophys Acta. 2012;1822(3):350–60. doi:10.1016/j.bbadis.2011.06.016.

Settivari R, Levora J, Nass R. The divalent metal transporter homologues SMF-1/2 mediate dopamine neuron sensitivity in caenorhabditis elegans models of manganism and parkinson disease. J Biol Chem. 2009;284(51):35758–68. doi:10.1074/jbc.M109.051409.

Sistrunk SC, Ross MK, Filipov NM. Direct effects of manganese compounds on dopamine and its metabolite Dopac: an in vitro study. Environ Toxicol Pharmacol. 2007;23(3):286–96. doi:10.1016/j.etap.2006.11.004.

Subbiah R, Veerapandian M, Yun KS. Nanoparticles: functionalization and multifunctional applications in biomedical sciences. Curr Med Chem. 2010;17(36):4559–77.

Sulston JE. Neuronal cell lineages in the nematode Caenorhabditis elegans. Cold Spring Harb Symp Quant Biol. 1983;48(Pt 2):443–52.

Sulston JE, Schierenberg E, White JG, Thomson JN. The embryonic cell lineage of the nematode *Caenorhabditis elegans*. Dev Biol. 1983;100(1):64–119.

Swain SC, Keusekotten K, Baumeister R, Sturzenbaum SR. C. elegans Metallothioneins: new insights into the phenotypic effects of cadmium toxicosis. J Mol Biol. 2004;341(4):951–9. doi:10.1016/j.jmb.2004.06.050.

Takeda A. Manganese action in brain function. Brain Res Brain Res Rev. 2003;41(1):79–87.

Tamm C, Duckworth JK, Hermanson O, Ceccatelli S. Methylmercury inhibits differentiation of rat neural stem cells via Notch signalling. Neuroreport. 2008;19(3):339–43. doi:10.1097/WNR.0b013e3282f50ca4.

Thanh NTK, Green LAW. Functionalisation of nanoparticles for biomedical applications. Nano Today. 2010;5:213–30. doi:10.1016/j.nantod.2010.05.003.

Tiernan CT, Edwin EA, Hawong HY, Rios-Cabanillas M, Goudreau JL, Atchison WD, Lookingland KJ. Methylmercury impairs canonical dopamine metabolism in rat undifferentiated pheochromocytoma (PC12) cells by indirect inhibition of aldehyde dehydrogenase. Toxicological sciences : an official journal of the Society of Toxicology. 2015;144(2):347–56. doi:10.1093/toxsci/kfv001.

Valentini S, Cabreiro F, Ackerman D, Alam MM, Kunze MB, Kay CW, Gems D. Manipulation of in vivo iron levels can alter resistance to oxidative stress without affecting ageing in the nematode C. elegans. Mech Ageing Dev. 2012;133(5):282–90. doi:10.1016/j.mad.2012.03.003.

Vanduyn N, Settivari R, Wong G, Nass R. SKN-1/Nrf2 inhibits dopamine neuron degeneration in a Caenorhabditis elegans model of methylmercury toxicity. Toxicol Sci. 2010;118(2):613–24. doi:10.1093/toxsci/kfq285.

Vanduyn N, Settivari R, Levora J, Zhou S, Unrine J, Nass R. The metal transporter SMF-3/DMT-1 mediates aluminum-induced dopamine neuron degeneration. J Neurochem. 2013;124(1):147–57. doi:10.1111/jnc.12072.

Vatamaniuk OK, Bucher EA, Ward JT, Rea PA. A new pathway for heavy metal detoxification in animals. Phytochelatin synthase is required for cadmium tolerance in Caenorhabditis elegans. J Biol Chem. 2001;276(24):20817–20. doi:10.1074/jbc.C100152200.

Verstraeten SV, Aimo L, Oteiza PI. Aluminium and lead: molecular mechanisms of brain toxicity. Arch Toxicol. 2008;82(11):789–802. doi:10.1007/s00204-008-0345-3.

Villeda-Hernandez J, Barroso-Moguel R, Mendez-Armenta M, Nava-Ruiz C, Huerta-Romero R, Rios C. Enhanced brain regional lipid peroxidation in developing rats exposed to low level lead acetate. Brain Res Bull. 2001;55(2):247–51.

Waalkes MP. Cadmium carcinogenesis in review. J Inorg Biochem. 2000;79(1–4):241–4.

Walton JR. Aluminum involvement in the progression of Alzheimer's disease. J Alzheimer's Dis JAD. 2013;35(1):7–43. doi:10.3233/JAD-121909.

Wan L, Nie G, Zhang J, Luo Y, Zhang P, Zhang Z, Zhao B. Beta-amyloid peptide increases levels of iron content and oxidative stress in human cell and Caenorhabditis elegans models of Alzheimer disease. Free Radic Biol Med. 2011;50(1):122–9. doi:10.1016/j.freeradbiomed.2010.10.707.

Wang B, Du Y. Cadmium and its neurotoxic effects. Oxidative Med Cell Longev. 2013;2013:898034. doi:10.1155/2013/898034.

Wang D, Xing X. Assessment of locomotion behavioral defects induced by acute toxicity from heavy metal exposure in nematode Caenorhabditis elegans. J Environ Sci. 2008;20(9):1132–7.

Wang D, Shen L, Wang Y. The phenotypic and behavioral defects can be transferred from zinc-exposed nematodes to their progeny. Environ Toxicol Pharmacol. 2007;24(3):223–30. doi:10.1016/j.etap.2007.05.009.

Wang W, Xu ZJ, Wu YQ, Qin LW, Li ZY, Wu ZX. Off-response in ASH neurons evoked by CuSO4 requires the TRP channel OSM-9 in Caenorhabditis elegans. Biochem Biophys Res Commun. 2015;461(3):463–8. doi:10.1016/j.bbrc.2015.04.017.

Wei X, Qi Y, Zhang X, Gu X, Cai H, Yang J, Zhang Y. ROS act as an upstream signal to mediate cadmium-induced mitophagy in mouse brain. Neurotoxicology. 2015;46:19–24. doi:10.1016/j.neuro.2014.11.007.

White JG, Southgate E, Thomson JN, Brenner S. The structure of the nervous system of the nematode Caenorhabditis elegans. Philos Trans R Soc Lond Ser B Biol Sci. 1986;314(1165):1–340.

White AR, Multhaup G, Galatis D, McKinstry WJ, Parker MW, Pipkorn R, Beyreuther K, Masters CL, Cappai R. Contrasting, species-dependent modulation of copper-mediated neurotoxicity by the Alzheimer's disease amyloid precursor protein. J Neurosci. 2002;22(2):365–76.

WHO. Guidelines for drinking-water quality in health criteria and other supporting information, vol. 2. Geneva: WHO; 1996.

Williams PL, Dusenbery DB. A promising indicator of neurobehavioral toxicity using the nematode Caenorhabditis elegans and computer tracking. Toxicol Ind Health. 1990;6(3–4):425–40.

Wright RO, Baccarelli A. Metals and neurotoxicology. J Nutr. 2007;137(12):2809–13.

Wu J, Basha MR, Brock B, Cox DP, Cardozo-Pelaez F, McPherson CA, Harry J, Rice DC, Maloney B, Chen D, Lahiri DK, Zawia NH. Alzheimer's disease (AD)-like pathology in aged monkeys after infantile exposure to environmental metal lead (Pb): evidence for a developmental origin and environmental link for AD. J Neurosci. 2008;28(1):3–9. doi:10.1523/JNEUROSCI.4405-07.2008.

Wu Q, Li Y, Tang M, Wang D. Evaluation of environmental safety concentrations of DMSA coated Fe2O3-NPs using different assay systems in nematode Caenorhabditis elegans. PLoS One. 2012;7(8):e43729. doi:10.1371/journal.pone.0043729.

Wu Q, Nouara A, Li Y, Zhang M, Wang W, Tang M, Ye B, Ding J, Wang D. Comparison of toxicities from three metal oxide nanoparticles at environmental relevant concentrations in nematode Caenorhabditis elegans. Chemosphere. 2013;90(3):1123–31. doi:10.1016/j.chemosphere.2012.09.019.

Wu Q, Zhao Y, Li Y, Wang D. Susceptible genes regulate the adverse effects of TiO2-NPs at predicted environmental relevant concentrations on nematode Caenorhabditis elegans. Nanomed Nanotechnol Biol Med. 2014;10(6):1263–71. doi:10.1016/j.nano.2014.03.010.

Wu T, He K, Zhan Q, Ang S, Ying J, Zhang S, Zhang T, Xue Y, Tang M. MPA-capped CdTe quantum dots exposure causes neurotoxic effects in nematode Caenorhabditis elegans by affecting the transporters and receptors of glutamate, serotonin and dopamine at the genetic level, or by increasing ROS, or both. Nanoscale. 2015;7(48):20460–73. doi:10.1039/c5nr05914c.

Xing XJ, Rui Q, Du M, Wang DY. Exposure to lead and mercury in young larvae induces more severe deficits in neuronal survival and synaptic function than in adult nematodes. Arch Environ Contam Toxicol. 2009;56(4):732–41. doi:10.1007/s00244-009-9307-x.

Ye H, Ye B, Wang D. Trace administration of vitamin E can retrieve and prevent UV-irradiation- and metal exposure-induced memory deficits in nematode Caenorhabditis elegans. Neurobiol Learn Mem. 2008;90(1):10–8. doi:10.1016/j.nlm.2007.12.001.

Ye B, Rui Q, Wu Q, Wang D. Metallothioneins are required for formation of cross-adaptation response to neurobehavioral toxicity from lead and mercury exposure in nematodes. PLoS One. 2010;5(11):e14052. doi:10.1371/journal.pone.0014052.

Yokel RA, Rhineheimer SS, Sharma P, Elmore D, McNamara PJ. Entry, half-life, and desferrioxamine-accelerated clearance of brain aluminum after a single (26)al exposure. Toxicol Sci. 2001;64(1):77–82.

Zeitoun-Ghandour S, Charnock JM, Hodson ME, Leszczyszyn OI, Blindauer CA, Sturzenbaum SR. The two Caenorhabditis elegans metallothioneins (CeMT-1 and CeMT-2) discriminate between essential zinc and toxic cadmium. FEBS J. 2010;277(11):2531–42. doi:10.1111/j.1742-4658.2010.07667.x.

Zhang Y, Ye B, Wang D. Effects of metal exposure on associative learning behavior in nematode Caenorhabditis elegans. Arch Environ Contam Toxicol. 2010;59(1):129–36. doi:10.1007/s00244-009-9456-y.

Zhao Y, Wu Q, Tang M, Wang D. The in vivo underlying mechanism for recovery response formation in nano-titanium dioxide exposed Caenorhabditis elegans after transfer to the normal condition. Nanomed Nanotechnol Biol Med. 2014;10(1):89–98. doi:10.1016/j.nano.2013.07.004.

Zhao Y, Wang X, Wu Q, Li Y, Wang D. Translocation and neurotoxicity of CdTe quantum dots in RMEs motor neurons in nematode Caenorhabditis elegans. J Hazard Mater. 2015;283:480–9. doi:10.1016/j.jhazmat.2014.09.063.

Zhu W, Xie W, Pan T, Xu P, Fridkin M, Zheng H, Jankovic J, Youdim MB, Le W. Prevention and restoration of lactacystin-induced nigrostriatal dopamine neuron degeneration by novel brain-permeable iron chelators. FASEB J. 2007;21(14):3835–44. doi:10.1096/fj.07-8386com.

CPSIA information can be obtained
at www.ICGtesting.com
Printed in the USA
LVOW05*2107210218
567424LV00001B/26/P